International Association of Fire Chiefs

National Fire Protection Association

Hazardous Materials

Awareness and Operations

SECOND EDITION

Rob Schnepp

Division Chief of Special Operations
Alameda County (CA) Fire Department

JONES & BARTLETT
LEARNING

Jones & Bartlett Learning
World Headquarters
5 Wall Street
Burlington, MA 01803
978-443-5000
info@jblearning.com
www.jblearning.com

National Fire Protection Association
1 Batterymarch Park
Quincy, MA 02169-7471
www.NFPA.org

International Association of Fire Chiefs
4025 Fair Ridge Drive
Fairfax, VA 22033
www.IAFC.org

Jones & Bartlett Learning books and products are available through most bookstores and online booksellers. To contact Jones & Bartlett Learning directly, call 800-832-0034, fax 978-443-8000, or visit our website, www.jblearning.com.

Substantial discounts on bulk quantities of Jones & Bartlett Learning publications are available to corporations, professional associations, and other qualified organizations. For details and specific discount information, contact the special sales department at Jones & Bartlett Learning via the above contact information or send an email to specialsales@jblearning.com.

05718-8

Production Credits
Chief Executive Officer: Ty Field
President: James Homer
Chief Product Officer: Eduardo Moura
VP, Publisher: Kimberly Brophy
VP, Sales, Public Safety Group: Matthew Maniscalco
Director of Sales, Public Safety Group: Patricia Einstein
Executive Editor: Bill Larkin
Associate Managing Editor: Janet Morris
Associate Director of Production: Jenny L. Corriveau
Senior Marketing Manager: Brian Rooney
VP, Manufacturing and Inventory Control: Therese Connell
Composition: diacriTech
Cover Design: Scott Moden
Rights and Photo Research Coordinator: Ashley Dos Santos
Cover Image: Courtesy of Rob Schnepp
Printing and Binding: Courier Companies
Cover Printing: Courier Companies

Library of Congress Cataloging-in-Publication Data
Schnepp, Rob, 1961-
Hazardous materials : awareness and operations / Rob Schnepp.
pages cm
ISBN 978-1-4496-4154-2 (pbk.)
1. Hazardous substances–Handbooks, manuals, etc. I. Title.
T55.3.H3S36 2015
363.17–dc23
2014017270

6048

Printed in the United States of America
18 17 16 15 14 10 9 8 7 6 5 4 3 2 1

Brief Contents

CHAPTER 1 Hazardous Materials: Overview 2

CHAPTER 2 Hazardous Materials: Properties and Effects 18

CHAPTER 3 Recognizing and Identifying the Hazards 44

CHAPTER 4 Estimating Potential Harm and Planning a Response 82

CHAPTER 5 Implementing the Planned Response 108

CHAPTER 6 Terrorism 132

CHAPTER 7 Fire Smoke: Responder Health and Safety 156

CHAPTER 8 Mission-Specific Competencies: Personal Protective Equipment 176

CHAPTER 9 Mission-Specific Competencies: Technical Decontamination 204

CHAPTER 10 Mission-Specific Competencies: Mass Decontamination 218

CHAPTER 11 Mission-Specific Competencies: Evidence Preservation and Sampling 234

CHAPTER 12 Mission-Specific Competencies: Product Control 252

CHAPTER 13 Mission-Specific Competencies: Victim Rescue and Recovery 272

CHAPTER 14 Mission-Specific Competencies: Response to Illicit Laboratories 300

CHAPTER 15 Mission-Specific Competencies: Air Monitoring and Sampling 312

APPENDIX A An Extract from NFPA 472, *Standard for Competencies of Responders to Hazardous Materials/Weapons of Mass Destruction Incidents* 334

APPENDIX B An Extract from NFPA 473, *Standard for Competencies for EMS Personnel Responding to Hazardous Materials/Weapons of Mass Destruction Incidents* 354

APPENDIX C OSHA 1910.120, Hazardous Waste Operations and Emergency Response 366

APPENDIX D ProBoard Assessment Methodology Matrices for NFPA 472 398

GLOSSARY 412

INDEX 423

Contents

CHAPTER 1 Hazardous Materials: Overview 2

Introduction 4
Hazardous Materials 5
Levels of Training: Regulations and Standards 6
- Awareness Level. 7
- Operations Level 9
- Technician/Specialist Level. 10
- Incident Commander. 11
- Other Governmental Agencies 11

The Difference Between a Hazardous Materials/WMD Incident and Other Types of Emergencies 12
- Preplanning 13

CHAPTER 2 Hazardous Materials: Properties and Effects. 18

Introduction 20
Physical and Chemical Changes 20
Critical Characteristics of Flammable Liquids 22
- Flash Point. 22
- Ignition Temperature 23
- Flammable Range. 23
- Vapor Pressure 24
- Boiling Point 25
- Vapor Density 25
- Specific Gravity 26
- Water Solubility 27
- Corrosivity (pH). 27
- Toxic Products of Combustion 28
- Radiation 29

Hazard, Exposure, and Contamination. . 31
- Hazard and Exposure. 31
- Contamination. 31
- Secondary Contamination 32

Types of Hazardous Materials: Weapons of Mass Destruction 32
- Nerve Agents 32
- Blister Agents 34
- Blood Agents 34
- Choking Agents 35
- Irritants (Riot Control Agents) 35
- Convulsants. 35

Harmful Substances' Routes of Entry into the Human Body 35
- Inhalation 36
- Absorption. 37
- Ingestion 37
- Injection 37

Chronic and Acute Health Effects. 37

CHAPTER 3 Recognizing and Identifying the Hazards 44

Introduction 46
Recognizing a Hazardous Materials/ Weapons of Mass Destruction Incident. 47
- Occupancy and Location 47
- Senses 47

Containers. 48
- Container Volume 49
- Nonbulk Storage Vessels 50

Transporting Hazardous Materials 54
- Roadway Transportation 54
- Railroad Transportation 56
- Pipelines 57

Transportation and Facility Markings . . . 57
- The Department of Transportation Marking System. 58
- The National Fire Protection Association 704 Marking System. 60
- Hazardous Materials Information System. 61
- Military Hazardous Materials/ Weapons of Mass Destruction Markings. 62

Chemical References 63
- The *Emergency Response Guidebook* 63

Other Reference Sources 66
Safety Data Sheets 66
Shipping Papers 66
CHEMTREC 73
National Response Center 73
Potential Terrorist Incidents 73
Chemical Agents 73
Biological Agents 74
Radiological Agents 74
Illicit Laboratories 76
Explosives 77
Secondary Devices 77

CHAPTER 4 **Estimating Potential Harm and Planning a Response 82**

Introduction 84
Estimating the Potential Harm or Severity of the Incident 84
Resources for Determining the Size of the Incident 86
Exposures 89
Reporting the Size and Scope of the Incident 91
Determining the Concentration of a Released Hazardous Material 91
Skin Contact Hazards 92
Plan an Initial Response 92
Response Objectives 93
Secondary Attacks and Devices 93
Defensive Actions 94
Personal Protective Equipment 94
Hazardous Materials: Types of PPE 96
Street Clothing and Work Uniforms 96
Structural Firefighting Protective Clothing 96
High-Temperature–Protective Clothing and Equipment 96
Chemical-Protective Clothing and Equipment 96
Chemical-Protective Clothing Ratings 98
Level A 98
Level B 98
Level C 99
Level D 100
Respiratory Protection 100
Physical Capability Requirements 101
Positive-Pressure Self-Contained Breathing Apparatus 101
Supplied-Air Respirators 102
Closed-Circuit SCBA 102
Air-Purifying Respirators 102
Powered Air-Purifying Respirators 103
Decontamination 103
Emergency Decontamination 103
Secondary Contamination 104

CHAPTER 5 **Implementing the Planned Response 108**

Introduction 110
Safety 110
Isolate 111
Notify 111
Response Safety Procedures 112
Establishing Control Zones 112
Performing Emergency Decontamination 114
Protective Actions 114
The Buddy System and Backup Personnel 115
Evacuation 116
Sheltering-in-Place 118
Search and Rescue 118
Safety Briefings 118
Excessive-Heat Disorders 119
Cold-Temperature Exposures 120
Personal Protective Equipment: Physical Capability Requirements 120
The Incident Command System 121
Command 121
General Staff Functions 124
Role of the Operations Level Responder 127

CHAPTER 6 **Terrorism 132**

Introduction 134
Responding to Terrorist Incidents 134
Potential Targets and Tactics 135
Classification of Terrorist Targets 135
Other Methods 138
Explosives and Incendiary Devices 138
Agents 140
Chemical Agents 140
Biological Agents 145
Radiological Agents 148
Operations 149
Initial Actions 149
Interagency Coordination 150
Decontamination 151
Mass Casualties 151

CHAPTER 7 **Fire Smoke: Responder Health and Safety 156**

Introduction 158
Overview of Fire Smoke 160

Carbon Monoxide and Hydrogen Cyanide: Silent Killers . . . 162
Smoke Inhalation Treatment . . . 165
Postfire Detection and Monitoring . . . 165
Why Use Detection Devices at the Fire Scene? . . . 166
Common Fire Scene Detection and Monitoring Technologies . . . 168
Electrochemical Sensors . . . 168
Photoionization (PID) Sensor . . . 169
Colorimetric Tubes . . . 170
Fire Scene Detection and Monitoring Practices . . . 170
Some Final Thoughts . . . 171
References . . . 172

CHAPTER 8 Mission-Specific Competencies: Personal Protective Equipment . . . 176

Introduction . . . 178
Hazardous Materials: Specific Personal Protective Equipment . . . 179
Chemical-Protective Clothing and Equipment . . . 180
Respiratory Protection . . . 184
Chemical-Protective Clothing Ratings . 185
Safety . . . 193
Chemical-Protective Equipment Performance Requirements . . . 193
Responder Safety . . . 196
Reporting and Documenting the Incident . . . 200

CHAPTER 9 Mission-Specific Competencies: Technical Decontamination . . . 204

Introduction . . . 206
Technical Decontamination . . . 207
Methods of Technical Decontamination . . . 208
The Technical Decontamination Process . . . 211
Performing Technical Decontamination . . . 213
Evaluating the Effectiveness of Technical Decontamination . . . 215
Reports and Documentation . . . 215

CHAPTER 10 Mission-Specific Competencies: Mass Decontamination . . . 218

Introduction . . . 220
Mass Decontamination . . . 220
Mass Decontamination Methods . . . 222
Evaluating the Effectiveness of Mass Decontamination . . . 224
The Role of Reference Sources in Mass Decontamination . . . 228
Crowd Control . . . 231
Reports and Documentation . . . 231
Evidence Preservation . . . 232

CHAPTER 11 Mission-Specific Competencies: Evidence Preservation and Sampling . . . 234

Introduction . . . 236
Analyzing the Incident . . . 237
Investigative Jurisdictions . . . 238
Types of Evidence . . . 238
Preservation of Evidence . . . 240
Chain of Custody . . . 242
Identifying Witnesses . . . 242
Taking Action . . . 243
Securing, Characterizing, and Preserving the Scene . . . 243
Documenting Personnel and Scene Activity . . . 244
Notifying Investigative Authority and Hazardous Devices Personnel . . . 246
Identifying Samples and Evidence to Be Collected . . . 246
Collecting Samples and Preventing Secondary Contamination . . . 246
Documentation of Sampling . . . 247
Sampling and Field Screening Protocols . . . 248
Labeling, Packaging, and Decontamination . . . 248

CHAPTER 12 Mission-Specific Competencies: Product Control . . . 252

Introduction . . . 254
Control Options . . . 254
Absorption and Adsorption . . . 255
Damming . . . 258
Diking . . . 259
Dilution . . . 259
Diversion . . . 260
Retention . . . 261
Remote Valve Shut-off . . . 262
Vapor Dispersion and Suppression . . . 263
Recovery . . . 267

CHAPTER 13 Mission-Specific Competencies: Victim Rescue and Recovery . . . 272

Introduction . . . 274
Tactical Considerations . . . 274
Entry Team Responsibilities . . . 275

Medical Care 278
Backup Team 278
Emergency Decontamination 279
Search, Rescue, and Recovery 279
Search and Rescue/Recovery Equipment 281
Search and Rescue/Recovery Methods. . 282
Sheltering-in-Place 283
Exit Assist 283
Simple Victim Carries 284
Emergency Drags 289

CHAPTER 14 Mission-Specific Competencies: Response to Illicit Laboratories . . 300

Introduction 302
Identifying Illicit Laboratories 302
Drug Laboratories 303
Weapons of Mass Destruction Laboratories. 305
Chemical Laboratories 306
Biological Laboratories. 306
Tasks and Operations. 306
Notifying Authorities 308
Determining Response Options 308

CHAPTER 15 Mission-Specific Competencies: Air Monitoring and Sampling 312

Introduction 314
Detection and Monitoring 316
Detection and Monitoring Terminology 316
Detection and Monitoring Concepts. 318
Detection and Monitoring Unknowns. 319
Types of Detectors and Monitors 321
Photo-Ionization Detector 321
Combustible Gas Indicator 321
Gas Chromatography. 322
Flame Ionization Detector 322
Oxygen Monitoring Device 322
Carbon Monoxide Detector 323
Hydrogen Sulfide Monitor 324
Multi-gas Meter 324
Colorimetric Tubes. 326
pH Paper 329
Chemical Test Strips. 329
Specialized Detection Devices 329
Radiation Detection Devices. 330
Personal Dosimeters. 331

APPENDIX A An Extract from NFPA 472, *Standard for Competencies of Responders to Hazardous Materials/Weapons of Mass Destruction Incidents* 334

APPENDIX B An Extract from NFPA 473, *Standard for Competencies for EMS Personnel Responding to Hazardous Materials/Weapons of Mass Destruction Incidents* 354

APPENDIX C OSHA 1910.120, Hazardous Waste Operations and Emergency Response 366

APPENDIX D ProBoard Assessment Methodology Matrices for NFPA 472 398

GLOSSARY 412

INDEX 423

Skill Drills

CHAPTER 5

Skill Drill 5-1 Performing Emergency Decontamination 115

CHAPTER 8

Skill Drill 8-1 Donning a Level A Ensemble 186
Skill Drill 8-2 Doffing a Level A Ensemble..........187
Skill Drill 8-3 Donning a Level B Nonencapsulated Chemical-Protective Clothing Ensemble.........................189
Skill Drill 8-4 Doffing a Level B Nonencapsulated Chemical-Protective Clothing Ensemble.........................191
Skill Drill 8-5 Donning a Level C Chemical-Protective Clothing Ensemble 192
Skill Drill 8-6 Doffing a Level C Chemical-Protective Clothing Ensemble 194
Skill Drill 8-7 Donning a Level D Chemical-Protective Clothing Ensemble 195

CHAPTER 9

Skill Drill 9-1 Performing Technical Decontamination on a Responder....................214

CHAPTER 10

Skill Drill 10-1 Performing Mass Decontamination on Ambulatory Victims 225
Skill Drill 10-2 Performing Mass Decontamination on Nonambulatory Victims 227

CHAPTER 11

Skill Drill 11-1 Collecting Samples and Preserving Evidence 241
Skill Drill 11-2 Securing, Characterizing, and Preserving the Scene 244
Skill Drill 11-6 Collecting Samples Using Equipment and Preventing Secondary Contamination 247
Skill Drill 11-7 Documentation of Sampling 248
Skill Drill 11-8 Labeling, Packaging, and Decontamination 249

CHAPTER 12

Skill Drill 12-1 Using Absorption/Adsorption to Manage a Hazardous Materials Incident 257
Skill Drill 12-2 Constructing an Overflow Dam 258
Skill Drill 12-3 Constructing an Underflow Dam..... 259
Skill Drill 12-4 Constructing a Dike 260
Skill Drill 12-5 Using Dilution to Manage a Hazardous Materials Incident........261
Skill Drill 12-6 Constructing a Diversion 261
Skill Drill 12-7 Using Retention to Manage a Hazardous Materials Incident....... 262
Skill Drill 12-8 Using Vapor Dispersion to Manage a Hazardous Materials Incident 264
Skill Drill 12-9 Using Vapor Suppression to Manage a Hazardous Materials Incident 265
Skill Drill 12-10 Performing the Rain-Down Method of Applying Foam 266
Skill Drill 12-11 Performing the Roll-In Method of Applying Foam 267
Skill Drill 12-12 Performing the Bounce-Off Method of Applying Foam 268

CHAPTER 13

Skill Drill 13-1 Performing a One-Person Walking Assist 284
Skill Drill 13-2 Performing a Two-Person Walking Assist.......................... 285
Skill Drill 13-3 Performing a Two-Person Extremity Carry.................. 286
Skill Drill 13-4 Performing a Two-Person Seat Carry 286
Skill Drill 13-5 Performing a Two-Person Chair Carry.. 287
Skill Drill 13-6 Performing a Cradle-in-Arms Carry ... 288
Skill Drill 13-7 Performing a Clothes Drag.......... 289
Skill Drill 13-8 Performing a Blanket Drag 290
Skill Drill 13-9 Performing a Standing Drag 291
Skill Drill 13-10 Performing a Webbing Sling Drag.... 292
Skill Drill 13-11 Performing a Fire Fighter Drag 293
Skill Drill 13-12 Performing a One-Person Emergency Drag from a Vehicle...... 294
Skill Drill 13-13 Performing a Long Backboard Rescue 296

CHAPTER 15

Skill Drill 15-3 Using a Multi-gas Meter............ 325
Skill Drill 15-4 Using Colorimetric Tubes........... 327

Acknowledgments

Jones & Bartlett Learning, the National Fire Protection Association, and the International Association of Fire Chiefs would like to thank all the authors, contributors, and reviewers of *Hazardous Materials Awareness and Operations, Second Edition.*

Author

Rob Schnepp Rob Schnepp has almost 30 years of fire service experience and is currently serving as the Division Chief of Special Operations for the Alameda County (CA) Fire Department. He is an instructor for the U.S. Defense Threat Reduction Agency, providing hazardous materials and weapons of mass destruction training internationally. Schnepp is a current member of the NFPA 472 Technical Committee on Hazardous Materials Response Personnel and the International Association of Fire Chiefs Hazardous Materials Committee, and is a former hazardous materials team manager for California Task Force 4, FEMA Urban Search and Rescue team.

With a variety of published works covering several fire service topics, including numerous magazine articles in *Fire Engineering* magazine, Schnepp is also an established author, receiving the Ben Franklin writing award for this textbook in 2012. He is a member of the Fire Engineering editorial advisory board and sits on the executive advisory board for the Fire Department Instructors Conference (FDIC).

Contributing Authors

Ed Allen Lt. Ed Allen has been with the Seminole County (FL) Sheriff's Office since 1987 and is currently the agency's Emergency Management Coordinator and SWAT Tactical Operations Coordinator. He is a Certified Emergency Manager and holds a BA in Behavioral Science. Lt. Allen serves on the Board of Directors for the National Tactical Officers Association and represents the organization on several national committees, including the NFPA 472 Technical Committee and the Inter-Agency Board.

Glen D. Rudner Glen Rudner is the General Manager of the Critical Incident Response Group at the Security and Emergency Response Training Center in Pueblo, Colorado. Prior to his move, he was a private consultant and a retired Hazardous Materials Response Officer for the Virginia Department of Emergency Management. His 35 years of experience in public safety include 12 years as a career fire fighter/hazardous materials specialist for the City of Alexandria (VA) Fire Department, as well as a former volunteer emergency medical technician (EMT), fire fighter, and officer. As a subcontractor, he has served as a consultant and assisted in the development of many training programs for local, state, and federal agencies. He also serves as Secretary for the National Fire Protection Association (NFPA) Hazardous Materials Committee, a member of the International Association of Fire Chiefs (IAFC) Hazardous Materials Committee, a member of the American Society of Testing and Materials (ASTM), and the Co-Chairman of the Ethanol Emergency Response Coalition.

Editorial Board

Ed Allen
Seminole County Sheriff's Office
Sanford, Florida

James A. Perkins, Jr.
International Association of Fire Chiefs
Fairfax, Virginia

Glen Rudner
Transportation Technology Center Inc. – SERTC
Pueblo, Colorado

Rob Schnepp
Alameda County Fire Department
San Leandro, California

David Trebisacci
National Fire Protection Association
Quincy, Massachusetts

Contributors and Reviewers

Benjamin Andersen
Loveland Fire Rescue Authority
Loveland, Colorado

Robert M. Barron
Prairie View Fire Department and Texas A&M Engineering Extension Service—Emergency Service Training Institute
College Station, Texas

Kevin R. Bartoe
Robins AFB Fire Emergency Services
Warner Robins, Georgia

Tim Beattie
University Department Chair, Keiser University
Sarasota, Florida

Hakim Benmoussa
City of Goodyear Fire Department Hazardous Materials Response Team
Goodyear, Arizona

Michael Blann
Lafayette Fire Department
Lafayette, Indiana

Jeffery L. Combs
Clark State Community College
Springfield, Ohio

Brian D. Daake
Beatrice Fire and Rescue
Beatrice, Nebraska

John T. Dean
Phoenix Fire Department
Phoenix, Arizona

Leo DeBobes
Suffolk County Community College
Selden, New York

Rick Emery
Lake Country Hazardous Materials Team (Retired)
Vernon Hills, Illinois

Michael L. Farrell
U.S. Capitol Police Hazardous Materials Response Team
Washington, District of Columbia

William Ferguson
Franklin County Public Safety
Rocky Mount, Virginia

Neil R. Fulton
Norwich Fire Department
Norwich, Vermont

Doug Goodings
Office of the Fire Marshal
North York, Ontario, Canada

Robert W. (Wayne) Green
Culpeper County Office of Emergency Services
Culpeper, Virginia

William D. Hadley, Jr
Louis F. Garland Department of Defense Fire Academy
Goodfellow AFB, San Angelo, Texas

John Hamilton
Southern Crescent Technical College and Griffin Fire Rescue
Griffin, Georgia

Kevin L. Hammons
Metro State University of Denver
Denver, Colorado

Stephen M. Hardesty
Howard County Fire and Rescue
Columbia, Maryland

Jeffry J. Harran
Lake Havasu City Fire Department
Lake Havasu City, Arizona

Jake Herring
Brampton Fire and Emergency Services
Brampton, Ontario, Canada

Judith A. Hoffmann
Hoffmann Consulting
South Bend, Indiana

Douglass Hogan
Texas A&M Engineering Extension Service (TEEX)
Doha, Qatar

Dominick Iannelli
Fairfax County Fire Department
Fairfax, Virginia

Jerry Janick
Illinois Fire Service Institute, Hazmat Program
Champaign, Illinois

Darin Keith
Rock Island Arsenal Fire and Emergency Services
Rock Island, Illinois

Richard Kosmoski
Middlesex County Fire Academy
Sayreville, New Jersey

Jason M. Krusen
Columbia Fire Department
Columbia, South Carolina

Steven Lague
Muskegon County Regional Hazmat Team
North Muskegon, Michigan

Jason Loeb
Illinois Fire Service Institute
Champaign, Illinois

Mike McEvoy
Saratoga County EMS
Saratoga County, New York

Tom McGowan
National Fire Protection Association
Quincy, Massachusetts

Patrick Nahgahgwon
Deerfield Township Fire Department
Mount Pleasant, Michigan

Paul E. Nelson
DeForest Area Fire and EMS
DeForest, Wisconsin

Patrick M. O'Connor
Providence Fire Department
Providence, Rhode Island

Mike Pannell
DeKalb County Fire and Rescue
Tucker, Georgia

Scott Pascu
Akron Fire Department
Akron, Ohio

Carl Raymond
Southern Crescent Technical College
Griffin, Georgia

Glen Rudner
Transportation Technology Center, Inc.—SERTC
Pueblo, Colorado

John Francis Ryan
Fort Gordon Fire and Emergency Services
Fort Gordon, Georgia

Mark J. Schmitt
Greensboro Fire Department
Greensboro, North Carolina

Mark D. Schuman
North Washington Fire Protection District
Denver, Colorado

Gary Seidel
Fire Chief (Retired)
Hillsboro, Oregon

Craig H. Shelley
City of Rutland Fire Department (Retired)
Rutland, Vermont

Bret Stohr
McChord Air Force Base
Tacoma, Washington

Lori P. Stoney
Homewood Fire and Rescue Service
Homewood, Alabama

Jimmy L. Talton
Reece City Volunteer Fire Department
Attalla/Reece City, Alabama

Tim Thomas
Indiana Department of Homeland Security
Indianapolis, Indiana

Ray Unrau
Saskatoon Fire and Protective Services
Saskatoon, Saskatchewan, Canada

Nick Zamiska
Brecksville Fire/SERT Hazmat
Brecksville, Ohio

Photographic Contributors

We would like to extend a huge *thank you* to Glen E. Ellman, the photographer for this project. Glen is a commercial photographer and fire fighter based in Fort Worth, Texas. His expertise and professionalism are unmatched!

Rob Schnepp provided many of the photographs that are new to this edition.

Thank you to the following organizations that opened up their facilities for these photo shoots:

Boca Raton Fire Rescue Services, Boca Raton, Florida
Hazardous Materials Response Team: Chief Scott Johnston, Paul Ceresa (Cid), Edward Guinn, William Puchalski, James Haag, Donnie Mullers, Philip Santa Maria, Michael Sklark, Scott Brooks, and Tom Carroll

The Mississippi State Fire Academy, Jackson, Mississippi
Reggie Bell—Executive Director
Daniel Cross—Instructor Chief
Marcus Collier—Instructor
Mike Word—Instructor
Tim Dennison, Brandon King, Benjamin Crabb, Bill Kitchings, Robert Gill, and the Fire Fighter Recruit class #104.

Hazardous Materials: Overview

Greg Henry/ShutterStock, Inc.

Courtesy of Rob Schnepp

© Photos.com

CHAPTER 1

Awareness Level Personnel

Knowledge Objectives

After studying this chapter, you will be able to:

- Define a hazardous material. (p. 5)
- Define weapons of mass destruction (WMD). (p. 5)
- Identify the location of the emergency response plan and/or standard operating procedures. (NFPA 4.4.1(1), p. 5)
- Describe the different levels of hazardous materials training: awareness, operations, technician, specialist, and incident commander. (p. 6–11)
- Understand the difference between the standards and federal regulations that govern hazardous material response activities. (p. 11–12)
- Explain the difference between hazardous materials incidents and other emergencies. (p. 12)
- Explain the need for a planned response to a hazardous materials incident. (p. 13–14)

Skills Objectives

This chapter has no skills objectives for Awareness Level personnel.

Operations Level Responders

Knowledge Objectives

After studying this chapter, you will be able to:

- Define a hazardous material. (p. 5)
- Define weapons of mass destruction (WMD). (p. 5)
- Describe the different levels of hazardous materials training: awareness, operations, technician, specialist, and incident commander. (p. 6–11)
- Understand the difference between the standards and federal regulations that govern hazardous material response activities. (p. 11–12)
- Explain the difference between hazardous materials incidents and other emergencies. (p. 12)
- Explain the need for a planned response to a hazardous materials incident. (p. 13–14)

Skills Objectives

This chapter has no skills objectives for Operations Level responders.

You Are the Responder

At 13:30 hours, your crew responds to a report of a suspicious odor at a small plastics manufacturing company. When you arrive, you see a cargo delivery truck parked at the loading dock, with the motor still idling. From where you stand, you can see one side of the trailer—it has a red diamond-shaped placard on it that reads "Flammable," with the number 3 at the bottom. The facility security guard reports that the truck has been there for several hours, and a site employee notified a supervisor that the driver was in a nearby break room complaining of nausea and a headache. Several of the employees stated that they can smell something like "paint thinner" in the air.

1. Is this an emergency? How would you go about sizing up the scene to better understand the problem? What other pieces of information would you want?
2. Does the placard on the side of the truck provide any clues about the nature of the cargo inside? If so, explain your findings.
3. Do you suspect criminal intent or some other illegal activity?

Introduction

Fire fighters, law enforcement personnel, and emergency medical services (EMS) personnel may be called upon to respond to a variety of incidents. These incidents may include structural fires, emergency medical calls, automobile accidents, confined-space rescues, water rescues, or acts of terrorism. These and other emergency situations may involve hazardous substances that threaten lives, property, and/or the environment. When an emergency clearly involves a hazardous material, or you suspect the presence of a hazardous materials release, the nature of the incident changes and so must the mentality of the responder. Generally speaking, hazardous materials incidents are handled in a more deliberate fashion than structural firefighting. There could be unknown substances present, or the setting of the release is such that it may take some time to get a full picture of the incident prior to taking action. For example, train derailments may involve chemical substances, in addition to having a large incident footprint. In any case, it does not mean that hazardous materials incidents are more or less complicated than fighting fires; it means that responders often take more time to get oriented to the problem and define a rational approach to solving the situation without getting exposed to the released substance.

With proper training and good situational awareness, responders can stay safe and make decisions that positively change the outcome of any incident. From a broad perspective, the goal of this text is to help you learn how to recognize the presence of a hazardous materials/weapons of mass destruction (WMD) incident; take initial actions including establishing scene control zones; implement the Incident Command System; use basic reference sources such as the *Emergency Response Guidebook (ERG)*; select personal protective clothing; implement product control measures when needed; perform appropriate decontamination; and ultimately, understand where you fit into a full-scale hazardous materials/WMD response.

In addition, you will be provided with information regarding mission-specific responsibilities in the following areas, based on the needs of the authority having jurisdiction (AHJ):

1. Operations level responders assigned to use personal protective equipment
2. Operations level responders assigned to perform mass decontamination
3. Operations level responders assigned to perform technical decontamination
4. Operations level responders assigned to perform evidence preservation and sampling
5. Operations level responders assigned to perform product control
6. Operations level responders assigned to perform air monitoring and sampling
7. Operations level responders assigned to perform victim rescue/recovery
8. Operations level responders assigned to respond to illicit laboratory incidents
9. Operations level responders assigned to perform improvised WMD dispersal device disablement/disruption and operations at improvised explosive laboratories[1]

Other National Fire Protection Association (NFPA) standards under development as of the writing of this text include:

- *NFPA 475: Recommended Practices for Organizing and Managing a Hazardous Materials/WMD Emergency Response Program.* The intent of the standard is to establish common criteria for the organization, management, programmatic elements, deployment of personnel, and resources for those entities responsible for the hazardous materials/WMD emergency preparedness function.
- *NFPA 1072: Standard for Hazardous Materials/Weapons of Mass Destruction Emergency Response Personnel Professional Qualifications:* This document is a hazardous

[1]The last mission-specific competency (9) will not be discussed in this text. The development of the competency occurred as a result of a request to the technical committee from the National Bomb Squad Commanders Advisory Board. The intent of the competency is to improve the certification process for bomb technicians.

materials ProQual standard that is based on job performance requirements (JPR). NFPA 1072 will cover the following levels: awareness, operations core, mission-specific competencies, hazardous materials technician, and incident commander.

The AHJ is the governing body that sets operational policy and procedures for the jurisdiction in which you operate. For example, the AHJ might identify a set of tasks that responders would be expected to perform in their course of duty and match the training competencies to address those tasks. Perhaps the AHJ would decide that operations level responders in that AHJ are required to perform victim rescue. Subsequently, those identified responders must be trained to meet all competencies at the awareness level, all core competencies at the operations level, all mission-specific competencies for personal protective equipment, and all mission specific competencies for victim rescue. There will be more on this topic later in the chapter.

Ultimately, you should be able to operate safely at a wide variety of hazardous materials/WMD incidents FIGURE 1-1. Regulations enforced by two U.S. federal agencies, the Occupational Safety and Health Administration (OSHA) and the Environmental Protection Agency (EPA), as well as consensus-based standards such as the National Fire Protection Association (NFPA) Standard 472, *Standard for Competence of Responders to Hazardous Materials/Weapons of Mass Destruction Incidents*, are in place to help you achieve this goal. The intent of this text is not to make you an expert on hazardous materials/WMD or on the regulations and standards that govern your response but to help you become a more informed responder.

Additionally, initial and ongoing actions may be guided by your local or organizational emergency response plans and/or standard operating procedures (SOPs). Every responder should have knowledge of and access to emergency response plans, and have been trained on appropriate SOPs prior to being considered a "responder."

FIGURE 1-1 The ability to recognize a potential hazardous materials/WMD incident is a critical first step to ensuring your safety.

Courtesy of George Roarty/VDEM

RESPONDER TIP

The *Emergency Response Guidebook (ERG)* is a preliminary action guide containing information regarding approximately 4000 chemicals. Once the incident progresses beyond the "initial response phase" (presence and/or identification of dangerous goods is confirmed, protective actions and area securement are initiated, and assistance of qualified personnel is requested), the *ERG* may no longer be the most appropriate source of information. Responders at the scene should seek additional specifics about any material in question by consulting the appropriate emergency response agency or using the emergency response number on a shipping document, if applicable, to gather more information.

Hazardous Materials

A hazardous material, as defined by the U.S. Department of Transportation (DOT), is any substance or material that is capable of posing an unreasonable risk to human health, safety, or the environment when transported in commerce, used incorrectly, or not properly contained or stored. The term *hazardous material* also includes hazardous substances, wastes, marine pollutants, and elevated-temperature materials.

As the title implies, the definition of a hazardous material as found in NFPA 472, *Standard for Competence of Responders to Hazardous Materials/Weapons of Mass Destruction Incidents*, includes the criminal use of hazardous materials, such as WMD. This definition encompasses illicit laboratories, environmental crimes, or industrial sabotage. It is for this reason that the text will refer to hazardous materials and WMD simultaneously.

From an emergency responder's perspective, a hazardous material can be almost anything, depending on the situation. Milk, for example, is not routinely regarded as a hazardous substance—but 5000 gallons of milk leaking into a creek does, in fact, pose an unreasonable risk to the environment. A large chlorine gas release also fits the definition, as would any substance used as a terrorist weapon. Regarding the threat of terrorism, it would be unwise to believe that your jurisdiction is immune to deliberate criminal acts. Such an event could happen anywhere, at any time, with any type of substance.

Hazardous materials can be found anywhere FIGURE 1-2. From hospitals to petrochemical plants, pure chemicals and chemical mixtures are used to create millions of consumer products. According to the Chemical Abstracts Service (CAS), which produces the largest databases on chemical information including the CAS Registry (https://www.cas.org/), approximately 83 million organic and inorganic substances are registered for use in commerce in the United States, with several thousand new ones being introduced each year. The bulk of the new chemical substances are industrial chemicals, household cleaners, and lawn care products. To that end, your response to almost any situation could include a hazardous materials/WMD component.

Responder Safety Tip

Responding personnel must be able to identify the substances involved in a release and place them in the proper context. In other words, ask yourself these questions: What is the substance? What are the conditions under which it was released? What is it going to do next?

FIGURE 1-2 A hazardous material can be found anywhere.

Manufacturing processes sometimes generate hazardous wastes. A hazardous waste is what remains after a process or manufacturing activity has used a particular substance and the material is no longer pure. Hazardous waste can be just as dangerous as pure chemicals. It can also comprise mixtures of several chemicals, which may make it difficult to determine how the substance will react when it is released or if it comes in contact with other chemicals. The waste generated by the illegal production of methamphetamine, for example, may produce a dangerous mixture of chemical wastes.

Levels of Training: Regulations and Standards

To understand where you fit in as a responder, you must first recognize some of the regulatory drivers that apply to hazardous materials response, beginning with the difference between a regulation and a standard. *Regulations* are issued and enforced by governmental bodies such as OSHA (part of the U.S. Department of Labor) and the EPA. Conversely, *standards* are issued by nongovernmental entities and are generally consensus based. A standard may be voluntary, meaning that an agency such as a fire department may not be required to adopt and follow the standard completely. For example, organizations such as the NFPA issue voluntary consensus-based standards that anyone can comment on before committee members agree to adopt them. The technical committee responsible for periodically revising any NFPA standard is required to meet regularly; revise, update, and possibly change a standard; and review and take action on any public comments during the revision process. Once the standard is finalized, agencies may *choose* to adopt it.

Responder Safety Tip

Being aware of your surroundings and observing the visual clues is your first defense against danger. Be aware of the hazards in your response area.

RESPONDER TIP

NFPA standards are voluntary and consensus based, which means that they can be adopted entirely or in part by an agency. This status is different from that of OSHA laws and regulations; compliance with OSHA mandates is not voluntary. Violation of an OSHA standard is a violation of law.

Each state in the United States has the right to adopt and/or supersede workplace health and safety regulations put forth by the federal agency OSHA. States that have adopted the OSHA regulations are called ***state-plan states***. California, for example, is a state-plan state; its regulatory body is called Cal-OSHA. About half of the states in the United States are state-plan states. States that have not adopted the OSHA regulations are non-plan states. Non-plan states are considered to be EPA states because they follow Title 40 of the Code of Federal Regulations (CFR), *Protection of the Environment, Part 311, Worker Protection*. The CFR is a collection of permanent rules published by the federal government. It includes 50 titles that represent broad areas of interest that are governed by federal regulation.

RESPONDER TIP

Responders should understand the relationship between OSHA regulations and NFPA standards when it comes to hazardous materials/WMD response. OSHA regulations are the law that governs hazardous materials/WMD responders; NFPA standards are guidelines that agencies choose to adopt. NFPA 472 clearly states: "First responders at the operational level also shall receive any additional training to meet applicable DOT, EPA, OSHA, and other state, local, or provincial occupational health and safety regulatory requirements."

NFPA standards governing hazardous materials/WMD response come from the Technical Committee on Hazardous Materials Response Personnel. This group includes more than 30 members from private industry, the fire service and law enforcement, professional organizations, and governmental agencies. Currently, two published standards are especially important to personnel who may be called upon to respond to hazardous materials/WMD incidents: NFPA 472, *Standard for Competence of Responders to Hazardous Materials/Weapons of Mass Destruction Incidents*, and NFPA 473, *Standard for Competencies for EMS Personnel Responding to Hazardous Materials/Weapons of Mass Destruction Incidents*. Extracts from each standard can be found in Appendix A and Appendix B. Generally speaking, NFPA 472 outlines training competencies for all hazardous materials responders; NFPA 473 spells out the competencies required for emergency medical services (EMS) personnel rendering medical care at hazardous materials/WMD incidents. NFPA 473 contains mission-specific competencies in a similar manner as NFPA 472. These mission-specific competencies include the following:

1. ALS Responder Assigned to a Hazardous Materials Team

2. ALS Responder Assigned to Provide Clinical Interventions at a Hazardous Materials/WMD Incident
3. ALS Responder Assigned to Treatment of Smoke Inhalation Victims

In the United States, NFPA standards as well as EPA and OSHA regulations are important to those personnel called upon to respond to hazardous materials/WMD incidents. The OSHA document containing the hazardous materials response competencies is commonly referred to as **HAZWOPER (HAZardous Waste OPerations and Emergency Response)**. The complete HAZWOPER regulation can be found in CFR, Title 29, 1910.120. See Appendix D of this text for the HAZWOPER regulation. The training levels found in HAZWOPER, much like the training levels found in the NFPA 472 standard, are identified as awareness, operations, technician, specialist (*specialist* is recognized only in the OSHA/HAZWOPER regulation), and incident commander. NFPA 472, since its first writing, has been updated every 5 years. This is a much more frequent revision cycle than that applied to the OSHA HAZWOPER regulation. This rapid revision cycle may be the source of some differences in the definitions of the levels of training represented here.

The following descriptions provide a broad overview, as found in NFPA 472 and the OSHA HAZWOPER regulation, of the different levels of hazardous materials/WMD responders and training competencies. When reading NFPA 472, it is important to understand that the standard is organized to first spell out the *tasks* that a responder (awareness, operations, technician, incident commander) may be called upon to perform on the scene. The subsequent *training competencies* follow. To be clear, NFPA 472 and the HAZWOPER regulations are not "how to respond" documents. They provide no direction on how to plug leaking containers, use detection and monitoring devices, or help you make a decision about what level of protection to wear in a certain situation. They are intended to provide guidance on the competencies associated with the various training levels. To stay focused on the intent of this text—which is to train responders to the awareness and operations level in accordance with the current edition of NFPA 472—we will discuss in detail only the tasks that awareness level personnel and operations level responders may be expected to perform. A general overview of the training competencies for the other response levels—technician, specialist, and incident commander—is provided, but not intended to be definitive or comprehensive.

Awareness Level

According to NFPA 472, **awareness level personnel** are those persons who, in the course of their normal duties, could encounter an emergency involving hazardous materials/WMD and who are expected to recognize the presence of the hazardous materials/WMD, protect themselves, call for trained personnel, and secure the area. According to NFPA 472, a person with awareness level training is not considered to be a *responder*. Instead, these individuals are now referred to as awareness level *personnel*. Persons receiving this level of training are not typically called to the scene to respond; rather, awareness level personnel, such as public works employees or fixed facility security personnel, function in support roles.

Tasks that awareness level personnel may be expected to perform on the scene include these duties:

- Analyzing the incident to detect the presence of hazardous materials/WMD
- Identifying the name, United Nations/North American Hazardous Materials Code (UN/NA) identification number, type placard, or other distinctive marking applied for the hazardous materials/WMD involved
- Collecting information from the current edition of the *ERG* about the hazard
- Initiating and implementing protective actions consistent with the emergency response plan, the standard operating procedures, and the current edition of the *ERG*
- Initiating the notification process

OSHA's HAZWOPER view of the awareness level is different, mostly because it does designate awareness level personnel as "responders." According to OSHA (1910.120(q) (6)(i)), "first responders at the awareness level are individuals who are likely to witness or discover a hazardous substance release and who have been trained to initiate an emergency response sequence by notifying the proper authorities of the release. They would take no further action beyond notifying the authorities of the release."

NFPA 472 Awareness Level Goals

4.1.2.2 When already on the scene of a hazardous materials/WMD incident, the awareness level personnel shall be able to perform the following tasks:

(1) Analyze the incident to determine both the hazardous material/WMD present and the basic hazard and response information for each hazardous material/WMD agent by completing the following tasks:

(a) Detect the presence of hazardous materials/WMD.

(b) Survey a hazardous materials/WMD incident from a safe location to identify the name, UN/NA identification number, type of placard, or other distinctive marking applied for the hazardous materials/WMD involved.

(c) Collect hazard information from the current edition of the DOT *Emergency Response Guidebook*.

(2) Implement actions consistent with the emergency response plan, the standard operating procedures, and the current edition of the DOT *Emergency Response Guidebook* by completing the following tasks:

(a) Initiate protective actions.

(b) Initiate the notification process.

VOICES OF EXPERIENCE

Over the last 30 years in which I have been involved in hazardous materials response, many changes have occurred. From the promulgation of the first regulations to the advent of the NFPA, it has been to the benefit of the first responder to have an understanding of how standards and regulations affect what we do.

One of the most interesting parts of the job is the exhilaration of a response, but *how* we respond is truly a result of what we have learned from the standards and regulations that we work with. We train to respond to hazardous materials/WMD incidents using the OSHA HAZWOPER regulation; NFPA 472, *Standard for Competence of Responders to Hazardous Materials/Weapons of Mass Destruction Incidents*; and NFPA 473, *Standard for Competencies for EMS Personnel Responding to Hazardous Materials/Weapons of Mass Destruction Incidents*. These regulations and standards are outgrowths of previous laws that established the basis for hazardous materials/WMD response, such as the Resource Conservation and Recovery Act of 1976 (RCRA), the Comprehensive Environmental Response, Compensation, and Liability Act of 1980 (CERCLA), and the Superfund Amendments and Reauthorization Act of 1986 (SARA).

Several years ago, while I was responding to an incident, I found that knowing the rules helped to assure we actually played by the rules. While working at an incident involving the spill of a nasty product, the state environmental and labor agencies began asking us questions about appropriate transportation, appropriate protection, cleanup, and disposal of the hazard. All of that information was listed on the material safety data sheet in the section that described spill, waste, and disposal considerations for the material. This information was available because of the RCRA, CERCLA, and SARA regulations. We had followed all of the rules and knew their importance to the responder. We must understand the *why* of *what* we do, and ensuring familiarity with and knowledge of the standards and regulations is a good way to protect ourselves throughout a hazardous materials/WMD incident.

Glen Rudner
Transportation Technology Center Inc—SERTC
Pueblo, Colorado

Based on the OSHA HAZWOPER regulation, first responders at the awareness level should have sufficient training or experience to objectively demonstrate competency in the following areas:

- An understanding of what hazardous substances are and the risks associated with them
- An understanding of the potential outcomes of an incident
- The ability to recognize the presence of hazardous substances
- The ability to identify the hazardous substances, if possible
- An understanding of the role of the first responder awareness individual in the emergency response plan
- The ability to determine the need for additional resources and to notify the communication center

Operations Level

According to NFPA 472, operations level responders are those persons who are tasked to respond to hazardous materials/WMD incidents for the purpose of taking action to protect nearby persons, the environment, or property from the effects of the release. These persons may also have competencies that are specific to their response mission, expected tasks, and equipment and training as determined by the AHJ–the mission specific competencies listed earlier in the chapter.

NFPA 472 significantly expands the scope of an operations level responder by separating the operations level suite of competencies into two distinct categories: core competencies and mission-specific competencies. The core competencies are based on those vital tasks that operations level personnel should perform on the scene of a hazardous materials/WMD incident. Some of those tasks are outlined here:

- Analyze the scene of a hazardous materials/WMD incident to determine the scope of the emergency
- Survey the scene to identify containers and materials involved
- Collect information from available reference sources
- Predict the likely behavior of a hazardous material
- Estimate the potential harm the substances might cause
- Plan a response to the release, including selection of the correct level of personal protective clothing
- Perform decontamination
- Preserve evidence
- Evaluate the status and effectiveness of the response

This abbreviated list should serve only as an illustration of the NFPA 472 core competencies. Keep in mind that the core training competencies are designed to provide the skills, knowledge, and abilities to safely accomplish the tasks listed. Core competency training is required of all operations level responders on the scene, no matter what their function. One of the goals of the NFPA 472 standard is to better match the expected tasks that may be required of the responder with the training that the responder should receive.

In addition to undertaking core competency training, an individual AHJ may find the need to do more training based on an identified or anticipated "mission-specific" need. To that end, those responders who are expected to perform additional missions, beyond the core competencies, shall be trained to carry out those mission-specific responsibilities. Remember, NFPA 472 allows each agency to pick and choose the training program that makes the most sense *for its jurisdiction*. These mission-specific competencies are *nonmandatory* and should be viewed as optional. To that end, operations level responders may end up performing a limited suite of technician level skills, but do not have the broader knowledge and abilities of a hazardous materials technician.

NFPA 472 provides a mechanism to ensure that operations level responders, including those with mission-specific training, do not go beyond their level of training and equipment. This is done by using technician level personnel to provide direct guidance to the operations level responder operating on a hazardous materials/WMD incident. Operations level responders are expected to work under the direct control of a hazardous materials technician or allied professional who has the ability to continuously assess and/or observe the actions of the operations level responder, *and* to provide immediate feedback. Guidance by a hazardous materials technician or an allied professional may be provided through direct visual observation or through assessments communicated from the operations level responder(s) to a technician. The concept of mission-specific competency training is also discussed in the chapter *Implementing the Planned Response*.

This text follows the format of NFPA 472 in that, after the core competencies have been covered, an individual chapter is devoted to eight of the nine mission-specific competencies (competencies for bomb technicians are not included in this text):

- Use of personal protective equipment
- Performing technical decontamination
- Performing mass decontamination
- Evidence preservation and sampling
- Performing product control
- Performing victim rescue and recovery operations
- Response to illicit laboratory incidents
- Performing air monitoring and sampling

According to the OSHA HAZWOPER regulation, first responders at the operations level are individuals who respond to releases or potential releases of hazardous materials incidents as part of their normal duties for the purpose of protecting nearby persons, property, or the environment from the effects of the release FIGURE 1-3. The OSHA HAZWOPER regulation mandates that the operations level responders must

RESPONDER TIP

The term *tasked to respond* may raise a question about your role as an operations level responder in an emergency response involving a hazardous material/WMD. To clarify this point, consider this scenario: If the local emergency response system you operate in—whether as a career professional or as a volunteer—is activated (e.g., 911 call), and your agency will respond to the scene to provide on-scene emergency services to the hazardous materials/WMD incident, you are viewed as a "responder." Personnel termed *responders* may include fire and rescue assets, law enforcement personnel, emergency medical services workers (basic or advanced life support providers), experts or other employees from private industry, and/or other allied professionals.

FIGURE 1-3 Operations level responders.

Courtesy of Rob Schnepp

be trained to respond in a defensive fashion, without actually trying to stop the release directly, while avoiding contact with the released substance. Their function is to contain the release from a safe distance, keep it from spreading, and prevent or reduce the potential for human exposures. First responders at the operations level are expected to have received at least 8 hours of training or to have had sufficient experience to objectively demonstrate competency in the following areas in addition to those listed for the awareness level:

- Knowledge of the basic hazard and risk assessment techniques
- Knowledge of how to select and use proper personal protective equipment provided
- An understanding of basic hazardous materials terms
- Knowledge of how to perform basic control, containment, and/or confinement operations within the capabilities of the resources and personal protective equipment available with their unit
- Knowledge of how to implement basic decontamination procedures
- An understanding of the relevant standard operating procedures and termination procedures

Technician/Specialist Level

NFPA 472 defines individuals at the technician level as follows: The hazardous materials technician is a person who responds to hazardous materials/WMD incidents using a risk-based response process with the ability to:

- Analyze a problem involving hazardous materials/WMD

NFPA 472 Operations Level Goals

5.1.2.2 When responding to hazardous materials/WMD incidents, operations level responders shall be able to perform the following tasks:

(1) Analyze a hazardous materials/WMD incident to determine the scope of the problem and potential outcomes by completing the following tasks:

(a) Survey a hazardous materials/WMD incident to identify the containers and materials involved, determine whether hazardous materials/WMD have been released, and evaluate the surrounding conditions.

(b) Collect hazard and response information from SDS; CHEMTREC/CANUTEC/SETIQ; local, state, and federal authorities; and shipper/manufacturer contacts.

(c) Predict the likely behavior of a hazardous material/WMD and its container.

(d) Estimate the potential harm at a hazardous materials/WMD incident.

(2) Plan an initial response to a hazardous materials/WMD incident within the capabilities and competencies of available personnel and personal protective equipment by completing the following tasks:

(a) Describe the response objectives for the hazardous materials/WMD incident.

(b) Describe the response options available for each objective.

(c) Determine whether the personal protective equipment provided is appropriate for implementing each option.

(d) Describe emergency decontamination procedures.

(e) Develop a plan of action, including safety considerations.

(3) Implement the planned response for a hazardous materials/WMD incident to favorably change the outcomes consistent with the emergency response plan and/or standard operating procedures by completing the following tasks:

(a) Establish and enforce scene control procedures, including control zones, emergency decontamination, and communications.

(b) Where criminal or terrorist acts are suspected, establish means of evidence preservation.

(c) Initiate an incident command system (ICS) for hazardous materials/WMD incidents.

(d) Perform tasks assigned as identified in the incident action plan.

(e) Demonstrate emergency decontamination.

(4) Evaluate the progress of the actions taken at a hazardous materials/WMD incident to ensure that the response objectives are being met safely, effectively, and efficiently by completing the following tasks:

(a) Evaluate the status of the actions taken in accomplishing the response objectives.

(b) Communicate the status of the planned response.

Source: Reproduced with permission from NFPA 472, *Competence of Responders to Hazardous Materials/Weapons of Mass Destruction Incidents*, Copyright© 2013, National Fire Protection Association. This reprinted material is not the complete and official position of the NFPA on the referenced subject, which is represented only by the standard in its entirety.

- Select appropriate decontamination procedures
- Control a release using specialized protective clothing and control equipment

These persons may have additional competencies that are specific to their response mission, expected tasks, and equipment and training as determined by the AHJ. A number of very detailed training competencies for technician level training are outlined in Chapter 7 of NFPA 472. Technician level personnel are integral to the NFPA 472 standard because they are, in many cases, intended to "supervise" the activities of on-scene operations level responders.

According to the OSHA HAZWOPER regulation, individuals at the technician level respond to hazardous material releases or potential releases for the purpose of stopping the release FIGURE 1-4. Conceptually, this is similar in intent to NFPA 472 in that a technician will likely assume a more proactive role than a first responder at the operations level. Technicians typically function at a higher level in terms of their cognitive approach to the response. They should be proficient at implementing a comprehensive risk-based approach to solving the problem. They will approach the point of release to plug, patch, or otherwise mitigate the problem.

FIGURE 1-4 Hazardous materials technicians.

Courtesy of Staff Sgt. Brandie Session/U.S. Air Force

According to the OSHA HAZWOPER standard, hazardous materials technicians should have received at least 24 hours of training equal to the first responder operations level. In addition, technicians should have the competency, knowledge, and understanding necessary to fulfill the following duties:

- Implement the employer's emergency response plan
- Classify, identify, and verify known and unknown materials by using field survey instruments and equipment
- Function within an assigned role in the Incident Command System
- Select and use proper specialized chemical personal protective equipment
- Understand hazard and risk assessment techniques
- Perform advanced control, containment, and/or confinement operations within the capabilities of the resources and personal protective equipment available with the unit
- Understand and implement decontamination procedures
- Understand termination procedures
- Understand basic chemical and toxicological terminology and behavior

The specialist level is identified only in the OSHA HAZWOPER standard. This level of responder receives more specialized training than a hazardous materials technician. Practically speaking, however, the two levels are not dramatically different.

Incident Commander

According to NFPA 472, the incident commander (IC) is the person responsible for all incident activities, including the development of strategies and tactics and the ordering and release of resources. The incident commander must receive any additional training necessary to meet applicable governmental occupational health and safety regulations and the specific needs of the jurisdiction.

The OSHA HAZWOPER standard requires specific IC Hazardous Materials level training for those assuming command of a hazardous materials incident requiring action beyond the operations level. Individuals trained as incident commanders should have at least operations level training as well as additional training specific to commanding a hazardous materials incident. Incident commanders, who will assume control of the incident scene beyond the first responder awareness level, must receive at least 24 hours of training equal to the first responder operations level and also have competency in the following areas:

- Know and implement the employer's incident command system
- Know how to implement the employer's emergency response plan
- Know and understand the hazards and risks associated with chemical protective clothing
- Know how to implement the local emergency response plan
- Know of the state emergency response plan and of the Federal Regional Response Team
- Know and understand the importance of decontamination procedures

For more complete information regarding the IC Hazardous Materials position, refer to Chapter 8 of NFPA 472.

In addition to the initial training requirements for all response levels listed earlier, OSHA regulations require annual refresher training of sufficient content and duration to ensure that responders maintain their competencies or that they demonstrate competency in those areas at least yearly. Consult your local agency for more specific information on refresher training, other hazardous materials laws, regulations, and regulatory agencies.

Other Governmental Agencies

In addition to OSHA and the EPA, several other governmental agencies are concerned with various aspects of hazardous materials/WMD response. The DOT, for example, promulgates and publishes laws and regulations that govern the transportation of goods by highway, rail, pipeline, air, and, in some cases, marine transport.

As discussed earlier in this chapter, the EPA regulates and governs issues relating to hazardous materials in the environment. The EPA's version of OSHA's HAZWOPER can be found in CFR Title 40, *Protection of the Environment, Part 311, Worker Protection.* In addition to creating standardized training for hazardous materials response and hazardous waste site operations, the Superfund Amendments and Reauthorization Act of 1986 (SARA) created a method and standard practice for a local community to understand and be aware of the chemical hazards in the community. Under SARA Title III, the Emergency Planning and Community Right to Know Act (EPCRA) requires a business that handles certain types and amounts of chemicals to report storage type, quantity, and storage methods to the fire department and the local emergency planning committee. This activity may be quite complex and should be undertaken only by those personnel who fully understand the requirements and the process.

Local emergency planning committees (LEPC) gather and disseminate information about hazardous materials to the public. These voluntary organizations are made up of members of industry, transportation, media, fire and police agencies, and the public at large; they are established to meet the requirements of EPCRA and SARA. Essentially, LEPCs ensure that local resources are adequate to respond to a chemical event in the community. Responders should be familiar with their local LEPC and know how their department works with this committee.

One piece of data collected by the LEPC is a material safety data sheet (MSDS). Note that this term is changing to safety data sheet (SDS), to be more in line with the effort to standardize terminology worldwide by way of the Globally Harmonized System of Classification and Labelling of Chemicals (GHS). In short, the GHS intends to create a standard methodology to define and classify hazards posed by chemical substances along with a standard method to communicate the hazards—the SDS. The intent of the document remains the same. More information on GHS can be found at http://www.osha.gov/dsg/hazcom/ghs.html.

The SDS is a detailed profile of a single chemical or mixture of chemicals provided by the manufacturer and/or supplier of a chemical **FIGURE 1-5**. The SDS details specific information about a chemical's physical and chemical properties, toxicological data, and appropriate first aid information in case of an accidental human exposure. These chemical-specific data sheets are an important source of information for responders when dealing with hazardous materials/WMD incidents. Responders also use the services of national response centers; local, state, and federal authorities; and shipper/manufacturer contacts when dealing with an incident.

Each state has a State Emergency Response Commission (SERC). The SERC acts as the liaison between local and state levels of authority. Its membership includes representatives from agencies such as the fire service, police services, and elected officials. The SERC is charged with the collection and dissemination of information relating to hazardous materials emergencies.

The Difference Between a Hazardous Materials/WMD Incident and Other Types of Emergencies

In most cases, fire fighters should not approach a hazardous materials/WMD incident with the same mindset used in structural firefighting. Similarly, law enforcement officers should be cognizant of the hazardous materials/WMD implications of some actions, such as responding to clandestine drug laboratory incidents. Most of the time, and certainly when lives are not at stake, response to a hazardous materials/WMD emergency may take more time than it would take to fight a structure fire or serve a search warrant to a high-risk suspect. Such responses will *not* require responders to rush headlong into the problem. This is not to say that hazardous materials/WMD incidents are doomed to always unfold slowly. It is more about the process of arriving at the best decision in the shortest amount of time. When it comes to chemical incidents, there is often a lot of information to process. Additionally, hazardous materials/WMD incidents are approached with more of a responder risk management strategy and responder health and safety mentality than most structure fires.

If a rescue is required, or if the situation is imminently dangerous in some other way and requires quick action, events may move quickly. For the most part, however, a hazardous materials/WMD emergency requires time, forethought, and planning to ensure that the incident is handled safely and effectively.

Responders must understand that actions taken at hazardous materials/WMD incidents are largely dictated by the chemicals or hazards involved; environmental influences such as wind, rain, and temperature; and the way the chemicals behave during the release. In short, the nature and circumstance of the response dictate the tactics and strategy.

Additionally, those personnel operating at the scene of a hazardous materials/WMD incident must be conscious of the potential or actual law enforcement aspect of the incident. Especially where terrorist or other criminal acts are suspected, responders should be mindful of evidentiary issues associated with the incident. Being mindful of potential evidence at a hazardous materials/WMD event may facilitate later efforts to identify, capture, and prosecute the person responsible for the act. To that end, every responder on the scene must be cognizant of the impact his or her presence and actions will have on potential evidence. Although evidence preservation should never impede the efforts to eliminate the problem or slow life-saving operations, every responder should be diligent in remembering that his or her actions and observations may play a vital role in the successful prosecution of a criminal suspect.

When responding to hazardous materials/WMD incidents, make a conscious effort to change your perspective. Slow down, think about the problem and available resources, and take well-considered actions to solve it.

SAFETY DATA SHEET
ANHYDROUS AMMONIA

TI

DISTRIBUTORS:

TANNER INDUSTRIES, INC.

DIVISIONS:

NATIONAL AMMONIA	NORTHEASTERN AMMONIA
HAMLER INDUSTRIES	BOWER AMMONIA & CHEMICAL

735 Davisville Road, Third Floor, Southampton, PA 18966; 215-322-1238

CORPORATE EMERGENCY TELEPHONE NUMBER: 800-643-6226 CHEMTREC: 800-424-9300

DESCRIPTION

CHEMICAL NAME: Ammonia, Anhydrous
CAS REGISTRY NO: 7664-41-7
SYNONYMS: Ammonia
CHEMICAL FAMILY: Inorganic Nitrogen Compound
FORMULA: NH_3
MOL. WT: 17.03 (NH_3)
COMPOSITION: 99+% Ammonia

STATEMENT OF HEALTH HAZARD

HAZARD DESCRIPTION:
Ammonia is an irritant and corrosive to the skin, eyes, respiratory tract and mucous membranes. Exposure to liquid or rapidly expanding gases may cause severe chemical burns and frostbite to the eyes, lungs and skin. Skin and respiratory related diseases could be aggravated by exposure.

Not recognized by OSHA as a carcinogen.
Not listed in the National Toxicology Program.
Not listed as a carcinogen by the International Agency for Research on Cancer.

EXPOSURE LIMITS FOR AMMONIA: Vapor

OSHA	50 ppm,	35 mg / m^3 PEL	8 hour TWA
NIOSH	35 ppm,	27 mg / m^3 STEL 15 minutes	
	25 ppm,	18 mg / m^3 REL	10 hour TWA
	300 ppm,	IDLH	
ACGIH	25 ppm,	18 mg / m^3 TLV	8 hour TWA
	35 ppm,	27 mg / m^3 STEL 15 minutes	

TOXICITY: LD 50 (Oral / Rat) 350 mg / kg

PHYSICAL DATA

BOILING POINT: -28°F at 1 Atm.
PH: N/A
SPECIFIC GRAVITY OF GAS (air = 1): 0.596 at 32°F
SPECIFIC GRAVITY OF LIQUID (water = 1): 0.682 at 28°F (Compared to water at 39°F).
PERCENT VOLATILE: 100% at 212°F
APPEARANCE AND ODOR: Colorless liquid or gas with pungent odor.
CRITICAL TEMPERATURE: 271.4°F
GAS SPECIFIC VOLUME: 20.78 Ft^3/Lb at 32°F and 1 Atm.
VAPOR DENSITY: 0.0481 Lb/Ft^3 at 32°F
LIQUID DENSITY: 38.00 Lb/Ft^3 at 70°F
APPROXIMATE FREEZING POINT: -108°F
WEIGHT (per gallon): 5.15 pounds at 60°F
VAPOR PRESSURE: 114 psig at 70°F
SOLUBILITY IN WATER (per 100 pounds of water): 86.9 pounds at 32°F, 51 pounds at 68°F
SURFACE TENSION: 23.4 Dynes / cm at 52°F
CRITICAL PRESSURE: 111.5 atm

FIGURE 1-5 An example of an SDS for anhydrous ammonia. Courtesy of Tanner Industries, Inc., Southhampton, PA

Preplanning

It is a mistake to assume that a response to a hazardous materials/WMD incident begins when the alarm sounds. In reality, the response begins with your initial training, continuing education, and preplanning activities at target hazards and other potential problem areas throughout the jurisdiction or response district FIGURE 1-6. Target hazards include any occupancy type or facility that presents a high potential for loss of

RESPONDER TIP

The goal of the emergency responder is to favorably change the outcome of the hazardous materials/WMD incident. This is accomplished by sound planning and by establishing safe and reasonable response objectives based on the level of training. Don't do it if you're not trained to do it!

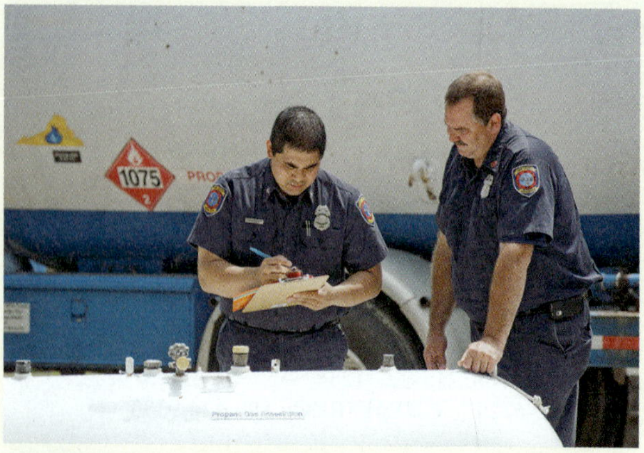

FIGURE 1-6 Conduct preincident planning activities at target hazards throughout the jurisdiction.

life or serious impact to the community resulting from fire, explosion, or chemical release.

Preplanning activities enable agencies to develop logical and appropriate response procedures for anticipated incidents. Planning should focus on the real threats that exist in your community or adjacent communities you could be assisting. Preplanning at major target hazards should include discussions and information sharing with the LEPC.

Once the threats have been identified, fire departments, police agencies, public health offices, and other governmental agencies should determine how they will respond and work together in case of a large-scale emergency. In many cases, the move toward interoperability before an incident will make the actual emergency response work run smoothly. Remember this important point: people making good decisions solve problems effectively. When people are acquainted with each other before the event happens, they tend to work together better. Get to know your peers around you at whatever level you operate.

Wrap-Up

Chief Concepts

- Governmental entities such as OSHA and the EPA issue and enforce regulations concerning hazardous materials emergencies.
- The consensus-based NFPA standards relating to hazardous materials/WMD incidents are available for those agencies that choose to adopt them.
- A hazardous material is any substance or material that is capable of posing an unreasonable risk to human health, safety, or the environment when transported in commerce, used incorrectly, or not properly contained or stored.
- The actions taken at hazardous materials/WMD incidents are largely dictated by the chemicals involved and the way they behave during the release.
- Two NFPA standards are important to responders who may be called upon to respond to hazardous materials/WMD incidents: NFPA 472, *Standard for Competence of Responders to Hazardous Materials/Weapons of Mass Destruction Incidents*, and NFPA 473, *Standard for Competencies for EMS Personnel Responding to Hazardous Materials/Weapons of Mass Destruction Incidents.*
- The OSHA HAZWOPER regulation is found in the CFR, Title 29, Standard 1910.120(q).
- The EPA's version of OSHA's HAZWOPER can be found in the CFR, Title 40, *Protection of the Environment, Part 311: Worker Protection.*
- The goals associated with the competencies of awareness level personnel are to recognize a potential hazardous materials emergency, to isolate the area, and to call for assistance. Awareness level personnel take protective actions only.
- NFPA 472 expands the scope of an operations level responder's duties by making a distinction between core competencies and mission-specific competencies.
- For those who choose to adopt the NFPA 472 standard, the core competencies are required for all operations level responders; each agency can then pick and choose to require any or all of the mission-specific responsibilities. The core competencies of operations level responders are defensive actions.
- The mission-specific responsibilities of operations level responders include personal protective equipment, technical and mass decontamination,

evidence preservation, product control, basic air monitoring, victim rescue, incidents occurring at illicit laboratories, and operations level responders assigned to perform improvised WMD dispersal device disablement/disruption and operations at improvised explosive laboratories.

- Hazardous materials technicians will approach the point of release so as to plug, patch, or otherwise mitigate a hazardous materials emergency.
- The hazardous materials incident commander is responsible for all incident activities.

Hot Terms

Authority having jurisdiction (AHJ) The governing body that sets operational policy and procedures for the jurisdiction you operate in.

Awareness level personnel Persons who, in the course of their normal duties, could encounter an emergency involving hazardous materials/weapons of mass destruction (WMD) and who are expected to recognize the presence of hazardous materials/WMD, protect themselves, call for trained personnel, and secure the area.

Chemical Abstracts Service (CAS) A division of the American Chemical Society. This resource provides hazardous materials responders with access to an enormous collection of chemical substance information—the CAS Registry.

Code of Federal Regulations (CFR) A collection of permanent rules published in the *Federal Register* by the executive departments and agencies of the U.S. federal government. Its 50 titles represent broad areas of interest that are governed by federal regulation. Each volume of the CFR is updated annually and issued on a quarterly basis.

Department of Transportation (DOT) The U.S. government agency that publicizes and enforces rules and regulations that relate to the transportation of many hazardous materials.

Emergency Planning and Community Right to Know Act (EPCRA) Legislation that requires a business that handles chemicals to report on those chemicals' type, quantity, and storage methods to the fire department and the local emergency planning committee.

Emergency Response Guidebook (ERG) A preliminary action guide for first responders operating at a hazardous materials incident in coordination with the U.S. Department of Transportation's (DOT) labels and placards marking system. The DOT and the Secretariat of Communications and Transportation of Mexico (SCT), along with Transport Canada, jointly developed the *Emergency Response Guidebook.*

Environmental Protection Agency (EPA) Established in 1970, the U.S. federal agency that ensures safe manufacturing, use, transportation, and disposal of hazardous substances.

Hazardous material Any substance or material that is capable of posing an unreasonable risk to human health, safety, or the environment when transported in commerce, used incorrectly, or not properly contained or stored.

Hazardous waste A substance that remains after a process or manufacturing plant has used some of the material and the substance is no longer pure.

HAZWOPER (HAZardous Waste OPerations and Emergency Response) The federal OSHA regulation that governs hazardous materials waste site and response training. Specifics to emergency response can be found in 29 CFR 1910.120(q).

Incident commander (IC) A level of training intended for those assuming command of a hazardous materials incident beyond the operations level. Individuals trained as incident commanders should have at least operations level training and additional training specific to commanding a hazardous materials incident.

Local emergency planning committee (LEPC) A committee made up of members of industry, transportation, the public at large, media, and fire and police agencies that gathers and disseminates information on hazardous materials stored in the community and ensures that there are adequate local resources to respond to a chemical event in the community.

Material safety data sheet (MSDS) *See* Safety data sheet (SDS).

National Fire Protection Association (NFPA) The association that develops and maintains nationally recognized minimum consensus standards on many areas of fire safety and specific standards on hazardous materials.

Occupational Safety and Health Administration (OSHA) The U.S. federal agency that regulates worker safety and, in some cases, responder safety. OSHA is a part of the U.S. Department of Labor.

Operations level responders Personnel who respond to hazardous materials/WMD incidents for the purpose of implementing or supporting actions to protect nearby persons, the environment, or property from the effects of the release.

Regulations Mandates issued and enforced by governmental bodies such as the U.S. Occupational Safety and Health Administration and the U.S. Environmental Protection Agency.

Safety data sheet (SDS) A form provided by manufacturers and compounders (blenders) of chemicals containing information about chemical composition, physical and chemical properties, health and safety hazards, emergency response, and waste disposal of a material.

Specialist level (OSHA/HAZWOPER only) A hazardous materials specialist who responds with, and provides support to, hazardous materials technicians. This individual's duties parallel those of the hazardous materials technician; however, the technician's duties require a more directed or specific knowledge of the various substances he or she may be called upon to contain. The hazardous materials specialist also acts as the incident-site liaison with federal, state, local, and other government authorities in regard to site activities.

Standards Guidelines issued by nongovernmental entities that are generally consensus based.

State Emergency Response Commission (SERC) The liaison between local and state levels that collects and disseminates information relating to hazardous materials emergencies. SERC includes representatives from agencies such as the fire service, police services, and elected officials.

Superfund Amendments and Reauthorization Act of 1986 (SARA) One of the first U.S. laws to affect how fire departments respond in a hazardous materials emergency.

Target hazards Any occupancy type or facility that presents a high potential for loss of life or serious impact to the community resulting from fire, explosion, or chemical release.

Technician level A person who responds to hazardous materials/WMD incidents using a risk-based response process by which he or she analyzes a problem involving hazardous materials/WMD, selects applicable decontamination procedures, and controls a release using specialized protective clothing and control equipment.

Weapons of mass destruction (WMD) Weapons whose use is intended to cause mass casualties, damage, and chaos. The NFPA includes the use of WMD in its definition of *hazardous materials*.

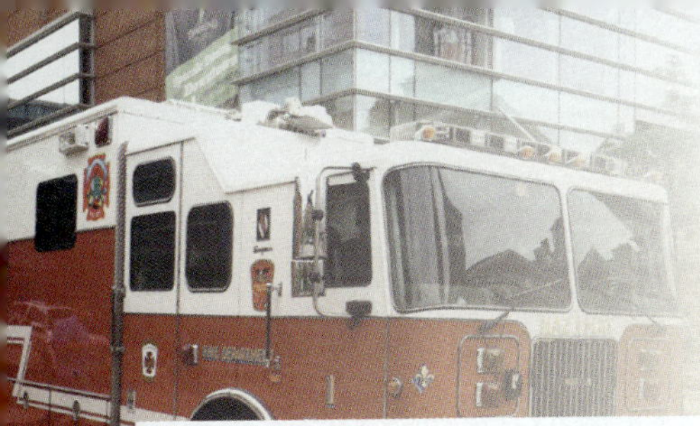

RESPONDER *in action*

Your volunteer fire company has just accepted five new members. You are tasked with providing them with core operations level hazardous materials training, so you spend several weeks preparing to instruct the class. The new members are given a preclass assignment to read through NFPA 472, *Standard for Competence of Responders to Hazardous Materials/Weapons of Mass Destruction Incidents* and to answer the following questions pertinent to the regulatory part of their training. Your instructional plan is to use this quiz to reinforce the main points of your introductory lecture.

1. Which consensus-based standard describes the hazardous materials training competencies for operations level responders?
   - **A.** NFPA 473, *Standard for Competencies for EMS Personnel Responding to Hazardous Materials/Weapons of Mass Destruction Incidents*
   - **B.** NFPA 472, *Standard for Competence of Responders to Hazardous Materials/Weapons of Mass Destruction Incidents*
   - **C.** Code of Federal Regulations 1910.150(q)
   - **D.** CFR Title 40, *Protection of the Environment, Part 311, Worker Protection*
2. NFPA 472 expands the scope of an operations level responder by separating the operations level competencies into two distinct categories. What are these two categories?
   - **A.** Technical and specialized competencies
   - **B.** Basic and advanced competencies
   - **C.** OSHA and NFPA competencies
   - **D.** Core and mission-specific competencies
3. According to the OSHA HAZWOPER regulation, which of the following competencies is *not* required of an operations level responder?
   - **A.** Knowledge of the basic hazard and risk assessment techniques
   - **B.** Knowledge of how to select and use proper personal protective equipment provided to the first responder
   - **C.** Knowledge of how to perform basic control, containment, and/or confinement operations within the capabilities of the resources and personal protective equipment available with the responder's unit
   - **D.** Knowledge of the classification, identification, and verification of known and unknown materials by using field survey instruments and equipment
4. What is the name of the federal document containing the hazardous materials response competencies? This regulation, issued in the late 1980s, standardized training for hazardous materials response and for hazardous waste site operations.
   - **A.** NFPA 471; paragraph 4, subsection (q)
   - **B.** NFPA 473; section 5.2.2.1
   - **C.** NFPA 472; CFR, book number 29, part 1910.120
   - **D.** HAZWOPER; CFR, Title 29, part 1910.120 subpart (q)

Hazardous Materials: Properties and Effects

Greg Henry/ShutterStock, Inc.

Courtesy of Rob Schnepp

© Photos.com

CHAPTER 2

Awareness Level Personnel

Knowledge Objectives

After studying this chapter, you will be able to:

- Explain the role of Awareness Level personnel during a hazardous materials/WMD incident. (NFPA 4.4.1(2), p. 31–32)
- Describe basic precautions when providing emergency care to victims of hazardous materials/WMD incidents. (NFPA 4.4.1(3)(a), p. 31–38)
- Identify common ignition sources found at a hazardous materials/WMD incident. (NFPA 4.4.1(3)(b), p. 22–23)
- Describe some ways in which hazardous materials are harmful to people and the environment. (NFPA 4.4.1(3)(c), p. 31–39)
- Describe the routes of exposure to hazardous materials for humans. (NFPA 4.4.1(3)(d), p. 35–37)

Skills Objectives

This chapter has no skills objectives for Awareness Level personnel.

Operations Level Responders

Knowledge Objectives

After studying this chapter, you will be able to:

- Describe the types of stress that can cause a container to release its contents, the ways in which containers can breach, and some of the dispersion patterns created by the release. (NFPA 5.2.3(2)(3)(4)(5), p. 20–22)
- Identify timeframes for estimating the duration that hazardous materials/WMD present risk of exposure. (NFPA 5.2.3(6), p. 37–39)
- Describe the following properties:
  - Boiling point (NFPA 5.2.3(1)(a)(i), p. 25)
  - Chemical reactivity (NFPA 5.2.3(1)(a)(ii), p. 21)
  - Corrosivity (pH) (NFPA 5.2.3(1)(a)(iii), p. 27)
  - Flammable (explosive) range [lower explosive limit (LEL) and upper explosive limit (UEL)] (NFPA 5.2.3(1)(a)(iv), p. 23–24)
  - Flash point (NFPA 5.2.3(1)(a)(v), p. 22–23)
  - Ignition (autoignition) temperature (NFPA 5.2.3(1)(a)(vi), p. 23)
  - Particle size (NFPA 5.2.3(1)(a)(vii), p. 36–37)
  - Persistence (NFPA 5.2.3(1)(a)(viii), p. 34)
  - Physical state (solid, liquid, gas) (NFPA 5.2.3(1)(a)(ix), p. 20)
  - Radiation (ionizing and non-ionizing) (NFPA 5.2.3(1)(a)(x), p. 29)
  - Specific gravity (NFPA 5.2.3(1)(a)(xi), p. 26)
  - Toxic products of combustion (NFPA 5.2.3(1)(a)(xii), (p. 28)
  - Vapor density (NFPA 5.2.3(1)(a)(xiii), p. 25)
  - Vapor pressure (NFPA 5.2.3(1)(a)(xiv), p. 24)
  - Water solubility (NFPA 5.2.3(1)(a)(xv), p. 27)
  - Physical change and chemical change (NFPA 5.2.3(1)(a)(ix), p. 20–22)
- Describe radiation (non-ionizing and ionizing) as well as the difference between alpha and beta particles, gamma rays, and neutrons. (NFPA 5.2.2(8), 5.2.3, p. 29–31)
- Describe the differences between the following pairs of terms:
  - Contamination and secondary contamination (NFPA 5.2.3(1)(b)(i), p. 31–32)
  - Exposure and contamination (NFPA 5.2.3(1)(b)(ii), p. 31–32)
  - Exposure and hazard (NFPA 5.2.3(1)(b)(iii) p. 31)
  - Infectious and contagious (NFPA 5.2.3(1)(b)(iv), p. 36)
  - Acute and chronic effects (NFPA 5.2.3(1)(b)(v), p. 37–38)
  - Acute and chronic exposures (NFPA 5.2.3(1)(b)(vi), p. 37–38)
- Describe the following types of weapons of mass destruction:
  - Blood agents (NFPA 5.2.3(9)(a), (p. 34)
  - Biological agents and biological toxins (NFPA 5.2.3(9)(b), (p. 32)
  - Choking agents (NFPA 5.2.3(9)(c), (p. 35)
  - Irritants (riot control agents) (NFPA 5.2.3(9)(d), (p. 35)
  - Nerve agents (NFPA 5.2.3(9)(e), (p. 32–34)
  - Radiological materials (NFPA 5.2.3(9)(f), (p. 32)
  - Vesicants (blister agents) (NFPA 5.2.3(9)(g), (p. 34)
- Describe the health and physical hazards of exposure to hazardous materials. (NFPA 5.2.3(7), p. 35–37)
- Describe the health hazards associated with the following:
  - Alpha, beta, gamma, and neutron radiation (NFPA 5.2.3(8)(a), p. 29–31)
  - Asphyxiant (NFPA 5.2.3(8)(b), p. 37)
  - Carcinogen (NFPA 5.2.3(8)(c), p. 37)
  - Convulsant (NFPA 5.2.3(8)(d), p. 35)
  - Corrosive (NFPA 5.2.3(8)(e), p. 27–28)
  - Highly toxic (NFPA 5.2.3(8)(f), p. 38)
  - Irritant (NFPA 5.2.3(8)(g), p. 35)
  - Sensitizer/allergen (NFPA 5.2.3(8)(h), p. 38)
  - Target organ effects (NFPA 5.2.3(8)(i), p. 37–38)
  - Toxic (NFPA 5.2.3(8)(j), p. 38)

Skills Objectives

This chapter has no skills objectives for Operations Level responders.

You Are the Responder

Just after midnight you receive a call for an explosion and a fire at a local landscaping company. Upon arrival at the site, you find a working fire in a storage shed located behind the main building. The wooden shed is fully involved, so you begin an indirect attack on the fire from the outside of the shed. As the fire is being knocked down, you receive an order over the radio to shut down the attack and move away from the building. About the same time, you experience an itchy feeling on the back of your neck. Other crew members are also complaining of itching and burning sensations around their wrists and necks. Some lower-floor residents of an adjacent multistory apartment complex are complaining of eye irritation and asking about a strange odor in the air.

1. Which types of chemicals might be found in this kind of occupancy?
2. Where could you obtain accurate technical information on the products stored in this building?
3. Which actions should be taken to address the complaints of burning and itching skin among fire fighters as well as the complaints of the residents of the apartment complex?

Introduction

To safely mitigate hazardous materials incidents, it is important to understand the chemical and physical properties of the substances involved. Chemical and physical properties are the characteristics of a substance that are measurable, such as vapor density, flammability, corrosivity, and water reactivity. However, you do not have to be a chemist to safely respond to hazardous materials incidents. In most cases, being an astute observer, referring to your emergency response plan and/or standard operating procedures, consulting the appropriate reference sources and correctly interpreting/understanding the visual clues presented to you will provide enough information to take basic actions at the incident.

This chapter will guide you in learning basic key terms that will help you digest the information contained in reference sources such as the safety data sheets (SDS).

FIGURE 2-1 The state of matter identifies the hazard as a solid, liquid, or gas. A. Solid. B. Liquid. C. Gas.

Physical and Chemical Changes

An important first step in understanding the hazard(s) associated with any chemical involves identifying the state of matter, or physical state, of the substance. The state of matter defines the substance as a solid, liquid, or gas **FIGURE 2-1**.

If you know the state of matter and other physical properties of the chemical, you can begin to predict what the substance will do if it escapes, or has escaped, from its containment vessel. For example, it would be vital to know if a released gas is heavier or lighter than air and how that particular physical property relates to the environmental factors at the time of the incident, along with the other characteristics of the emergency scene. Imagine a release of a heavy gas such as propane in the setting of a trench rescue or other below-grade emergency: The physical properties of propane could be a complicating factor to performing a safe rescue in such a scenario.

Another critical part of comprehending the nature of the release comes from identifying the reason(s) the containment vessel failed. In many cases, responders focus on the fact that a substance *is* being released rather than understanding *why* the product is escaping its container. It's one thing to notice that a container is leaking or generating a cloud from a puncture, crack, split, or tear. It's equally important, however, to figure out what type of stress caused the vessel to breach or release its contents in the first place. Is the product release caused by a thermal influence—heat from inside or outside the container that may be the root cause of a catastrophic container failure? Maybe an errant forklift driver struck a drum, or a valve failure occurred. In these examples, physical damage allows the release of the container contents. In other cases, a chemical reaction inside a container may cause the entire container to breach **FIGURE 2-2**.

Responders must also determine and/or estimate the duration of the event and link that to the other information gathered. For example, you should think of all parameters of the event as clues, and use those clues to form a hypothesis and action plan. What's leaking? How did it get out of its container? How long might the event last and what happens if we

FIGURE 2-2 Four examples of container failure. To understand the nature of the release, it's important to determine the reason behind the failure. Look at the four examples and think about potential causes of the breach.

Courtesy of Rob Schnepp

do nothing? If we choose to intervene, are we trained to handle the incident? Do we have the proper personal protective equipment (PPE)? What is the likelihood of a positive outcome—safely resolving the problem? Incidents may last anywhere from seconds and minutes to several days and, in extreme cases, months or years. Tactics and strategies may vary greatly depending on the projected duration of the event. At any hazardous materials incident, it's important to link together all the bits of information to make an informed decision about how to handle the problem.

Chemicals can undergo a physical change when they are subjected to environmental influences such as heat, cold, and pressure. For example, when water is frozen, it undergoes a physical change from a liquid to a solid. The actual chemical make-up of the water (H_2O) is still the same, but the state in which it exists is different.

Like water, propane gas is subject to physical change based on environmental influences such as heat and cold. If a propane cylinder is exposed to heat, the compressed liquefied propane inside changes phase, becoming gaseous propane—which, in turn, increases the pressure inside the vessel. If this uncontrolled expansion takes place faster than the relief valve can vent, a BLEVE could occur. A BLEVE (boiling liquid expanding vapor explosion) occurs when pressurized liquefied materials (propane or butane, for example) inside a closed vessel are exposed to a source of high heat. In this case, the chemical make-up of propane has not changed; rather, the propane has physically changed state from liquid to gas.

The expansion ratio is a description of the volume increase that occurs when a compressed liquefied gas (e.g., propane, chlorine, oxygen) material changes to a gas. For example, propane has an expansion ratio of 270 to 1. This means that for every 1 volume of liquid propane, 270 times that amount of propane exists as vapor (the gas phase of a substance). If released into the atmosphere, 1 gallon of liquid propane would vaporize to 270 gallons of propane vapor.

Chemical reactivity (also known as chemical change) describes the ability of a substance to undergo a transformation at the molecular level, usually with a release of some form of energy. Physical change is essentially a change in state. By contrast, a chemical change results in an alteration of the chemical nature of the material. Steel rusting and wood burning are examples of chemical change.

Responder Safety Tip

The classic example of a BLEVE is a fire occurring below or adjacent to a propane tank. Once the liquid inside the vessel begins to boil, large volumes of vapor are generated within the vessel. Ultimately, the vessel fails catastrophically if it is unable to relieve the pressure through the safety valve. The ensuing explosion throws fragments of the vessel in all directions. When the pressurized liquefied material is flammable, a tremendous fireball is generated. Additionally, an overpressure blast wave is created as a result of the rapidly expanding vapor released by the vessel failure.

FIGURE 2-3 Responders should take great care when handling incidents involving compressed liquefied gases. Courtesy of Rob Schnepp

If fire is impinging on a vessel that contains pressurized liquefied materials, responders should carefully evaluate the risks of attempting to fight the fire FIGURE 2-3. BLEVEs have claimed the lives of many fire fighters throughout the years—and this history should serve as a reminder of the dangers of these types of emergency situations.

Responder Safety Tip

It is important to understand that chemical change is different from physical change. The reactive nature of any substance can be influenced in many ways, such as by mixing it with another substance (adding an acid to a base) or by applying heat.

To relate the concepts of physical and chemical change to a hazardous materials incident, think back to the initial case study. Assume for a moment that the owner of the landscape company wadded up some rags soaked with linseed oil and left them in a corner of the shed. The rags spontaneously ignited and started a fire that ultimately caused the failure of a small propane cylinder (the explosion that occurred prior to your arrival). The fire also melted the fusible plug on a chlorine cylinder (at approximately 160°F [71°C]), causing a release of chlorine gas; chlorine gas is 2.5 times heavier than air.

Looking closely at each event in this sequence, you can see the effect of physical and chemical changes. Linseed oil is an organic material that generates heat as it decomposes (chemical change). That heat, in the presence of oxygen in the surrounding air, ignited the rags, which in turn ignited other combustibles. The surrounding heat caused the propane inside the tank to expand (physical change) until it overwhelmed the ability of the relief valve to handle the build-up of pressure. The cylinder ultimately exploded. At the same time, the escaping chlorine gas mixed with the moisture in the air and formed an acidic mist (a chemical change). A slight breeze carried the smoke and mist toward the apartment complex. The fire fighters began to feel itchy and burning sensations where the acidic mist contacted their moist skin. Because chlorine vapors are heavier than air, most of the residents complaining about eye irritation are located on the bottom floor of the complex.

Critical Characteristics of Flammable Liquids

When looking at the fire potential of a flammable liquid, several important aspects must be considered. Among them are flash point, ignition temperature, and flammable range. More important than memorizing the definitions for these terms, however, responders must understand the relationships between these and other physical characteristics of a flammable liquid. Keep in mind that when it comes to combustion (burning), all materials must be in a gaseous or vapor state prior to flaming combustion: Solids do not burn, and liquids do not burn—they give off a gas or vapor that is ultimately ignited. Think of a log burning in a fireplace. If you look closely, the fire is not directly on the log, but rather appears slightly above the surface. The reason for this phenomenon is that the wood has to be heated to the point where it produces enough "wood gas" to support combustion. The log, then, does not burn—the gas produced by heating the log is what starts and sustains the process of combustion.

Conceptually, liquid fuels such as gasoline, diesel fuel, and other hydrocarbon-based dissolving solutions (solvents) behave the same way as the log. One way or another, vapor production must occur before there can be fire. This vapor production must be factored in when estimating the probability of fire during a release.

Flash Point

Flash point is an expression of the minimum temperature at which a liquid or solid gives off sufficient vapors such that, when an ignition source (e.g., a flame, electrical equipment, lightning, or even static electricity) is present, the vapors will result in a flash fire. The flash fire involves only the vapor phase of the liquid (like the example of the log) and will go out once the vapor fuel is consumed.

To illustrate how such an event occurs, consider the flash point of gasoline at –45°F (–43°C). When the temperature of gasoline reaches –45°F (–43°C) as a result of heat from an external source or from the surrounding environment, it gives off sufficient flammable vapors to support combustion. In nearly all circumstances, when gasoline is spilled or otherwise released, the temperature of the external environment is well above the flash point of gasoline, creating the potential for ignition. Diesel fuel, by comparison, has a much higher flash point than gasoline—in the range of approximately 120°F (49°C) to 140°F (60°C), depending on the fuel grade. In either case, once the temperature of the liquid surpasses its flash point, the fuel will give off sufficient flammable vapors to support combustion FIGURE 2-4.

FIGURE 2-4 Responders should always be mindful of ignition sources at flammable/combustible liquid incidents.

Additionally, liquids with low flash points—such as gasoline, ethyl alcohol, and acetone—typically have higher vapor pressures and higher ignition temperatures. Vapor pressure and ignition temperature are explained in more detail later in this chapter. Consider the case of gasoline, whose flash point is 45°F (–43°C). At a vapor pressure in excess of 275 mm Hg (compared to the vapor pressure of ethyl alcohol of 40 mm Hg at approximately 68°F [20°C]), gasoline has an ignition temperature of approximately 475°F (246°C). By contrast, flammable/combustible liquids with high flash points typically have lower ignition temperatures and lower vapor pressures. A particular grade of diesel—#2 grade diesel, for example—may have a flash point of 125°F (52°C) and a corresponding ignition temperature of approximately 500°F (260°C); this is clearly a much narrower range than that described in the gasoline example. The vapor pressure of #2 grade diesel fuel is very low—much lower than the vapor pressure of water. TABLE 2-1 lists several

Table 2-1 Flash Point, Vapor Pressure, and Ignition Temperature

	Flash Point	Vapor Pressure	Ignition Temperature
Water	N/A	25 mm Hg at 68°F (20°C)	N/A
Gasoline	–45°F (–43°C)	275–400 mm Hg at 70°F (21°C)	475°F (246°C)
Acetone	–4°F (–20°C)	400 mm Hg at 104°F (40°C)	869°F (465°C)
#2 grade diesel	125°F (52°C)	<2 mm Hg at 68°F (20°C)	500°F (260°C)

other examples of flash points of flammable/combustible liquids, illustrating the relationships among flash point, ignition temperature, and vapor pressure.

A direct relationship also exists between temperature and vapor production. Simply put, when the temperature increases, the vapor production of any flammable liquid increases, leading to a higher concentration of vapors produced. Therefore, even liquids with low flash points can be expected to produce a significant amount of flammable vapors at all but the lowest ambient temperatures.

Flash point is merely one aspect to consider when flammable/combustible liquids are released from a container. The fire point is another definition that should be appreciated. Fire point is the temperature at which sustained combustion of the vapor will occur. It is usually only slightly higher than the flash point for most materials.

Ignition Temperature

Ignition (autoignition) temperature is another important temperature landmark for flammable/combustible liquids. From a technical perspective, you can think of ignition temperature as the minimum temperature at which a fuel, when heated, will ignite in the presence of air and continue to burn.

From a responder's perspective, it is important to realize that when a liquid fuel is heated beyond its ignition temperature, from any type of heat, it will ignite without an external ignition source. Think of a pan full of cooking oil on the stove. For illustrative purposes, assume that the ignition temperature of the oil is 300°F (148°C). What would happen if the burner was set on high and left unattended, so that the oil was heated past 300°F (148°C)? Once the temperature of the oil exceeds its ignition temperature, it will ignite; there is no need for an external ignition source. In fact, this scenario is a common cause of stove fires.

Flammable Range

Flammable range is another important term to understand. Defined broadly, flammable range is an expression of a fuel/air mixture, defined by upper and lower limits, that reflects an amount of flammable vapor mixed with a given volume of air. Gasoline will serve as our example. The flammable range for gasoline vapors is 1.4 percent to 7.6 percent. The two percentages, called the lower explosive limit (LEL) (1.4 percent) and the upper explosive limit (UEL) (7.6 percent), define the boundaries of a fuel/air mixture necessary for gasoline to burn properly. If a given gasoline/air mixture falls between the LEL and the UEL, and that mixture comes in contact with an ignition source, a flash fire will occur.

The concept of automobile carburetion capitalizes on the notion of flammable range—the carburetor is the place where gasoline and air are mixed. When the mixture of gasoline vapors and air occurs in the right proportions, the car runs smoothly. When there is too much fuel and not enough air, the mixture is considered to be too "fuel rich." When there is too much air and not enough fuel, the carburetion is too "fuel lean." In either of these two cases, optimal combustion is not achieved and the motor does not run correctly.

Understanding flammable range, as it relates to hazardous materials response, is based on the same line of thinking. If gasoline is released from a rolled-over cargo tank, the vapors from the spilled liquid will mix with the surrounding air. If the mixture of vapors and air falls between the UEL and the LEL, and those vapors reach an ignition source, a flash fire will occur. Ultimately, the entire volume of gasoline will likely burn. As with carburetion, if too much or too little fuel is present, the mixture will not adequately support combustion.

Generally speaking, the wider the flammable range, the more dangerous the material **TABLE 2-2**. This relationship reflects the fact that the wider the flammable range, the more opportunity there is for an explosive mixture to find an ignition source.

Table 2-2	Flammable Ranges of Common Gases
	Flammable Ranges of Common Gases
Hydrogen	4.0%–75%
Natural gas	5.0%–15%
Propane	2.5%–9.0%

Vapor Pressure

For our purposes, the definition of vapor pressure will pertain to liquids held inside any type of closed container. When liquids are held in a closed 55-gallon drum or a 4-liter glass bottle, for example, some amount of pressure (in the headspace above the liquid) will develop inside **FIGURE 2-5**. All liquids, even water, will develop a certain amount of pressure in the airspace between the top of the liquid and the container.

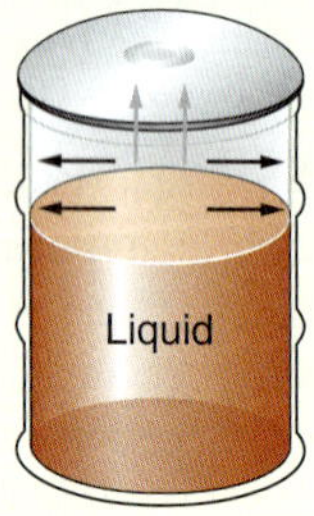

FIGURE 2-5 Vapor pressure.

The key point to understanding vapor pressure is this: *The vapors released from the surface of any liquid must be contained if they are to exert pressure.* Essentially, the liquid inside the container will vaporize until the molecules given off by the liquid reach equilibrium with the liquid itself. Equilibrium is a balancing act between the liquid and the vapors—some molecules turn to vapor, whereas others leave the vapor phase and return to the liquid phase.

Carbonated beverages illustrate this concept. Inside the basic cola drink is the formula for the soda and a certain amount of carbon dioxide (CO_2) molecules. The bubbles in the soda are composed of CO_2. When the soda sits in an unopened can or plastic bottle on the shelf, the balancing act of CO_2 is happening: CO_2 from the liquid becomes gaseous CO_2 above the liquid; at the same time, some of the CO_2 dissolves back into the liquid. The pressure inside the bottle can be verified by feeling the rigidity of the container—until you open it. Once the lid is removed, the CO_2 is no longer in a closed vessel, and it escapes into the atmosphere. With the pressure released, the sides of the bottle can be squeezed easily. Ultimately, the soda goes flat because all of the CO_2 has escaped.

With the soda example in mind, consider the technical definition of vapor pressure: The vapor pressure of a liquid is the pressure exerted by its vapor until the liquid and the vapor are in equilibrium. Again, this process occurs inside a closed container, and temperature has a direct influence on the vapor pressure. For example, if heat impinges on a drum of acetone, the pressure above the liquid will increase and perhaps cause the drum to fail. If the temperature is dramatically reduced, the vapor pressure will drop. This is true for any liquid held inside a closed container.

What happens if the container is opened or spilled onto the ground to form a puddle? The liquid still has a vapor pressure but it is no longer confined to a container. *In this case, we can conclude that liquids with high vapor pressures will evaporate much more quickly than will liquids with low vapor pressures.* Vapor pressure directly correlates to the speed with which a material will evaporate once it is released from its container.

RESPONDER TIP

Two specific points of temperature 68°F (20°C) and pressure (14.7 psi) are commonly referred to as *normal temperature and pressure* (NTP). These values represent standard conditions under which chemical and physical properties can be tested and documented. Another term, *standard temperature and pressure* (STP), as defined by the International Union of Pure and Applied Chemistry (IUPAC), describes an ideal testing environment for a variety of chemical and physical processes: 14.7 psi at 32°F (0°C).

For example, motor oil has a low vapor pressure. When it is released, it will stay on the ground a long time. Chemicals such as isopropyl alcohol and diethyl ether exhibit the opposite behavior: When either of these materials is released and collects on the ground, it will evaporate rapidly. If ambient air temperature or pavement temperatures are elevated, their evaporation rates will increase even further. Wind speed, shade, humidity, and the surface area of the spill also influence how fast the chemical will evaporate.

When consulting reference sources, be aware that the vapor pressure may be expressed in pounds per square inch (psi), atmospheres (atm), torr, or millimeters of mercury (mm Hg); most references give the vapor pressures of substances at a temperature of 68°F (20°C). Each expression of pressure is a valid point of reference in emergency response. The term *millimeters of mercury* (mm Hg) is commonly found in reference books; it is defined as the pressure exerted at the base of a column of fluid that is exactly 1 millimeter in height. The *torr* unit is named after the Italian physicist Evangelista Torricelli, who discovered the principle of the mercury barometer in 1644. In his honor, the torr was equated to 1 mm Hg. Another common expression of pressure is bar. The conversion from bar to psi is: 1 bar = 14.7 psi.

Certain conversion factors allow calculations from one reference point to another, but some of these factors are very complex. For the purposes of simplicity and emergency response, the important point is to have some frame of reference to understand the concept of vapor pressure and to recognize how that concept will affect the release of chemicals into the environment.

To understand the relationships among the various units of pressure, use the following comparison:

14.7 psi = 1 atm = 760 torr = 760 mm Hg = 1 bar

Another method of comparison is to take the values from the reference books and compare those values to the behavior of substances you may be familiar with. The following example compares three common substances (using mm Hg): water, motor oil, and isopropyl alcohol. The vapor pressure of water at room temperature is approximately 25 mm Hg. Standard 40-weight motor oil has a vapor pressure of less than 0.1 mm Hg at 68°F (20°C)—it is practically vaporless at room temperature. Isopropyl alcohol, by contrast, has a high vapor pressure of 30 mm Hg at room temperature. Again, temperature has a direct correlation to vapor pressure, but all things being equal, these three substances give you a good starting point for making comparisons.

Boiling Point

Boiling point is the temperature at which a liquid will continually give off vapors in sustained amounts and, if held at that temperature long enough, will turn completely into a gas. The boiling point of water, for example, is 212°F (100°C). At this temperature, water molecules have enough kinetic energy (energy in motion) to overcome the downward force of the surrounding atmospheric pressure (at sea level, a pressure of 14.7 psi is exerted on every surface of every object, including the surface of the water) FIGURE 2-6. At temperatures less than 212°F (100°C), there is insufficient heat to create enough kinetic energy to allow the water molecules to escape.

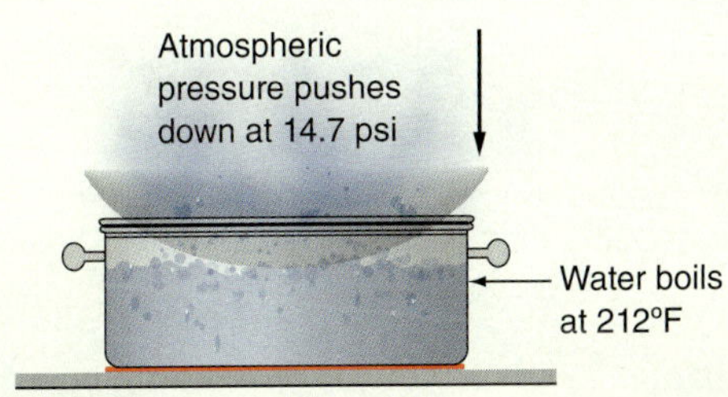

FIGURE 2-6 The concept of boiling point versus atmospheric pressure.

An illustration of boiling point, which demonstrates how heated liquids inside closed containers might create a deadly situation, can be found by looking at a completely benign and common example—popcorn. Fundamentally, popcorn pops when the trapped water inside the kernel of corn is heated beyond its boiling point. When the popcorn kernels are heated, the small amount of water inside each kernel eventually exceeds its boiling point and expands to 1700 times its original water volume. The kernel then "pops." A kernel of corn has no relief valve, so a rapid build-up of pressure will cause it to breach. Unpopped kernels are those with insufficient water inside the kernel to permit this expansion.

Responder Safety Tip

Flammable liquids with low boiling points are dangerous because they have the potential to produce large volumes of flammable vapor at relatively low temperatures.

RESPONDER TIP

The molecular weight of a flammable liquid dictates the behavior of the substance once it is released. A molecule of motor oil, for example, is much larger than a molecule of isopropyl alcohol. Some general conclusions about the relationships among vapor pressure, boiling point, flash point, ignition temperature, and heat output can be linked based on the molecular size of a chemical compound. Specifically, as the molecular weight of a substance increases:

⇓ Vapor pressure decreases.
⇑ Boiling point increases.
⇑ Flash point increases.
⇓ Ignition temperature decreases.
⇑ Heat output increases.

Vapor Density

In addition to identifying the flammability of a particular vapor or gas, responders must determine whether the vapor or gas is heavier or lighter than air. In essence, you must know the vapor density of the substance in question. Vapor density is the weight of an airborne concentration of a vapor or gas as compared to an equal volume of dry air FIGURE 2-7.

Basically, vapor density is a question of comparison—what will the gas or vapor do when it is released in the air? Will it collect in low spots in the topography or somewhere inside a building, or will it float upward into the ventilation system or upward into the air? These are important response considerations for hazardous materials incidents and may influence the severity of the incident.

You can find a chemical's vapor density by consulting a good reference source such as an SDS. The vapor density will be expressed in numerical fashion, such as 1.2, 0.59, or 4.0. The vapor density of propane, for example, is 1.51. Air has a set vapor density value of 1.0. Gases such as propane or chlorine are heavier than air; therefore, they have a vapor density value greater than 1.0. Substances such as acetylene, natural gas, and hydrogen are lighter than air and will have a vapor density value of less than 1.0.

FIGURE 2-7 Cylinder A. Vapor density less than 1. Cylinder B. Vapor density greater than 1.

In the absence of reliable reference sources in the field, there is a mnemonic you can use to remember a number of lighter-than-air gases: 4H MEDIC ANNA. This mnemonic translates as follows:

H: Hydrogen
H: Helium
H: Hydrogen cyanide
H: Hydrogen fluoride
M: Methane
E: Ethylene
D: Diborane
I: Illuminating gas (methane/ethane mixture)
C: Carbon monoxide
A: Ammonia
N: Neon
N: Nitrogen
A: Acetylene

If you encounter a leaking propane cylinder, for example, just refer back to 4H MEDIC ANNA. Propane is not on that list, so by default, you can assume that propane is heavier than air. You won't find chlorine or butane on the list, either: Both are heavier than air. Again, if the substance is not on this list, it is most likely heavier than air.

Specific Gravity

Specific gravity is to liquids what vapor density is to gases and vapors—namely, a comparison value. In this case, the comparison is between the weight of a liquid chemical and the weight of water. Water is assigned a value of 1.0 as its specific gravity. Any material with a specific gravity value less than 1.0 will float on water FIGURE 2-8. Any material with a specific gravity value greater than 1.0 will sink and remain below the surface of water. A comprehensive reference source will provide the specific gravity of the chemical in question.

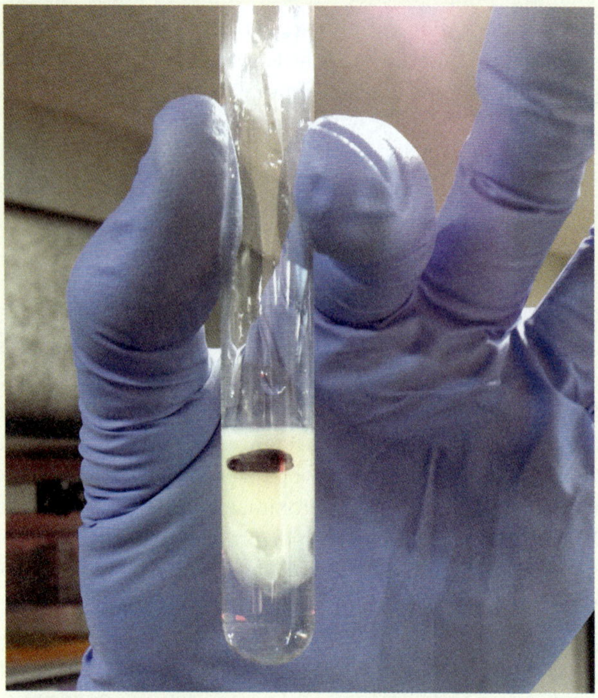

FIGURE 2-8 An example of a substance floating on water.

Courtesy of Rob Schnepp

Most flammable liquids will float on water. Gasoline, diesel fuel, motor oil, and benzene are all examples of liquids that float on water FIGURE 2-9. Carbon disulfide, by comparison, has a specific gravity of approximately 2.6. Consequently, if water is gently applied to a puddle of carbon disulfide, the water will rest on top and cover the puddle completely FIGURE 2-10.

FIGURE 2-9 Gasoline floating on water.

Courtesy of Rob Schnepp

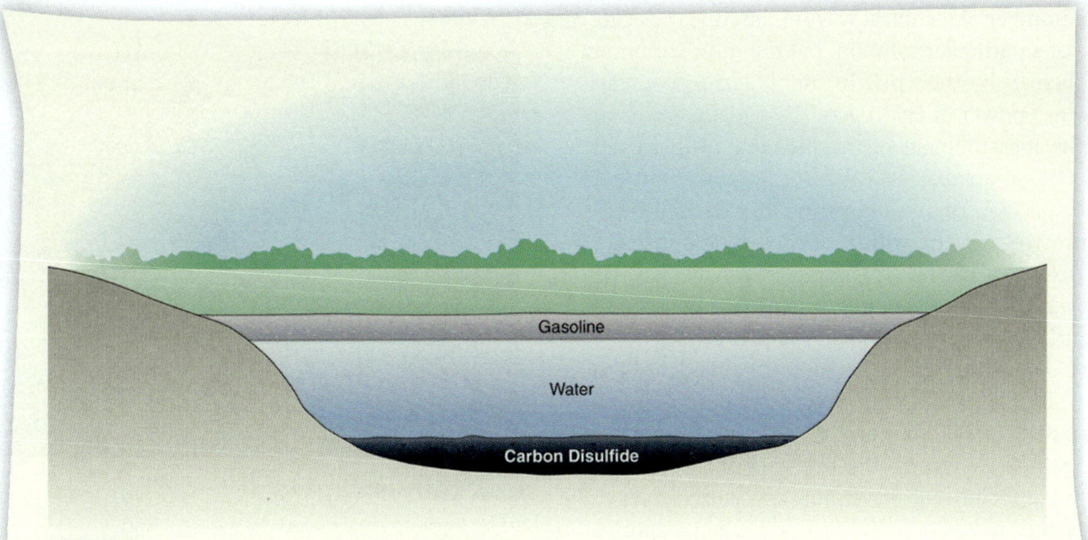

FIGURE 2-10 Gasoline will float on water, whereas carbon disulfide will not.

Responder Safety Tip

It is important to carefully consider the application of water in all cases involving spilled or released chemicals. The results could be something worse than the initial problem.

Water Solubility

When discussing the concept of specific gravity, it is also necessary to determine if a chemical will mix with water. Water solubility describes the ability of a substance to dissolve in water. Water is the predominant agent used to extinguish fire, but when you are dealing with chemical emergencies, it may not always be the best and safest choice to mitigate the situation. That's because water is a tremendously aggressive solvent and has the ability to react violently with certain chemicals.

Concentrated sulfuric acid, metallic sodium, and magnesium are just a few examples of substances that will adversely react with water. If water is applied to burning magnesium, for example, the heat of the fire will break apart the water molecule, creating an explosive reaction. Adding water to sulfuric acid would be like throwing water on a pan full of hot oil—popping and spattering may occur.

In other circumstances, water can be a friendly ally when you are attempting to handle a chemical emergency. Depending on the chemical involved, fog streams operated from handlines may knock down vapor clouds. Additionally, heavier than water flammable liquids can be extinguished by gently applying water to the surface of the liquid. In some cases, water may be an effective way to dilute a chemical, thereby rendering it less hazardous.

Corrosivity (pH)

Corrosivity is the ability of a material to cause damage (on contact) to skin, eyes, or other parts of the body. The technical definition, as found in the U.S. Code of Federal Regulations, includes the language, "destruction or irreversible damage to living tissue at the site of contact." Such materials are often also damaging to clothing, rescue equipment, and other physical objects in the environment.

Corrosives are a complex group of chemicals that should not be taken lightly. The tens of thousands of corrosive chemicals used in general industry, semiconductor manufacturing, and biotechnology can be further categorized into two classes: acid and base FIGURE 2-11.

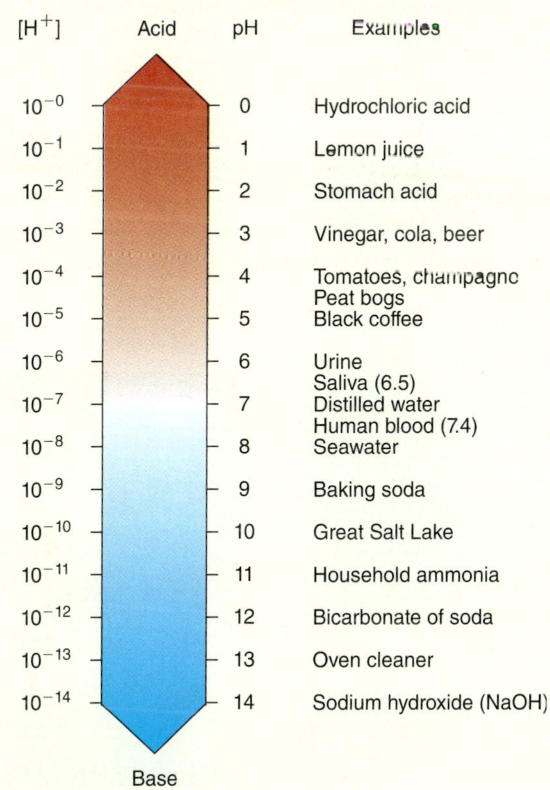

FIGURE 2-11 The pH scale.

There are a number of technical ways to describe the acidity or alkalinity of a particular solution, but the most common way to define them is by their pH. In simple terms, you can think of pH as the "power of Hydrogen." Essentially, pH is an expression of the measurement of the presence of dissolved hydrogen ions (H^+) in a substance. When thinking about pH from a field perspective, this value can be viewed as a measurement of corrosive strength, which can loosely be translated into a certain degree of hazard. In simple terms, we can use pH to judge how aggressive a particular corrosive might be. To assess that hazard, you must understand the pH scale.

How do we determine pH in the field? One method is to use specialized pH test paper. You can also obtain this information from the SDS, or call for a specialized hazardous materials response team to determine the pH level of a chemical. Hazardous materials technicians have more specialized ways of determining pH and may be a useful resource when responders are handling corrosive incidents.

Common acids, such as sulfuric, hydrochloric, phosphoric, nitric, and acetic (vinegar) acids, have a predominant amount of hydrogen ions (H^+) in the solution and, therefore, will have pH values less than 7 FIGURE 2-12. Chemicals that are considered to be bases, such as sodium and potassium hydroxide, sodium carbonate, and ammonium hydroxide, have a predominant amount of hydroxide (OH^-) ions in the solution and will have pH values greater than 7 FIGURE 2-13.

The middle of the pH scale (7) is where a chemical is considered to be neutral—that is, neither acidic nor basic. A pH of 7 is considered "neutral" because the concentration of hydrogen ions (H^+) in such a material is exactly equal to the concentration of hydroxide (OH^-) ions produced by dissociation of the water. At this value, a chemical will not harm human tissue.

Generally, pH values of 2.5 or less, and those of 12.5 or greater, are considered to be strong. In practical terms, this designation means that strong corrosives (acids and bases) will react more aggressively with metallic substances such as steel and iron; they will cause more damage to unprotected skin; they will react more adversely when contacting other chemicals; and they may react violently with water.

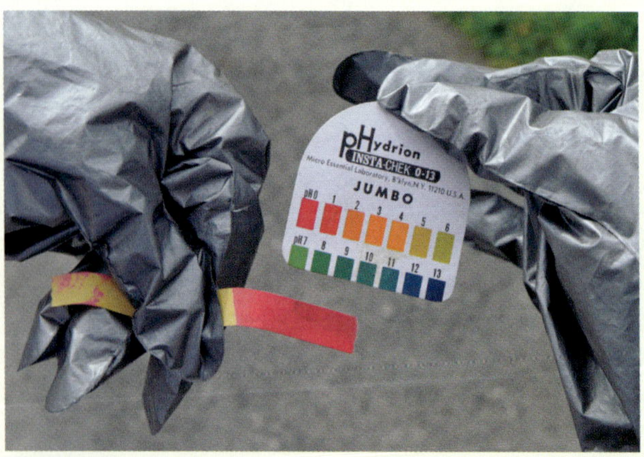

FIGURE 2-12 An acid.

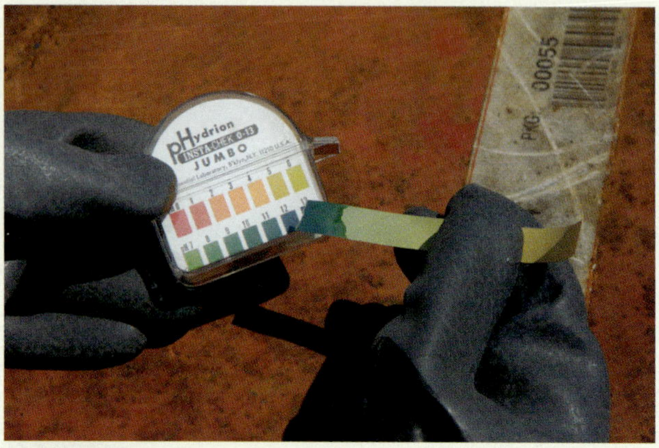

FIGURE 2-13 A base.

At the awareness operations level, it is critical to understand that incidents involving corrosives may not be as straightforward as other types of chemical emergencies. The materials themselves are more complicated and, in fact, the tactics employed to deal with them could be outside your scope of responsibilities.

Toxic Products of Combustion

Toxic products of combustion are the hazardous chemical compounds released when a material decomposes under heat. Recall that the process of combustion is a chemical reaction and, like other reactions, will generate a given amount of gases (many of them toxic) and particulates as by-products of the decomposition of the fuel. Have you really thought about what's in the smoke you may be breathing or taken a moment to think about the toxic gases liberated during a residential structure fire?

An easy way to think about this issue is to apply a phrase long used in the world of chemistry: garbage in, garbage out. This phrase reflects the idea that whatever objects are involved in the fire (chairs, tables, and sofas, for example) decompose and release a host of chemical by-products.

Notable substances found in most fire smoke include soot (carcinogen), carbon monoxide, carbon dioxide, polycyclic aromatic hydrocarbons, benzene (carcinogen),water vapor, formaldehyde (carcinogen), cyanide compounds, chlorine compounds, and many oxides of nitrogen. Each of these substances is unique in its chemical make-up and most are toxic to humans, even in small doses.

For example, carbon monoxide affects the ability of the human body to transport oxygen. When it is present in the body in excessive amounts, the red blood cells cannot get oxygen to the other cells of the body and, subsequently, a person will die from tissue asphyxiation. Cyanide compounds also adversely affect oxygen uptake in the body and are often a cause or contributing cause of smoke related illness and death. Formaldehyde is found in many plastics and resins; it is one of the many components of smoke that cause eye and lung irritation. The oxides of nitrogen, which include nitric oxide, nitrous oxide, and nitrogen dioxide, are deep lung irritants that may cause a serious medical condition called pulmonary edema (fluid build-up in the lungs). There is more on this topic in the *Fire Smoke* chapter.

Radiation

Most fire fighters have not been trained on the finer points of handling incidents involving radioactive materials and, therefore, have many misconceptions about radiation. Radiation is energy transmitted through space in the form of electromagnetic waves or energetic particles.

Fundamentally, you should understand that you cannot escape radiation. It is all around you, every day. During the course of your life, you will receive radiation from the sun and the soil, when taking a plane ride, or by having an x-ray. Radiation has been around since the beginning of time. Our focus here, however, is not background radiation—the kind of radiation you received from the sun or the soil—rather, it is on the occupational exposures encountered in the field. For the most part, the health hazards posed by radiation are a function of two factors:

- The amount of radiation absorbed by your body has a direct relationship to the degree of damage done.
- The exposure time to the radiation will ultimately affect the extent of the injury.

The periodic table of elements illustrates all the known elements that are found in all the chemical compounds on the face of the Earth **FIGURE 2-14**. Those elements, in turn, are made up of atoms. In the nucleus of those atoms are protons (positive [+] electrical charge) and neutrons (no electrical charge). Orbiting the nucleus are electrons (negative [–] electrical charge).

All stable atoms of any given element will have the same number of protons and neutrons in their nucleus. Those same elements, however, are prone to having an imbalance in the numbers of protons and neutrons. This variation in the number of neutrons creates a radioactive isotope of the element. Carbon-14, for example, is a radioactive isotope of carbon because of the imbalance in the numbers of protons and neutrons found in the nucleus of carbon-14. You will notice on the periodic table of elements that stable carbon (C) has an atomic mass of 12. Carbon-14 illustrates a different mass, which shows its imbalance of protons and neutrons. Other examples include sulfur-35 and phosphorus-32, which are radioactive isotopes of sulfur and phosphorus, respectively.

RESPONDER TIP

Isotopes are atoms with the *same number* of protons but *different numbers* of neutrons.

Radioactivity is the natural and spontaneous process by which unstable atoms (isotopes) of an element decay to a different state and emit or radiate excess energy in the form of particles or waves. Radioactive isotopes give off energy from the nucleus of an unstable atom in an attempt to reach a stable state. Typically, you will encounter a combination of alpha, beta, and gamma radiation. Each of these forms of energy given off by a radioactive isotope will vary in intensity, and consequently determine your efforts to reduce the exposure potential **FIGURE 2-15**.

Small radiation detectors are available that can be worn on turnout gear. These detectors sound an alarm when dangerous levels of radiation are encountered and alert responders to leave the scene and call for more specialized assistance (See the mission-specific chapter on detection and monitoring for more information on radiation detection devices).

Alpha Particles

As mentioned previously, radiation stems from an imbalance in the number of protons and neutrons in the nucleus of an atom. Alpha radiation is a reflection of that instability. This form of radiation energy is produced when an electrically charged particle is given off by the nucleus of an unstable atom. Alpha particles have weight and mass; as a consequence, they cannot travel very far (less than a few inches) from the nucleus of the atom. For the purpose of comparison, alpha particles are like dust particles. Typical alpha emitters include americium (found in smoke detectors), polonium (identified in cigarette smoke), radium, radon, thorium, and uranium.

You can protect yourself from alpha emitters by staying several feet away from their source, and by protecting your respiratory tract with a P100 filter on an air-purifying respirator or by using a self-contained breathing apparatus (SCBA).

RESPONDER TIP

Radioactive *contamination* occurs when radioactive material is deposited on an object or a person. The release of radioactive materials into the environment can cause air, water, surfaces, soil, plants, buildings, people, or animals to become contaminated. A contaminated person has radioactive materials on or inside their body.

Radioactive materials give off a form of energy that travels in waves or particles. When a person is exposed to radiation, the energy penetrates the body. For example, when a person has an X-ray, he or she is said to be *exposed* to radiation.

Reproduced from Centers for Disease Control and Prevention (CDC), *Fact Sheet: Radiation Emergencies*, May 2005.

Beta Particles

Beta particles are more energetic than alpha particles, and therefore pose a greater health hazard. Essentially, beta particles are like electrons, except that a beta particle is ejected from the nucleus of an unstable atom. Depending on the strength of the source, beta particles can travel 10 to 15 feet in the open air. The beta particles themselves are not radioactive; rather, the radiation energy is generated by the speed at which the particles are emitted from the nucleus. Due to this phenomenon, beta radiation can break chemical bonds at the molecular level and cause damage to living tissue. This breaking of chemical bonds creates an ion; therefore, beta particles are considered ionizing radiation. Ionizing radiation has the capability to cause changes in human cells, which may ultimately lead to a mutation of the cell and become the root cause of cancer. Examples of ionizing radiation include X-rays and gamma rays.

PERIODIC TABLE OF THE ELEMENTS

Key	
Element	hydrogen
Atomic Number	1
Symbol	H
*Atomic Mass	1.01

Metals
Metalloids
Nonmetals

	1 IA	2 IIA	3 IIIB	4 IVB	5 VB	6 VIB	7 VIIB	8 VIII	9 VIII	10 VIII	11 IB	12 IIB	13 IIIA	14 IVA	15 VA	16 VIA	17 VIIA	18 VIIIA
																		helium 2 **He** 4.00
2	lithium 3 **Li** 6.94	beryllium 4 **Be** 9.01											boron 5 **B** 10.81	carbon 6 **C** 12.01	nitrogen 7 **N** 14.01	oxygen 8 **O** 16.00	fluorine 9 **F** 19.00	neon 10 **Ne** 20.18
3	sodium 11 **Na** 22.99	magnesium 12 **Mg** 24.31											aluminum 13 **Al** 26.98	silicon 14 **Si** 28.09	phosphorus 15 **P** 30.97	sulfur 16 **S** 32.07	chlorine 17 **Cl** 35.45	argon 18 **Ar** 39.95
4	potassium 19 **K** 39.10	calcium 20 **Ca** 40.08	scandium 21 **Sc** 44.96	titanium 22 **Ti** 47.88	vanadium 23 **V** 50.94	chromium 24 **Cr** 52.00	manganese 25 **Mn** 54.94	iron 26 **Fe** 55.85	cobalt 27 **Co** 58.93	nickel 28 **Ni** 58.69	copper 29 **Cu** 63.55	zinc 30 **Zn** 65.39	gallium 31 **Ga** 69.72	germanium 32 **Ge** 72.61	arsenic 33 **As** 74.92	selenium 34 **Se** 78.96	bromine 35 **Br** 79.90	krypton 36 **Kr** 83.80
5	rubidium 37 **Rb** 85.47	strontium 38 **Sr** 87.62	yttrium 39 **Y** 88.91	zirconium 40 **Zr** 91.22	niobium 41 **Nb** 92.91	molybdenum 42 **Mo** 95.94	technetium 43 **Tc** (99)	ruthenium 44 **Ru** 101.07	rhodium 45 **Rh** 102.91	palladium 46 **Pd** 106.42	silver 47 **Ag** 107.87	cadmium 48 **Cd** 112.41	indium 49 **In** 114.82	tin 50 **Sn** 118.71	antimony 51 **Sb** 121.75	tellurium 52 **Te** 127.60	iodine 53 **I** 126.90	xenon 54 **Xe** 131.29
6	cesium 55 **Cs** 132.91	barium 56 **Ba** 137.33	lanthanum 57 **La** 138.91	hafnium 72 **Hf** 178.49	tantalum 73 **Ta** 180.95	tungsten 74 **W** 183.85	rhenium 75 **Re** 186.21	osmium 76 **Os** 190.2	iridium 77 **Ir** 192.22	platinum 78 **Pt** 195.08	gold 79 **Au** 196.97	mercury 80 **Hg** 200.59	thallium 81 **Tl** 204.38	lead 82 **Pb** 207.2	bismuth 83 **Bi** 208.98	polonium 84 **Po** (209)	astatine 85 **At** (210)	radon 86 **Rn** (222)
7	francium 87 **Fr** (223)	radium 88 **Ra** (226)	actinium 89 **Ac** (227)	rutherfordium 104 **Rf** (261)	dubnium 105 **Db** (262)	seaborgium 106 **Sg** (263)	bohrium 107 **Bh** (262)	hassium 108 **Hs** (265)	meitnerium 109 **Mt** (266)	ununnilium 110 **Uun** (269)	unununium 111 **Uuu** (272)	ununbium 112 **Uub** (277)	ununtrium 113 **Uut** (284.18)	flerovium 114 **Fl** (289.19)	ununpentium 115 **Uup** (288.19)	livermorium 116 **Lv** (293)	ununseptium 117 **Uus** (294)	ununoctium 118 **Uuo** (294)

Lanthanide Series	cerium 58 **Ce** 140.12	praseodymium 59 **Pr** 140.91	neodymium 60 **Nd** 144.24	promethium 61 **Pm** (147)	samarium 62 **Sm** 150.36	europium 63 **Eu** 151.97	gadolinium 64 **Gd** 157.25	terbium 65 **Tb** 158.93	dysprosium 66 **Dy** 162.50	holmium 67 **Ho** 164.93	erbium 68 **Er** 167.26	thulium 69 **Tm** 168.93	ytterbium 70 **Yb** 173.04	lutetium 71 **Lu** 174.97
Actinide Series	thorium 90 **Th** 232.04	protactinium 91 **Pa** (231)	uranium 92 **U** 238.03	neptunium 93 **Np** (237)	plutonium 94 **Pu** (244)	americium 95 **Am** (243)	curium 96 **Cm** (247)	berkelium 97 **Bk** (247)	californium 98 **Cf** (251)	einsteinium 99 **Es** (252)	fermium 100 **Fm** (257)	mendelevium 101 **Md** (258)	nobelium 102 **No** (259)	lawrencium 103 **Lr** (260)

*Note: For radioactive elements, the mass number of an important isotope is shown in parentheses; for thorium and uranium, the atomic mass of the naturally occurring radioisotopes is given.

FIGURE 2-14 The periodic table of elements.

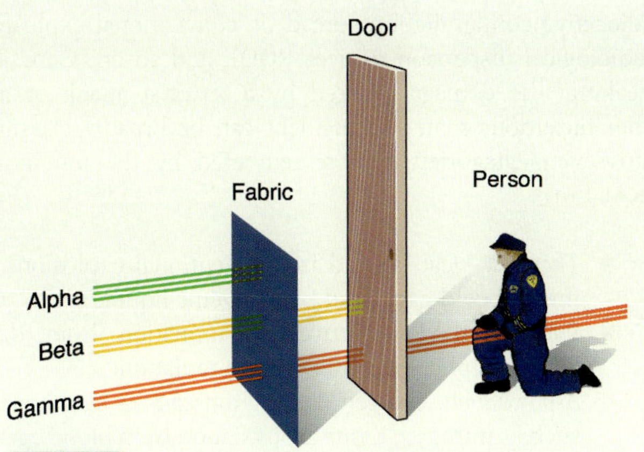

FIGURE 2-15 Alpha, beta, and gamma radiation.

Non-ionizing radiation comes from electromagnetic waves, which are capable of causing a disturbance of activity at the atomic level but do not have sufficient energy to break bonds and create ions. Typical non-ionizing waves include sound waves, radio waves, and microwaves.

Beta particles can redden (erythema) and burn skin; they can also be inhaled. Beta particles that are inhaled can directly damage the cells of the human body. Most solid objects can stop these particles, and your SCBA should provide adequate respiratory protection against them. Common beta emitters include tritium (luminous dials on gauges), iodine (medical treatment), and cesium.

Responder Safety Tip

To reduce the effects of a radiation exposure, responders should understand the concept of time–distance–shielding (TDS). Think of TDS in this way: The less time you spend in the sun, the less chance you have of suffering a sunburn (*time*); the closer you stand to a fire, the hotter you will get (*distance*); and if you come inside during a rainstorm, you will stop getting wet (*shielding*). TDS directly correlates to ALARA (as low as reasonably achievable)—a principle of radiation protection philosophy that requires that exposures to ionizing radiation shall be kept as low as reasonably achievable.

Gamma Rays and Neutrons

Gamma radiation is the most energetic radiation responders may encounter. Gamma radiation differs from alpha and beta radiation in that it is not a particle ejected from the nucleus; rather, it is pure electromagnetic energy. A gamma ray has no mass and no electrical charge and travels at the speed of light. Gamma rays can pass through thick solid objects (including the human body) very easily and generally follow the emission of a beta particle. If the nucleus still has too much energy after ejecting beta particles, it may release a photon (a packet of pure energy) that takes the form of a gamma ray.

Like beta particles, gamma radiation is a form of ionizing radiation and can be deadly. Structural firefighting gear that includes SCBA will not protect you from gamma rays; if fire fighters are near the source of this type of radiation, they will be exposed. Typical sources of gamma radiation include cesium (cancer treatment and soil density testing at construction sites) and cobalt (medical instrument sterilization).

Neutrons are also penetrating particles found in the nucleus of the atom. Neutrons themselves are not radioactive, but exposure to neutrons can create radiation, such as gamma radiation.

Hazard, Exposure, and Contamination

When responders operate at hazardous materials incidents, it is vital to identify the potential hazards and risks to minimize the potential for exposure. This begins by keen observation of the scene to determine whether a hazardous substance is released or has the potential to escape its container. At some point, early in the incident, it is vital to determine whether the pool of responders has the right training, equipment, and protective gear to positively influence the outcome of the incident. In some cases, a "no-go" situation may exist if the problem exceeds the ability of the responders to solve it safely within the boundaries of their training. It is as important to determine what can't be done as it is to decide on what actions to take. Additionally, responders must establish a safe perimeter around the problem, keeping it secure from accidental entrance and possible human exposures. In fact, the most important initial actions taken at a hazardous materials incident revolve around identifying the problem and taking actions to limit the spread of contamination and/or human exposures.

Hazard and Exposure

A *hazardous material* is defined as a material capable of posing an unreasonable risk to health, safety, or the environment—that is, a material capable of causing harm. The same source defines exposure as the process by which people, animals, the environment, and equipment are subjected to or come into contact with a hazardous material.

Contamination

When a chemical has been released and physically comes in contact with people, the environment, and everything around it, either intentionally or unintentionally, the residue of that chemical is called contamination. In some cases, the process of removing such contaminants is complex and requires a significant effort to complete. In other cases, the contamination is easily eliminated. The chapter *Mission-Specific Competencies: Technical Decontamination* covers the finer points of removing contaminants and decontaminating the responders and their gear. For now, simply understand that when a chemical escapes its container, it's possible it will get into or onto something and you will need a system to safely and efficiently remove that chemical or otherwise reduce the hazard. This could include exposed persons, PPE, or tools and equipment.

Secondary Contamination

Secondary contamination, also known as cross-contamination, occurs when a person or object transfers the contaminant or the source of contamination to another person (responder or civilian) or object by direct contact. Responders may become contaminated and subsequently handle tools and equipment, touch door handles or other responders, and spread the contamination. Additionally, contaminated victims who are improperly handled by unprotected responders also run the risk of spreading contamination FIGURE 2-16.

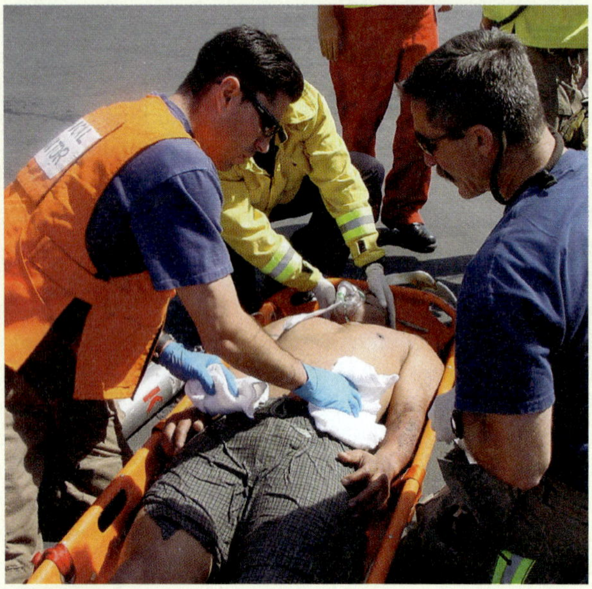

FIGURE 2-16 Responders should ensure contaminated victims are completely decontaminated prior to rendering care.

Courtesy of Rob Schnepp

The possibility of spreading contamination brings up a point that all responders should understand: The cleaner responders stay during the response, the less decontamination they will have to do later. Many responders have the misperception that PPE is worn to enable you to contact the product. In fact, quite the opposite is true: PPE, including specialized chemical protective gear, is worn to protect you in the event that you cannot avoid product contact. The less contamination encountered or spread, the easier the decontamination will be.

Types of Hazardous Materials: Weapons of Mass Destruction

In recent years, weapons of mass destruction (WMD) have become a highly relevant topic for responders. WMD represent a real threat in the United States, and each responder should have a basic knowledge of those potential threats. To that end, this section briefly addresses the threats posed by nerve, blister, blood, and choking agents and other potential hazards associated with various irritant materials.

WMD events can be complex, and the harm inflicted during such incidents may be the result of several factors—nerve agents dispersed by explosives or other dispersal devices, radioactive contamination spread by conventional explosives (radiological dispersion devices RDD), and so on. Generally speaking, the damages caused by a terrorist attack or any other hazardous materials incident can be broadly classified into seven categories, best remembered by the mnemonic TRACEMP:

- *Thermal:* Heat created from intentional explosions or fires, or cold generated by cryogenic liquids
- *Radiological:* Radioactive contamination from dirty bombs; alpha, beta, and gamma radiation
- *Asphyxiation:* Oxygen deprivation caused by materials such as nitrogen; tissue asphyxiation from blood agents
- *Chemical:* Injury and death caused by the intentional release of toxic industrial chemicals, nerve agents, vesicants, poisons, or other chemicals
- *Etiological:* Illness and death resulting from biohazards such as anthrax, plague, and smallpox; hazards posed by bloodborne pathogens
- *Mechanical:* Property damage and injury caused by explosions, falling debris, shrapnel, firearms, explosives, and slips, trips, and falls
- *Psychogenic:* The mental harm from being potentially exposed to, contaminated by, and even just being in close proximity to an incident of this nature

Nerve Agents

Nerve agents pose a significant threat to civilians and responders alike. The main threat stems from the ability of a nerve agent to enter the body through the lungs or the skin and then systemically affect the function of the human body. Nerve agents, like many organophosphate pesticides, disrupt the central nervous system, possibly causing death or serious impairment.

The human central nervous system is laid out much like the electrical wiring in a house. The brain serves as the main panel. The wires—that is, the nerves—go to muscles, organs, glands, and other critical areas of the body. When a nerve agent enters the body, it affects the body's ability to transact nerve impulses at the junction points between nerves and muscles, nerves and vital organs, and nerves and glands. Subsequently, those critical areas will not function properly, and associated bodily functions will be affected. For example, sarin is a water-like liquid nerve agent that is primarily a vapor hazard.

Signs and symptoms of nerve agent exposure may include pinpoint pupils, tearing of the eyes, twitching muscles, loss of bowel and bladder control, and slow or rapid heartbeat. Mild exposures may result in mild symptoms and include some or all of the symptoms listed previously. Significant exposures may result in rapid death, serious impairment, or a more pronounced manifestation of symptoms. The bottom line is that an exposure to a nerve agent is not an automatic death sentence. Much like the case of radiation exposure, the dose absorbed is directly correlated with the degree of damage done and the symptoms presented by the victim.

VOICES OF EXPERIENCE

It was so cold on that January morning in central Canada that a mere 5 minutes outside transformed my turnout gear into 25 pounds of crackling cellophane. The cold on this particular morning had the added complication of a thick blanket of "ice fog." The low visibility caused by the thick fog was the reason we found ourselves responding to a two-vehicle highway accident.

On arrival at the scene, we quickly determined that a semi-trailer carrying a full load of a highly volatile petroleum-based liquid had accidentally driven too close to another large vehicle. The two vehicles had rubbed the sides of their trailers together at a speed of about 30 miles per hour, causing about 20 feet of "tearing and creasing" along the side of the trailer containing the volatile liquid. At the scene, we noticed a puddle of product with about a 10-foot radius on the ground.

Different atmospheric conditions at this accident would have spurred us to different initial actions. In this case, it was so cold that the liquid was unable to generate any significant vapors that could cause immediate concerns about ignition outside the product's immediate area. Our captain recognized that the cold weather affected this product in specific ways, and therefore decided on situation-specific actions and priorities that would provide reasonable safety for all responders and bystanders, prevent the leaking product from spreading, and enable us to work to safely offload the remaining product to a suitable replacement container. On a hot day, the liquid would have been actively producing dangerous vapors, changing our initial actions. In that case, our priorities would have involved removal of ignition sources through isolation of the area as well as aggressive vapor suppression.

Changing factors, such as weather conditions and the environment immediately surrounding the affected area, will affect the properties and effects of hazardous materials. It is critical to gather as much information as possible about the hazard to accurately determine your initial actions and priorities.

Ray Unrau
Saskatoon Fire and Protective Services
Saskatoon, Saskatchewan

Recognition of the signs and symptoms of a nerve agent exposure is vital in determining the presence of such a threat. Nerve agents in general are reported to have fruity odors. In many cases, responders will not have the specialized detection devices that are necessary to identify the presence of a nerve agent exposure; the symptoms presented by the victims should tip them off, however. To that end, the mnemonic SLUDGEM briefly summarizes some of the more common signs and symptoms of nerve agent exposure:

S: Salivation
L: Lacrimation (tearing)
U: Urination
D: Defecation
G: Gastric disturbance
E: Emesis (vomiting)
M: Miosis (constriction of the pupil)

Many times, nerve agents are incorrectly referred to as nerve gas. Nerve agents are liquids, not gases, and each one evaporates at a different rate. Sarin, for example, is a water-like liquid that evaporates more easily than VX, which is a thick, oily liquid.

VX is the most toxic of the weapons-grade nerve agents. It also has the most persistence (continued presence). Other nerve agents include soman and tabun. The organophosphate and carbamate classes of pesticides are similar to the weapons-grade nerve agents but are less toxic in general terms.

Blister Agents

Blister agents (also known as vesicants) have the ability to cause blistering of the skin. Common blister agents include sulfur mustard and lewisite, which interact in unique ways with the human body.

Sulfur mustard was first used as a weapon in World War I. Chemical casualties suffered horrific injuries to the skin and lungs; in some cases, those exposures proved fatal. Although sulfur mustard is commonly referred to as mustard gas, it is more typically found in a liquid state. This chemical is colorless and odorless when pure. When mixed with other substances or otherwise impure, however, it can look brownish in color and smell like garlic or onions.

Exposures to sulfur mustard may not be immediately apparent. A victim may not experience pain or any other sign indicating current contact with this agent. Indeed, redness and blistering may not appear until 2 to 24 hours after exposure. Once the blistering does occur, skin decontamination will not reduce its effects FIGURE 2-17. If sulfur mustard exposure is suspected, immediate skin decontamination is indicated. Early skin decontamination may reduce the effects of exposure.

Sulfur mustard can cause injuries similar to second- and third-degree thermal burns. In cases involving mild skin exposures, victims typically make a complete recovery. Inhalation exposures are much more serious, with the damage to lung tissues sometimes resulting in permanent respiratory function impairment. Additionally, sulfur mustard is very persistent in the environment. When it is released in this way, it can remain intact, sticking to the ground or other surfaces, for several days.

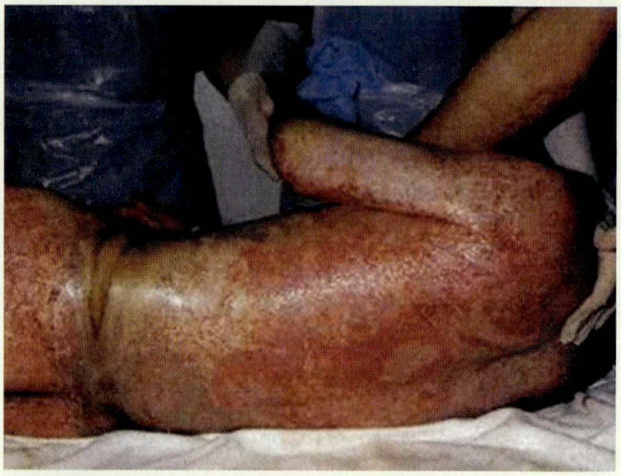

FIGURE 2-17 Blister agent exposure.

Courtesy of Dr. Saeed Keshavarz/RCCI (Research Center of Chemical Injuries)

Lewisite has many of the same characteristics as sulfur mustard. Lewisite, however, contains arsenic; thus, when it is absorbed by the skin, this blister agent may produce some symptoms specific to arsenic poisoning, such as vomiting and low blood pressure. Unlike sulfur mustard, lewisite exposures to the skin will produce immediate pain. Lewisite is a colorless-to-dark-brown oily liquid that smells of geraniums. If a lewisite exposure is suspected, immediate and aggressive skin decontamination is indicated. Like sulfur mustard, lewisite does not occur naturally in the environment and is considered to be a chemical warfare agent.

Blood Agents

Blood agents are chemicals that, when absorbed by the body, interfere with the transfer of oxygen from the blood to the cells. Cyanide compounds, including hydrogen cyanide and cyanogen chloride, prevent the body from using oxygen. Victims exposed to high levels of cyanide, for example, could be placed on supplemental oxygen but still not survive the exposure. The cause of death in such cases is not a lack of breathable oxygen. It is the inability of the body to use the available oxygen effectively.

The main route of exposure to cyanide compounds is through the lungs, but many of these agents can also be ingested or absorbed through the skin. Hydrogen cyanide has a reported odor of bitter almonds, but approximately 60 percent of the general population cannot detect the presence of hydrogen cyanide by smell. Thus, because the odor threshold and the amount of cyanide that would be lethal are very close, odor alone cannot be used to detect dangerous amounts of hydrogen cyanide. By contrast, cyanogen chloride, because of the chlorine atom within the chemical compound, has a much more irritating and pungent odor.

Typical signs and symptoms of cyanide exposure include vomiting, dizziness, watery eyes, and deep and rapid breathing. High concentrations lead to convulsions, inability to breathe, loss of consciousness, and death.

Choking Agents

Choking agents are predominantly designed to inhibit breathing and are typically intended to incapacitate rather than to kill; nevertheless, death and serious injuries from these agents are possible. The nature of choking agents (extremely irritating odor) alerts potential victims to their presence and allows for escape from the environment when possible. Commonly recognized choking agents include chlorine, phosgene, and chloropicrin.

Each of these substances is associated with a specific odor. Chlorine smells like a swimming pool, whereas phosgene and chloropicrin have been reported to smell like freshly mown grass or hay. In either case, the odors are noticeable and, in many cases, strong enough and irritating enough to alert victims to the presence of an unusual situation.

In the event of a significant exposure, the main threat from these materials is pulmonary edema (also known as dry drowning, because the victim's lungs fill with fluid). When inhaled, choking agents damage sensitive lung tissue and cause fluid to be released. The excess fluid results in chemically induced pneumonia, which could end up being fatal. Additionally, choking agents act as skin irritants; they can cause mild to moderate skin irritation and significant burning when victims are exposed to them in high concentrations.

Irritants (Riot Control Agents)

Irritants (also known as riot control agents) include substances, such as mace, that can be dispersed to briefly incapacitate a person or groups of people. Chiefly, irritants cause pain and a burning sensation to exposed skin, eyes, and mucous membranes. The onset of symptoms occurs within seconds after contact and lasts several minutes to several hours, usually with no lasting effects. Irritants can be dispersed from canisters, handheld sprayers, and grenades.

From a terrorism perspective, irritants may be employed to incapacitate rescuers or to drive a group of people into another area where a more dangerous substance can be released. Of all the groups of WMD, irritants pose the least amount of direct danger in terms of toxicity. Exposed patients can be decontaminated with clean water, and the residual effects of the exposure are rarely significant. Responders are likely to encounter these agents when they respond to calls in which law enforcement personnel have used these agents for riot control or suspect apprehension. It is important for responders to know which specific agents are used by local law enforcement agencies for such purposes and to identify any specific decontamination techniques recommended for use with those agents.

Convulsants

Chemicals classified as convulsants are capable of causing convulsions or seizures when they are absorbed by the body. Chemicals that fall into this category include nerve agents such as sarin, soman, tabun, VX, and the organophosphate and carbamate classes of pesticides. These substances interfere with the central nervous system and disrupt normal transmission of neuromuscular impulses throughout the body. Examples of pesticides that are capable of functioning as convulsants include parathion, aldicarb, diazinon, and fonofos.

When dealing with these materials, it is important to identify their presence and to avoid breathing in their vapors or allowing the liquid to contact your skin. In some cases, even a small exposure can be fatal.

Harmful Substances' Routes of Entry into the Human Body

For a chemical or other harmful substance to injure a person, it must first get into the individual's body. Throughout the course of his or her service career, a responder will inevitably be bombarded by harmful substances such as diesel exhaust, bloodborne pathogens, smoke, and accidentally and intentionally released chemicals. Fortunately, most of these exposures are not immediately deadly. Unfortunately, repetitive exposures to these materials may have negative health effects after a 20-year career. To protect yourself now and give yourself the best shot at a healthy retirement, it is important to understand some basic concepts about toxicology. Chemical substances can enter the human body in four ways FIGURE 2-18:

- *Inhalation:* Through the lungs
- *Absorption:* By permeating the skin
- *Ingestion:* Via the gastrointestinal tract
- *Injection:* Through cuts or other breaches in the skin

Toxicology is the study of the adverse effects of chemical or physical agents on living organisms. The following sections discuss these agents' potential routes of entry into the human body and methods used to protect against these agents.

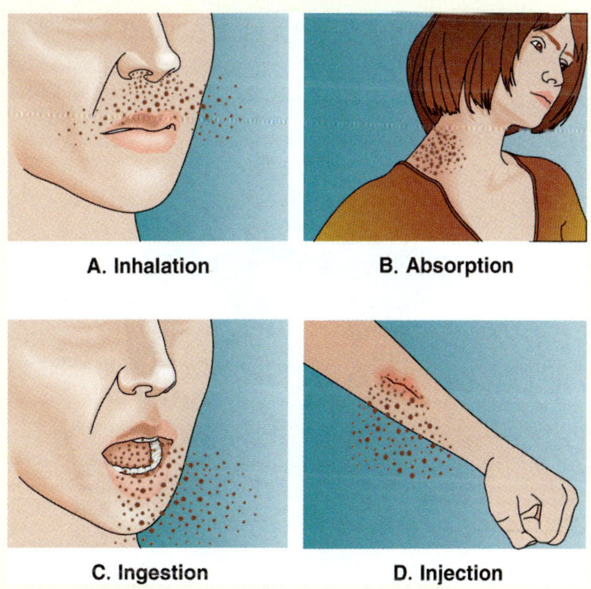

FIGURE 2-18 The four ways a chemical substance can enter the body.

Inhalation

Inhalation exposures occur when harmful substances are brought into the body through the respiratory system. The lungs are a direct point of access to the bloodstream, so they can quickly transfer an airborne substance into the circulatory system and onward to the rest of the body. In addition, the lungs cannot be decontaminated, so any exposure will result in some type of harm that cannot be addressed in the same way as an exposure to skin.

The respiratory system is vulnerable to attack from a wide range of substances, from corrosive materials such as chlorine and ammonia to solvent vapors such as gasoline and acetone, or superheated air from a fire, or any other material finding its way into the air. In addition to gases and vapors, small particles of dust, fiberglass insulation, asbestos, or soot from a fire can become lodged in sensitive lung tissue, causing substantial irritation.

Given these facts, it is imperative that responders wear appropriate respiratory protection when operating in the presence of airborne contamination. Fortunately, fire fighters have ready access to an excellent form of respiratory protection—namely, the positive-pressure, open-circuit SCBA. This equipment is by far the single most important piece of PPE that fire fighters have at their disposal. Sometimes fire fighters may need to use other forms of respiratory protection depending on the specific respiratory hazard they are facing while performing a given task. For example, full-face and half-face air-purifying respirators (APRs) offer specific degrees of protection if the chemical hazard present is known and the appropriate filter canister is used FIGURE 2-19.

Air-purifying respirators do not provide oxygen, however. Thus, if the oxygen content of the work area is low, full- or half-face respirators are not a viable option. According to the Occupational Safety and Health Administration (OSHA), any work environment containing less than 19.5% oxygen is considered to be oxygen deficient and will require the use of an SCBA or a supplied air respirator.

FIGURE 2-19 Air-purifying respirators offer some degree of protection against airborne chemical hazards.

Courtesy of Sperian Respiratory Protection

Respirators are lighter than SCBA, are more comfortable to wear, and usually allow longer work periods because they are not dependent on a limited source of breathing air. SCBAs certainly offer a higher level of respiratory protection, but in the right circumstances, APRs may represent a viable form of respiratory protection.

When considering protection against airborne contamination, it is important to understand the origin, concentration, and potential impact of the contamination relative to the oxygen levels in the area. In short, the appropriate type of respiratory protection is determined by looking at the overall situation, including the nature of the contaminant. In some cases, the anticipated particle size of the contamination may dictate the level of respiratory protection employed TABLE 2-3.

Table 2-3	Particle Sizes of Common Types of Respiratory Hazards
Fume	<1 micron
Smoke	≤ 1 micron
Dust	≥ 1 micron
Fog	>40 microns
Mist	<40 microns

Anthrax spores offer an excellent example to illustrate this point. Weaponized anthrax spores typically vary in size from 0.5 micron to 1 micron. Based on that size range, a typical full-face APR with a nuisance dust filter, or a surgical-type mask, would not offer sufficient protection against this hazard. Anyone operating in an area contaminated with anthrax should wear SCBA or, at a minimum, a full-face APR with P100 filtration (which filters out greater than 99% of 0.3-micron or larger particles).

Anthrax exposures also illustrate another important pair of terms and definitions that responders should understand: infectious and contagious. Anthrax is a pathogenic microorganism capable of causing an illness (infectious). A person with an illness caused by an anthrax exposure, however, is not capable of passing it along to another person (contagious). In other words, anthrax is not contagious. Conversely, smallpox can be both infectious and contagious, which is why this pathogen poses such a high risk in the event of an outbreak.

Now consider this scenario: A container of gaseous helium is leaking inside a poorly ventilated storage room. Helium is nonflammable; the main threat it poses is the possibility of oxygen deficiency. In this case, SCBA is the appropriate type of respiratory protection based on the anticipated hazard. In many cases, when hazardous materials technicians respond to incidents, they will use air-monitoring devices to characterize the work area prior to entry. When this step is taken, choices of respiratory protection are based on more definitive information.

Particle size also determines where the inspired contamination will eventually end up. The larger particles that make up visible mists will be captured in the nose and upper airway, for example, whereas smaller particles may work their way deeper into the lung TABLE 2-4.

Table 2-4	Location of Respiratory Trapping by Particle Size
<7 microns	Nose
5–7 microns	Larynx
3–5 microns	Trachea and bronchi
2–3 microns	Bronchi
1–2.5 microns	Respiratory bronchioles
0.5–1 micron	Alveoli

Respiratory protection is one of the most important components of any PPE. In all cases where airborne contamination is encountered, take the time to understand the nature of the threat, evaluate the respiratory protection available, and decide if it provides adequate protection. When protecting the lungs, "good enough" is not an option. The chapter *Mission-Specific Competencies: Personal Protective Equipment*, addresses the concept of chemical PPE and describes how respiratory protection fits into the "big picture" of remaining safe at chemical incidents.

Absorption

The skin is the largest organ in the body and is susceptible to the damage inflicted by a number of substances. In addition to serving as the body's protective shield against heat, light, and infection, the skin helps to regulate body temperature, stores water and fat, and serves as a sensory center for painful and pleasant stimulation. Without this important organ, human beings would not be able to survive.

When discussing chemical exposures, however, absorption is not limited only to the skin. Absorption is the process by which substances travel through body tissues until they reach the bloodstream. The eyes, nose, mouth, and, to a certain degree, the intestinal tract are also part of the equation. The eyes, for example, will absorb a high amount of liquid and vapor that come into contact with these sensitive tissues. This absorption is particularly problematic because the eyes connect directly to the optic nerve, which allows the chemical to follow a direct route to the brain and the central nervous system.

Although the skin functions as a shield for the body, that shield can be pierced by a number of chemicals. Aggressive solvents such as methylene chloride (found in paint stripper), for example, can be readily absorbed through the skin. A secondary hazard associated with this chemical occurs when the body attempts to metabolize the substance after it is absorbed. A by-product of that metabolism reaction is carbon monoxide, a cellular asphyxiant. Asphyxiants are substances that prevent the body (at the cellular level) from using oxygen, thereby causing suffocation. In this scenario, the initial chemical is broken down to form another substance that is potentially a greater health hazard than the original chemical. Methylene chloride is also suspected to be a human cancer-causing agent (carcinogen).

Absorption hazards are not limited to solvents. Hydrofluoric acid, for example, poses a significant threat to life when it is absorbed through the skin. This unique corrosive has the ability to bind with certain substances in the body (predominantly calcium). Secondary health effects occurring after exposure can include muscular pain and potentially lethal cardiac arrhythmias.

In the field, responders must constantly evaluate the possibility of chemical contact with their skin and eyes. In many cases, structural firefighting turnout gear provides little or no protection against liquid chemicals. Consult the *Emergency Response Guidebook* or an SDS for response guidance when deciding whether turnout gear is appropriate for the hazard you are facing. In the event the turnout gear does not offer adequate protection, responders may have to increase their level of protection to include specialized chemical-protective clothing.

Ingestion

In addition to absorption through the skin, chemicals can be brought into the body through the gastrointestinal tract, by the process of ingestion. The water, nutrients, and vitamins the body requires are predominantly absorbed in this manner. For example, at a structure fire, fire fighters generally have an opportunity to rotate out of the building for rest and refreshment and may not take the time to wash up prior to eating or drinking. This leads to a high probability of spreading contamination from the hands to the food and subsequently to the intestinal tract. If you do not think about every situation where you might become exposed, you may put yourself in harm's way.

Injection

Chemicals brought into the body through open cuts and abrasions qualify as injection exposures. To protect yourself from this route of exposure, begin by realizing when you will work in a compromised state. Any cuts or open wounds should be addressed before reporting for duty. If they are significant, you may be excluded from operating in contaminated environments. Open wounds act as a direct portal to the bloodstream and subsequently to muscles, organs, and other body systems. If a chemical substance comes in contact with this open portal, the health effects could be immediate and pronounced. Remember—intact skin is a good protective shield. Do not go into battle if your shield is not up to the task.

Chronic and Acute Health Effects

A chronic health hazard (also known as a chronic health effect) is an adverse health effect that occurs gradually over time after long-term exposure to a hazard. Chronic health effects may appear either after long-term or chronic exposures or following multiple short-term exposures that occur over a shorter period. The exposures can involve all routes of entry into the body and may result in such health problems as cancer, permanent loss of lung function, or repetitive skin rashes. For example, inhaling asbestos fibers for years without respiratory protection can result in a form of lung cancer called asbestosis. This is an example of a target organ effect—chronic asbestos exposures target the lungs as the site of their damage.

Chemicals that pose a hazard to health after relatively short exposure periods are said to cause acute health effects. Such an exposure period is considered to be subchronic, with effects occurring either immediately after a single exposure or as long as several days or weeks after an exposure. Essentially, acute exposures are "right now" exposures that produce some observable conditions such as eye irritation, coughing, dizziness, and skin burns.

Skin irritation and burning after a dermal exposure to sulfuric acid would be classified as an acute health effect. Other chemicals, such as formaldehyde, are also capable of causing acute health effects. Among other things, formaldehyde (also classified as a human carcinogen) is a sensitizer that is capable of causing an immune system response when it is inhaled or absorbed through the skin. This reaction is similar to other allergic reactions, in that the initial exposure produces mild health effects, but subsequent exposures cause much more severe reactions such as breathing difficulties and skin irritation. OSHA defines a sensitizer as a chemical that causes a substantial proportion of exposed people or animals to develop an allergic reaction in normal tissue after repeated exposure to the chemical.

Toxicity is a measure of the degree to which something is toxic or poisonous. This term can also refer to the adverse effect(s) a substance may have on a whole organism, such as a human (or a bacterium or a plant), or to a substructure such as a cell or a specific organ such as the liver, kidneys, or lungs. To understand the risks posed by a material's toxicity, emergency responders must consider the physical and chemical properties of the substance causing the illness or injury, as well as the dose–response relationship that exists when a person is exposed to any substance. This important correlation refers to the response (signs, symptoms, illness, or injury) that a specific dose might provoke from the human body. The magnitude of the response depends on several factors, including the concentration of the hazardous substance and the duration of the exposure. The actual dose is the amount taken up by a person through one of the four routes of entry.

Cyanide, for example, is a harmful substance that can enter the body by all routes of entry. If cyanide gets into the body via the lungs, for example, it would be useful to understand at what airborne concentration (dose) the exposure might produce adverse health effects (response). To fully understand the threat, responders must be familiar with several health-related terms and definitions as they relate to airborne and dermal exposures. For example, the lethal dose (LD) of a material is a single dose that causes the death of a specified number of the group of test animals exposed by any route other than inhalation. The lethal concentration (LC) is defined as the concentration of a material in air that, on the basis of laboratory tests (inhalation route), is expected to kill a specified number of the group of test animals when administered over a specified period of time.

Several subcategories of LD and LC exist, each depicting a benchmark concentration of a particular exposure that will be harmful and/or fatal to a certain percentage of the test population. These values are commonly found in SDS, electronic databases, and printed reference books. In most cases, these values are derived from animal studies, so the results may not exactly correspond to the levels/concentrations that would be harmful to humans.

The following overall descriptions of these benchmark values offer relative guidelines when it comes to determining potential levels of human toxicity. As mentioned earlier, many chemical substances on the market today have established LD or LC values. Typically, if the substance is a vapor or gas, it will be expressed as an LC value. If the substance poses a dermal threat or is harmful when ingested, its toxicity will be expressed as an LD value.

For any given substance, the LD is the lowest dosage per unit of body weight (typically stated in terms of milligrams per kilogram [mg/kg] of body weight) of a substance known to have resulted in fatality in a particular animal species. The median lethal dose (LD_{50}) of a toxic material is the dose required to kill half (50 percent) of the members of a tested population.

For example, the LD_{50} of sodium cyanide (based on testing done on laboratory rats) is approximately 6.4 mg/kg. LD_{50} figures are frequently used as a general indicator of a substance's toxicity. The LD_{hi} or LD_{100} is the absolute dose of a toxic material required to kill all (100 percent) of the members of a tested population.

The LC_{lo} is the lowest lethal concentration of a material reported to cause death in a particular animal species, when administered via the inhalation route; it is the lowest lethal concentration for gases, dusts, vapors, and mists. The LC_{50} is the concentration of a material in air that is expected to kill 50 percent of the group of a particular animal species when administered via the inhalation route. Some literature reports the LC_{50} of carbon monoxide to be approximately 3700 parts per million (ppm) for a 1-hour exposure time frame. Carbon monoxide levels have been measured in working structure fires to approach 10,000 ppm—well above the LC_{50}. The LC_{hi} or LC_{100} is the absolute concentration of a toxic material required to kill all (100 percent) of the members of a tested population when the substance is administered via the inhalation route.

The values expressed here figure prominently in OSHA's view when it comes to identifying the relative toxicity of a particular substance. The following information reflects OSHA's view of exposure levels considered to be toxic and highly toxic.

The label *toxic*, as defined by OSHA 29 CFR 1910.1200, is assigned to any chemical that falls in one of these three categories:

- A chemical that has an LD_{50} of more than 50 mg/kg but not more than 500 mg/kg of body weight when administered orally to albino rats weighing between 200 and 300 g each
- A chemical that has an LD_{50} of more than 200 mg/kg but not more than 1000 mg/kg of body weight when

administered by continuous contact for 24 hours (or less if death occurs within 24 hours) with the bare skin of albino rabbits weighing between 2 and 3 kg each

- A chemical that has an LC_{50} in air of more than 200 ppm but not more than 2000 ppm by volume of gas or vapor, or more than 2 mg/L but not more than 20 mg/L of mist, fume, or dust, when administered by continuous inhalation for 1 hour (or less if death occurs within 1 hour) to albino rats weighing between 200 and 300 g each

OSHA 29 CFR 1910.1200 describes *highly toxic* materials as follows:

- A chemical that has an LD_{50} of 50 mg/kg or less of body weight when administered orally to albino rats weighing between 200 and 300 g each
- A chemical that has an LD_{50} of 200 mg/kg or less of body weight when administered by continuous contact for 24 hours (or less if death occurs within 24 hours) with the bare skin of albino rabbits weighing between 2 and 3 kg each
- A chemical that has an LC_{50} in air of 200 ppm by volume or less of gas or vapor, or 2 mg/L or less of mist, fume, or dust, when administered by continuous inhalation for 1 hour (or less if death occurs within 1 hour) to albino rats weighing between 200 and 300 g each

Wrap-Up

Chief Concepts

- The most fundamental of all actions is the ability to observe the scene and understand the problem you are facing. Think before you act—it could save your life!
- An important first step in understanding the hazards of any chemical is identifying the state of matter and defining whether the substance is a solid, liquid, or gas.
- A critical step in comprehending the nature of the release is identifying the reason(s) why the containment vessel failed.
- Chemical change is not the same thing as physical change. Physical change is a change in state; chemical change describes the ability of a substance to undergo a transformation at the molecular level, usually with a release of some form of energy.
- There are many critical characteristics of flammable liquids, and you should be familiar with each of them.
- When responders respond to hazardous materials incidents, they must fully understand the hazards to minimize the potential for exposure.
- Avoid contamination whenever possible—it will reduce the likelihood of harmful exposures.
- Chemicals can enter the body in four ways: inhalation, absorption, ingestion, and injection.
- Nerve agents, blister agents, blood agents, choking agents, irritants, and convulsants are examples of hazardous materials that may be used as WMD.
- Irritants such as mace cause immediate irritation of the skin, eyes, and mucous membranes.
- Equipment such as APR and SCBA can protect lungs from select particulates such as smoke, dust, and fumes.
- Chronic health effects may occur after years of exposure to hazardous materials. Wear all protective gear to minimize the impacts of repeated exposures.

Hot Terms

Absorption The process by which substances travel through body tissues until they reach the bloodstream.

Acid A material with a pH value less than 7.

Acute exposure A "right now" exposure that produces observable signs such as eye irritation, coughing, dizziness, and skin burns.

Acute health effects Health problems caused by relatively short exposure periods to a harmful substance that produces observable conditions such as eye irritation, coughing, dizziness, and skin burns.

Alpha particle A type of radiation that quickly loses energy and can travel only 1 or 2 inches from its source. Clothing or a sheet of paper can stop this type of energy. Alpha particles are not dangerous to plants, animals, or people unless the alpha-emitting substance has entered the body.

Anthrax An infectious disease spread by the bacterium *Bacillus anthracis*; typically found around farms, infecting livestock.

Asphyxiant A material that causes the victim to suffocate.

Base A material with a pH value greater than 7.

Beta particle A type of radiation that is capable of travelling 10 to 15 feet from its source. Heavier materials, such as metal and glass, can stop this type of energy.

BLEVE Boiling liquid/expanding vapor explosion; an explosion that occurs when pressurized liquefied materials (propane or butane, for example) inside a closed vessel are exposed to a source of high heat.

Blister agent A chemical that causes the skin to blister; also known as a vesicant.

Blood agent Chemicals that interfere with the utilization of oxygen by the cells of the body. Cyanide is an example of a blood agent.

Boiling point The temperature at which a liquid will continually give off vapors in sustained amounts and, if held at that temperature long enough, will eventually turn completely into a gas.

Carcinogen A cancer-causing agent.

Chemical and physical properties Measurable characteristics of a chemical, such as its vapor density, flammability, corrosivity, and water reactivity.

Chemical reactivity The ability of a chemical to undergo an alteration in its chemical make-up, usually accompanied by a release of some form of energy.

Choking agent A chemical designed to inhibit breathing and typically intended to incapacitate rather than kill its victims.

Chronic exposure Long-term exposure, occurring over the course of many months or years.

Chronic health hazard An adverse health effect occurring after a long-term exposure to a substance.

Combustible The process of burning from a gaseous or vaporous state.

Contagious Capable of transmitting a disease.

Contamination The process of transferring a hazardous material from its source to people, animals, the environment, or equipment, all of which may act as carriers for the material.

Convulsant A chemical capable of causing convulsions or seizures when absorbed by the body. Chemicals that fall into this category include nerve agents such as sarin, soman, tabun, VX, and the organophosphate and carbamate classes of pesticides.

Corrosivity The ability of a material to cause damage (on contact) to skin, eyes, or other parts of the body.

Cyanide compound Blood agent that prevents the body from using the available oxygen effectively.

Expansion ratio A description of the volume increase that occurs when a liquid changes to a gas.

Exposure The process by which people, animals, the environment, and equipment are subjected to or come into contact with a hazardous material.

Fire point The temperature at which sustained combustion will occur. The fire point is usually only slightly higher than the flash point for most materials.

Flammable range An expression of a fuel/air mixture, defined by upper and lower limits, that reflects an amount of flammable vapor mixed with a given volume of air.

Flash point The minimum temperature at which a liquid or a solid releases sufficient vapor to form an ignitable mixture with air.

Gamma radiation A type of radiation that can travel significant distances, penetrating most materials and passing through the body. Gamma radiation is the most destructive type of radiation to the human body.

Hazard A material capable of posing an unreasonable risk to health, safety, or the environment; a material capable of causing harm.

Ignition (autoignition) temperature The minimum temperature at which a fuel, when heated, will ignite in air and continue to burn. Also called the *autoignition temperature*.

Infectious Capable of causing an illness by entry of a pathogenic microorganism.

Ingestion Exposure to a hazardous material by swallowing the substance.

Inhalation Exposure to a hazardous material by breathing the substance into the lungs.

Injection Exposure to a hazardous material by the substance entering cuts or other breaches in the skin.

Ionizing radiation Electromagnetic waves of such intensity that chemical bonds at the atomic level can be broken (creating an ion).

Irritant A substance (such as mace) that can be dispersed to briefly incapacitate a person or groups of people. Irritants cause pain and a burning sensation to exposed skin, eyes, and mucous membranes.

Lethal concentration (LC) The concentration of a material in air that, on the basis of laboratory tests (inhalation route), is expected to kill a specified number of the group of test animals when administered over a specified period of time.

Lethal dose (LD) A single dose that causes the death of a specified number of the group of test animals exposed by any route other than inhalation.

Lewisite A blister-forming agent that is an oily, colorless-to-dark-brown liquid with an odor of geraniums.

Lower explosive limit (LEL) The minimum amount of gaseous fuel that must be present in the air for the air/fuel mixture to be flammable or explosive. Also referred to as the lower flammable limit (LFL).

Nerve agent A toxic substance that attacks the central nervous system in humans.

Neutrons Penetrating particles found in the nucleus of the atom that are removed through nuclear fusion or fission. Although neutrons are not radioactive, exposure to them can create radiation.

Non-ionizing radiation Electromagnetic waves capable of causing a disturbance of activity at the atomic level but that do not have sufficient energy to break bonds and create ions.

Persistence The continued presence of a substance in the environment. Persistent chemicals are typically dense, oily substances with high molecular weights and low vapor pressures.

pH An expression of the amount of dissolved hydrogen ions (H^+) in a solution.

Physical change A transformation in which a material changes its state of matter—for instance, from a liquid to a solid.

Plague An infectious disease caused by the bacterium *Yersinia pestis*, which is commonly found on rodents.

Pulmonary edema Fluid build-up in the lungs.

Radiation The combined process of emission, transmission, and absorption of energy traveling by electromagnetic wave propagation between a region of higher temperature and a region of lower temperature.

Radioactive isotope A variation of an element created by an imbalance in the numbers of protons and neutrons in an atom of that element.

Radioactivity The spontaneous decay or disintegration of an unstable atomic nucleus accompanied by the emission of radiation.

Sarin A liquid nerve agent that is primarily a vapor hazard.

Secondary contamination The process by which a contaminant is carried out of the hot zone and contaminates people, animals, the environment, or equipment. Also referred to as cross-contamination.

Sensitizer A chemical that causes a large percentage of people or animals to develop an allergic reaction after repeated exposure.

Smallpox A highly infectious disease caused by the *Variola* virus.

Solvent A substance that dissolves another substance so as to make a solution.

Specific gravity The weight of a liquid as compared to water.

State of matter The physical state of a material—solid, liquid, or gas.

Sulfur mustard A clear, yellow, or amber oily liquid that may be dispersed in an aerosol form. It causes blistering of exposed skin.

Target organ effect A situation in which specific bodily organs are typically affected by exposures to certain chemical substances.

Toxicity A measure of the degree to which something is toxic or poisonous. Toxicity can also refer to the adverse effects a substance may have on a whole organism, such as a human (or a bacterium or a plant), or to a substructure such as a cell or a specific organ.

Toxicology The study of the adverse effects of chemical or physical agents on living organisms.

Toxic products of combustion Hazardous chemical compounds that are released when a material decomposes under heat.

TRACEMP An acronym to help remember the effects and potential exposures to a hazardous materials incident: thermal, radiation, asphyxiant, chemical, etiologic, mechanical, and psychogenic.

Upper explosive limit (UEL) The maximum amount of gaseous fuel that can be present in the air if the air/fuel mixture is to be flammable or explosive.

Vapor The gas phase of a substance, particularly of those substances that are normally liquids or solids at room temperature.

Vapor density The weight of an airborne concentration (vapor or gas) as compared to an equal volume of dry air.

Vapor pressure For our purpose, the pressure associated with liquids held inside any type of closed container.

VX A thick, oily liquid that is the most toxic of the weapons-grade nerve agents.

Water solubility The ability of a substance to dissolve in water.

RESPONDER *in action*

Your crew arrives at a biotechnology research company, where you find an 8-liter spill of acetonitrile in a laboratory.

The scientist working in the laboratory dropped two glass containers, each containing 4 liters, shattering both on the tile floor. The chemical saturated both legs of the scientist's pants up to the knee. You find him sitting on a chair in an adjacent break room, still dripping wet from the spill. No decontamination has been performed. He is coughing and complains of a severe headache. One of the scientist's co-workers hands you an SDS. Answer the following questions based on the selected information found on the SDS.

Chemical name: ACETONITRILE

Appearance: Clear, colorless liquid

Odor: Similar to ether

Flash point: 36°F (2°C)

Flammable range:

Lower explosive limit (LEL): 4.4 percent

Upper explosive limit (UEL): 16.0 percent

Solubility: Miscible in water.

Specific gravity: 0.79

pH: No information found.

Vapor density (air = 1): 1.4

Vapor pressure (mm Hg): 73 @ 68°F (20°C)

Evaporation rate (BuAc = 1): 5.79

Health effects: Effects of overexposure are often delayed, possibly due to the slow formation of cyanide anions (negatively charged ions or atoms) in the body. These cyanide anions prevent the body from using oxygen and can lead to internal asphyxiation. Early symptoms may include nose and throat irritation, flushing of the face, and chest tightness.

1. Based on the information found on the SDS, would the victim's signs and symptoms lead you to believe he requires immediate decontamination and medical treatment? If so, choose the following answer that best represents your next course of action.

A. The victim is not suffering any health effects related to the exposure. He should simply change his clothes and return to work.

B. You should assist the victim with removing his contaminated pants only, and direct him to the nearest safety shower. Once decontamination is complete, he is eligible to return to work.

C. The victim should remove all of his own clothing, be accompanied to the nearest safety shower, decontaminated for 15 minutes, and transported to the hospital for medical evaluation. You should keep a safe distance from the victim and avoid touching him. As long as he is alert and able to follow commands, he is still able to help himself.

D. This chemical is extremely toxic. You and your crew should withdraw from the laboratory and set up a controlled area outside of the building for decontamination. You will need to research the chemical fully, don the appropriate PPE, and then reenter the laboratory later to address the medical needs of the victim.

2. The ambient temperature in the laboratory is 77°F (25°C). The flash point of acetonitrile is 36°F (2°C). Which of the following choices best describes the threat of a flash fire?

A. The flash point of acetonitrile is well below the temperature of the room. If the vapors reach an uncontrolled ignition source, a flash fire could occur.

B. The flash point of acetonitrile is so low that it cannot ignite under these conditions.

C. In this scenario, acetonitrile would not give off any flammable vapors.

D. Eight liters of acetonitrile would create a small puddle and, therefore, would not create much flammable vapor.

RESPONDER *in action* (continued)

3. The vapor density of acetonitrile is 1.4. Based on this value, you can expect that the vapors will be ________ than air.
   - **A.** lighter
   - **B.** darker
   - **C.** more reactive
   - **D.** heavier
4. The SDS shows the flammable range for acetonitrile to be 4.4 percent to 16 percent. Which of the following definitions for flammable range is most accurate?
   - **A.** An expression of a fuel/air mixture, defined by upper and lower flammable limits, that reflects an amount of flammable vapor mixed with a given volume of air.
   - **B.** A description of the volume increase that occurs when a liquid material changes to a gas.
   - **C.** The minimum temperature at which a fuel, when heated, will ignite in air and continue to burn.
   - **D.** The ability of a chemical to undergo a change in its chemical make-up, usually with the release of some form of energy.

Recognizing and Identifying the Hazards

Greg Henry/ShutterStock, Inc.

©Mark Winfrey/ShutterStock, Inc.

© Photos.com

CHAPTER 3

Awareness Level Personnel

Knowledge Objectives

After studying this chapter, you will be able to:

- Define *hazardous material* and *weapon of mass destruction (WMD)*. (NFPA 4.2.1(1), p. 46–47)
- Identify the UN/DOT hazard classes, as well as the primary hazards associated with each (NFPA 4.2.1(2)(3), p. 64–66)
- Describe the difference between hazardous materials/WMD incidents and other emergencies. (NFPA 4.2.1(4), p. 47)
- Describe occupancies that may contain hazardous materials. (NFPA 4.2.1(5), p. 47)
- Describe some ways you might detect the presence of a hazardous materials release. (NFPA 4.2.1(11), p. 47)
- Describe the limitations of using your senses to detect the presence of hazardous materials. (NFPA 4.2.1(12), p. 47–48)
- Describe specific containers and container shapes that might indicate hazardous materials. (NFPA 4.2.1(6), p. 48–53)
- Describe how to identify the product, owner, and emergency telephone number on a pipeline marker. (NFPA 4.2.1(7), p. 57)
- Describe how to identify a placard, label, or other marking used to indicate the presence of a hazardous material. (NFPA 4.2.1(9), 4.2.2, 4.4.1(4), p. 58–60)
- Describe the NFPA 704 hazard identification system. (NFPA 4.2.1(8), p. 60–61)
- Describe safety data sheets (SDS) and shipping papers. (NFPA 4.2.1(10), p. 66–71)
- Describe how to identify criminal or terrorist activity involving chemical, biological, or radiological agents. (NFPA 4.2.1(13), (14), (15), (16), (17), p. 73–76)
- Describe how to identify an illicit laboratory, and explosive and secondary devices. (NFPA 4.2.1(18), (19), (20), p. 76–77)

Skills Objectives

After studying this chapter, you will be able to:

- Use the *Emergency Response Guidebook*. (NFPA 4.2.3, 4.4.1, p. 63–66)

Operations Level Responders

Knowledge Objectives

After studying this chapter, you will be able to:

- Describe occupancies that may contain hazardous materials. (p. 47)
- Describe specific containers and container shapes that might indicate/contain hazardous materials. (NFPA 5.2.1.1, 5.2.1.1.5, 5.2.1.1.6, 5.2.1.2, 5.2.1.2.2, p. 48–53)
- Describe shipping and storage vessels that could hold hazardous materials. (NFPA 5.2.1.1.1, 5.2.1.1.2, 5.2.1.1.3, 5.2.1.1.4, p. 48–57)
- Describe various modes of transportation used to transport hazardous materials. (NFPA 5.2.1.1.1, p. 54–57)
- Describe how to identify the product, owner, and emergency telephone number on a pipeline marker. (NFPA 5.2.1.3.1, p. 57)
- Describe how to identify information and its significance on a pesticide label or a radioactive material label. (NFPA 5.2.1.3.2, 5.2.1.3.3, p. 51, 74–76)
- Describe the NFPA 704 hazard identification system. (p. 60–61)
- Describe safety data sheets (SDS) and shipping papers. (NFPA 5.2.1.2.1, 5.2.2, 5.2.3(1), p. 66–71)
- Describe assistance provided by CHEMTREC/CANUTEC/SETIQ. (NFPA 5.2.2, 5.2.3 (1), p. 73)
- Recognize the surrounding conditions to take note of during a hazardous materials/WMD incident survey. (NFPA 5.2.1.4, p. 46–48)
- Describe ways to verify information obtained from a hazardous materials/WMD incident survey. (NFPA 5.2.1.5, p. 57–73)
- Describe how to identify criminal or terrorist activity involving chemical, biological, or radiological agents. (NFPA 5.2.1.6, 5.2.2, p. 73–76)
- Describe how to identify an illicit laboratory, and explosive and secondary devices. (p. 76–77)
- Identify the corresponding UN/DOT hazard class and division, given the following:
  - Blood agents (NFPA 5.2.3(9)(a), p. 66)
  - Biological agents and biological toxins (NFPA 5.2.3(9)(b), p. 66)
  - Choking agents (NFPA 5.2.3(9)(c), p. 66)
  - Irritants (riot control agents) (NFPA 5.2.3(9)(d), p. 66)
  - Nerve agents (NFPA 5.2.3(9)(e), p. 66)
  - Radiological materials (NFPA 5.2.3(9)(f), p. 66)
  - Vesicants (blister agents) (NFPA 5.2.3(9)(g), p. 66)

Skills Objectives

After studying this chapter, you will be able to:

- Use the *Emergency Response Guidebook*. (NFPA 5.2.3 (1), p. 63–66)

You Are the Responder

During an odor investigation in a light-industrial area of your district, you are directed by a police officer to an abandoned open-head steel 55-gallon drum just off the city street. The officer directs you to a local shopkeeper who made the 911 call. The shopkeeper reports that the drum must have been dropped off overnight and that he smelled an unusual odor when he walked by the drum. You notice a green label on the side drum reading "non-hazardous waste."

1. What type of material is most likely stored in the drum (e.g., corrosive, solvent, explosive)? Does the description of the drum provide any other useful information? If so, what?
2. What does the label on the drum signify, and does the report of an unusual odor make sense given the information you have?
3. How would you go about obtaining more information on the potential contents of the drum?

Introduction

Scene size-up is important in any emergency situation, but especially during hazardous materials incidents. The ability to "read" the scene is a critical skill, and responders must interpret the available clues and weave them together to make informed decisions and operate safely. It is not enough to simply scan the emergency scene—you must train yourself to stop for a moment and pay attention. Looking at something is nothing more than pointing your eyes in the right direction; seeing, by contrast, is taking in the visual clues and piecing them together to form a conclusion—this is the basis for situational awareness **FIGURE 3-1**. The opening scenario, for example, contains a few clues. An open-head drum (the type where the entire lid is secured by a bolted clasp-type ring that circles the entire head of the drum) typically contains solid materials or those types of substances that cannot be poured or pumped easily. Additionally, steel drums don't usually contain corrosives. The hazardous waste label provides some preliminary information. Also, the shopkeeper mentioned detecting an odor. Each point does not tell the whole story, but at least provides a place from which to start your size-up.

FIGURE 3-1 Pay attention: situational awareness is key.

Courtesy of Rob Schnepp

RESPONDER TIP

Remember to pay attention to the scene and think before you act: It could save your life.

In some cases, it may be possible to detect the presence of a hazardous materials emergency based on information relayed in the initial dispatch or perhaps from persons on the scene, or based on your own knowledge of the response area. In other cases, you may have to put on your detective hat and search for clues that may indicate whether or not a hazardous substance is present. At all times, departmental standard operating procedures (SOPs) and your level of training, along with your information gathering efforts at the scene, should guide your initial and ongoing actions. Size-up is a continual process and is never complete until the incident is over.

RESPONDER TIP

A weapon of mass destruction (WMD), as defined in NFPA 472, *Standard for Competence of Responders to Hazardous Materials/Weapons of Mass Destruction Incidents*, includes (1) any destructive device, such as any explosive, incendiary, or poison gas, bomb, grenade, rocket having a propellant charge of more than 4 ounces, missile having an explosive or incendiary charge of more than one-quarter ounce (7 grams), mine, or device similar to the above; (2) any weapon involving toxic or poisonous chemicals; (3) any weapon involving a disease organism; or (4) any weapon that is designed to release radiation or radioactivity at a level dangerous to human life.

More than 4 billion tons of hazardous materials are shipped annually in the United States by rail, highway, sea, and air. While it is not possible to know each and every chemical by name, it is possible to identify many of the more common commodities by way of available identification systems (placards, labels, and other signage) and/or using other methods (detection devices, eyewitness accounts, visible indicators) to identify the presence of a hazardous material. This chapter provides you with guidance on interpreting visual clues, which may signal the possible presence of a hazardous materials incident.

Recognizing a Hazardous Materials/ Weapons of Mass Destruction Incident

A hazardous material, as defined by the U.S. Department of Transportation (DOT), is any substance or material that is capable of posing an unreasonable risk to human health, safety, or the environment when transported in commerce, used incorrectly, or not properly contained or stored. Recognizing the presence of a hazardous materials/weapons of mass destruction (WMD) event, determining the identity of the material(s), and understanding the hazards involved often require some detective work. You must train yourself to take the time to look at the whole scene so that you can identify the critical visual indicators and fit them into what is known about the problem. Some of the key differences between hazardous materials/WMD emergencies and other types of emergencies include:

- A hazardous materials/WMD event typically moves more slowly than other emergencies and usually does not require responders to rush headlong into the problem.
- Actions taken at hazardous materials/WMD incidents are largely dictated by the chemicals or hazards involved; the circumstances of the release and the risks associated with taking action.
- Personnel operating at the scene of a hazardous materials/WMD incident must be conscious of the law enforcement aspect of the incident.

On any hazardous materials incident, your first action should always be to approach the scene from a safe location and direction. The traditional rules of staying uphill and upwind are a good place to start. If possible, it may make sense to use binoculars and view the scene from a safe distance, again looking for labels, placards, type of container, or other clues that could help you understand the scene. Be sure to question anyone involved in the incident—a wealth of information may be available to you if you simply ask the right person. Take enough time to assess the scene and interpret other clues such as dead animals near the release, discolored pavement, dead grass, visible vapors or puddles, or other indicators that may help identify the presence of a hazardous material. A wet area near unidentified containers on an asphalt parking lot may not initially seem significant, but perhaps it's a hot day, and the "wet" area still looks wet an hour after the containers were noticed, well past the point that water would have evaporated. Perhaps the substance is a hydrocarbon and the wet-looking pavement is actually solvent that has permeated into the asphalt FIGURE 3-2.

A thorough and thoughtful size up will help clarify the problem you are facing. Once you have a basic idea of what happened or have determined that danger may be present, you can begin to formulate a plan for addressing the incident.

FIGURE 3-2 Notice the details—and think about what they mean.

Courtesy of Rob Schnepp

Occupancy and Location

A wide variety of chemicals are stored in warehouses, hospitals, laboratories, industrial occupancies, residential garages, bowling alleys, home improvement centers, garden supply stores, restaurants, and scores of other facilities or businesses in your response area. So many different chemicals exist in so many different locations that you could encounter almost anything during any type of emergency situation.

The chapter, *Hazardous Materials: Overview*, discusses the importance of preplanning a response to occupancies that use hazardous materials. Identifying the kinds and quantities of hazardous materials used and stored by local facilities should be an integral part of any comprehensive community response plan. However, once a chemical emergency has occurred at a particular occupancy, responders should locate some key personnel. Generally speaking, most fixed facilities that use and/or store a significant amount of chemicals will have an Environmental Health and Safety (EH&S) department. In many cases, EH&S departments will employ certified industrial hygienists, chemists and chemical engineers, and/or certified safety professionals to ensure safe work practices at the site. These industry experts may be a valuable source of information for understanding the chemical inventory of the facility, the ventilation systems, high-hazard chemical storage areas, and specialized areas such as clean rooms or refrigeration systems. Additionally, representatives of the EH&S department have the ability to connect the responders with site security, maintenance workers, and other important employees at the facility. It's important to find the right people with keys, unique site knowledge, and instructions on how to get around the facility.

Senses

Another possible way to detect the presence of hazardous materials is to use your senses, although doing this must be employed carefully to avoid becoming contaminated or exposed. The senses that are typically safe to employ on a regular basis are those of sight and sound. Generally, the farther you are from the incident when you notice a problem, the safer you will be.

Using any of your senses that bring you in close proximity to the chemical should be done with caution or avoided. When it comes to hazardous materials incidents, "leading with your nose" is not a good tactic—using binoculars from a distance is.

Clues that are seen or heard may also provide valuable information from a distance, enabling you to take precautionary steps. Vapor clouds at the scene, for example, are a signal to move yourself and others away to a place of safety; the sound of an alarm from a toxic gas sensor in a chemical storage room or laboratory may also serve as a warning to retreat. Some highly vaporous and odorous chemicals—chlorine and ammonia, for example—may be detected by smell a long way from the actual point of release. These and other clues may alert the operations level responder to the presence of a hazardous atmosphere.

Containers

In basic terms, a container is any vessel or receptacle that holds a material. Often the container type, size, and material of construction provide important clues about the nature of the substance inside. Nevertheless, responders should not rely solely on the type of container when making a determination about hazardous materials. The reason for the release may be that an incompatible substance was placed in the wrong type of container, and that's why you are there.

Red phosphorus from a drug laboratory, for example, might be found in an unmarked plastic container. In this case, there may be no legitimate markings to alert a responder to the possible contents. Gasoline or waste solvents may be stored in 55-gallon steel drums with two capped openings (2" and ¾") on the top FIGURE 3-3.

Sulfuric acid, at 97 percent concentration, could be found in a polyethylene drum that might be colored black, red, white, or blue. In most cases, there is no correlation between the color of the drum and the possible contents. The same sulfuric acid might also be found in a 1-gallon amber glass container. Hydrofluoric acid, by contrast, is incompatible with silica (glass) and would be stored in a plastic container.

FIGURE 3-3 An abandoned steel drum. Notice the configuration of the holes on top.

Courtesy of Rob Schnepp

Steel or polyethylene drums, bags, high-pressure gas cylinders, railroad tank cars, plastic buckets, aboveground and underground storage tanks, cargo tanks, and pipelines are all representative examples of how hazardous materials are used, stored, and shipped FIGURE 3-4.

Some very recognizable chemical containers, such as 55-gallon drums and compressed gas cylinders, can be found in almost every type of manufacturing facility. Materials stored in a cardboard drum are usually in solid form. Stainless steel containers may hold particularly dangerous chemicals, and cold liquids are kept in Thermos-like Dewar containers designed to maintain the appropriate temperature FIGURE 3-5.

FIGURE 3-4 Drums may be constructed of many different types of materials, including cardboard, polyethylene, or stainless steel.

Courtesy of Rob Schnepp

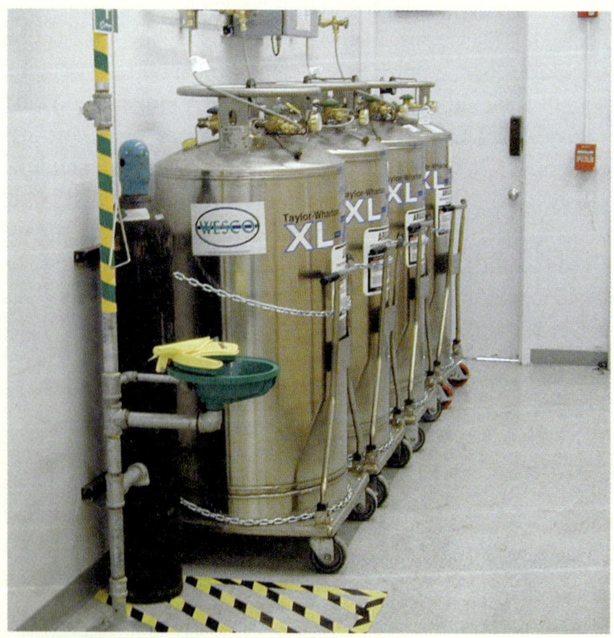

FIGURE 3-5 A series of Dewar containers stored adjacent to a compressed gas cylinder.

Courtesy of Rob Schnepp

One way to distinguish containers is to divide them into two categories based on their capacity: bulk and nonbulk storage containers.

RESPONDER TIP

When you think about all the places a hazardous materials incident could occur, do not limit your thinking. Explore your response district—you may be surprised at how many different kinds of clues you might find.

Container Volume

Bulk storage containers, or large-volume containers, are defined by their internal capacity based on the following measures:

- Liquids greater than 119 gallons
- Bulk packaging with a net mass greater than 882 pounds
- Gases greater than 100 lbs water capacity

Bulk storage containers include fixed tanks, highway cargo tanks, rail tank cars, totes, and intermodal tanks.

In general, bulk storage containers are found in occupancies that rely on and need to store large quantities of a particular chemical. Most manufacturing facilities have at least one type of bulk storage container TABLE 3-1. Often these bulk storage containers are surrounded by a supplementary containment system to help control an accidental release. Secondary containment is an engineered method to control spilled or released product if the main containment vessel fails. A 5000-gallon vertical storage tank, for example, may be surrounded by a series of short walls commonly referred to as *dikes* or *berms* that form a catch basin around the vessel.

Secondary containment basins typically can hold the entire volume of the tank, along with a percentage of the water flowed from hose lines or sprinkler systems in the event of fire and, in the case of outdoor storage, a certain amount of rainfall. Many storage vessels, including 55-gallon drums, may have secondary containment systems. If facilities that use bulk storage containers in your response area have secondary containment systems, it will be easier to handle leaks when they occur.

Large-volume horizontal tanks are also common at fixed facilities. When stored aboveground, these tanks are referred to as aboveground storage tanks (ASTs); if they are placed underground, they are known as underground storage tanks (USTs).

These tanks can hold a few hundred gallons to several million gallons of product and are usually made of aluminum, steel, or plastic. ASTs and USTs can be pressurized or nonpressurized. Nonpressurized horizontal tanks are usually made of steel or aluminum. Because it is difficult to relieve the internal pressure in these tanks, they pose a danger when they are exposed to fire. Typically, these containers hold flammable or combustible materials, such as gasoline, oil, or diesel fuel FIGURE 3-6.

Pressurized horizontal tanks have rounded ends and large vents or pressure-relief stacks. The most common aboveground pressurized tanks contain liquid propane and liquid ammonia; such containers can hold a few hundred gallons to several thousand gallons of product. These tanks usually have a small vapor space—called the *headspace*—above the liquid. In most cases, 10 to 15 percent of the total container capacity is vapor.

Intermodal tanks are both shipping and storage vessels. They hold between 5000 and 6000 gallons of product and can be either pressurized or nonpressurized. Intermodal (IM or

FIGURE 3-6 Example of an above-ground diesel storage tank at a fixed facility.
Courtesy of Rob Schnepp

Table 3-1 Common Bulk Storage Vessels, Locations, and Contents

Tank Shape	Common Locations	Hazardous Materials Commonly Stored
Underground tanks	Residential, commercial	Fuel oil and combustible liquids
Covered floating roof tanks	Bulk terminal and storage	Highly volatile flammable liquids
Cone roof tanks	Bulk terminal and storage	Combustible liquids
Open floating roof tanks	Bulk terminal and storage	Flammable and combustible liquids
Dome roof tanks	Bulk terminal and storage	Combustible liquids
High-pressure horizontal tanks	Industrial storage and terminal	Flammable gases, chlorine, ammonia
High-pressure spherical tanks	Industrial storage and terminal	Liquid propane gas, liquid nitrogen gas
Cryogenic liquid storage tanks	Industrial and hospital storage	Oxygen, liquid nitrogen gas

IMO) tanks can also be used to ship and store gaseous substances that have been chilled until they liquefy (cryogenic liquids), such as liquid nitrogen and liquid argon. In most cases, an IM tank is shipped to a facility, where it is stored and used, and then returned to the shipper for refilling. Intermodal tanks can be shipped by all methods of transportation—air, sea, or land.

Typically, a box-like steel framework, constructed to facilitate efficient stacking and shipping, surrounds an IM tank. Several types of IM tanks are available:

- IM-101 portable tanks (IMO type 1 internationally) have a 6300-gallon capacity, with internal working pressures between 25.4 pounds per square inch (psi) and 100 psi. These containers typically carry mild corrosives, food-grade products, and flammable liquids FIGURE 3-7A.
- IM-102 portable tanks (IMO type 2 internationally) have a 6300-gallon capacity, with internal working pressures between 14.7 psi and 25.4 psi. They primarily carry nonhazardous materials but may also contain flammable liquids and corrosives FIGURE 3-7B.
- Pressure intermodal tanks (IMO type 5 internationally or DOT Spec 51) are high-pressure vessels with internal pressures in the range of 100–600 psi. These intermodal containers commonly hold liquefied compressed gases such as propane and butane FIGURE 3-7C.
- Cryogenic intermodal tanks (IMO type 7 internationally) are low-pressure containers in transport but can be pressurized to 600 psi. This kind of container commonly carries cryogenic materials that have temperatures less than –150°F (66°C), such as liquefied oxygen, nitrogen, or helium.
- Tube modules consist of several high-pressure tubes attached to a frame. The tubes are individually specified and have working pressures that range as high as 5000 psi. The products commonly carried include hydrogen and oxygen.

Nonbulk Storage Vessels

Essentially, nonbulk storage vessels are all types of containers other than bulk containers. Nonbulk storage vessels can hold a few ounces to 119 gallons of product and include vessels such as drums, bags, compressed gas cylinders, cryogenic containers, and more. Nonbulk storage vessels hold commonly used commercial and industrial chemicals such as solvents, industrial cleaners, and compounds. This section describes the most commonly encountered types of nonbulk storage vessels.

Drums

Drums are easily recognizable, barrel-like containers. They are used to store a wide variety of substances, including food-grade materials, corrosives, flammable liquids, and grease. Drums may be constructed of low-carbon steel, polyethylene, cardboard, stainless steel, nickel, or other materials. Generally, the nature of the chemical dictates the construction of the storage drum. Steel utility drums, for example, hold flammable liquids, cleaning fluids, oil, and other noncorrosive chemicals. Polyethylene drums are used for corrosives such as acids, bases, oxidizers, and other materials that cannot be stored in steel containers. Cardboard drums hold solid materials such as soap flakes, sodium hydroxide pellets, and food-grade materials. Stainless steel or other heavy-duty drums generally hold materials too aggressive (i.e., too reactive) for either plain steel or polyethylene.

FIGURE 3-7 A. IM-101 portable tanks (IMO type 1 internationally). B. IM-102 portable tanks (IMO type 2 internationally). C. Pressure intermodal tanks (IMO type 5 internationally).

Courtesy of UBH International Ltd

Closed-head drums have a permanently attached lid with one or more small openings. The opening is called a bung. Typically, these openings are threaded holes sealed by caps that can be removed only by using a special tool called a bung wrench FIGURE 3-8. Closed-head drums usually have one 2" bung and one ¾" bung. The larger bung is used to pump product from the drum; the smaller bung functions as a vent.

An open-head drum has a removable lid fastened to the drum with a ring FIGURE 3-9. The ring is tightened with a clasp or a threaded nut-and-bolt assembly.

FIGURE 3-9 An open-head drum has a lid that is fastened with a ring that is tightened with a clasp or a nut-and-bolt assembly.

Courtesy of Globalindustrial.com

Bags

Bags are commonly used to store solids and powders such as cement powder, sand, pesticides, soda ash, and slaked lime. Storage bags may be constructed of plastic, paper, or plastic-lined paper. Bags come in different sizes and weights, depending on their contents.

Pesticide bags must be labeled with specific information FIGURE 3-10. Responders can learn a great deal from the label, including the following details:

- Name of the product
- Active ingredients
- Hazard statement
- Total amount of product in the container
- Manufacturer's name and address
- Environmental Protection Agency (EPA) registration number, which provides proof that the product was registered with the EPA
- EPA establishment number, which shows where the product was manufactured
- Signal words to indicate the relative toxicity of the material:
  - *Danger—Poison:* Highly toxic by all routes of entry
  - *Danger:* Severe eye damage or skin irritation
  - *Warning:* Moderately toxic
  - *Caution:* Minor toxicity and minor eye damage or skin irritation
- Practical first-aid treatment description
- Directions for use
- Agricultural use requirements

FIGURE 3-8 A bung wrench is used to operate the openings on the top of a closed-head drum.

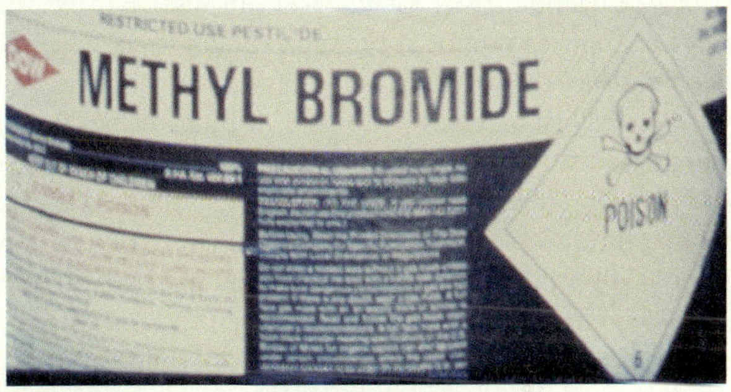

FIGURE 3-10 A pesticide bag must be labeled with the appropriate information.

Courtesy of the USDA

- Precautionary statements such as mixing directions or potential environmental hazards
- Storage and disposal information
- Classification statement on who may use the product

In addition, every pesticide label must carry the statement, "Keep out of reach of children." In Canada, the pest control product (PCP) number can also be found on the pesticide label.

Carboys

Some corrosives and other types of chemicals are transported and stored in vessels called carboys **FIGURE 3-11**. A carboy is a glass, plastic, or steel container that holds 5 to 15 gallons of product. Glass carboys are often placed in a protective wood, foam, fiberglass, or steel box to help prevent breakage. For example, nitric acid, sulfuric acid, and other strong acids are often transported and stored in thick glass carboys protected by a wooden or Styrofoam crate to shield the glass container from damage during normal shipping.

Cylinders

Several types of cylinders are used to hold liquids and gases. Uninsulated compressed gas cylinders are used to store substances such as nitrogen, argon, helium, and oxygen. They come in a range of sizes and have variable internal pressures. An oxygen cylinder used for medical purposes, for example, has a pressure reading of approximately 2000 psi when full. By comparison, the very large compressed gas cylinders found at a fixed facility may have pressure readings of 5000 psi or greater.

The high pressures exerted by these cylinders create a potential for danger. If the cylinder is punctured or the valve assembly fails or the cylinder falls over and damages the valve, causing a rapid release of compressed gas, it will turn the cylinder into an unpredictable missile. Also, if the cylinder is heated rapidly, it could explode with tremendous force, spewing product and metal fragments over long distances. Compressed gas cylinders have pressure-relief valves, but those valves may not be sufficient to relieve the pressure created during a fast-growing fire **FIGURE 3-12**.

A propane cylinder is another type of compressed gas cylinder. Propane cylinders have lower pressures (200–300 psi) and contain a liquefied gas. Liquefied gases such as propane are subject to the phenomenon known as BLEVE (boiling liquid/expanding vapor explosion). BLEVEs occur when pressurized liquefied materials (propane or butane, for example) inside a closed vessel are exposed to a source of high heat.

The low-pressure Dewar container is another commonly encountered cylinder type. As mentioned previously, Dewars are Thermos-like vessels designed to hold cryogenic liquids (cryogens) such as helium, liquid nitrogen, and liquid argon **FIGURE 3-13**.

Typical cryogens include oxygen, helium, hydrogen, argon, and nitrogen. Under normal atmospheric conditions, each of these substances is a gas. A complex process turns them into liquids that can be stored and used for long periods of time. Nitrogen, for example, becomes a liquid at –320°F (–160°C) and must be kept at that temperature if it is to remain in a liquid state.

Cryogens pose a substantial threat if the Dewar container fails to maintain the low temperature of the cryogenic liquid. Cryogens have large expansion ratios—even larger than the expansion ratio of propane (270:1). Cryogenic helium, for example, has an expansion ratio of approximately 750:1. If one volume of liquid helium is warmed to room temperature and vaporized in a totally enclosed container, it can generate a pressure of more than 14,500 psi. To counter this possibility, cryogenic containers usually have two pressure-relief devices: a pressure-relief valve and a frangible (easily broken) metal disk.

FIGURE 3-11 A carboy is used to transport and store corrosive chemicals.

Courtesy of EMD Chemicals, Inc.

FIGURE 3-12 Compressed gas cylinders in storage—what clues do you see?

Courtesy of Rob Schnepp

FIGURE 3-13 A small cryogenic Dewar container.

Courtesy of Cryofab, Inc.

FIGURE 3-14 A. Ton chlorine containers in use at a water treatment facility. B. A ton chlorine container after a leak.

Courtesy of Jason Krusen

Responder Safety Tip

Cryogens are stored in a liquid state—beware of skin exposures. Significant injuries, similar to those associated with thermal burns, can occur when skin comes in contact with one of these liquids.

Ton Containers

Ton containers commonly hold compressed liquefied gases such as chlorine and sulfur dioxide, although you may encounter some select refrigerant fluorocarbon gases such as trichlorofluoromethane (Freon-11) and dichlorodifluoromethane (Freon-12) stored in ton containers.

Regardless of the material inside, these vessels will hold 2000 lbs. of product (hence the name, *ton container*) and are 8 feet in length and 3 feet in diameter. All ton containers other than those containing phosphine have pressure relief valves called fusible plugs. FIGURE 3-14 shows ton containers at a water treatment facility and a ton container after a leak.

Intermediate Bulk Containers (IBC)

These containers are so named because the volumes stored inside fall between what is typically found in drums or bags and what is in cargo tanks. Their capacities are greater than 119 gallons but less than 793 gallons. IBCs are either flexible or rigid containers and are often referred to as "super sacks" or "totes." Super sacks are much larger than a normal bag and may hold solid material anywhere from 500 lbs. to several thousand pounds FIGURE 3-15. The fabric of construction ranges from woven cloth to polyethylene, and they may be palletized for transport. Solid materials transported and stored in super sacks may include oxidizers, corrosives, and flammable solids.

Totes can be found as a high-density polyethylene tank inside a rigid stainless steel frame or perhaps a square metal tank around 4 ft. square and 6 ft. tall. Both types have bottom-mounted discharge valves and are commonly found at fixed facilities. Totes may hold a wide variety of substances such as solvents, corrosives, or select oxidizers. Totes typically have capacities ranging from 119 to 793 gallons.

Shipping and storing totes can be hazardous. These containers often are stacked atop one another and moved with a forklift, such that a mishap with the loading or moving process can compromise the tote FIGURE 3-16. Because totes have no secondary containment system, any leak has the potential to create a large puddle. Additionally, the steel webbing around the tote makes it difficult to access and patch leaks.

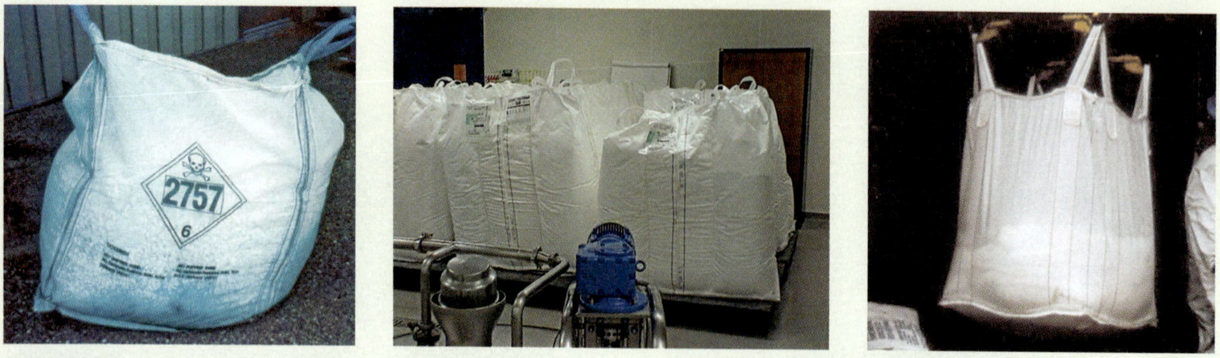

FIGURE 3-15 A super sack.

Courtesy of Hildebrand and Noll Associates, Inc.

FIGURE 3-16 Totes waiting for use at a fixed facility.

Courtesy of Rob Schnepp

Transporting Hazardous Materials

Hazardous materials may be transported by air, sea, or land, or a combination of any or all of these modes of transport. Although transportation of hazardous materials occurs most often on the roadway, when another primary mode of transport is used, roadway vehicles often transport the shipments from the rail station, airport, or dock to the point where it will be used. For this reason, responders must become familiar with all types of chemical transport vehicles they might encounter during a transportation emergency.

Roadway Transportation

As mentioned previously, the most common method of hazardous material transport is over land, by roadway transportation vehicles. According to the Code of Federal Regulations, 49 CFR 171.8(2), or local jurisdictional regulations (for example, Transport Canada), a cargo tank is bulk packaging that is permanently attached to or forms a part of a motor vehicle, or is not permanently attached to any motor vehicle, and that, because of its size, construction, or attachment to a motor vehicle, is loaded or unloaded without being removed from the motor-vehicle. The U.S. Department of Transportation (DOT) does not view tube trailers (which consist of several individual cylinders banded together and affixed to a trailer) as cargo tanks.

One of the most common and reliable transportation vessels is the MC-306/DOT 406 cargo tank FIGURE 3-17. These tanks frequently carry liquid food-grade products, gasoline, or other flammable and combustible liquids. The oval-shaped tank is pulled by a diesel tractor and can carry between 6000 and 10,000 gallons of product. The MC-306/DOT 406 is non-pressurized (its working pressure is between 2.65 and 4 psi), usually made of aluminum or stainless steel, and loaded and offloaded through valves at the bottom of the tank. These cargo tanks have several safety features, including full rollover protection and remote emergency shut-off valves FIGURE 3-18.

A vehicle that is similar to the MC-306/DOT 406 is the MC-307/DOT 407 chemical hauler. It has a round or horseshoe-shaped tank and is capable of holding 6000 to 7000 gallons of liquid FIGURE 3-19. The MC-307/DOT 407, which is also a tractor-drawn tank, is used to transport flammable liquids, mild corrosives, and poisons. This type of cargo tank may be insulated (horseshoe) or uninsulated (round), and have a higher internal working pressure than the MC-306/DOT 406—in some cases up to 35 psi. Cargo tanks that transport corrosives may have a rubber lining to prevent corrosion of the tank structure.

The MC-312/DOT 412 corrosives cargo tank is commonly used to carry corrosives such as concentrated sulfuric acid, phosphoric acid, and sodium hydroxide FIGURE 3-20. This cargo tank has a smaller diameter than the MC-306/DOT 406 or the MC-307/DOT 407 and is often identifiable by the presence of several heavy-duty reinforcing rings around the tank. The rings provide structural stability during transportation and in the event of a rollover. The inside of an MC-312/DOT 412 tanker operates at approximately 15 to 25 psi and holds approximately 6000 gallons. These cargo tanks have substantial rollover protection to reduce the potential for damage to the top-mounted valves.

The MC-331 pressure cargo tank carries materials such as ammonia, propane, Freon, and butane FIGURE 3-21. The liquid volume inside the tank varies, ranging from the 1000-gallon delivery truck to the full-size 11,000-gallon cargo tank. The MC-331 cargo tank has rounded ends, typical of a pressurized

FIGURE 3-17 The MC-306/DOT 406 flammable liquid tank typically hauls flammable and combustible liquids.

Courtesy of Polar Tank Trailer L.L.C.

FIGURE 3-18 The MC-306/DOT 406 cargo tank has a remote emergency shut-off valve as a safety feature.

Courtesy of Glen Rudner

FIGURE 3-19 The MC-307/DOT 407 chemical hauler carries flammable liquids, mild corrosives, and poisons.

Courtesy of Polar Tank Trailer L.L.C.

FIGURE 3-20 The MC-312/DOT 412 corrosives tanker is commonly used to carry corrosives such as concentrated sulfuric acid, phosphoric acid, and sodium hydroxide.

Courtesy of National Tank Truck Carriers Association

FIGURE 3-21 The MC-331 pressure cargo tank carries materials such as ammonia, propane, Freon, and butane.

Courtesy of Rob Schnepp

FIGURE 3-22 The MC-338 cryogenic tank maintains the low temperatures required for the cryogens it carries.

Courtesy of Jack B. Kelly, Inc.

vessel, and is commonly constructed of steel or stainless steel with a single tank compartment. The MC-331 operates at approximately 300 psi, with typical internal working pressures being in the vicinity of 250 psi. These cargo tanks are equipped with spring-loaded relief valves that traditionally operate at 110 percent of the designated maximum working pressure. A significant explosion hazard arises if an MC-331 cargo tank is impinged on by fire, however. The nature of most materials carried in MC-331 tanks means a threat of explosion exists because of the inability of the relief valve to keep up with the rapidly building internal pressure. Responders must use great care when dealing with this type of transportation emergency.

The MC-338 cryogenic tank operates much like the Dewar container described earlier and carries many of the same substances FIGURE 3-22. This low-pressure tank relies on tank insulation to maintain the low temperatures required for the cryogens it carries. A boxlike structure containing the tank control valves is typically attached to the rear of the tank. Special training is required to operate valves on this and any other tank. An untrained individual who attempts to operate the valves may disrupt the normal operation of the tank, thereby compromising its ability to keep the liquefied gas cold and creating a potential explosion hazard. Cryogenic cargo tanks have a relief valve near the valve control box. From time to time, small puffs of white vapor will be vented from this valve. Responders should understand that this is a normal occurrence—the valve is working to maintain the proper internal pressure. In most cases, this vapor is not indicative of an emergency situation.

Tube trailers carry compressed gases such as hydrogen, oxygen, helium, and methane FIGURE 3-23. Essentially, they are high-volume transportation vehicles that are made up of several individual cylinders banded together and affixed to a trailer. The individual cylinders on the tube trailer are much like the smaller compressed gas cylinders discussed earlier in this chapter. These large-volume cylinders operate at working pressures of 3000 to 5000 psi. One trailer may carry several different gases in individual tubes. Typically, a valve control box is found toward the rear of the trailer, and each individual cylinder has its own relief valve. These trailers can frequently be seen at construction sites or at facilities that use large quantities of compressed gases.

Dry bulk cargo tanks are commonly seen on the road; they carry dry bulk goods such as powders, pellets, fertilizers, or grain FIGURE 3-24. These tanks are not pressurized, but may use pressure to offload the product. Dry bulk cargo tanks are generally *V*-shaped with rounded sides that funnel the contents to the bottom-mounted valves.

FIGURE 3-23 A tube trailer.

Courtesy of Jack B. Kelly, Inc.

FIGURE 3-24 A dry bulk cargo tank carries dry goods such as powders, pellets, fertilizers, or grain.

Courtesy of Polar Tank Trailer L.L.C.

Railroad Transportation

Railroads move almost 2 million carloads of freight each year in the United States, with relatively few accidents. Even so, responders should recognize that when rail incidents do occur, they can create unique and significant hazards. Railway cars include passenger cars, freight cars, and tank cars, some of which are capable of carrying more than 30,000 gallons of product. Hazardous materials incidents involving railroad transportation have the potential to be large-scale emergencies.

Operations level responders should recognize three basic types of rail tank cars: nonpressurized, pressurized, and special use. Each has a distinctive profile that can be recognized from a distance. Additionally, rail tank cars are usually labeled on both sides with, among other things, the owner of the car, the car's capacity, and the specification. With dedicated haulers, the chemical name is often clearly visible on both sides of the rail tank car.

Nonpressurized (general-service) rail tank cars typically carry general industrial chemicals and consumer products such as corn syrup, flammable and combustible liquids, and mild corrosives. The new style of the nonpressurized rail tank car has visible valves and piping without protective housing on top though the next generation tank car will have protective housing that has all the valves inside FIGURE 3-25. Older nonpressurized railcars may have a dome that covers the top-mounted valves and bottom outlet valves. Nonpressurized rail tank cars may hold volumes ranging from 4000 to 40,000 gallons.

Pressurized rail tank cars transport materials such as propane, ammonia, ethylene oxide, and chlorine. These cars have internal working pressures ranging from 100 to 500 psi and are equipped with top-mounted fittings for loading and unloading. These fittings are protected by a sturdy and easily identified protective housing that sits atop the rail tank car FIGURE 3-26. Unfortunately, the high volumes carried in these cars can generate long-duration, high-pressure leaks that may prove difficult to control. For example, a liquid or vapor release from the valve arrangements on a chlorine tank car requires a special kit and specific training to stop the leak. Additionally, pressurized rail tank cars may be insulated or uninsulated, depending on the material being transported.

FIGURE 3-25 A nonpressurized rail tank car has visible valves and piping.

Courtesy of private source

FIGURE 3-26 Pressurized rail tank cars have top-mounted fittings for loading and unloading.

Courtesy of private source

FIGURE 3-27 Special-use rail cars include boxcars, flat cars, cryogenic, and corrosive tank cars. Tube cars are no longer in service.

© Mark Gibson/Index Stock Images, Inc./Alamy Images

Cryogenic tank cars are the most common special-use railcars emergency responders may encounter, although there are not many of them. Cryogenic railcars illustrate an important concept for the emergency responder: The hazard will be unique to the particular railcar and its contents FIGURE 3-27.

The train's engineer or conductor will have information about all the cars on the train and should be located and consulted during a railway emergency. The engineer will have knowledge about the availability of a specialized response team from the railroad and the location and contents of the railcars, and will be able to help responders understand the scope and impact of the derailment. These types of incidents can have significant effects on communities adjacent to the incident and may pose a significant risk to life, property, or the environment. Details about the shipping papers unique to railway transportation are presented later in this chapter.

Pipelines

Of all the various methods used to transport hazardous materials, the high-volume pipeline is the one that is most rarely involved in emergencies. A pipeline is a length of pipe—including pumps, valves, flanges, control devices, strainers, and/or similar equipment—for conveying fluids and gases over potentially long distances. Like rail incidents, pipeline incidents may present responders with challenges and hazards not typically

FIGURE 3-28 A pipeline warning sign provides information about the pipe's contents, the owner's name, and contact information.

© Photodisc

encountered at most hazardous materials incidents. In many areas, large-diameter pipelines transport natural gas, gasoline, diesel fuel, and other products from delivery terminals to distribution facilities. Pipelines are often buried underground, but may be aboveground in remote areas. Additionally, subject matter experts from the company that owns the pipeline may be required to assist hazardous materials responders for the local jurisdiction. These incidents, like rail incidents, could have far-reaching implications and present responders with a challenging set of circumstances.

The pipeline right-of-way is an area, patch of land, or roadway that extends a certain number of feet on either side of the pipe itself. The company that owns the pipeline maintains this area. The company is also responsible for placing warning signs at regular intervals along the length of the pipeline. Pipeline warning signs include a warning symbol, the pipeline owner's name, and an emergency contact phone number FIGURE 3-28.

Again, pipeline emergencies are complicated events that require specially trained responders. If you suspect an emergency involving a pipeline, contact the owner of the line immediately. The company will dispatch a crew to assist with the incident.

Information about the pipe's contents and owner is also often found at the vent pipes. These inverted *J*-shaped tubes provide pressure relief or natural venting during maintenance and repairs. Vent pipes are clearly marked and are located approximately 3 feet above the ground.

Transportation and Facility Markings

The presence of labels, placards, and other markings on buildings, packages, boxes, and containers often enables responders to identify a released chemical. When used correctly, marking systems indicate the presence of a hazardous material and provide clues about the substance. This section provides an introduction to the various marking systems currently being used. It does not cover all the intricacies and requirements for every marking system; however, it will acquaint you with the most common systems.

The Department of Transportation Marking System

The U.S. Department of Transportation (DOT) marking system is an identification system characterized by labels, placards, and markings FIGURE 3-29. This marking system is used when materials are being transported from one location to another in the United States. The same marking system is also used in Canada by Transport Canada.

Placards are diamond-shaped indicators (10¾" on each side) that are placed on all four sides of highway transport vehicles, railroad tank cars, and other forms of transportation carrying hazardous materials FIGURE 3-30. Labels are smaller versions (4" diamond-shaped indicators) of placards; they are placed on the four sides of individual boxes and smaller packages being transported FIGURE 3-31.

FIGURE 3-29 The DOT uses labels, placards, and markings (such as these found in the *Emergency Response Guidebook*) to give responders a general idea of the hazard inside a particular container or cargo tank.

Courtesy of the U.S. Department of Transportation

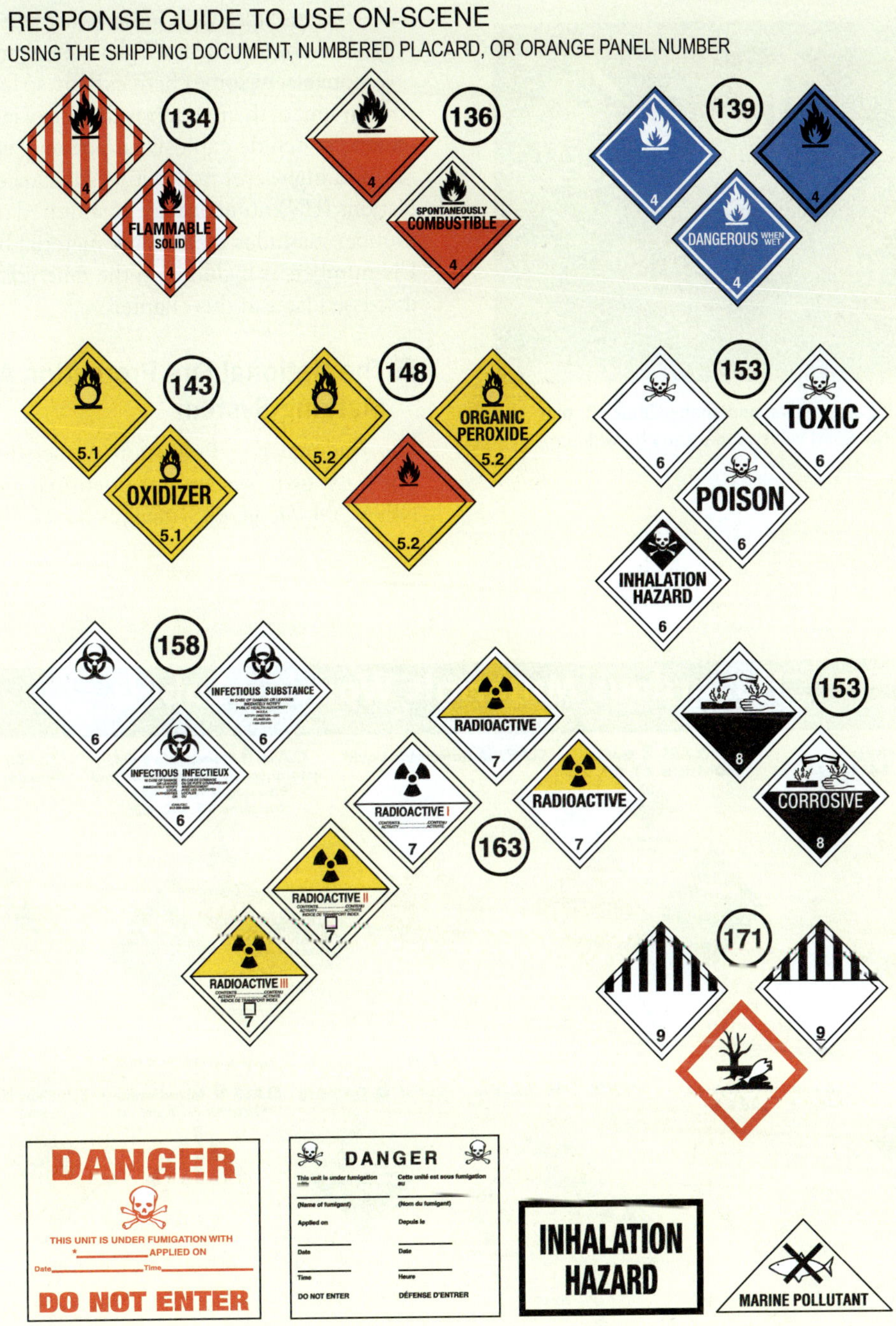

FIGURE 3-29 *(Continued)*

Placards, labels, and markings are intended to give responders a general idea of the hazard inside a particular container or cargo tank. A placard identifies the broad hazard class (flammable, poison, corrosive) to which the material inside belongs. A label on a box inside a delivery truck, for example, relates only to the potential hazard inside that particular package.

Other Considerations

The DOT system does not require that all chemical shipments be marked with placards or labels. In most cases, the package or cargo tank must contain a certain amount of hazardous material before a placard is required. For example, the "1000-pound rule" applies to explosives, flammable and nonflammable gases, flammable/combustible liquids, flammable solids, air-reactive solids, oxidizers

FIGURE 3-30 A placard is a large diamond-shaped indicator that is placed on all sides of transport vehicles that carry hazardous materials.

and organic peroxides, poison solids, corrosives, and miscellaneous (class 9) materials. Placards are required for these materials only when the shipment weighs more than 1000 pounds.

Conversely, some chemicals are so hazardous that shipping any amount of them requires the use of labels or placards. These materials include explosives, poison gases, water-reactive solids, and high-level radioactive substances. A four-digit United Nations (UN) number may be required on some placards. This number identifies the specific material being shipped; a list of UN numbers is included in the *Emergency Response Guidebook*, discussed later in this chapter.

The National Fire Protection Association 704 Marking System

The National Fire Protection Association (NFPA) has developed its own system for identifying hazardous materials. NFPA 704, *Standard System for the Identification of the Hazards*

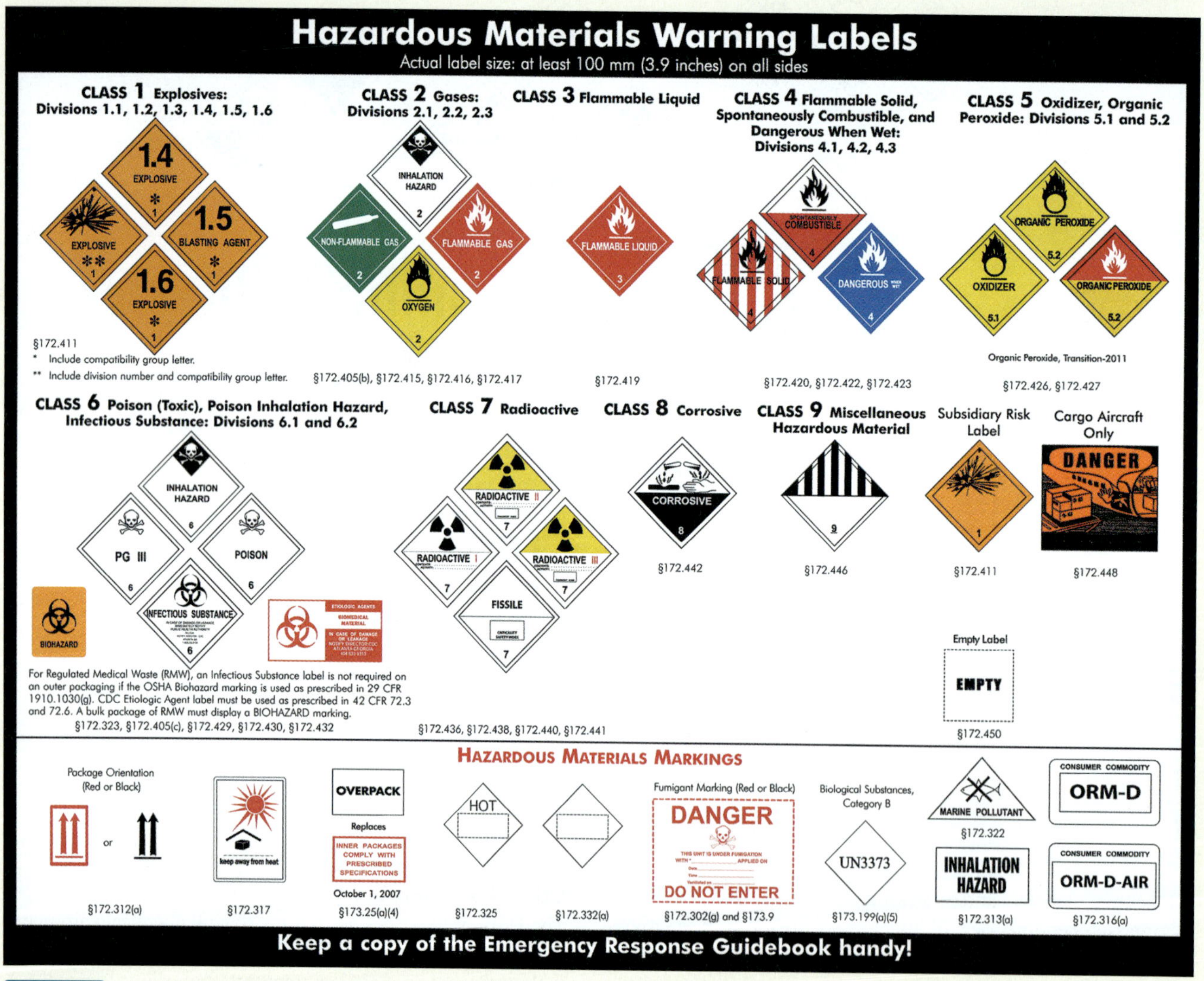

FIGURE 3-31 A label is a smaller version of the placard and is placed on boxes or smaller packages that contain hazardous materials.

Courtesy of the U.S. Department of Transportation

of Materials for Emergency Response, outlines a marking system characterized by a set of diamonds that are found on the outside of buildings, on doorways to chemical storage areas, and on fixed storage tanks. This marking system is designed for fixed-facility use. Responders can use the NFPA diamonds to understand the broad hazards posed by chemicals stored in a building or part of a building.

The NFPA 704 hazard identification system uses a diamond-shaped symbol of any size, which is itself broken into four smaller diamonds, each representing a particular property or characteristic of a substance or group of substances **FIGURE 3-32**. The blue, red, and yellow diamonds will each contain a numerical rating in the range of 0–4, with 0 being the least hazardous and 4 being the most hazardous **TABLE 3-2**.

The blue diamond (at the nine o'clock position) indicates the health hazard posed by a particular material alone or perhaps within a group of other chemicals. Responders must understand that when an NFPA diamond represents a series of hazards posed by a number of different substances, the most severe characteristic of any of the substances may be used to represent the hazard within any of the four colored diamonds. For example, if any one of the substances in a grouping of chemicals could be fatally toxic, that single substance causes a 4 to appear in the blue diamond. All other substances could be much less hazardous, but the one causing the 4 represents the health hazard for the group. The same logic holds true for the flammability and reactivity diamonds. The top red diamond indicates flammability. The yellow diamond (at the three o'clock position) indicates reactivity. The bottom white diamond will not have a number, but may contain special symbols. Among the symbols used are a burning *O* (oxidizing capability), a three-bladed trefoil (radioactivity), and a *W* with a slash through it (water reactive). For complete information on the NFPA 704 system, consult NFPA 704, *Standard System for the Identification of the Hazards of Materials for Emergency Response*.

To accurately compare and contrast the DOT marking system and NFPA 704 system, remember this important difference:

- The DOT hazardous materials marking system is used when materials are being transported from one location to another.
- The NFPA 704 hazard identification system is designed for fixed-facility use.

Hazardous Materials Information System

Since 1983, the Hazardous Materials Information System (HMIS) hazard communication program has helped employers comply with the Hazard Communication Standard established by the U.S. Occupational Safety and Health Administration (OSHA). The HMIS is similar to the NFPA 704 marking system and uses a numerical hazard rating with similarly colored horizontal columns **FIGURE 3-33**.

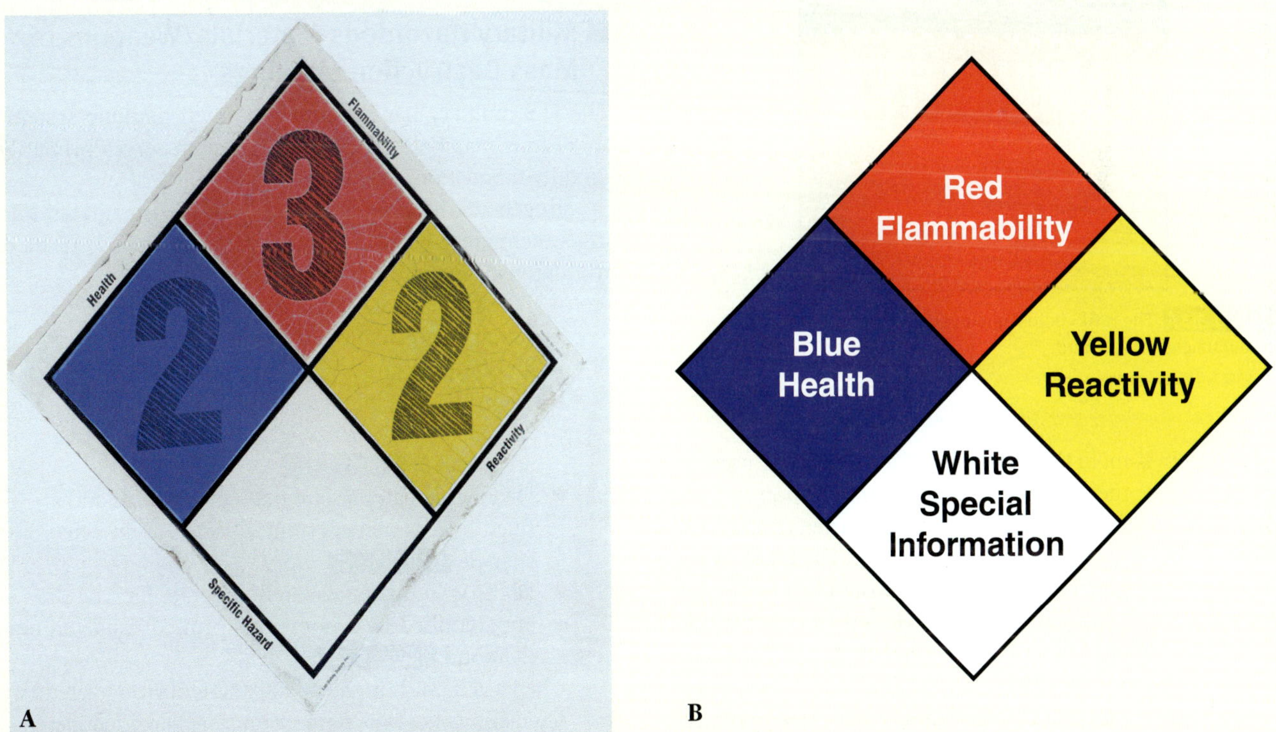

FIGURE 3-32 A. Example of a placard using the NFPA 704 hazard identification system that is used for fixed-facility use. B. Each color used in the diamond represents a particular property or characteristic.

Table 3-2 Hazard Levels in the NFPA Hazard Identification System

Flammability Hazards (Red Diamond)	Reactivity Hazards (Yellow Diamond)
4 Materials that will rapidly or completely vaporize at atmospheric pressure and normal ambient temperature, or that are readily dispersed in air and that will burn readily. Liquids with a flash point below 73°F (22°C) and a boiling point below 100°F (38°C). 3 Liquids and solids that can be ignited under almost all ambient temperature conditions. Liquids with a flash point below 73°F (22°C) and a boiling point above 100°F (38°C) or liquids with a flash point above 73°F (22°C) but not exceeding 100°F (38°C) and a boiling point below 100°F (38°C). 2 Materials that must be moderately heated or exposed to relatively high ambient temperatures before ignition can occur. Liquids with a flash point above 100°F (38°C) but not exceeding 200°F (93°C). 1 Materials that must be preheated before ignition can occur. Liquids that have a flash point above 200°F (93°C). 0 Materials that will not burn.	4 Materials that in themselves are readily capable of detonation or of explosive decomposition or reaction at normal temperatures and pressures. 3 Materials that in themselves are capable of detonation or explosive decomposition or reaction but require a strong initiating source, or that must be heated under confinement before initiation, or that react explosively with water. 2 Materials that readily undergo violent chemical change at elevated temperatures and pressures, or that react violently with water, or that may form explosive mixtures with water. 1 Materials that in themselves are normally stable, but can become unstable at elevated temperatures and pressures. 0 Materials that in themselves are normally stable, even under fire exposure conditions, and are not reactive with water.
Health Hazards (Blue Diamond)	**Special Hazards (White Diamond)**
4 Materials that on very short exposure could cause death or major residual injury.	*ACID* Acid
3 Materials that on short exposure could cause serious temporary or residual injury.	*ALK* Alkali
2 Materials that on intense or continued, but not chronic, exposure could cause incapacitation or possible residual injury.	*COR* Corrosive
1 Materials that on exposure would cause irritation but only minor residual injury.	*OX* Oxidizer
0 Materials that on exposure under fire conditions would offer no hazard beyond that of ordinary combustible material.	*W* Reacts with water

FIGURE 3-33 The HMIS helps employers comply with the Hazard Communication Standard.

The HMIS is more than just a label; it is a method used by employers to give their personnel necessary information to work safely around chemicals and includes training materials to inform workers of chemical hazards in the workplace. The HMIS is not required by law but rather is a voluntary system that employers choose to use to comply with OSHA's Hazard Communication Standard. In addition to describing the chemical hazards posed by a particular substance, the HMIS provides guidance about the personal protective equipment that employees need to use to protect themselves from workplace hazards. Letters and icons specify the different levels and combinations of protective equipment.

Responders must understand the fundamental difference between the NFPA 704 marking system and the HMIS. NFPA 704 is intended for responders; the HMIS is intended for the employees of a facility. Although the HMIS is not a response information tool, it can give clues about the presence and nature of the hazardous materials found in the facility.

Military Hazardous Materials/Weapons of Mass Destruction Markings

The U.S. military has developed its own marking system for hazardous materials. The military system serves primarily to identify detonation, fire, and special hazards.

In general, hazardous materials within the military marking system are divided into four categories based on the relative detonation and fire hazards:

- Division 1 materials are considered mass detonation hazards and are identified by a number 1 printed inside an orange octagon FIGURE 3-34A.
- Division 2 materials have explosion-with-fragment hazards and are identified by a number 2 printed inside an orange X FIGURE 3-34B.
- Division 3 materials are mass fire hazards and are identified by a number 3 printed inside an inverted orange triangle FIGURE 3-34C.
- Division 4 materials are moderate fire hazards and are identified by a number 4 printed inside an orange diamond FIGURE 3-34D.

Chemical hazards in the military system are depicted by colors. Toxic agents (such as sarin or mustard) are identified by the color red. Harassing agents (such as tear gas and smoke producers) are identified by yellow. White phosphorus is identified by white. Specific personal protective gear requirements are identified using pictograms. Military shipments containing hazardous materials/WMD are not required, by exception, to be placarded.

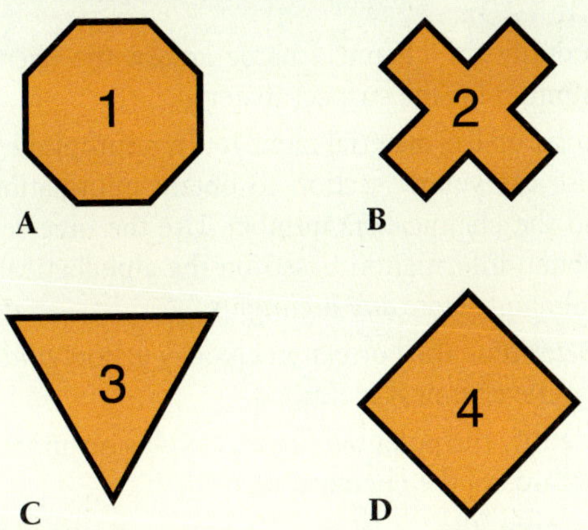

FIGURE 3-34 A. Mass detonation hazards. B. Explosion-with-fragment hazards. C. Mass fire hazards. D. Moderate fire hazards.

Chemical References

Numerous reference materials are available to the responder, including the DOT's *Emergency Response Guidebook*. The following sections describe these resources.

The *Emergency Response Guidebook*

The DOT *Emergency Response Guidebook (ERG)* offers a certain amount of guidance for responders operating at a hazardous materials incident FIGURE 3-35. This guide, which is intended to help responders decide which preliminary action to take, provides information on approximately 4000 chemicals. Responders at the scene should seek additional specifics about any material in question by consulting the appropriate emergency response agency or using the emergency response number on a shipping document, if applicable, to gather more information.

The U.S. DOT and the Secretariat of Communications and Transportation of Mexico (SCT), along with Transport Canada, jointly developed the *ERG*.

RESPONDER TIP

The same information, organized differently, is found in both the blue and yellow sections of the *ERG*.

Using the *ERG*

When the *ERG* refers to a *small spill*, it means a leak from one small package, a small leak in a large container (up to a 55-gallon drum), a small cylinder leak, or any small leak, even one in a large package. A *large spill* is a large leak or spill from a larger container or package, a spill from a number of small packages, or anything from a 1-ton cylinder, tank truck, or railcar.

The *ERG* is divided into four colored sections: yellow, blue, orange, and green.

- *Yellow section:* More than 4000 chemicals are found in this section, listed numerically by their four-digit UN number/identification (ID) number. Entry number 1005, for example, identifies "ammonia, anhydrous." Use the yellow section when the UN/ID number is known or can be identified. The entries include the name of the chemical and the emergency action guide number. For example:

ID No.	Guide No.	Name of Material
1005	125	Ammonia, anhydrous

- *Blue section:* The same chemicals listed in the yellow section are found here, listed alphabetically by name. The entry will include the emergency action guide number and the identification number.

Name of Material	Guide No.	ID No.
Ammonia, anhydrous	125	1005

- *Orange section:* This section is organized by guide number FIGURE 3-36. The general hazard class, fire/explosion hazards, health hazards, and basic emergency actions, based on hazard class, are provided.
- *Green section:* This section is organized numerically by UN/ID number and provides the initial isolation distances for certain materials FIGURE 3-37. Chemicals included in this section consist of the chemicals highlighted from the yellow and blue sections.[*] The green section includes water-reactive materials that produce toxic gases (calcium phosphide and trichlorosilane, for example); toxic inhalation hazards (TIH), which are gases or volatile liquids that are extremely toxic to humans; chemical warfare agents (CWA); and dangerous water-reactive materials (WRM). Examples include such substances as anhydrous ammonia, sarin, and sodium cyanide. These gases or volatile liquids are extremely toxic to humans and pose a hazard to health during transport. Any material listed in the green section is extremely hazardous.

Responder Safety Tip

Any substance highlighted, in either the yellow or blue section of the *ERG*, is either a toxic inhalation hazard, a chemical warfare agent, or a water-reactive material.

The nine chemical families, and their respective divisions recognized in the *ERG*, are outlined here:

- DOT Class 1—Explosives
  Division 1.1 Explosives with a mass explosion hazard
  Division 1.2 Explosives with a projection hazard
  Division 1.3 Explosives with predominantly a fire hazard
  Division 1.4 Explosives with no significant blast hazard
  Division 1.5 Very insensitive explosives with a mass explosion hazard
  Division 1.6 Extremely insensitive articles

[*]The yellow and blue sections of the *ERG* contain the same information; it is simply organized differently. Therefore, any substance highlighted in the blue section will also be highlighted in the yellow section, and vice versa.

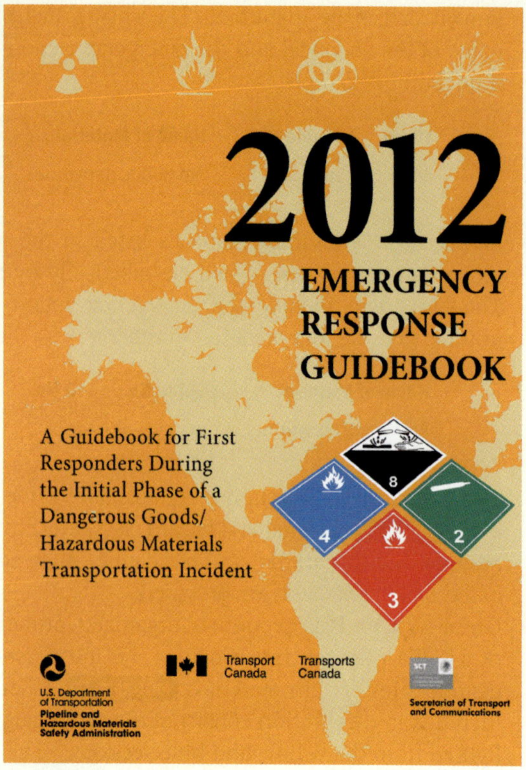

FIGURE 3-35 The *Emergency Response Guidebook* is a reference used as a base for your initial actions at a hazardous materials incident.

Courtesy of the U.S. Department of Transportation

To use the *ERG*, follow the steps in SKILL DRILL 3-1 (NFPA 4.2.3, 4.4.1, 5.2.3):

1. Identify the chemical name and/or the chemical ID number for the suspect material.
2. Look up the material name in the appropriate section. Use the yellow section to obtain information based on the chemical ID number. Use the blue section to obtain information based on the alphabetical chemical name. *Note any highlights.*
3. Determine the correct emergency action guide to use for the chemical identified.
4. Identify the potential fire and explosion and/or health hazards of the chemical identified.
5. Identify the isolation distance and the protective actions required for the chemical identified.
6. Identify the emergency response actions of the chemical identified.

The *ERG* organizes chemicals into nine basic hazard classes, or families; the members of each family exhibit similar properties. There is also a "Dangerous" placard, which indicates that more than one hazard class is contained in the same load FIGURE 3-38.

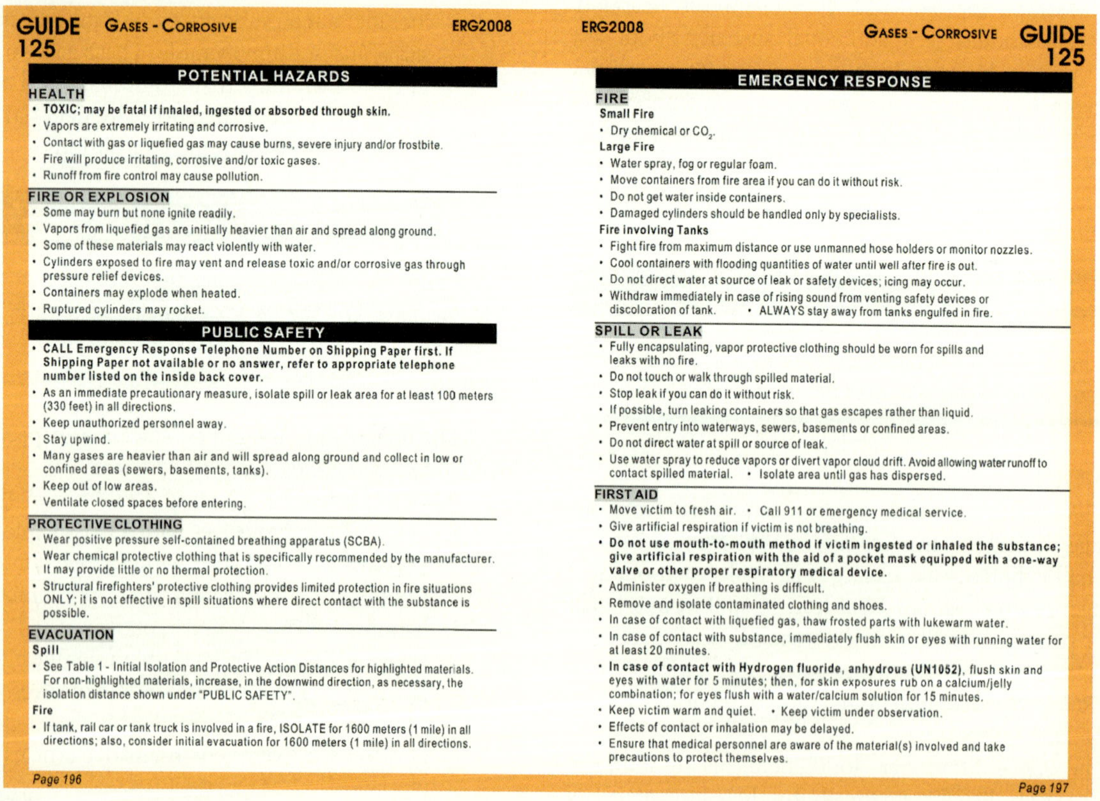

GUIDE 125 GASES - CORROSIVE ERG2008

POTENTIAL HAZARDS

HEALTH

- **TOXIC; may be fatal if inhaled, ingested or absorbed through skin.**
- Vapors are extremely irritating and corrosive.
- Contact with gas or liquefied gas may cause burns, severe injury and/or frostbite.
- Fire will produce irritating, corrosive and/or toxic gases.
- Runoff from fire control may cause pollution.

FIRE OR EXPLOSION

- Some may burn but none ignite readily.
- Vapors from liquefied gas are initially heavier than air and spread along ground.
- Some of these materials may react violently with water.
- Cylinders exposed to fire may vent and release toxic and/or corrosive gas through pressure relief devices.
- Containers may explode when heated.
- Ruptured cylinders may rocket.

PUBLIC SAFETY

- **CALL Emergency Response Telephone Number on Shipping Paper first. If Shipping Paper not available or no answer, refer to appropriate telephone number listed on the inside back cover.**
- As an immediate precautionary measure, isolate spill or leak area for at least 100 meters (330 feet) in all directions.
- Keep unauthorized personnel away.
- Stay upwind.
- Many gases are heavier than air and will spread along ground and collect in low or confined areas (sewers, basements, tanks).
- Keep out of low areas.
- Ventilate closed spaces before entering.

PROTECTIVE CLOTHING

- Wear positive pressure self-contained breathing apparatus (SCBA).
- Wear chemical protective clothing that is specifically recommended by the manufacturer. It may provide little or no thermal protection.
- Structural firefighters' protective clothing provides limited protection in fire situations ONLY; it is not effective in spill situations where direct contact with the substance is possible.

EVACUATION

Spill

- See Table 1 - Initial Isolation and Protective Action Distances for highlighted materials. For non-highlighted materials, increase, in the downwind direction, as necessary, the isolation distance shown under "PUBLIC SAFETY".

Fire

- If tank, rail car or tank truck is involved in a fire, ISOLATE for 1600 meters (1 mile) in all directions; also, consider initial evacuation for 1600 meters (1 mile) in all directions.

Page 196

ERG2008 GASES - CORROSIVE **GUIDE 125**

EMERGENCY RESPONSE

FIRE

Small Fire

- Dry chemical or CO_2.

Large Fire

- Water spray, fog or regular foam.
- Move containers from fire area if you can do it without risk.
- Do not get water inside containers.
- Damaged cylinders should be handled only by specialists.

Fire involving Tanks

- Fight fire from maximum distance or use unmanned hose holders or monitor nozzles.
- Cool containers with flooding quantities of water until well after fire is out.
- Do not direct water at source of leak or safety devices; icing may occur.
- Withdraw immediately in case of rising sound from venting safety devices or discoloration of tank. • ALWAYS stay away from tanks engulfed in fire.

SPILL OR LEAK

- Fully encapsulating, vapor protective clothing should be worn for spills and leaks with no fire.
- Do not touch or walk through spilled material.
- Stop leak if you can do it without risk.
- If possible, turn leaking containers so that gas escapes rather than liquid.
- Prevent entry into waterways, sewers, basements or confined areas.
- Do not direct water at spill or source of leak.
- Use water spray to reduce vapors or divert vapor cloud drift. Avoid allowing water runoff to contact spilled material. • Isolate area until gas has dispersed.

FIRST AID

- Move victim to fresh air. • Call 911 or emergency medical service.
- Give artificial respiration if victim is not breathing.
- **Do not use mouth-to-mouth method if victim ingested or inhaled the substance; give artificial respiration with the aid of a pocket mask equipped with a one-way valve or other proper respiratory medical device.**
- Administer oxygen if breathing is difficult.
- Remove and isolate contaminated clothing and shoes.
- In case of contact with liquefied gas, thaw frosted parts with lukewarm water.
- In case of contact with substance, immediately flush skin or eyes with running water for at least 20 minutes.
- **In case of contact with Hydrogen fluoride, anhydrous (UN1052)**, flush skin and eyes with water for 5 minutes; then, for skin exposures rub on a calcium/jelly combination; for eyes flush with a water/calcium solution for 15 minutes.
- Keep victim warm and quiet. • Keep victim under observation.
- Effects of contact or inhalation may be delayed.
- Ensure that medical personnel are aware of the material(s) involved and take precautions to protect themselves.

Page 197

FIGURE 3-36 The orange section of the *ERG*.

Courtesy of the U.S. Department of Transportation

Page 300

TABLE 1 - INITIAL ISOLATION AND PROTECTIVE ACTION DISTANCES

ID No.	NAME OF MATERIAL	SMALL SPILLS (From a small package or small leak from a large package) First ISOLATE in all Directions: Meters	(Feet)	Then PROTECT persons Downwind during- DAY: Kilometers	(Miles)	NIGHT: Kilometers	(Miles)	LARGE SPILLS (From a large package or from many small packages) First ISOLATE in all Directions: Meters	(Feet)	Then PROTECT persons Downwind during- DAY: Kilometers	(Miles)	NIGHT: Kilometers	(Miles)
1005 1005	Ammonia, anhydrous Anhydrous ammonia	30 m	(100 ft)	0.1 km	(0.1 mi)	0.2 km	(0.1 mi)	150 m	(500 ft)	0.8 km	(0.5 mi)	2.3 km	(1.4 mi)
1008 1008	Boron trifluoride Boron trifluoride, compressed	30 m	(100 ft)	0.1 km	(0.1 mi)	0.6 km	(0.4 mi)	300 m	(1000 ft)	1.9 km	(1.2 mi)	4.8 km	(3.0 mi)
1016 1016	Carbon monoxide Carbon monoxide, compressed	30 m	(100 ft)	0.1 km	(0.1 mi)	0.1 km	(0.1 mi)	150 m	(500 ft)	0.7 km	(0.5 mi)	2.7 km	(1.7 mi)
1017	Chlorine	60 m	(200 ft)	0.4 km	(0.3 mi)	1.6 km	(1.0 mi)	600 m	(2000 ft)	3.5 km	(2.2 mi)	8.0 km	(5.0 mi)
1023 1023	Coal gas Coal gas, compressed	30 m	(100 ft)	0.1 km	(0.1 mi)	0.1 km	(0.1 mi)	60 m	(200 ft)	0.3 km	(0.2 mi)	0.4 km	(0.3 mi)
1026 1026	Cyanogen Cyanogen gas	30 m	(100 ft)	0.2 km	(0.1 mi)	0.9 km	(0.5 mi)	150 m	(500 ft)	1.0 km	(0.7 mi)	3.5 km	(2.2 mi)
1040 1040	Ethylene oxide Ethylene oxide with Nitrogen	30 m	(100 ft)	0.1 km	(0.1 mi)	0.2 km	(0.1 mi)	150 m	(500 ft)	0.8 km	(0.5 mi)	2.5 km	(1.6 mi)
1045 1045	Fluorine Fluorine, compressed	30 m	(100 ft)	0.1 km	(0.1 mi)	0.3 km	(0.2 mi)	150 m	(500 ft)	0.8 km	(0.5 mi)	3.1 km	(1.9 mi)
1048	Hydrogen bromide, anhydrous	30 m	(100 ft)	0.1 km	(0.1 mi)	0.4 km	(0.3 mi)	300 m	(1000 ft)	1.5 km	(1.0 mi)	4.5 km	(2.8 mi)
1050	Hydrogen chloride, anhydrous	30 m	(100 ft)	0.1 km	(0.1 mi)	0.4 km	(0.2 mi)	60 m	(200 ft)	0.3 km	(0.2 mi)	1.4 km	(0.9 mi)
1051	AC **(when used as a weapon)**	100 m	(300 ft)	0.3 km	(0.2 mi)	1.1 km	(0.7 mi)	1000 m	(3000 ft)	3.8 km	(2.4 mi)	7.2 km	(4.5 mi)
1051 1051 1051	Hydrocyanic acid, aqueous solutions, with more than 20% Hydrogen cyanide Hydrogen cyanide, anhydrous, stabilized Hydrogen cyanide, stabilized	60 m	(200 ft)	0.2 km	(0.1 mi)	0.6 km	(0.4 mi)	400 m	(1250 ft)	1.6 km	(1.0 mi)	4.1 km	(2.5 mi)
1052	Hydrogen fluoride, anhydrous	30 m	(100 ft)	0.1 km	(0.1 mi)	0.5 km	(0.3 mi)	300 m	(1000 ft)	1.7 km	(1.1 mi)	3.6 km	(2.2 mi)

FIGURE 3-37 The green section of the *ERG*.

Courtesy of the U.S. Department of Transportation

FIGURE 3-38 A "Dangerous" placard indicates that more than one hazard is contained within the same load.

Courtesy of Rob Schnepp

- DOT Class 2—Gases
  Division 2.1 Flammable gases
  Division 2.2 Nonflammable, nontoxic gases
  Division 2.3 Toxic gases
- DOT Class 3—Flammable liquids (and combustible liquids in the United States)
- DOT Class 4—Flammable solids, spontaneously combustible materials, and dangerous-when-wet materials/water-reactive substances
  Division 4.1 Flammable solids
  Division 4.2 Spontaneously combustible materials
  Division 4.3 Water-reactive substances/dangerous when wet
- DOT Class 5—Oxidizing substances and organic peroxides
  Division 5.1 Oxidizing substances
  Division 5.2 Organic peroxides
- DOT Class 6—Toxic substances and infectious substances
  Division 6.1 Toxic substances
  Division 6.2 Infectious substances
- DOT Class 7—Radioactive materials
- DOT Class 8—Corrosive substances
- DOT Class 9—Miscellaneous hazardous materials/products, substances or organisms

TABLE 3-3 provides examples of substances emergency responders should be aware of, along with their UN hazard class designations.

Fire fighters, police, and other emergency services personnel—all of whom may be the first to arrive at the scene of a transportation incident involving a hazardous material—are the primary audience for both the *ERG* and the *Fire Fighter's Handbook of Hazardous Materials*.

Table 3-3 Corresponding UN/DOT Hazard Classes and Divisions

Common Name	UN/DOT Hazard Class/Division
Blood Agents	
Hydrogen cyanide	6.1
Cyanogen chloride	2.3
Biological Agents and Toxins	
Smallpox	6.2
Ricin	6.2
Anthrax	6.2
Choking Agents	
Chlorine	2.3
Phosgene	2.3
Irritants (Riot Control Agents)	
Pepper spray, mace	2.2 (subsequent risk 6.1)
Tear gas	6.1
Nerve Agents	
Sarin	6.1
Soman	6.1
Tabun	6.1
VX	6.1
Radiological Materials	7
Vesicants (Blister Agents)	
Mustard	6.1
Lewisite	6.1
Nitrogen mustard	6.1

RESPONDER TIP

The responder should always attempt to identify hazardous materials using more than one source. The *ERG* is only one source and should not be used exclusively.

Other Reference Sources

Labels and placards may be helpful in identifying hazardous materials, but other sources of information are also available. Among these reference sources are safety data sheets (SDS), shipping papers, and staffed national resource centers. First responding fire fighters, law enforcement personnel, or representatives of other allied agencies should place a high priority on identifying the released material and finding a reliable source of information about the chemical and physical properties of the released substance. The most appropriate source will depend on the situation; use your best judgment in making your selection.

Safety Data Sheets

A common source of information about a particular chemical is the SDS specific to that substance FIGURE 3-39. Essentially, an SDS provides basic information about the chemical make-up of a substance, the potential hazards it presents, appropriate first aid in the event of an exposure, and other pertinent data for safe handling of the material. An SDS will typically include the following details:

- The name of the chemical, including any synonyms
- Physical and chemical characteristics of the material
- Physical hazards of the material
- Health hazards of the material
- Signs and symptoms of exposure
- Routes of entry
- Permissible exposure limits
- Responsible-party contact
- Precautions for safe handling (including hygiene practices, protective measures, and procedures for cleaning up spills or leaks)
- Applicable control measures, including personal protective equipment
- Emergency and first-aid procedures
- Appropriate waste disposal

When responding to a hazardous materials incident at a fixed facility, responders should ask the site representative for an SDS for the spilled material. All facilities that use or store chemicals are required by law to have an SDS on file for each chemical used or stored in the facility. Many sites, but especially those that stock many different chemicals, may keep this information archived on a computer database. Although the SDS is not a definitive response tool, it is a key piece of the puzzle. Responders should investigate as many sources as possible (preferably at least three) to gather emergency information about a released substance. An SDS can also be obtained from staffed national resource centers or on the transporting vehicle.

Shipping Papers

Shipping papers are required whenever materials are transported from one place to another. They include the names and addresses of the shipper and the receiver, identify the material being shipped, and specify the quantity and weight of each part of the shipment. Shipping papers for road and highway transportation are called bills of lading or freight bills and are located in the cab of the vehicle FIGURE 3-40. Drivers transporting chemicals are required by law to have a set of shipping papers on their person or within easy reach inside the cab at all times.

A bill of lading may provide additional information about a hazardous substance, such as its packaging group designation. The packaging group designation is another system used by shippers to identify special handling requirements or hazards. Some DOT hazard classes require shippers to assign packaging groups based on the material's flash point and toxicity. A packaging group designation may signal that the material poses a greater hazard than similar materials in a hazard class. There are three packaging group designations:

- *Packaging group I:* High danger
- *Packaging group II:* Medium danger
- *Packaging group III:* Minor danger

SAFETY DATA SHEET
ANHYDROUS AMMONIA

TI

DISTRIBUTORS:

TANNER INDUSTRIES, INC.

DIVISIONS:

NATIONAL AMMONIA	NORTHEASTERN AMMONIA
HAMLER INDUSTRIES	BOWER AMMONIA & CHEMICAL

735 Davisville Road, Third Floor, Southampton, PA 18966; 215-322-1238

CORPORATE EMERGENCY TELEPHONE NUMBER: 800-643-6226 CHEMTREC: 800-424-9300

DESCRIPTION

CHEMICAL NAME: Ammonia, Anhydrous
CAS REGISTRY NO: 7664-41-7
SYNONYMS: Ammonia
CHEMICAL FAMILY: Inorganic Nitrogen Compound
FORMULA: NH_3
MOL. WT: 17.03 (NH_3)
COMPOSITION: 99+% Ammonia

STATEMENT OF HEALTH HAZARD

HAZARD DESCRIPTION:
Ammonia is an irritant and corrosive to the skin, eyes, respiratory tract and mucous membranes. Exposure to liquid or rapidly expanding gases may cause severe chemical burns and frostbite to the eyes, lungs and skin. Skin and respiratory related diseases could be aggravated by exposure.

Not recognized by OSHA as a carcinogen.
Not listed in the National Toxicology Program.
Not listed as a carcinogen by the International Agency for Research on Cancer.

EXPOSURE LIMITS FOR AMMONIA: Vapor

OSHA	50 ppm,	35 mg / m^3 PEL	8 hour TWA
NIOSH	35 ppm,	27 mg / m^3 STEL 15 minutes	
	25 ppm,	18 mg / m^3 REL	10 hour TWA
	300 ppm,	IDLH	
ACGIH	25 ppm,	18 mg / m^3 TLV	8 hour TWA
	35 ppm,	27 mg / m^3 STEL 15 minutes	

TOXICITY: LD 50 (Oral / Rat) 350 mg / kg

PHYSICAL DATA

BOILING POINT: -28°F at 1 Atm.
PH: N/A
SPECIFIC GRAVITY OF GAS (air = 1): 0.596 at 32°F
SPECIFIC GRAVITY OF LIQUID (water = 1): 0.682 at 28°F (Compared to water at 39°F).
PERCENT VOLATILE: 100% at 212°F
APPEARANCE AND ODOR: Colorless liquid or gas with pungent odor.
CRITICAL TEMPERATURE: 271.4°F
GAS SPECIFIC VOLUME: 20.78 Ft^3/Lb at 32°F and 1 Atm.
VAPOR DENSITY: 0.0481 Lb/Ft^3 at 32°F
LIQUID DENSITY: 38.00 Lb/Ft^3 at 70°F
APPROXIMATE FREEZING POINT: -108°F
WEIGHT (per gallon): 5.15 pounds at 60°F
VAPOR PRESSURE: 114 psig at 70°F
SOLUBILITY IN WATER (per 100 pounds of water): 86.9 pounds at 32°F, 51 pounds at 68°F
SURFACE TENSION: 23.4 Dynes / cm at 52°F
CRITICAL PRESSURE: 111.5 atm

Revision: September 2005 Page 1 of 4 Prepared By: JRP

FIGURE 3-39 An example of an SDS for anhydrous ammonia.

Courtesy of Tanner Industries, Inc., Southhampton, PA

STRAIGHT BILL OF LADING
ORIGINAL - NOT NEGOTIABLE

BOL/Reference No.
RSI82715

CARRIER: NORFOLK SOUTHERN

Date: 12/23/2008

Shipper: RSI LOGISTICS, INC (OKEMOS, MI US)

The property described below, in apparent good order, except as noted (contents and condition of packages unknown), marked, consigned, and destined as indicated below, which said carrier (the word carrier being understood throughout this contract as meaning any person or corporation in possession of the property under the contract) agrees to carry to its usual place of delivery at said destination, if on its route, otherwise to deliver to another carrier on the route to said destination. It is mutually agreed, as to each carrier of all or any said property, that every service to be performed hereunder shall be subject to all the terms and conditions of the Uniform Domestic Straight Bill of Lading set forth (1) in Official, Southern, Western and Illinois Freight Classification in effect on the date hereof, if this is a rail or a rail-water shipment, or (2) in the applicable motor carrier classification or tariff if this is a motor carrier shipment

Shipper hereby certifies that he is familiar with all the terms and conditions or the said bill of lading, including those on the back thereof, set forth in the classification or tariff which governs the transportation of this shipment, and the said terms and conditions are hereby agreed to by the shipper and accepted for himself and his assigns.

Consignee Information: CONSIGNEE DEER PARK, TX Address: City: DEER PARK, TX US	
Route: NS-ESTL-BNSF	
Origin Switch Route:	
Destination Switch Route: HUSTN-PTRA	Rail Car No: GATX290861

For assistance in any transportation emergency involving chemicals, phone CHEMTREC, day or night, Toll Free 1-800-424-9300

DESCRIPTION		*WEIGHT
ONE TANK CAR	Contains: Methyl Esters STCC#2899415 BIODIESEL-15, Biodiesel Sales Order Contract No: RSI82715 Sales Order Contract No: AAT122308-4 Purchase Order Contract No: AAT122308-4	(Sub. To Correction) 204400 Lbs.

SEAL NUMBERS:

Gross

Tare

Net

Weighed By: ____________________

If charges are to be prepaid, write or stamp here, "To be Prepaid"
Prepaid

Subject to Section 7 of the conditions of applicable bill of lading, if this shipment is to be delivered to the consignee without recourse on the consignor, the consignor shall sign the following statement:: *The carrier shall not make delivery of this shipment without payment of freight and all other lawful charges.*

Not In Effect

* This is to certify that the above named materials are properly classified, described, packaged, marked, and labeled, and are in proper condition for transportation, according to the applicable regulations of the Department of Transportation.

FIGURE 3-40 A bill of lading or freight bill.

Courtesy of RSI Logistics, Inc.

Shipping papers for railroad transportation are called waybills FIGURE 3-41; a list of the contents in every car on the train is called the consist or train list. The conductor, engineer, or a designated member of the train crew will have a copy of the consist in mainline use. If the incident happens in a railyard, then you may find a waybill, as well.

On a marine vessel, shipping papers are called the dangerous cargo manifest FIGURE 3-42. The manifest is generally kept in a tube-like container in the wheelhouse in the custody of the captain or master.

For air transport, the air bill is the shipping paper FIGURE 3-43. It is kept in the cockpit and is the pilot's responsibility.

WAYBILL NO.

WAYBILL
NON-NEGOTIABLE

SHIP DATE:

SHIPPER NUMBER	SHIPPER REFERENCE NUMBER	CONSIGNEE REFERENCE NUMBER	P.O. NUMBER
SHIPPER		CONSIGNEE	
STREET ADDRESS		STREET ADDRESS	
STREET ADDRESS		STREET ADDRESS	
CITY, STATE AND ZIP CODE		CITY, STATE AND ZIP CODE	
CONTACT	PHONE NUMBER	CONTACT	PHONE NUMBER

3rd PARTY NUMBER

3rd PARTY NAME

STREET ADDRESS

CITY, STATE AND ZIP CODE

CONTACT NAME

SHIPPING CO. LIABILITY IS LIMITED TO $50 PER SHIPMENT OR 50 CENTS PER POUND (U.S. DOLLARS), WHICHEVER IS HIGHER SUBJECT TO A MAXIMUM LIABILITY OF $25,000, UNLESS A HIGHER VALUE IS DECLARED AND APPLICABLE CHARGES (FOR DECLARING A VALUE ON WAYBILL) ARE PAID PRIOR TO SHIPPING (SUBJECT TO THE TERMS AND CONDITIONS ON REVERSE SIDE, AND THE SERVICE CONDITIONS FOUND IN THE SHIPPING CO. SERVICE CONDITIONS POLICY).

DECLARED VALUE

$

(SUBJECT TO CORRECTION)

NO. OF PIECES	TYPE	HM	KIND OF PACKAGE, DESCRIPTION OF ARTICLES, SPECIAL MARKS & EXCEPTIONS	ACTUAL WEIGHT	LENGTH	WIDTH	HEIGHT

Special Instructions:

☐ SATURDAY DELIVERY ☐ SUNDAY DELIVERY ☐ APPOINTMENT DELIVERY ☐ INSIDE DELIVERY ☐ RESIDENTIAL DELIVERY

Services Requested:

☐ SAME DAY/ NEXT FLIGHT OUT ☐ NEXT DAY AM ☐ NEXT DAY PM ☐ 2ND DAY ☐ ECONOMY DEFERRED (3-5 DAYS)

FOR CHARTER AIR, TIME DEFINITE, OR GUARANTEED SERVICE, PLEASE CALL 1-800-000-XXXX FOR AVAILABILITY.

QUOTE NUMBER | DIM WEIGHT

SHIPPER CERTIFICATION: Shipper certifies by its signature, its agreement to all of the foregoing terms and conditions, and further certifies that the above named materials are properly classified, described, packaged, marked and labeled, and are in proper condition for transportation according to the applicable regulations of the DOT.

SHIPPER REPRESENTATIVE

SIGNATURE X ____________________ Print Name X ____________________ Date ____________

PICKED UP BY:	RECEIVED BY:	RECEIVED BY CONSIGNEE IN GOOD ORDER UNLESS NOTED BELOW:
DRIVER SIGNATURE ____________	CONSIGNEE SIGNATURE ____________	# S/W SKIDS DEL'D INTACT _____
PLEASE PRINT ____________	PLEASE PRINT ____________	# SKIDS DEL'D: _____
COMPANY ____________	COMPANY ____________	☐ GOOD ORDER ☐ SHORT ☐ OVER ☐ DAMAGED
DATE ____________	DATE ____________	DESCRIBE EXCEPTIONS:
TIME ____________	TIME ____________	

All rules as contained in Shipping Co. Services Conditions Policy will apply. Terms and conditions stated on any Bill of Lading used to transfer goods for carriage, other than an Shipping Co. Waybill, will be null and void. Quotes are based on the information provided and are only an estimate. Final charges are based on actual shipment pieces, weight, dimensions, and services performed as a requirement for delivery. Any changes in actual shipment details will affect the final charges.

Shipping Co. is a certified participant in compliance with the Transportation Security Administration Regulations, Part 109, a Federal Security program.

1 - SHIPPER'S COPY

FIGURE 3-41 A waybill.

Courtesy of private source

CARGO MANIFEST

AIR	AIRCRAFT DATA			DEST CODE	REF	DESTINATION	MISSION DATA			ALW WT	ALW CU	MANIFEST ID	PAGE NO.
	CARRIER	A/C NO.	A/C MODEL				NO.	SU	DATE			STA / FY / TY / NO.	
SUR-FACE	POE	DATE SAILED	VOYAGE DOCUMENT NO.	POD	REF	VESSEL NAME	STATUS	SUST	TRUCK NO.	REMARKS			PAGE NO.

DOC ID	VEHICLE TRAILER OR CNTNR NUMBER	YR / MAKE / CNTNR NUMBER / COMMODTY DESCRIP	COM CODE	CAR-GO EXC	VOYAGE DOC NO. / AIRDM / TRNS PT	PORT OF DISCH	TYPE PACK	TRANSPORTATION CONTROL NUMBER	CONSIGNEE / SERV / ACTVTY ADDRESS	PRIORITY	NAME / AMMO LOT NO./NOMEN / DIMENSIONS / RDD / PROJ / STOW LOC / TRANS ACCT	IDENTIFICATION NO. OR REMARKS	PIECES	WEIGHT	CUBE
												TOTALS	0	0.00	0.00

ITEMS HAVE BEEN LOADED:

ITEMS HAVE BEEN RECEIVED EXCEPT AS CIRCLED NOTED ON REVERSE SIDE

DATE	SIGNATURE OF LOADING AGENT	DATE	SIGNATURE OF UNLOADING AGENT	DATE	SIGNATURE OF RECEIVING AGENT

DD FORM 1385, NOV 78 REPLACES EDITION OF 1 APR 66 WHICH MAY BE USED Adobe Professional 7.0

FIGURE 3-42 A dangerous cargo manifest.

Courtesy of the Department of Defense

STAPLE DOCUMENTS ABOVE PERFORATION

Set your tabulator stops here

Line-up here

Shipper's Name and Address	Shipper's Account Number	Not Negotiable **Air Waybill** Issued by
		Copies 1, 2 and 3 of this Air Waybill are originals and have the same validity.
Consignee's Name and Address	Consignee's Account Number	It is agreed that the goods described herein are accepted in apparent good order and condition (except as noted) for carriage SUBJECT TO THE CONDITIONS OF CONTRACT ON THE REVERSE HEREOF. ALL GOODS MAY BE CARRIED BY ANY OTHER MEANS INCLUDING ROAD OR ANY OTHER CARRIER UNLESS SPECIFIC CONTRARY INSTRUCTIONS ARE GIVEN HEREON BY THE SHIPPER, AND SHIPPER AGREES THAT THE SHIPMENT MAY BE CARRIED VIA INTERMEDIATE STOPPING PLACES WHICH THE CARRIER DEEMS APPROPRIATE. THE SHIPPER'S ATTENTION IS DRAWN TO THE NOTICE CONCERNING CARRIER'S LIMITATION OF LIABILITY. Shipper may increase such limitation of liability by declaring a higher value for carriage and paying a supplemental charge if required.
Issuing Carrier's Agent Name and City		Accounting Information
Agent's IATA Code	Account No.	
Airport of Departure (Addr. of First Carrier) and Requested Routing		Reference Number / Optional Shipping Information

To	By First Carrier (Routing and Destination)	to	by	to	by	Currency	CHGS Code	WT/VAL PPD	WT/VAL COLL	Other PPD	Other COLL	Declared Value for Carriage	Declared Value for Customs

Airport of Destination	Requested Flight/Date		Amount of Insurance	INSURANCE - If carrier offers insurance, and such insurance is requested in accordance with the conditions thereof, indicate amount to be insured in figures in box marked "Amount of Insurance".

Handling Information

These commodities, technology or software were exported from the United States in accordance with the Export Administration Regulations. Ultimate destination

Diversion contrary to U.S. law prohibited.

SCI

No. of Pieces RCP	Gross Weight	kg / lb	Rate Class / Commodity Item No.	Chargeable Weight	Rate / Charge	Total	Nature and Quantity of Goods (incl. Dimensions or Volume)

Prepaid	Weight Charge	Collect	Other Charges
	Valuation Charge		
	Tax		
	Total Other Charges Due Agent		Shipper certifies that the particulars on the face hereof are correct and that **insofar as any part of the consignment contains dangerous goods, such part is properly described by name and is in proper condition for carriage by air according to the applicable Dangerous Goods Regulations.**
	Total Other Charges Due Carrier		Signature of Shipper or his Agent
Total Prepaid		Total Collect	
Currency Conversion Rates		CC Charges in Dest. Currency	Executed on (date) at (place) Signature of Issuing Carrier or its Agent
For Carriers Use only at Destination		Charges at Destination	Total Collect Charges

APPERSON K0419 (10/03) WHSE. #05640

FIGURE 3-43 An air bill.

Courtesy of Apperson, Inc.

VOICES OF EXPERIENCE

In 2006, I had a little more than a year under my belt as a company officer and even less time as the Captain of HazMat 1, so when we were dispatched to a chlorine leak, I was a tad nervous. It was the middle of the day when we received the call, and the wastewater treatment facility was a good ride from our station, so I had a little time to mentally prepare—or so I thought. The questions began to come across the radio as soon as we left the station. This quickly ceased, though, because E-3 would be arriving and report the fixed monitoring system was activated. They quickly followed with a report of a worker acting hysterical near the gate of the facility.

My first major hazardous materials incident as an officer now included an exposure. As we arrived on scene, most of the first alarm units were there and staged a significant distance away from the scene. Fortunately, the individual that was thought to have been exposed to chlorine was merely upset about the release he had caused and was panicking. We were provided an accountability of the workers at the facility, which helped steer the incident.

Because the worker familiar with what had occurred prior to our arrival was upset, it was difficult to get helpful information from him. We knew the leak was coming from the cylinder on the dock and was extremely active. The worker did inform us that he was moving a cylinder with the cylinder hoist that was positioned at the manifold. He thought that he had hit the valves against the header of the manifold, causing the leak. By the look of the release, his story supported the large release we were seeing.

The first team went in with the Chlorine B Kit prepared to install the dome over the damaged valve. The team faced two issues from the start: First, the entry team was near the speaker for the fixed alarm system, which hindered communications both for the entry team and via the radio back to Operations. The second pressing issue was the lack of visibility at the cylinder head due to the large vapor cloud. A positive pressure fan was taken up to the team and set up to help disperse the vapor cloud, but this added to the noise issue.

The team realized that the cylinder was partially raised in the hoist and was leaning towards the valves. Both the liquid and vapor lines were broken, causing a significant liquid release. The crew observed no damage to the valves; they lowered the cylinder back into the cradle and started to close the valves. The crew met its time commitment and returned to decon. The second team now made entry and was prepared to stop the leak. During this task, the team reported back to Operations that one of their gloves was damaged, but it was the outer glove only. They were able to quickly control the valve and return.

During decon, it appeared that the glove had been ripped, but further inspection revealed that the glove was cracked. The cryogenic-like action of the rapid release of chlorine had caused a rather large buildup of frozen material on the head of the cylinder. The team had to chip away this material, and during this process the glove made contact and froze. Unknown to the member, he pulled his hand away from the material and inadvertently ripped the outer glove.

That day we learned that, with a large liquid leak of chlorine freezing, it is necessary to address and plan for cryogenic-like issues. Communications in noisy areas with alarms or in production areas is difficult, so some type of hearing protection and a voice-activated communications system will do wonders.

Jason Krusen
Special Operations Chief
Columbia Fire Department
Columbia, South Carolina

RESPONDER TIP

Standard operating procedures, the emergency response plan, and the *ERG* will help you to identify the various protective actions and notifications that must be made for the types of responses anticipated within the jurisdiction. Standard operating procedures should define points of contact for local, state, and federal resources that might be called upon for assistance during hazardous materials emergencies.

The responsible person in each situation should maintain the information about hazardous cargo and provide it in an emergency situation. When encountering a chemical release on any of these modes of transportation, take time to find the responsible person and track down these valuable pieces of information. Additionally, the person who finds it for you, or has it in his or her possession, is likely to be a great resource to connect you with subject matter experts from the company or may have some level of knowledge about the material or the mode of transportation that may prove valuable.

CHEMTREC

Located in Arlington, Virginia, the Chemical Transportation Emergency Center (CHEMTREC), which is operated by the American Chemistry Council, is a clearinghouse of technical chemical information. Since 1971, this emergency call center has served as an invaluable information resource for first responders of all disciplines who are called upon to respond to chemical incidents. CHEMTREC has the ability to provide responders with technical chemical information via telephone, fax, or other electronic media. It also offers a phone conferencing service that will put a responder in touch with thousands of shippers, subject matter experts, and chemical manufacturers.

When calling CHEMTREC, be sure to have the following basic information ready:

- The name of the chemical(s) involved in the incident (if known)
- Name of the caller and callback telephone number
- Location of the actual incident or problem
- Shipper or manufacturer of the chemical (if known)
- Container type
- Railcar or vehicle markings or numbers
- The shipping carrier's name
- Recipient of material
- Local conditions and exact description of the situation

When speaking with CHEMTREC personnel, be very clear about the name of the substance and spell out the name if necessary; if using a third party, such as a dispatcher, it is vital that you confirm all spellings to avoid misunderstandings. One number or letter out of place could throw off all subsequent research. When in doubt, be sure to obtain clarification.

The Canadian equivalent of CHEMTREC is the Canadian Transport Emergency Centre (CANUTEC), which is located in Ottawa. This organization serves Canadian responders (in French and English) in much the same way that CHEMTREC serves responders in the United States. CANUTEC may be called 24 hours a day for emergency situations.

The Mexican equivalent of CHEMTREC and CANUTEC is the Emergency Transportation System for the Chemical Industry, Mexico (SETIQ). SETIQ may be called 24 hours a day.

Phone numbers for all of these agencies can be found in the *ERG*.

National Response Center

Whenever a significant hazardous materials incident occurs, the National Response Center (NRC) must be notified. The NRC is operated by the U.S. Coast Guard and serves as a central notification point, rather than a guidance center. Once the NRC is notified, it will alert the appropriate state and federal agencies. The NRC must be notified if any spilled hazard discharges into the environment. The toll-free number for the NRC is 1-800-424-8802.

The NRC has established complex reporting requirements for different chemicals based on the reportable quantity (RQ) for that chemical. The shipper or the owner of the chemical has the ultimate legal responsibility to make this call, but by doing so themselves, response agencies will have their reporting bases covered.

Potential Terrorist Incidents

The threat of terrorism has changed the way public safety agencies operate. In small towns and major metropolitan areas, the possibility exists that on any day, any agency could find itself in the eye of a storm involving intentionally released chemical substances, biological agents (disease-causing bacteria or other viruses that attack the human body), or an attack on buildings or people using explosives. No area of any country is immune to such attacks. In this day and age, responders must recognize that they are a part of a much larger response mechanism in the United States; they must also understand that any incident involving terrorism will require the assistance and cooperation of countless local, state, and federal resources. Responders within every jurisdiction should think about the locations of potential targets for terrorists; the general and specific hazards posed by chemical, biological, and radiological agents (materials that emit radioactivity); possible indicators of illicit laboratories; and basic operational guidelines for dealing with explosive events and identifying the possible indicators of secondary devices.

Potential targets for terrorist activities include both natural landmarks and human-made structures. These sites can be classified into three broad categories: infrastructure targets, symbolic targets, and civilian targets. The following sections will help you build a foundation to become a more informed responder when it comes to dealing with potential terrorist incidents.

Chemical Agents

Indicators of possible criminal or terrorist activity involving chemical agents may vary depending on the complexity of the operation, and there is no single indicator that may tip you off to the presence of such illicit activities. There may be overt indicators such as chemical-type gloves, chemical suits, respirators, and marked or unmarked containers made of various materials in a variety of shapes and sizes. For example, glass containers are prevalent at such locations, but may or may not appear to be obvious hazards or threatening in any way. The chemicals may provide unexplained odors that are out of character for the surroundings. Residual chemicals (liquid, powder, or gas form) may also be found in the area. Chemistry books or other reference materials may be seen, as well as materials that are used to manufacture chemical weapons (such as scales, thermometers, or torches). There may or may not be some type of easily identifiable signature such as an odor, liquid or solid residue or dead insects or foliage. The main point is that you must always be on the lookout for items that may appear out of context with the setting—pay attention to your surroundings at all times. The routine medical or garage fire may quickly bring you up close and personal with a very dangerous situation.

Persons working around the illicit materials may themselves become exposed and exhibit symptoms of chemical exposure—for example, irritation to the eyes, nose, and throat; difficulty breathing; tightness in the chest; nausea and vomiting; dizziness; headache; blurred vision; blisters or rashes; disorientation; or even convulsions.

Biological Agents

Indicators of incidents that may potentially involve biological agents may include chemicals or production equipment such as Petri dishes, vented hoods, Bunsen burners, pipettes, microscopes, and incubators. Reference manuals, such as microbiology or biology textbooks or handwritten research or directions—maybe in a foreign language—may be present. Containers used to transport biological agents may include metal cylindrical cans or red plastic boxes or bags, probably without specific biological hazard labels. Personal protective equipment, including respirators, chemical or biological suits, and latex gloves, may be on the scene, as well as excessive amounts of antibiotics as a means to protect those working with the agents. Other potential indicators may include abandoned spray devices and unscheduled or unusual sprays being disseminated (especially if outdoors at night).

Persons working in the area of the lab may eventually exhibit symptoms consistent with the biological weapons with which they are working. Biological agents have a delayed onset of symptoms (usually days to weeks after the initial exposure). In fact, the biggest difference between a chemical incident and a biological incident is typically the speed of onset of the health effects from the involved agents. Because most biological agents are odorless and colorless, there are usually no outward indicators that the agents have spread. Again, there may or may not be overt indicators of criminal activities—be aware and attentive!

Radiological Agents

Indicators of radiological agents may include production or containment equipment, such as lead or stainless steel containers (with or without labels), and explosives that may be used to disperse the radioactive source, along with containers (e.g., pipes), caps, fuses, gunpowder, timers, wire, and detonators. Personal protective equipment present may include radiological protective suits and respirators. Radiation monitoring equipment such as Geiger counters or radiation pagers may be present or perhaps similar radiation detection devices used by responders to alert them to the presence of a potential threat. As with other locations where nefarious activities may be present, responders must put the potential threat in context. There may or may not be a valid explanation for the presence of something suspicious.

A major difference between an illicit location where radioactive substances are present and a legitimate operation may be the way the substances are packaged and stored. Legitimate sources are well marked, tracked, and regulated. When radiological agents are shipped, the package type is dictated by the degree of radiation activity inside the package—that is, the labeling is driven by the *amount of radiation that can be measured outside the package*. Three varieties of labels are found on radioactive packages: White I **FIGURE 3-44**, Yellow II **FIGURE 3-45**, and Yellow III **FIGURE 3-46**.

Additionally, shippers of radiological materials are required to include a transport index (TI) number on the package label. This number indicates the highest amount of radiation that can be measured 1 meter away from the surface of the package.

Responders must be able to recognize situations where radioactive materials might be encountered. Industries that routinely use radioactive materials include food testing labs, hospitals, medical research centers, biotechnology facilities, construction sites, and medical laboratories. For the most part, there will be some visual indicators (signs or placards) that indicate the presence of radioactive substances, but this is not always the case.

The key is to be able to suspect, recognize, and understand when and where you may encounter radioactive sources. If you suspect a radiation incident at a fixed facility, you should

FIGURE 3-44 A White I label.

Courtesy of the U.S. Department of Transportation

FIGURE 3-45 A Yellow II label.

Courtesy of the U.S. Department of Transportation

FIGURE 3-46 A Yellow III label.

Courtesy of the U.S. Department of Transportation

initially consult with the radiation safety officer of the facility. This person is responsible for the use, handling, and storage procedures for all radioactive material at the site. He or she likely will be a tremendous resource to you and will know exactly what is being used at the facility. If the incident is not at a fixed site, the presence of radiation may never be apparent. Radioactive isotopes are not detected by sight, smell, taste, or any of the other senses. Therefore, if you have any suspicion that the incident involves radiation, it will be necessary to call a hazardous materials team or some other resource with radiation detection capabilities.

Significant incidents involving radiation are rare, largely due to the comprehensiveness of the regulations for using, storing, and transporting significant radioactive sources. This is not to say that these incidents will not happen; nevertheless, the regulations have helped considerably in keeping the number of incidents low. Most of the incidents you may encounter will involve low-level radioactive sources and can be handled safely. These low-level sources are typically found in Type A packaging. This packaging method is unique to radioactive substances and contains materials such as radiopharmaceuticals and other low-level emitters.

Radiological Packaging

The most common types of containers and packages used to store radioactive materials are divided into five major categories: excepted range radioactive packaging, industrial radioactive packaging, and Type A FIGURE 3-47, Type B FIGURE 3-48, and Type C FIGURE 3-49 packaging.

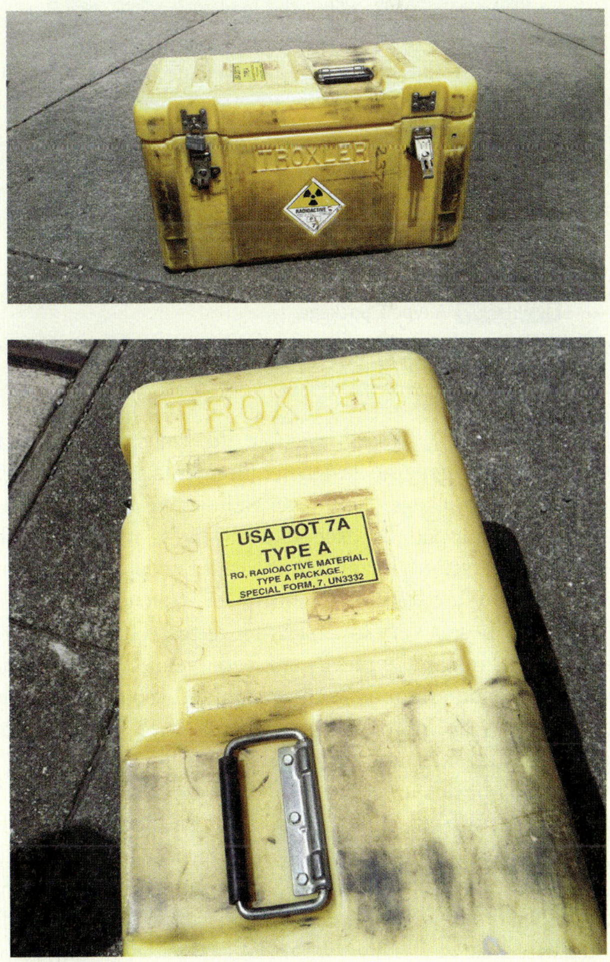

FIGURE 3-47 A Type A package.

Courtesy of Rob Schnepp

FIGURE 3-48 Type B packaging.

Courtesy of the U.S. Department of Energy

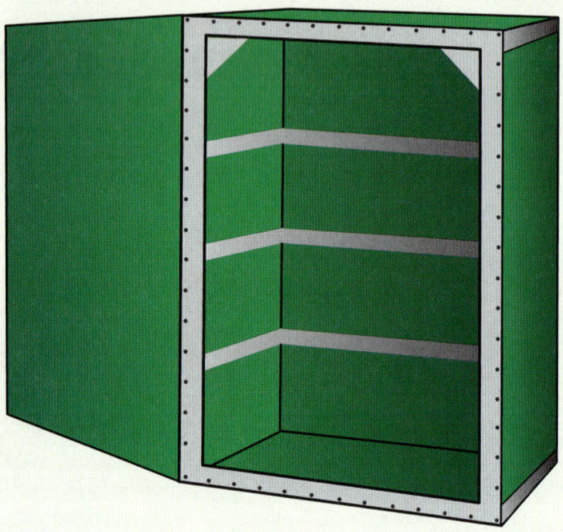

FIGURE 3-49 A Type C package.

Excepted packaging is used to transport materials that meet only general design requirements for any hazardous material package. Low-level radioactive substances are commonly shipped in excepted packages, which may be constructed out of heavy cardboard. Excepted packaging is authorized for limited quantities of radioactive material that would pose a very low hazard if released in an accident. Examples of material typically shipped in excepted packaging include consumer goods such as smoke detectors. Excepted packaging is excepted (excluded) from specific packaging, labeling, and shipping paper requirements; however, it is required to have the letters "UN" and the appropriate four-digit UN identification number marked on the outside of the package.

Industrial packaging is used in certain shipments of low activity material and contaminated objects, which are usually categorized as radioactive waste. Most low-level radioactive waste is shipped in these packages. DOT regulations require that these packages allow no identifiable release of the material to the environment during normal transportation and handling. There are three categories of industrial packages: IP-1, IP-2, and IP-3. The category of package will be marked on the exterior of the package.

Type A packaging is designed to protect the internal radiological contents during normal transportation and in the event of a minor accident. Such packaging is characterized by having an inner containment vessel made of glass, plastic, or metal and external packaging materials made of polyethylene, rubber, or vermiculite. Examples of material typically shipped in Type A packages include nuclear medicines (radiopharmaceuticals), radioactive waste, and radioactive sources used in industrial applications. Type A packaging and its radioactive contents must meet standard testing requirements designed to ensure that the package retains its containment integrity and shielding under normal transport conditions and may be transported by air. Type A packages must withstand moderate degrees of heat, cold, reduced air pressure, vibration, impact, water spray, drop, penetration, and stacking tests. The consequences of a release of the material in one of these packages would not be significant because the quantity of material in this package is so limited.

Type B packaging is far more durable than Type A packaging and is designed to prevent a release in the case of extreme accidents during transportation. More dangerous radioactive sources might be found in Type B packaging. Some of the tests that Type B containers must undergo include heavy fire, pressure from submersion, and falls onto spikes and unyielding surfaces. Type B packages include small drums and heavily shielded casks weighing more than 100 metric tons. This type of containment vessel contains materials such as spent nuclear fuel, high-level radioactive waste, and high concentrations of other radioactive material such as cesium and cobalt. Type B packages are designed to protect their contents from greater exposure; the amount of protection is based on the potential severity of the hazard. These package designs must withstand all Type A tests, and a series of tests that simulate severe or "worst-case" accident conditions. Accident conditions are simulated by performance testing and engineering analysis. Life-endangering amounts of radioactive material are required to be transported in Type B packages. Type B packages are often used for air transportation.

Type C packaging is used for transporting high-activity radioactive substances by air. Dangerous radioactive sources are shipped in Type C packaging. Type C packaging is not certified for use in the United States and is not part of the transportation regulations. This type of packaging is only referenced in international regulations.

Illicit Laboratories

Many indicators of possible criminal or terrorist activity involving illicit laboratories may be evident to responders. For example, terrorist paraphernalia may include terrorist training manuals, ideological propaganda, and documents indicating affiliation with known terrorist groups. Locations with certain characteristics are also commonly sites of illicit (clandestine) laboratories—for example, basements with unusual or multiple vents, buildings with heavy security,

FIGURE 3-50 Items typically found at clandestine drug laboratories are everyday items.

buildings with obscured windows, and buildings with odd or unusual odors. Personnel working in illegal laboratory settings may exhibit a certain degree of unusual or suspicious behavior; for instance, they may be nervous and have a high level of anxiety. In addition, they may be very protective of the laboratory area and not want to allow anyone to access the area for any reason, or they may rush people out of the area as soon as possible.

Equipment that may be present in illicit laboratory areas includes surveillance materials (such as videotapes, photographs, maps, blueprints, or time logs of the target hazard locations), non-weapon supplies (such as identification badges, uniforms, and decals that would be used to allow the terrorist to access target hazards), and weapon-related supplies (such as timers, switches, fuses, containers, wires, projectiles, and gunpowder or fuel). Security weapons such as guns, knives, and booby trap systems may also be present.

Drug laboratories are by far the most common type of clandestine laboratory encountered by responders. These laboratories are typically very primitive. Materials used to manufacture the drugs often consist of everyday items (jars, bottles, glass cookware, coolers, and tubing) that have been modified to produce the illicit drugs FIGURE 3-50.

Specific chemicals and materials found at the scene may include large quantities of cold tablets (ephedrine or pseudoephedrine), hydrochloric or sulfuric acid, paint thinner, drain cleaners, iodine crystals, table salt, aluminum foil, and camera batteries. The strong smell of urine or unusual chemical smells such as ether, ammonia, or acetone are very common indicators of clandestine drug manufacturing. Illicit drug laboratories should be considered significant hazardous materials scenes, because the inexperienced chemists who run them take many shortcuts and disregard typical safety protocols so as to increase production.

The examples listed here are intended to give you an idea of some of the things that might tip you off to the presence of suspicious or illicit activity. It is not an exhaustive list, and you may encounter an illicit lab without any of these indicators. Remember, the people carrying out such activities are not interested in being obvious in their behaviors or actions.

Explosives

Indicators of possible criminal or terrorist activity involving explosives typically include materials that fit into four major categories—protective equipment, production and containment materials, explosive materials, and support materials. Protective equipment may include rubber gloves, goggles and face shields, and maybe even fire extinguishers. Production and containment equipment may include funnels, spoons, threaded pipes, caps, fuses, timers, wires, detonators, and concealment containers such as briefcases, backpacks, or other innocuous-looking packages. Explosive materials may include gunpowder, gasoline, fertilizer, solvents, oxidizers, and similar materials. Support materials may include explosive reference manuals, Internet-based reference materials, and military information. If you encounter a situation where a bomb-making operation is active, it is important to note what you see, get yourself out of danger, secure an appropriate amount of real estate, and call for your local bomb squad. This is not a situation in which untrained responders should be investigating.

Secondary Devices

By definition, a secondary device is some form of explosive or incendiary device designed to harm those responders summoned to the scene for some other reason. Terrorists who want to injure responding personnel with a secondary device or attack will typically make the initial attack very dramatic so as to draw responders into close proximity of the scene. The secondary attack usually takes place as the responders begin to treat victims of the initial attack.

Indicators of potential secondary devices may include "trip devices" such as timers, wires, or switches. Common concealment containers, such as briefcases, backpacks, boxes, or other common packages, may also be present; uncommon concealment containers may include pressure vessels (propane tanks) or industrial chemical containers (chlorine storage containers). Personnel may watch the site of the primary devices, as part of preparing to manually activate the secondary devices. Responders can use the EVADE acronym to help think critically about the presence of secondary devices. This acronym can be found in NFPA 473:

- Evaluate the scene for likely areas where secondary devices can be placed.
- Visually scan operating areas for a secondary device before providing patient care.
- Avoid touching or moving anything that can conceal an explosive device.
- Designate and enforce scene control zones.
- Evacuate victims, other responders, and nonessential personnel as quickly and safely as possible.

Wrap-Up

Chief Concepts

- Responders must interpret visual clues effectively to improve their ability to mitigate an incident.
- Responders should be able to recognize the various container profiles and understand the general classifications of materials that may be stored inside each type of container.
- Responders should be able to name, understand, and locate the various types of shipping papers on various modes of transportation.
- Roadway vehicles often transport shipments from the rail station, airport, or dock to the point where the materials will ultimately be used. For this reason, responders must become familiar with all types of chemical transport vehicles.
- When used correctly, various marking systems indicate the presence of a hazardous material and provide clues about the substance. The DOT, NFPA, HMIS, and military have all developed marking systems specific to their level of response.
- Responders should be able to demonstrate proficiency when using the *Emergency Response Guidebook.*
- Responders should know how to obtain SDS documentation from various sources, including their own department, the scene of the incident itself, or the manufacturer of the material.
- Responders should become familiar with the locations of potential terrorist targets in their jurisdiction; the general and specific hazards of chemical, biological, and radiological agents; indicators of illicit laboratories; and basic operational guidelines for dealing with explosive events and identifying the possible indicators of secondary devices.

Hot Terms

Aboveground storage tank (AST) A tank that can hold anywhere from a few hundred gallons to several million gallons of product. ASTs are usually made of aluminum, steel, or plastic.

Air bill The shipping papers on an airplane.

Bill of lading The shipping papers used for transport of chemicals over roads and highways. Also referred to as a *freight bill.*

Biological agents Disease-causing bacteria, viruses, and other agents that attack the human body.

Bulk storage container A large-volume container that has an internal volume greater than 119 gallons for liquids and a capacity greater than 882 pounds for solids and greater than 882 pounds for gases.

Bung A small opening in a closed-head drum.

Canadian Transport Emergency Centre (CANUTEC) A national call center located in Ottawa, Canada. This organization serves Canadian responders (in French and English) in much the same way CHEMTREC serves responders in the United States.

Carboy A glass, plastic, or steel nonbulk storage container, ranging in volume from 5 to 15 gallons.

Cargo tank Bulk packaging that is permanently attached to or forms a part of a motor vehicle, or is not permanently attached to any motor vehicle, and that, because of its size, construction, or attachment to a motor vehicle, is loaded or unloaded without being removed from the motor vehicle.

Chemical Transportation Emergency Center (CHEMTREC) A U.S. national call center that provides basic chemical information. It is operated by the American Chemistry Council.

Consist A list of the contents of every car on a train. Also called a train list.

Container Any vessel or receptacle that holds material, including storage vessels, pipelines, and packaging.

Cryogenic liquid (cryogen) A gaseous substance that has been chilled to the point where it has liquefied; a liquid having a boiling point lower than −150°F (−101°C) at 14.7 psi (an absolute pressure of 101 kPa).

Cylinder A portable, nonbulk, compressed gas container used to hold liquids and gases. Uninsulated compressed gas cylinders are used to store substances such as nitrogen, argon, helium, and oxygen. They have a range of sizes and internal pressures.

Dangerous cargo manifest The shipping papers on a marine vessel, generally located in a tube-like container.

Dewar container A container designed to preserve the temperature of the cold liquid held inside.

Drum A barrel-like nonbulk storage vessel used to store a wide variety of substances, including food-grade materials, corrosives, flammable liquids, and grease. Drums may be constructed of low-carbon steel, polyethylene, cardboard, stainless steel, nickel, or other materials.

Dry bulk cargo tank A tank designed to carry dry bulk goods such as powders, pellets, fertilizers, or grain. Such tanks are generally *V*-shaped with rounded sides that funnel toward the bottom.

Emergency Response Guidebook (ERG) A preliminary action guide for first responders operating at a hazardous materials incident in coordination with the U.S. Department of Transportation's (DOT's) labels and placards marking system. The DOT and the Secretariat of Communications and Transportation of Mexico (SCT), along with Transport Canada, jointly developed the *Emergency Response Guidebook*.

Emergency Transportation System for the Chemical Industry, Mexico (SETIQ) A national response center that is the Mexican equivalent of CHEMTREC and CANUTEC.

Excepted packaging Packaging used to transport materials that meets only general design requirements for any hazardous material. Low-level radioactive substances are commonly shipped in excepted packages, which may be constructed out of heavy cardboard.

Freight bill The shipping papers used for transport of chemicals along roads and highways. Also referred to as a *bill of lading*.

Hazardous material Any substance or material that is capable of posing an unreasonable risk to human health, safety, or the environment when transported in commerce, used incorrectly, or not properly contained or stored.

Hazardous Materials Information System (HMIS) A color-coded marking system by which employers give their personnel the necessary information to work safely around chemicals. The Workplace Hazardous Materials Information System (WHMIS) is the Canadian hazard communication standard.

Industrial packaging Packaging used to transport materials that present a limited hazard to the public or the environment. Contaminated equipment is an example of such material, because it contains a non-life-endangering amount of radioactivity. Industrial packaging is classified into three categories, based on the strength of the packaging.

Intermodal tank A bulk container that serves as both a shipping and storage vessel. Such tanks hold between 5000 and 6000 gallons of product and can be either pressurized or nonpressurized. Intermodal tanks can be shipped by all modes of transportation—air, sea, or land.

Label A smaller version of a placard (4-inch, diamond-shaped markings). Labels are placed on all four sides of individual boxes and smaller packages that are being transported.

MC-306/DOT 406 cargo tank Such a vehicle typically carries between 6000 and 10,000 gallons of a product such as gasoline or other flammable and combustible materials. The tank is nonpressurized.

MC-307/DOT 407 chemical hauler A tank with a rounded or horseshoe-shaped tank capable of holding 6000 to 7000 gallons of flammable liquid, mild corrosives, and poisons. The tank has a high internal working pressure.

MC-312/DOT 412 corrosive tank A tank that often carries aggressive (highly reactive) acids such as concentrated sulfuric and nitric acid. It is characterized by several heavy-duty reinforcing rings around the tank and holds approximately 6000 gallons of product.

MC-331 pressure cargo tank A tank that carries materials such as ammonia, propane, Freon, and butane. This type of tank is commonly constructed of steel and has rounded ends and a single open compartment inside. The liquid volume inside the tank varies, ranging from a1000-gallon delivery truck to a full-size 11,000-gallon cargo tank.

MC-338 cryogenic tanker A low-pressure tank designed to maintain the low temperature required by the cryogens it carries. A boxlike structure containing the tank control valves is typically attached to the rear of the tanker.

National Response Center (NRC) An agency maintained and staffed by the U.S. Coast Guard; it should always be notified if a hazard discharges into the environment.

NFPA 704 hazard identification system A hazardous materials marking system designed for fixed-facility use. It uses a diamond-shaped symbol of any size, which is itself broken into four smaller diamonds, each representing a particular property or characteristic of the material.

Nonbulk storage vessel Any container other than bulk storage containers such as drums, bags, compressed gas cylinders, and cryogenic containers. Nonbulk storage vessels hold commonly used commercial and industrial chemicals such as solvents, industrial cleaners, and compounds.

Nonpressurized (general-service) rail tank car A railcar equipped with a tank that typically holds general industrial chemicals and consumer products such as corn syrup, flammable and combustible liquids, and mild corrosives.

Pipeline A length of pipe—including pumps, valves, flanges, control devices, strainers, and/or similar equipment—for conveying fluids and gases.

Pipeline right-of-way An area, patch, or roadway that extends a certain number of feet on either side of a pipeline and that may contain warning and informational signs about hazardous materials carried in the pipeline.

Placard Signage required to be placed on all four sides of highway transport vehicles, railroad tank cars, and other forms of hazardous materials transportation; the sign identifies the hazardous contents of the vehicle, using a standardization system with 10¾-inch diamond-shaped indicators.

Pressurized rail tank car A railcar used to transport materials such as propane, ammonia, ethylene oxide, and chlorine.

Radiological agents Materials that emit radioactivity (the spontaneous decay or disintegration of an unstable atomic nucleus accompanied by the emission of radiation).

Secondary containment Any device or structure that prevents environmental contamination when the primary container or its appurtenances fail. Examples of secondary containment mechanisms include dikes, curbing, and double-walled tanks.

Secondary device An explosive or incendiary device designed to harm emergency responders who have responded to an initial event.

Shipping papers A shipping order, bill of lading, manifest, or other shipping document serving a similar purpose; it usually includes the names and addresses of both the shipper and the receiver as well as a list of the shipped materials along with their quantity and weight.

Signal words Information on a pesticide label that indicates the relative toxicity of the material.

Special-use railcar A boxcar, flat car, cryogenic tank car, or corrosive tank car.

Toxic inhalation hazard (TIH) Any gas or volatile liquid that is extremely toxic to humans.

Train list See *consist*.

Tube trailer A high-volume transportation device made up of several individual compressed gas cylinders banded together and affixed to a trailer. Tube trailers carry compressed gases such as hydrogen, oxygen, helium, and methane. One trailer may carry several different gases in individual tubes.

Type A packaging Packaging that is designed to protect its internal radiological contents during normal transportation and in the event of a minor accident.

Type B packaging Packaging that is far more durable than Type A packaging and is designed to prevent a release of the radiological hazard in the case of extreme accidents during transportation. Type B containers must undergo a battery of tests including those involving heavy fire, pressure from submersion, and falls onto spikes and rocky surfaces.

Type C packaging Packaging used when radioactive substances must be transported by air.

Underground storage tank (UST) A type of tank that can hold anywhere from a few hundred gallons to several million gallons of product. USTs are usually made of aluminum, steel, or plastic.

U.S. Department of Transportation (DOT) marking system A unique system of labels and placards that is used when materials are being transported from one location to another in the United States. The same marking system is used in Canada by Transport Canada.

Vent pipes Inverted *J*-shaped tubes that allow for pressure relief or natural venting of the pipeline for maintenance and repairs.

Waybill Shipping papers for railroad transport.

© Greg Henry/ShutterStock, Inc.

RESPONDER in action

It is a rainy afternoon when your engine company is dispatched to a motor vehicle accident. Upon arrival, you see a large tractor trailer rig on its side, with a single tank compartment that has rounded ends. The tank is painted white and appears to be a pressurized vessel.

1. Which type of container is this likely to be?
   - **A.** MC-306/DOT 406
   - **B.** MC-307/DOT 407
   - **C.** MV-312
   - **D.** MC-331
2. Which documents should be found in the cab of the vehicle?
   - **A.** Bill of lading
   - **B.** Waybill
   - **C.** Dangerous cargo manifest
   - **D.** Consists
3. Which section of the *Emergency Response Guidebook* (*ERG*) includes the name of the chemical first and then the guide number and ID number?
   - **A.** Green section
   - **B.** Orange section
   - **C.** Yellow section
   - **D.** Blue section
4. In which section of the *Emergency Response Guidebook* should you look when you can identify a UN/ID number?
   - **A.** Green section
   - **B.** Orange section
   - **C.** Yellow section
   - **D.** Blue section

Estimating Potential Harm and Planning a Response

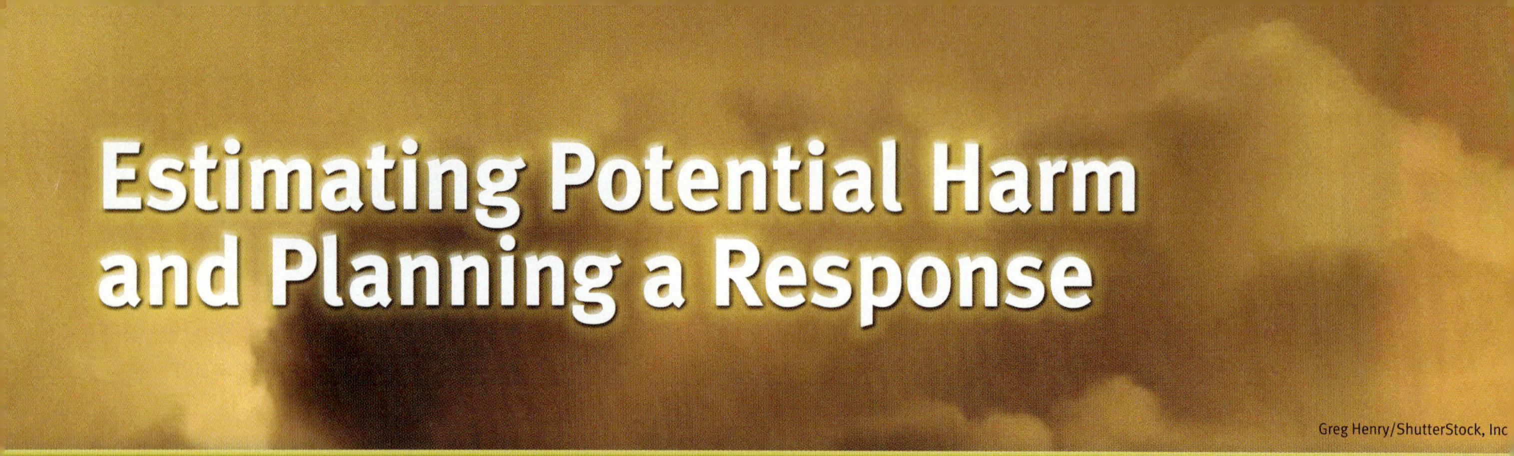

Greg Henry/ShutterStock, Inc

© Courtesy of Captain David Jackson, Saginaw Township Fire Department

CHAPTER 4

Awareness Level Personnel

Knowledge Objectives

After studying this chapter, you will be able to:

- Describe resources to determine the size of a hazardous materials/WMD incident. (NFPA 4.4.1, p. 86–88)
- Describe how to report the size and scope of an incident. (NFPA 4.4.1(8)(9)(10), p. 91)
- Identify protective actions. (NFPA 4.4.1(6)(7)(11), p. 89–90)
- Identify skin-contact hazards encountered at hazardous materials/WMD incidents. (p. 92)
- Describe how to plan initial actions within the scope of awareness personnel. (NFPA 4.4.1(12), p. 92–93)
- Describe the potential for secondary attacks/devices. (p. 93–94)
- Identify the purpose, advantages, and limitations of the following items (NFPA 4.4.1(5)):
  - Street clothing and work uniforms (p. 96)
  - Structural firefighting protective clothing (p. 96)
  - High-temperature–protective clothing and equipment (p. 96)
  - Chemical-protective clothing and equipment (p. 96–97)
- Discuss the levels of hazardous materials/WMD personal protective equipment (PPE). (NFPA 4.4.1(5), p. 98–100)

Skills Objectives

This chapter has no skills objectives for Awareness Level personnel.

Operations Level Responders

Knowledge Objectives

After studying this chapter, you will be able to:

- Describe how to estimate the potential harm or the severity of a hazardous materials/weapons of mass destruction (WMD) incident. (NFPA 5.1.2.2, 5.2.4, p. 84–86)
- Describe resources to determine the scope of a hazardous materials/WMD incident. (NFPA 5.1.2.2, 5.2.4, p. 91)
- Describe exposure protection. (NFPA 5.2.4, 5.3.1, p. 89–90)
- Describe resources available for determining the concentrations of a released hazardous material. (NFPA 5.2.4, p. 91–92)
- Identify skin-contact hazards encountered at hazardous materials/WMD incidents. (NFPA 5.3.3(2)(a), p. 92)
- Describe how to plan an initial response. (NFPA 5.3.1, 5.3.2, p. 92–94)
- Describe the potential for secondary attacks/devices. (NFPA 5.3.1(4), p. 93–94)
- Describe personal protective equipment (PPE) used for hazardous materials/WMD incidents, as well as the care and maintenance of PPE provided by the AHJ. (NFPA 5.3.3(2), 5.4.4(6)(7), p. 94–103)
- Identify the purpose, advantages, and limitations of the following items:
  - Structural firefighting protective clothing (p. 96)
  - Chemical-protective clothing and equipment (p. 96–97)
  - High-temperature–protective clothing and equipment (p. 96)
- Discuss the importance of respiratory protection in a hazardous materials/WMD incident. (NFPA 5.3.3(1), p. 100–103)
- Describe the importance of having a plan in place to decontaminate a victim. (NFPA 5.3.2, 5.3.4, p. 103–104)

Skills Objectives

This chapter has no skills objectives for Operations Level responders.

You Are the Responder

Your paramedic ambulance and local engine company have been called to a semiconductor fabrication facility for a report of shortness of breath. You arrive at the reception area and are led to an employee break room. As you approach the seated patient, you notice an odor of ammonia. The laboratory manager is standing next to the patient—a laboratory technician—who is doubled over in the chair. The laboratory manager tells you that the technician was splashed in the face with approximately 100 ml of ammonium hydroxide while pouring chemicals into an instrument. He states that he helped the laboratory technician to an eye-wash station immediately after the incident and then escorted him to the break room. The laboratory manager tells you he has seen this sort of injury before. In his opinion, the laboratory technician doesn't need to go to the hospital—he just needs some oxygen. The paramedic begins an assessment and finds that the laboratory technician has reddened skin over his entire face and is complaining of shortness of breath. You are the officer on the engine company and your four-person crew is trained at the first responder operations level in the core competencies of NFPA 472.

1. What are the initial response priorities for the engine crew and ambulance once the ammonia exposure is discovered?
2. What are the initial objectives for treating/transporting the patient?
3. Are the ambulance and engine crew at risk of being exposed or contaminated to potentially toxic levels of ammonium hydroxide from the patient?

Introduction

It is important to have a set of basic response priorities to guide your decision making at the scene of a hazardous materials/WMD incident. To that end, the first response priority should be to ensure your own safety while operating at the scene. You're no good to anyone if you've become part of the problem! At a minimum, you must arrive at the scene in a safe manner and make sure that you and your crew do not become a liability during the course of the incident.

After ensuring your own safety, your next objective should be to address the potential life safety of those persons affected or potentially affected by the incident. In the scenario described in the chapter-opening vignette, your crew responded for one type of emergency—shortness of breath—and ended up facing a completely different kind of problem—a medical issue related to a chemical exposure. This new set of parameters will require you to quickly shift gears. You now have an exposed victim and a potentially exposed laboratory manager. Are there additional response personnel and/or equipment required to handle the situation? You can smell the ammonia: Does that mean you and other personnel are in danger, too? Has the patient been adequately decontaminated to the point you can safely render care? Who performed the decontamination? What type of decontamination was done? Should you back out of the area without treating the victim and call for a hazardous materials/WMD response team? Can the exposed person wait for a team of specially trained hazardous materials responders to perform decontamination? Do you have enough information and a decision-making process to weigh the risks and benefits of safely rendering patient care to a chemically exposed person?

Such real-world challenges present a set of complicated questions—quickly. This chapter will help you to estimate and plan for the challenges you may encounter at a hazardous materials/WMD incident.

Estimating the Potential Harm or Severity of the Incident

Hazardous materials/WMD incident response priorities should be based on the need to protect and/or reduce the threat to life, property, critical systems, and the environment.

RESPONDER TIP

Remember—it is important to separate the people from the problem as soon and safely as possible.

In some cases, the people and the problem are one and the same, as in the chapter-opening scenario. Once the threat to life has been handled, the incident becomes a matter of reducing the impact to the property that may be affected and minimizing environmental complications.

The damage that a hazardous material/WMD will inflict on a human being is a function of the physical and chemical properties of the released substance, as well as the conditions under which it was released along with the duration of the exposure. Among other factors, characteristics such as the concentration of the material, the temperature of the material at the time of its release, and the pressure under which the substance was released will affect both the release parameters and the potential health effects on those exposed to the material. Additionally, the age, gender, genetics, and underlying medical conditions of the exposed person will have some bearing on patient outcome. Chemical exposures are complicated events because of the number of variables that may be present.

When it comes to rendering medical care to persons exposed to harmful substances, guidance can be found in NFPA 473, *Standard for Competencies for EMS Personnel Responding to Hazardous Materials/Weapons of Mass Destruction Incidents*. This standard centers on the notion that all EMS responders, regardless of their scope of practice, should share a basic set of hazardous materials response skills to work safely on a hazardous materials/WMD scene and deliver effective patient care. In most cases, that care is performed in a safe area (cold zone) away from the hazard, after decontamination. There are very few circumstances where definitive advanced life support must be rendered in the hot zone. EMS responders, however, should understand the nature of the incident and look at the scene with a critical eye to pick up the clues that might assist with defining the nature of the exposure.

To have some frame of reference for the degree of harm a substance may inflict, responders should have a basic understanding of some commonly used terms and definitions. Two main organizations establish and publish the toxicological data typically used by hazardous materials/WMD responders: the American Conference of Governmental Industrial Hygienists (ACGIH) and the Occupational Safety and Health Administration (OSHA).

For more than 60 years, the ACGIH has been a respected and trusted source for occupational health and industrial hygiene guidelines and information. Its best-known committee, the Threshold Limit Values for Chemical Substances (TLV-CS) Committee, was established in 1941. This committee first introduced the concept of threshold limit value (TLV) in 1956. The TLV is the point at which a hazardous material/WMD begins to affect a person. Today, TLVs have been published for more than 600 chemical substances. Any values indicated with a TLV are established by the ACGIH.

OSHA was created in 1971 with three goals: to improve worker safety, to conduct research, and to publish toxicological data. OSHA's permissible exposure limit (PEL), for example, is conceptually the same as ACGIH's TLV term. The PEL is the established standard limit of exposure to a hazardous material. You might see both of these terms in a reference source, for example. Nevertheless, there is an important distinction between the ACGIH and OSHA standards—ACGIH sets guidelines; OSHA standards are the law.

It is common to see toxicological values expressed in units of parts per million (ppm), parts per billion (ppb), and, in some cases, parts per trillion (ppt). Typically, the amount of airborne contamination encountered with releases of gases such as arsine, chlorine, and ammonia will be expressed in this manner. Arsine, for example, has an OSHA-established PEL of 0.05 ppm. This is a very small amount compared to the OSHA PEL for chlorine, which is 1 ppm. Comparatively speaking, arsine is a far more toxic substance than chlorine.

Another way to express contamination levels for substances other than gases, such as fibers and dusts, is through units of milligrams per cubic meter (mg/m^3). For the purpose of simplicity and better understanding of this text, keep in mind that you may see toxicological data expressed in several different ways. Regardless, the lower the value, the more toxic the product.

The threshold limit value/short-term exposure limit (TLV/STEL) is the maximum concentration of a hazardous material that a person can be exposed to in 15-minute intervals, up to four times per day, without experiencing irritation or chronic or irreversible tissue damage. A minimum 1-hour rest period should separate any exposures to this concentration of the material. The lower the TLV/STEL concentration,

RESPONDER TIP

One part per million (ppm) means there is one particle of a given substance for 999,999 other particles. Here's an illustration to put this unit of measurement into perspective: 1 ppm equates to 1 second in 280 hours.

One part per billion is equivalent to one particle in 999,999,999 other particles. Thus, 1 ppb equates to 1 second in 32 years.

One part per trillion (ppt) is equivalent to 1 second in 320 centuries.

the more toxic the substance. The threshold limit value/time-weighted average (TLV/TWA) is the maximum airborne concentration of material that a worker could be exposed to for 8 hours a day, 40 hours a week, with no ill effects. As with the TLV/STEL, the lower the TLV/TWA, the more toxic the substance. The threshold limit value/ceiling (TLV/C) is the maximum concentration of a hazardous material that a worker should not be exposed to, even for an instant. Again, the lower the TLV/C, the more toxic the substance.

The threshold limit value/skin indicates that direct or airborne contact with a material could result in possible and significant exposure from absorption through the skin, mucous membranes, and eyes. This designation is intended to suggest that appropriate measures be taken to minimize skin absorption so that the TLV/TWA is not exceeded.

As mentioned earlier, the permissible exposure limit (PEL) is the standard limit of exposure to a hazardous material as established and enforced by OSHA. The recommended exposure level (REL) is a value established by the National Institute for Occupational Safety and Health (NIOSH), and is comparable to OSHA's PEL. NIOSH is part of the U.S. Department of Health and Human Services and is charged with ensuring that individuals have a safe and healthy work environment by providing information, training, research, and education in the field of occupational safety and health. The PEL and REL limits are comparable to ACGIH's TLV/TWA. These three terms (PEL, REL, and TLV/TWA) measure the maximum, time-weighted concentration of material to which 95 percent of healthy adults can be exposed without suffering any adverse effects over a 40-hour workweek.

Responder Safety Tip

Identifying and measuring the levels of airborne contamination require specific detection and monitoring instruments along with training to interpret the results.

The designation immediately dangerous to life and health (IDLH) means that an atmospheric concentration of a toxic, corrosive, or asphyxiant substance poses an immediate threat to life or could cause irreversible or delayed adverse health effects. Three types of IDLH atmospheres are distinguished: toxic, flammable, and oxygen deficient. Individuals exposed to atmospheric concentrations below the IDLH value (in theory) could escape from the atmosphere without experiencing irreversible damage to their health, even if their respiratory protection fails. Individuals who may be exposed to atmospheric concentrations equal to or higher than the IDLH value must use positive-pressure self-contained breathing apparatus (SCBA), or equivalent protection.

With the appropriate equipment, responders will be able to measure concentrations of specific chemicals. Once these exposure values are understood, they can be applied at the scene of a hazardous materials/WMD emergency. For example, exposure guidelines can be used to identify three basic atmospheres or environments that might be encountered at a hazardous materials/WMD emergency. Green environments could be considered low hazard and low risk. These environments would not require any PPE beyond a normal work uniform. Yellow environments are transitional whereby the hazard is increasing, and some level of PPE is required (at least splash protection and some level of respiratory protection). Red environments are those that pose a high hazard and high risk; therefore, you must wear the highest level of skin and respiratory protection. An exposure to unprotected skin and/or lungs could be fatal. These are above IDLH environments where toxic chemicals are present. You can certainly work around lethal concentrations of released chemicals safely if the correct type and level of PPE is used. These are rough guidelines and serve only to illustrate an actual or potential level of risk.

In many cases, responders have no control over the hazard—the genie may be out of the bottle upon your arrival. You have some control over your risk by understanding why and determining how you will interact with the problem. How will various levels of PPE change the risk? What is the risk of taking no action? Do I have the proper type and level of protection that allows for safe entry into a contaminated atmosphere?

- *Safe atmosphere:* No harmful hazardous materials effects exist, so personnel can handle routine emergencies without donning specialized personal protective equipment (PPE).
- *Unsafe atmosphere:* A hazardous material that is no longer contained has created an unsafe condition or atmosphere. A person who is exposed to the material for long enough will probably experience some form of acute or chronic injury.
- *Dangerous atmosphere:* Serious, irreversible injury or death may occur in the environment.

All exposure guidelines share a common goal: to ensure the safety and health of people exposed to a hazardous material.

Responder Safety Tip

Initial isolation zones and protective actions are just that—initial. You must constantly evaluate the conditions and adjust tactics and strategy accordingly.

Resources for Determining the Size of the Incident

It is vital to understand the incident as a whole. To do so, responders must plug in results obtained from detection and monitoring devices, reference sources, bystander information, the current environmental conditions surrounding the incident, and other information to get a clear picture of what is going on and what is likely to happen next. Sometimes a decision must be made to evacuate or rescue people in danger. In those instances, responders may need to consult printed and electronic reference sources for guidance on evacuation distances and other safety information. A number of computer programs can be used to model and predict the direction and size of vapor clouds. When used properly, these computer programs can be a valuable source of information for predicting the size, shape, and direction of movement of vapor clouds. Again, it is important to identify the health hazards posed by the substance to accurately set safe parameters around the entire incident.

A valuable resource to consult for evacuation distances is the *Emergency Response Guidebook* (*ERG*), which is updated every 3 or 4 years to address new substances and/or new technologies. This reference book identifies and outlines predetermined evacuation distances and basic action plans for chemicals, based on spill size estimates FIGURE 4-1.

Responder Safety Tip

To reduce the effects of a radiation exposure, responders should understand the concept of time–distance–shielding (TDS). When radiation source is suspected or confirmed, responders should take action to reduce the amount of time they are exposed to the source, remain as far away as necessary, and place some barrier between themselves and the source. Identifying the presence of a radioactive source may require using a radiation detector. In the event you do not have such a device, or if you have a reasonable suspicion that the incident may involve a radioactive source, employ basic tactics that use the concept of TDS. Think of TDS in this way: The less time you spend in the sun, the less chance you have of suffering a sunburn (*time*); the closer you stand to a fire, the hotter you will get (*distance*); and if you come inside during a rainstorm, you will stop getting wet (*shielding*). These basic illustrations are analogous to the TDS concept for reducing the health effects of a radiation exposure.

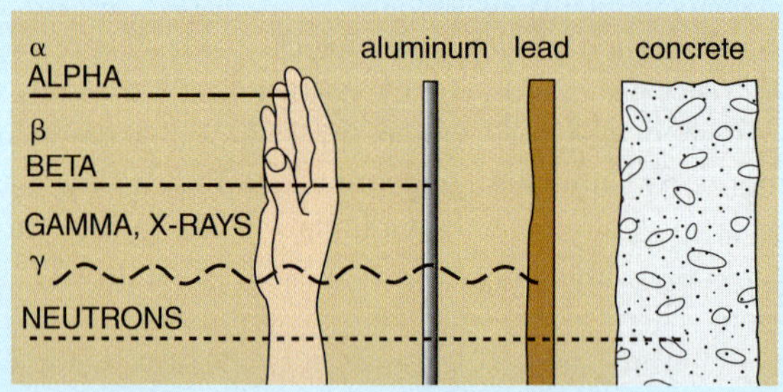

© Jones & Bartlett Learning

All responders should be equipped with the latest version of the *ERG* and take time to become familiar with it.

A good way to practice is to imagine a particular chemical and credible location or condition in which the chemical might be released. Identify the unique United Nations/North American Hazardous Materials Code (UN/NA) identification number for the chemical, check whether it is highlighted and found in the green section of the book, and determine the recommended emergency actions and PPE that might be required to handle the incident. Also, it is useful to imagine the release occurring in several different areas of your jurisdiction. Take a few minutes to think about initial isolation distances or other protective actions in each case. Part of being

HOW TO USE TABLE 1 - INITIAL ISOLATION AND PROTECTIVE ACTION DISTANCES

(1) The responder should already have:

- Identified the material by its ID Number and Name; (if an ID Number cannot be found, use the Name of Material index in the blue-bordered pages to locate that number.)
- Found the three-digit guide for that material in order to consult the emergency actions recommended jointly with this table;
- **Noted the wind direction.**

(2) Look in Table 1 (the green-bordered pages) for the ID Number and Name of the Material involved in the incident. Some ID Numbers have more than one shipping name listed—look for the specific name of the material. (If the shipping name is not known and Table 1 lists more than one name for the same ID Number, use the entry with the largest protective action distances.)

(3) Determine if the incident involves a SMALL or LARGE spill and if DAY or NIGHT. Generally, a SMALL SPILL is one which involves a single, small package (e.g., a drum containing up to approximately 208 liters (55 US gallons)), a small cylinder, or a small leak from a large package. A LARGE SPILL is one which involves a spill from a large package, or multiple spills from many small packages. DAY is any time after sunrise and before sunset. NIGHT is any time between sunset and sunrise.

(4) Look up the INITIAL ISOLATION DISTANCE. Direct all persons to move, in a crosswind direction, away from the spill to the distance specified—in meters and feet.

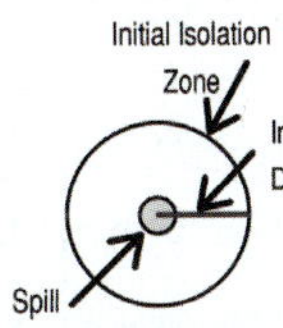

(5) Look up the initial PROTECTIVE ACTION DISTANCE shown in Table 1. For a given material, spill size, and whether day or night, Table 1 gives the downwind distance—in kilometers and miles— for which protective actions should be considered. For practical purposes, the Protective Action Zone (i.e., the area in which people are at risk of harmful exposure) is a square, whose length and width are the same as the downwind distance shown in Table 1.

(6) Initiate Protective Actions to the extent possible, beginning with those closest to the spill site and working away from the site in the downwind direction. When a water-reactive TIH producing material is spilled into a river or stream, the source of the toxic gas may move with the current or stretch from the spill point downstream for a substantial distance.

The shape of the area in which protective actions should be taken (the Protective Action Zone) is shown in this figure. The spill is located at the center of the small circle. The larger circle represents the INITIAL ISOLATION zone around the spill.

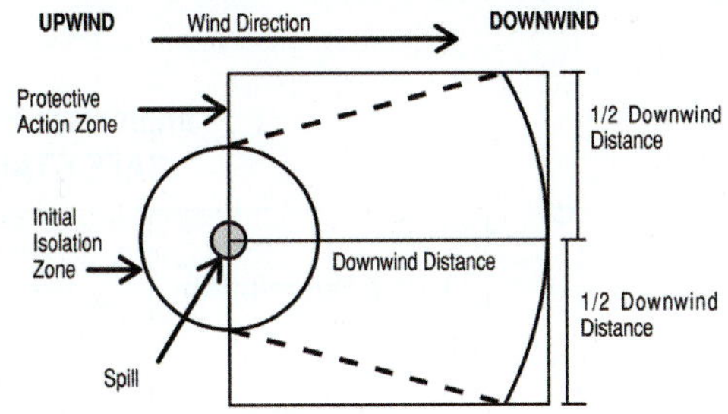

NOTE 1: See "Introduction To Green Tables – Initial Isolation And Protective Action Distances" under "Factors That May Change the Protective Action Distances" (page 285)

NOTE 2: See Table 2 – Water-Reactive Materials which Produce Toxic Gases for the list of gases produced when these materials are spilled in water.

Call the emergency response telephone number listed on the shipping paper or the appropriate response agency as soon as possible for additional information on the material, safety precautions and mitigation procedures.

Page 290

Page 291

Page 292

TABLE 1 - INITIAL ISOLATION AND PROTECTIVE ACTION DISTANCES

			SMALL SPILLS (From a small package or small leak from a large package)						LARGE SPILLS (From a large package or from many small packages)					
			First ISOLATE in all Directions		Then PROTECT persons Downwind during-				First ISOLATE in all Directions		Then PROTECT persons Downwind during-			
					DAY		NIGHT				DAY		NIGHT	
ID No.	Guide	NAME OF MATERIAL	Meters	(Feet)	Kilometers	(Miles)	Kilometers	(Miles)	Meters	(Feet)	Kilometers	(Miles)	Kilometers	(Miles)
1005 *	**125**	Ammonia, anhydrous	30 m	(100 ft)	0.1 km	(0.1 mi)	0.2 km	(0.1 mi)	150 m	(500 ft)	0.8 km	(0.5 mi)	2.0 km	(1.3 mi)
1005 *	**125**	Anhydrous ammonia												
1008	**125**	Boron trifluoride	30 m	(100 ft)	0.1 km	(0.1 mi)	0.5 km	(0.4 mi)	300 m	(1000 ft)	1.7 km	(1.1 mi)	4.8 km	(3.0 mi)
1008	**125**	Boron trifluoride, compressed												
1016	**119**	Carbon monoxide	30 m	(100 ft)	0.1 km	(0.1 mi)	0.2 km	(0.1 mi)	200 m	(600 ft)	1.2 km	(0.8 mi)	4.8 km	(3.0 mi)
1016	**119**	Carbon monoxide, compressed												
1017 *	**124**	Chlorine	60 m	(200 ft)	0.4 km	(0.2 mi)	1.5 km	(1.0 mi)	500 m	(1500 ft)	3.0 km	(1.9 mi)	7.9 km	(4.9 mi)
1023	**119**	Coal gas	60 m	(200 ft)	0.2 km	(0.1 mi)	0.2 km	(0.1 mi)	100 m	(300 ft)	0.4 km	(0.2 mi)	0.5 km	(0.3 mi)
1023	**119**	Coal gas, compressed												
1026	**119**	Cyanogen	30 m	(100 ft)	0.1 km	(0.1 mi)	0.5 km	(0.3 mi)	60 m	(200 ft)	0.4 km	(0.2 mi)	1.7 km	(1.0 mi)
1026	**119**	Cyanogen gas												
1040 *	**119P**	Ethylene oxide	30 m	(100 ft)	0.1 km	(0.1 mi)	0.2 km	(0.1 mi)	150 m	(500 ft)	0.9 km	(0.5 mi)	2.0 km	(1.3 mi)
1040 *	**119P**	Ethylene oxide with Nitrogen												
1045	**124**	Fluorine	30 m	(100 ft)	0.1 km	(0.1 mi)	0.2 km	(0.1 mi)	100 m	(300 ft)	0.5 km	(0.3 mi)	2.3 km	(1.4 mi)
1045	**124**	Fluorine, compressed												
1048	**125**	Hydrogen bromide, anhydrous	30 m	(100 ft)	0.1 km	(0.1 mi)	0.3 km	(0.2 mi)	200 m	(600 ft)	1.2 km	(0.8 mi)	3.9 km	(2.4 mi)
1050 *	**125**	Hydrogen chloride, anhydrous	30 m	(100 ft)	0.1 km	(0.1 mi)	0.3 km	(0.2 mi)	60 m	(200 ft)	0.3 km	(0.2 mi)	1.3 km	(0.8 mi)
1051	**117**	AC **(when used as a weapon)**	60 m	(200 ft)	0.3 km	(0.2 mi)	1.0 km	(0.6 mi)	1000 m	(3000 ft)	3.7 km	(2.3 mi)	8.4 km	(5.3 mi)

FIGURE 4-1 Instructions and example pages from the Initial Isolation and Protective Action Distances table found in the *ERG*.

Courtesy of the U.S. Department of Transportation

a prepared responder is to conduct some level of preplanning of your own relative to the credible release scenarios that may exist.

In the orange section of the *ERG*, for example, evacuation distances assume that the listed chemical is involved in a fire or is a result of a large spill. The green section, by comparison, is based on toxicity concerns. Materials are listed in this section because they present an extraordinary threat from a health and safety perspective. To that end, responders should understand that the "initial isolation" distances (the distance all persons should be considered for evacuation in all directions) and the "protective action" distances (the downwind distance over which some form of protective actions might be required) are based on the nature of the material, the environmental conditions of the release, and the size of the release. A small spill, for example, means that a spill involves less than 200 liters (approximately 52 gallons) for liquids or less than 300 kilograms (approximately 661 pounds) of a solid. The ERG also offers suggested stand-off distances for improvised explosive devices (IEDs) and potential BLEVE situations **FIGURE 4-2**.

Improvised Explosive Device (IED)
SAFE STAND OFF DISTANCE

A.

	Threat Description	Explosives Mass (TNT equivalent)[1]		Building Evacuation Distance[2]		Outdoor Evacuation Distance[3]	
High Explosives (TNT Equivalent)	Pipe Bomb	5 lbs	2.3 kg	70 ft	21 m	850 ft	259 m
	Suicide Belt	10 lbs	4.5 kg	90 ft	27 m	1,080 ft	330 m
	Suicide Vest	20 lbs	9 kg	110 ft	34 m	1,360 ft	415 m
	Briefcase/Suitcase Bomb	50 lbs	23 kg	150 ft	46 m	1,850 ft	564 m
	Compact Sedan	500 lbs	227 kg	320 ft	98 m	1,500 ft	457 m
	Sedan	1,000 lbs	454 kg	400 ft	122 m	1,750 ft	534 m
	Passenger/Cargo Van	4,000 lbs	1,814 kg	640 ft	195 m	2,750 ft	838 m
	Small Moving Van/ Delivery Truck	10,000 lbs	4,536 kg	860 ft	263 m	3,750 ft	1,143 m
	Moving Van/Water Truck	30,000 lbs	13,608 kg	1,240 ft	375 m	6,500 ft	1,982 m
	Semitrailer	60,000 lbs	27,216 kg	1,570 ft	475 m	7,000 ft	2,134 m

B.

	Threat Description	LPG Mass/ Volume1		Fireball Diameter4		Safe Distance5	
Liquefied Petroleum Gas (LPG - Butane or Propane)	Small LPG Tank	20 lbs/5 gal	9 kg/19 L	40 ft	12 m	160 ft	48 m
	Large LPG Tank	100 lbs/25 gal	45 kg/95 L	69 ft	21 m	276 ft	84 m
	Commercial/ Residential LPG Tank	2,000 lbs/500 gal	907 kg/1,893 L	184 ft	56 m	736 ft	224 m
	Small LPG Truck	8,000 lbs/2,000 gal	3,630 kg/7,570 L	292 ft	89 m	1,168 ft	356 m
	Semitanker LPG	40,000 lbs/10,000 gal	18,144 kg/37,850 L	499 ft	152 m	1,996 ft	608 m

[1] Based on the maximum amount of material that could reasonably fit into a container or vehicle. Variations possible.

[2] Governed by the ability of an unreinforced building to withstand severe damage or collapse.

[3] Governed by the greater of fragment throw distance or glass breakage/falling glass hazard distance. These distances can be reduced for personnel wearing ballistic protection. Note that the pipe bomb, suicide belt/vest, and briefcase/suitcase bomb are assumed to have a fragmentation characteristic that requires greater standoff distances than an equal amount of explosives in a vehicle.

[4] Assuming efficient mixing of the flammable gas with ambient air.

[5] Determined by U.S. firefighting practices wherein safe distances are approximately 4 times the flame height. Note that an LPG tank filled with high explosives would require a significantly greater standoff distance than if it were filled with LPG.

Page 372

FIGURE 4-2 The ERG offers suggested stand-off distances for IEDs (A) and potential BLEVE situations (B).

Courtesy of the U.S. Department of Transportation

BLEVE – SAFETY PRECAUTIONS

Use with caution. The following table gives a summary of tank properties, critical times, critical distances and cooling water flow rates for various tank sizes. This table is provided to give responders some guidance but it should be used with caution.

Tank dimensions are approximate and can vary depending on the tank design and application.

Minimum time to failure is based on ***severe torch fire impingement*** on the vapour space of a tank in good condition, and is approximate. Tanks may fail earlier if they are damaged or corroded. Tanks may fail minutes or hours later than these minimum times depending on the conditions. It has been assumed here that the tanks are not equipped with thermal barriers or water spray cooling.

Minimum time to empty is based on an engulfing fire with a properly sized pressure relief valve. If the tank is only partially engulfed then time to empty will increase (i.e., if tank is 50% engulfed then the tanks will take twice as long to empty). Once again, it has been assumed that the tank is not equipped with a thermal barrier or water spray.

Tanks equipped with thermal barriers or water spray cooling significantly increase the times to failure and the times to empty. A thermal barrier can reduce the heat input to a tank by a factor of ten or more. This means it could take ten times as long to empty the tank through the Pressure Relief Valve (PRV).

Fireball radius and emergency response distance is based on mathematical equations and is approximate. They assume spherical fireballs and this is not always the case.

Two safety distances for public evacuation. The minimum distance is based on tanks that are launched with a small elevation angle (i.e., a few degrees above horizontal). This is most common for horizontal cylinders. The preferred evacuation distance has more margin of safety since it assumes the tanks are launched at a 45 degree angle to the horizontal. This might be more appropriate if a vertical cylinder is involved.

It is understood that these distances are very large and may not be practical in a highly populated area. However, it should be understood that the risks increase rapidly the closer you are to a BLEVE. Keep in mind that the farthest reaching projectiles tend to come off in the zones 45 degrees on each side of the tank ends.

Water flow rate is based on $\sqrt[5]{\text{capacity (USgal)}}$ = USgal/min needed to cool tank metal.

Warning: the data given are approximate and should only be used with extreme caution. For example, where times are given for tank failure or tank emptying through the pressure relief valve – these times are typical but they can vary from situation to situation. Therefore, never risk life based on these times.

Page 366

FIGURE 4-2 *(Continued)*

Exposures

When considering the potential consequences of a hazardous materials/WMD incident, the on-scene crews must consider how exposures might be affected. In firefighting, the term *exposures* typically refers to those areas adjacent to the fire that might become involved if the fire is left unchecked. For our purposes here, exposures include any people, property, structures, or environments that are subject to influence, damage, or injury as a result of contact with a hazardous material/WMD. The number of exposures is determined by the location of the incident, the physical and chemical properties of the released substance, and the amount of progress that has been made in protecting those exposures (such as people, property, and the environment) by isolating the release site or by taking protective actions such as evacuation or sheltering-in-place. Incidents in urban areas are likely to have a greater potential for exposures; consequently, more resources will likely

FIGURE 4-3 What material is likely to be in the container, and how much? What resources would you use to determine the evacuation distances?
Courtesy of Rob Schnepp

FIGURE 4-4 Law enforcement can be a great asset to ensuring your restricted area stays restricted.
Courtesy of Rob Schnepp

be needed to protect those exposures from the hazardous materials/WMD.

Isolation of the hazard area is one of the first actions responders must take at a hazardous materials/WMD incident. The general philosophy of isolation revolves around the concept of life safety and separating the people from the problem: *Responders and civilians alike must be kept a safe distance from the release site—a vital first step in beginning to establish safe work zones and identify the areas of high hazard.*

Imagine arriving on scene as the incident commander and encountering the situation shown in **FIGURE 4-3**. First arriving units put master streams in service to cool the container. It is the middle of the day in a residential area; there are several blocks of occupied homes. What material is likely to be in the container, and how much? Would you evacuate or shelter-in-place? What resources could you use to determine the evacuation distances?

Isolating the hazard may be accomplished in several ways. Law enforcement officers may be posted a safe distance from the release to create a secure perimeter.

Other public safety personnel such as fire fighters may serve the same function, although it is important not to waste the skills of a cache of trained hazardous materials/WMD responders by assigning them to guard doors, other points of ingress or egress from a building, or other contaminated areas. In many cases, responders will stretch a length of barrier tape across roadways, doors, or other access points. Care must be taken, however, not to rely solely on this method of scene control. Quite often, areas marked with barrier tape are not respected by public safety responders or the general public. If and when barrier tape is used, it should still be backed up by a human presence **FIGURE 4-4**. Also, keep in mind that the precise type of isolation efforts undertaken will be driven by the nature of the released chemical and the environmental conditions.

Once the hazard is isolated, access is denied to all but a small group of responders who are trained and equipped to enter the contaminated atmosphere. Isolating a contaminated atmosphere is always conjoined in some way with denial of entry (i.e., restriction of access) to the site. Practically speaking, one action should not exist without the other. Typically, site access control is established so as to control the movement of personnel into and out of a contaminated area.

Evacuation is the removal/relocation of those individuals who may be affected by an approaching release of a hazardous material. If the threat will be sustained over a long period of time, it may be advisable to evacuate people from a predicted or anticipated hazard area, making sure to evacuate those in the most danger first.

RESPONDER TIP

When conducting an evacuation, response personnel should concentrate on keeping themselves from becoming contaminated.

Evacuation efforts should not require personnel to wear PPE or enter contaminated atmospheres. Think of it this way: If you are wearing PPE to move people from one area to another, you may have shifted gears from conducting an evacuation to performing a rescue. The latter responsibilities may or may not be within the scope of your training.

Sheltering-in-place is a method of safeguarding people located near or in a hazardous area by temporarily keeping them in a cleaner atmosphere, usually inside structures. In some cases—for example, with a transitory problem such as a mobile vapor cloud—it is advisable to use a shelter-in-place strategy. This method is desirable only when the population being protected in place can care for themselves, control the air, and the structure can be sealed. Awareness level personnel, for example, could be expected to initiate some form of protective action such as directing civilians away from the contaminated area or directing certain populations of civilians to follow a shelter-in-place approach. *Remember that the NFPA 472 standard does not consider awareness level personnel to be responders.*

Reporting the Size and Scope of the Incident

Reporting the estimated physical size of the area affected by a hazardous materials/WMD incident is accomplished by using information available at the scene. If a vehicle is transporting a known amount of material, for example, an estimate of the size of the release might be made by subtracting the amount remaining in the container from the maximum capacity of the container. This can be computed by looking at the shipping papers to see whether deliveries have been made. To "see" into containers such as railroad tank cars, steel drums, or cargo tanks and estimate their remaining contents, responders may use thermal imagers (TIs) FIGURE 4-5.

For example, you might use a TI to investigate a steel drum discovered in a vacant lot. A quick look at the scene may reveal some wet-looking soil around the base of the drum. By using the TI, you might be able to determine the percentage of liquid remaining, and make an educated guess about how much could have leaked. Of course, these estimations may be quite rough, especially when it is unknown how much a given vessel may have contained prior to a release; however, it does allow for some "worst case scenario" estimations.

Depending on its size, the extent of the release may be expressed in units as small as square feet or as large as square miles. There are no hard and fast rules here: Be as accurate and as clear as possible when communicating with other responders or assisting agencies, or when contacting call centers such as CHEMTREC, CANUTEC, or SETIQ for assistance. Remember that the safety of responders is paramount to maintaining an effective response to any hazardous materials/WMD incident.

Determining the Concentration of a Released Hazardous Material

Concentration, from the perspective of a chemist, refers to the amount of solute in a given amount of solution. From a practical perspective, a concentrated solution of any kind contains a large amount of solute for a given amount of solution FIGURE 4-6. A dilute solution, by contrast, contains a small amount of solute for a given solution FIGURE 4-7.

FIGURE 4-5 Thermal imagers allow responders to estimate how much material remains in a container.
Courtesy of Bullard

FIGURE 4-6 A concentrated solution of any kind contains a large amount of solute for a given amount of solution.
© Jones & Bartlett Learning

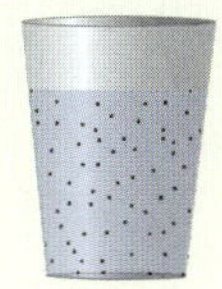

FIGURE 4-7 A dilute solution contains a small amount of solute for a given solution.
© Jones & Bartlett Learning

Consider a natural gas release. If a gas heater were to fail in some way, resulting in a sustained release of natural gas (the solute), a high concentration of gas could build up inside a tightly sealed house (with air being the solution in this instance). The consequence of that accumulation could be an explosion if the gas were to reach the proper proportions and find an ignition source. If the windows or doors of the house were opened, however, the concentration of natural gas would decrease, perhaps becoming so "dilute" that it would pose no significant fire or health hazard. Typically, concentrations of gases are expressed as percentages—think of flammable range as an example. The flammable range of natural gas is 5 percent to 15 percent, meaning that the concentration of vapors must be between these two values if combustion is to occur.

In the preceding example, responders may be called upon to determine the airborne concentration of natural gas within the house. When this step is necessary, the use of specialized detection and monitoring equipment may be required. Using detection and monitoring equipment properly requires some technical expertise, a lot of common sense, and a commitment to continual training. It is a mistake for responders to believe that they can simply turn on a machine, point it in some direction, and expect it to solve the problem. A reading from a gas detector, taken out of context, may cause an entire response to head off in the wrong direction, leading to an unsafe decision or a series of inefficient tactics. The responder must interpret the information the instrument is providing and make decisions based on the information. Always remember that using a detector/monitor entails more than just reading the screen or waiting for an alarm to sound. For more detail regarding the use of these devices, see the chapter, *Mission-Specific Competencies: Air Monitoring and Sampling*.

When responding to incidents involving corrosives, it is also important to know the concentration of the released substance. Concentration, when discussing corrosives, is an expression of how much of the acid or the base is dissolved

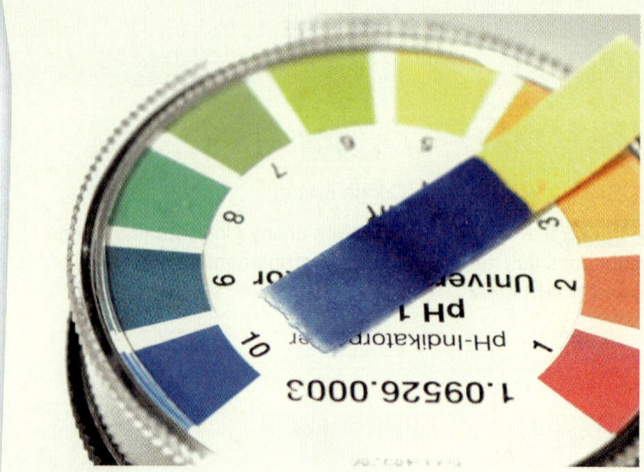

FIGURE 4-8 Litmus paper (pH strips) is used to determine the hazardous material's pH.

in a solution (usually water). Again, this value is generally expressed as a percentage. Sulfuric acid at a concentration of 97 percent is considered to be "concentrated," whereas the sulfuric acid found in car batteries (approximately 30 percent) is considered to be "dilute." In practical terms, responders should understand whether a released corrosive is concentrated or diluted—but should not confuse *concentration* with the *strength* of the solution. The words "strong" and "weak" do not correspond with "concentrated" and "dilute," respectively. The strength of a corrosive refers to the degree of ionization that occurs in a solution, which is determined by the solution's pH. A strong acid such as hydrochloric acid (HCl) is strong even if it's found in a dilute concentration. Acetic acid (vinegar) is considered to be a weaker acid, and it remains a weak acid even when it occurs in high concentrations. In general, strong corrosives will react more vigorously with incompatible materials such as organic substances and will be more aggressive when they come into contact with metallic objects such as metal shelving, shovels, and other items. (For more information on corrosives, refer to the chapter, *Hazardous Materials: Properties and Effects*.)

To measure pH in the field, hazardous materials/WMD responders can use litmus paper, sometimes referred to as pH paper **FIGURE 4-8**. Several styles of pH paper are in use today; refer to the tools used in your own jurisdiction to determine the specific styles of pH paper you may be called upon to use. Although specialized laboratory instruments are also used to measure pH, these kinds of tools are rarely deployed in the field.

Skin Contact Hazards

Many hazardous substances on the market today have the ability to produce harmful effects on the unprotected or inadequately protected human body. The skin can absorb harmful toxins without any sensation to the skin itself. Given this fact, responders should not rely on pain or irritation as a warning sign of absorption. Some poisons (the nerve agent VX, for example) are so concentrated that just a few drops placed on the skin may result in death.

Skin absorption is enhanced by abrasions, cuts, heat, and moisture—all of which allow materials to enter the body more easily. This relationship can create critical problems for responders who are working at incidents that involve any form of chemical or biological agents. Responders with large open cuts, rashes, or abrasions should be prohibited from working in areas where they may be exposed to hazardous materials/WMD. Smaller cuts or abrasions should be covered with nonporous dressings.

The rate of absorption can vary depending on the body part that is exposed. For example, chemicals can be absorbed through the skin on the scalp much faster than they are absorbed through the skin on the forearm. The high absorbency rate associated with the eyes makes them one of the fastest means of exposure. For example, a chemical may quickly enter the body through this route when it is splashed directly into the eyes or carried from a fire by toxic smoke particles, or when the eyes are exposed to gases or vapors.

Chemicals such as corrosives will immediately damage skin or body tissues upon contact. Acids, for example, have a strong affinity for moisture and can create significant skin and respiratory tract burns. In contrast, alkaline materials dissolve the fats and lipids that make up skin tissue and change solid tissue into a soapy-like liquid. This process is similar to the way caustic cleaning solutions dissolve grease and other materials in sinks and drains. As a result, alkaline burns are often much deeper and more destructive than are acid burns.

Plan an Initial Response

Planning a response boils down to understanding the nature of an incident and determining a course of action that will favorably change the outcome. On the surface, this seems like a straightforward, uncomplicated task. In truth, the decision to act can be a weighty one, fraught with many pitfalls and dangers. When planning an initial hazardous materials/WMD incident response, it is important to be mindful of the safety of the responding personnel. The responders are there to isolate, contain, and/or remedy the problem—not to become part of it. Proper incident planning will keep responders safe and provide a means to control the incident effectively, preventing further harm to persons or property.

Tactical control objectives are actions that may or may not involve the actual stopping of the leak or release of a hazardous material. In some cases, tactical objectives may include preventing further injury and controlling or containing the spread of the hazardous material. An example of a tactical control objective may include identifying and securing potential ignition sources when flammable liquids and gases have been released. Common examples of ignition sources include open flames from pilot lights or other sources, arcs occurring when electrical switches are turned on or off, static electricity, and/or smoking materials.

The information obtained from the initial call for help is used to determine the safest, most effective, and fastest route to the hazardous materials/WMD scene. Choose a route that approaches the scene from an upwind and upgrade direction, so that natural wind currents blow the hazardous material vapors away from arriving responders FIGURE 4-9. A route that places the responders uphill as well as upwind of the site is also desirable, so that a liquid or vapor hazardous material flows away from responders.

Responders need to know as much as possible about the material involved. Is the material a solid, a liquid, or a gas? Is it contained in a drum, a barrel, or a pressurized tank? Is the spill still in progress (dynamic) or has it ceased (static)? The response to a spill of a solid hazardous material will differ from the response to a liquid-release incident or a vapor-release incident. A solid may be easily contained, whereas a released gas can be widespread and constantly moving, depending on the gas characteristics and weather conditions FIGURE 4-10.

The characteristics of the affected area near the location of the spill or leak are also important factors in planning the response to an incident. If an area is heavily populated, evacuation procedures may be established very early in the course of the incident. If the area is sparsely populated and rural, isolating the area from anyone trying to enter the location may be the top priority. A high-traffic area such as a
would necessitate immediate rerouting of traffic,
during rush hours.

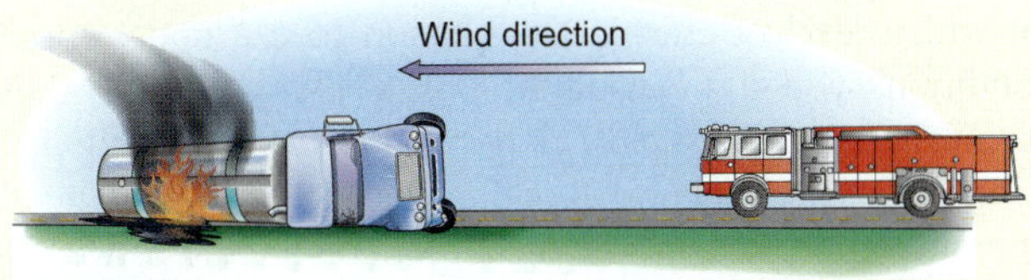

FIGURE 4-9 Approach a hazardous materials incident cautiously; choose a route that approaches from an upwind direction.
© Jones & Bartlett Learning

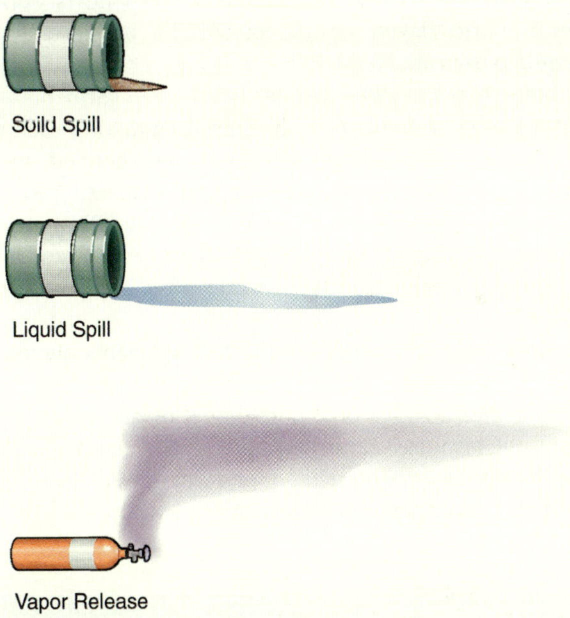

FIGURE 4-10 The response to a spill of a solid hazardous material will differ from the response to a liquid-release or vapor-release incident.
© Jones & Bartlett Learning

Response Objectives

Response objectives should be measurable, flexible, and time sensitive; they should also be based on the chosen strategy. Some examples might include the following:

- A three-person team will construct a dirt berm around the drain at the south end of the leaking tanker so as to protect the adjacent waterway. This task will need to be completed in the next 30 minutes.
- A two-person team, wearing a Level B ensemble, will immediately enter the steel door on the east side of the building and shut down the ventilation system.

In some cases, several response objectives may be developed to solve a problem. To be effective and meaningful, however, those objectives need to be tied to the reason you chose to take action in the first place. Again, if you don't understand why you are taking action, reevaluate the situation so that you can better understand the problem.

Secondary Attacks and Devices

In today's political climate, first responders of all disciplines should be aware of the potential for secondary attacks or devices and understand the need for acknowledging them in response objectives. Implementing a secondary attack on responders is a tactic that many terrorist organizations use to make their attacks even more dramatic. For this reason, responders must be able to assess the potential for secondary attacks and devices. This may be much easier said than done, however, because the indicators of a secondary attack or device(s) can be difficult to locate.

Terrorists who want to injure responding personnel with a secondary device or attack will typically make the initial attack very dramatic to draw responders into close proximity to the scene. Indeed, the primary attack may purposely injure members of the public to draw responders into the scene. As the responders begin to treat victims, the secondary attack then takes place.

All responders should maintain good situational awareness at every emergency scene, but especially in cases where one detonation has already taken place. In particular, their actions should include a brief scan of areas where secondary devices are likely to be placed. You may seek the assistance of someone familiar with the area—that person may be more attuned to noticing unusual items. Avoid touching or moving anything that is not critical to the operation or anything that may reasonably conceal an explosive device. The NFPA 473, *Standard for Competencies for EMS Personnel Responding to Hazardous Materials/Weapons of Mass Destruction Incidents*, EVADE mnemonic serves as a useful reminder of some key points when it comes to secondary devices:

- Evaluate the scene for areas where secondary devices are likely to be placed.
- Visually scan operating areas for a secondary device before providing patient care.

- Avoid touching or moving anything that may conceal an explosive device.
- Designate and enforce scene control zones.
- Evacuate victims, other responders, and nonessential personnel as quickly and safely as possible.

Defensive Actions

Examples of defensive actions that can be taken include diking and damming, stopping the flow of a substance remotely from a valve or shut-off, diluting or diverting the material, or suppressing or dispersing vapor. These and other actions are covered in detail in the chapter, *Mission-Specific Competencies: Product Control.*

Personal Protective Equipment

The determination of which PPE is needed is based on the hazardous material involved, the specific hazards present, and the physical state of the material, along with a consideration of the tasks to be performed by the operations level responder. The chapter, *Mission-Specific Competencies: Personal Protective Equipment*, discusses proper PPE for a hazardous materials/WMD incident in detail.

In the realm of hazardous materials/WMD response, the selection and use of chemical-protective clothing may have the greatest direct impact on responder health and safety. Without the proper PPE, responders place themselves at risk of suffering harmful exposures. Corrosives such as concentrated sulfuric acid and hydrochloric acid, as well as caustic substances such as sodium hydroxide, for example, can damage the skin. The human body is also susceptible to the adverse effects of solvents such as methylene chloride and toluene, which may penetrate the skin and cause systemic health effects. Some poisons—the nerve agent VX, for example—are highly toxic and can be fatal in small doses.

Skin protection, however, is just one factor to be considered when discussing PPE. Anyone planning to work in a contaminated atmosphere must also place a high priority on respiratory protection. Respiratory protection is so important that it can be viewed as the defining element of PPE. The head-to-toe ensemble is not complete until the hazards have been identified and the respiratory protection has been properly matched to both the hazard and the garment. Keep in mind that PPE is not intended to function as an impenetrable suit of armor: It has limitations!

The National Fire Protection Association (NFPA) publishes protective clothing standards to provide guidance on the performance of certain types of chemical-protective garments. NFPA 1991, for example, is the *Standard on Vapor-Protective Ensembles for Hazardous Material Emergencies.* NFPA 1992, *Standard on Liquid Splash-Protective Ensembles and Clothing for Hazardous Materials Emergencies*, covers a different type of chemical-protective garment. The NFPA also acknowledges the importance of chemical-protective garments as they relate to WMD response. To obtain guidance in this area, responders may reference NFPA 1994, *Standard on Protective Ensembles for First Responders to CBRN Terrorism Incidents*. (CBRN stands for chemical, biological, radiological, and nuclear.)

The NFPA does not "certify" any garments. This is a common misperception in the hazardous materials/WMD response industry. Instead, the intent of the NFPA clothing standards is to provide guidance on manufacturing quality and performance standards. A third-party testing laboratory carries out the testing and "certifies" the garment in question. In short, the NFPA publishes performance standards (durability, flammability, chemical resistance, and cold temperature); third-party laboratories then test manufacturers' garments to determine if they meet the NFPA standards. Much like other NFPA committees, the Technical Committee for Chemical Protective Clothing consists of end users (responders), manufacturers, government representatives, and other recognized experts in the field.

When it comes to the selection and use of PPE, responders must understand how standards and regulations influence their decision making in the field. The NFPA protective clothing standards do not tell you when, or under which conditions, to wear a particular level of chemical protection. Rather, these standards are performance documents for the garments only—they are not intended to guide responders. For guidance on which level of chemical protection to use under specific conditions, you may consult the OSHA HAZWOPER (*HAZardous Waste OPerations and Emergency Response)* regulation, 29 CFR 1910.120; Appendix B of the OSHA HAZWOPER regulation offers guidance on which components should be worn for certain levels of protection, and under which conditions the various levels of protection should be chosen.

In addition to the performance standards, responders must be aware of the procedures for cleaning, disinfecting, and inspecting PPE. These procedures may vary from manufacturer to manufacturer, so it is important for all responders working in an Authority Having Jurisdiction (AHJ) to understand what is required to maintain the PPE.

Some types of PPE—primarily reusable garments—are required to be tested at regular intervals and after each use. Individual manufacturers will have well-defined procedures for the maintenance, testing, and inspection of their particular equipment. Prior to purchasing any PPE, the AHJ should understand what is required in terms of maintenance and upkeep, including the cleaning and disinfection of the PPE.

Most types of chemical-protective garments should be stored in a cool dry place, free of significant temperature swings and/or high levels of humidity. If repairs are required, consult the manufacturer prior to performing any work. There is a risk that the garment will not perform as expected if it has been modified or repaired incorrectly.

RESPONDER TIP

The NFPA does not "certify" chemical-protective garments of any kind but rather sets performance standards for the garments.

VOICES OF EXPERIENCE

My engine company was dispatched to a small fire in a garbage dump. As we arrived on the scene, we could see a small fire burning about 200 yards away. We dismounted and walked closer to determine whether to pull a line or try to hit the fire with the master stream. My engine carried 1000 gallons of water, so we felt confident that we could handle a refuse fire.

We were approximately 20 feet away when bright red smoke began billowing from the burning pile of junk. I had never seen anything like it before. We immediately pulled back, and I ordered the crew to don SCBA. You do not have to be a chemist to understand that junk fire smoke is not typically bright red.

Common sense prevailed, and we exercised caution by moving the engine back toward the entrance road to stay upwind of the red smoke. I called dispatch and reported that we had a possible hazardous materials fire. I was told to stand by for a hazardous materials officer. When the hazardous materials officer arrived, it was decided that we should let the fire burn out. We later learned that someone had piled old tires around abandoned drums of unknown liquid and added a few gallons of gasoline. It was a cheap (and fast) way to dispose of the chemicals.

This was my first experience as a company officer dealing with a hazardous materials incident. At that time, we did not know a lot about hazardous materials. I was not sure if we should use water on this fire, so I elected to do nothing. After estimating and predicting the risks involved, we planned our actions accordingly. We understood that this incident had significant potential to become a much larger event.

This call convinced me that I needed to know a lot more about hazardous materials. Shortly after the incident, I registered for a class on hazardous materials chemistry. I was later appointed to be the coordinator to start the hazardous materials team for our county.

Rick Emery
Lake County Hazardous Materials Team (Retired)
Vernon Hills, Illinois

Hazardous Materials: Types of PPE

Several types and levels of PPE may be selected for use at a hazardous materials/WMD incident. These levels are spelled out in great detail in OSHA's HAZWOPER regulations and in local jurisdictional regulations. This section reviews the protective qualities of various ensembles, from the lowest level of protection to the greatest, and discusses the selection criteria for each level.

Street Clothing and Work Uniforms

At the lower end of the PPE spectrum is normal street clothing or work uniforms, which offer the least amount of protection in a hazardous materials/WMD emergency FIGURE 4-11. Work uniforms may prevent a "nuisance" powder from coming into direct contact with the skin but offer no chemical protection. Typically, those personnel performing support functions away from the areas of contamination wear normal work uniforms.

Structural Firefighting Protective Clothing

The next level of protection is provided by structural firefighting protective equipment FIGURE 4-12. Structural firefighting gear is not recognized as a chemical-protective ensemble, though it does have a place at a hazardous materials/WMD incident. Many support functions can be carried out in structural fire fighter's gear. In some cases (such as during incidents involving chemicals with low toxicity and/or high flammability), structural gear may be safer than traditional types of chemical-protective clothing. A full set of structural fire fighter's gear includes a helmet, a bunker coat, bunker pants, boots, gloves, a hood, SCBA, and a personal alert safety system (PASS) device.

FIGURE 4-11 A Nomex jumpsuit.
Courtesy of The DuPont Company

FIGURE 4-12 Standard structural firefighting gear.
© Jones & Bartlett Learning. Photographed by Glen E. Ellman

High-Temperature–Protective Clothing and Equipment

High-temperature–protective equipment is a level above structural fire fighter's gear FIGURE 4-13. This type of personal protective equipment shields the wearer during short-term exposures to high temperatures. Sometimes referred to as a proximity or entry suit, high-temperature–protective equipment allows the properly trained fire fighter to work in extreme fire conditions. It provides protection against high temperatures only—it is not designed to protect the fire fighter from hazardous materials/WMD.

RESPONDER TIP

A fire department should issue PPE that properly fits each fire fighter, regardless of gender, size, or shape.

Chemical-Protective Clothing and Equipment

Chemical-protective clothing is unique in that it is designed to prevent chemicals from coming in contact with the body. Not all chemical-protective clothing is the same, and each type/brand/style may offer varying degrees of resistance. There is no single chemical-protective garment on the market that will protect you from everything. Manufacturers supply compatibility charts with

FIGURE 4-13 High-temperature–protective equipment protects the wearer from high temperatures during a short exposure.

FIGURE 4-14 Vapor-protective clothing.

all protective equipment, these charts are intended to assist you in choosing the right chemical-protective clothing. You must match the anticipated chemical hazard to these charts to determine the resistance characteristics of the garment. Time, temperature, and resistance to cuts, tears, and abrasions are all factors that affect the chemical resistance of materials. Other requirements include flexibility, temperature resistance, shelf life, and sizing criteria.

Chemical resistance is the ability of the garment to resist damage or become compromised as a result of direct contact with a chemical. Chemical-resistant materials are specifically designed to inhibit or resist the passage of chemicals into and through the material by the processes of penetration, permeation, or degradation.

Penetration is the flow or movement of a hazardous chemical through closures (e.g., zippers), seams, porous materials, pinholes, or other imperfections in the material. Although liquids are most likely to penetrate a material, solids (e.g., asbestos) can also penetrate protective clothing materials.

Permeation is the process by which a hazardous chemical moves through a given material on the molecular level. It differs from penetration in that permeation occurs through the material itself rather than through openings in the material.

Degradation is the physical destruction or decomposition of a clothing material owing to chemical exposure, general use, or ambient conditions (e.g., storage in sunlight). It may be evidenced by visible signs such as charring, shrinking, swelling, color changes, or dissolving. Materials can also be tested for weight changes, loss of fabric tensile strength, and other properties to measure degradation.

Chemical-protective clothing can be constructed as a single-piece or multiple-piece garment. For example, a single-piece garment completely encloses the wearer and is referred to as an encapsulated suit. A variety of materials are used to manufacture encapsulated suits and nonencapsulated multiple-piece garments. Some common suit materials include butyl rubber, Tyvek, Saranex, polyvinyl chloride, and Viton, which are used either singly or in multiple layers of several materials. Special chemical-protective clothing is adequate for some chemicals, yet useless for other chemicals; no single material provides satisfactory protection from all chemicals.

Fully encapsulating protective clothing offers full body protection from highly contaminated environments and requires supplied-air respiratory protection devices such as SCBA. NFPA 1991, *Standard for Vapor-Protective Ensembles for Hazardous Materials Emergencies*, sets the performance standards for these types of garments, more correctly referred to as vapor-protective clothing FIGURE 4-14. NFPA 1991 garments are tested for permeation resistance against 21 standard chemicals, several toxic industrial chemicals, and two chemical weapons.

Liquid splash–protective clothing is designed to protect the wearer from chemical splashes FIGURE 4-15. It does not provide total body protection from gases or vapors, however, and should not be used for incidents involving liquids that emit vapors known to affect or be absorbed through the skin. NFPA 1992, *Standard on Liquid Splash-Protective Ensembles and Clothing for Hazardous Materials Emergencies*, is the performance document for liquid-splash garments and ensembles. This type of equipment is tested for penetration resistance against a test battery of five chemicals. The tests include no gases, because this level of protection is not considered to be vapor protection.

FIGURE 4-15 Liquid splash–protective clothing must be worn when there is the danger of chemical splashes.
© Jones & Bartlett Learning. Photographed by Glen E. Ellman

FIGURE 4-16 The Level A ensemble envelops the wearer in a totally encapsulating suit.
© Jones & Bartlett Learning. Courtesy of MIEMSS

Chemical-Protective Clothing Ratings

Chemical-protective clothing is rated for its effectiveness in several different ways. The U.S. Environmental Protection Agency (EPA) defines levels of protection using an alphabetic system.

Level A

The Level A ensemble consists of a fully encapsulating garment that completely envelops both the wearer and the respiratory protection, gloves, boots, and communications equipment. The Level A ensemble should be used when the hazardous material identified requires the highest level of protection for the skin, eyes, and respiratory tract. Typically, this level is indicated when the operating environment is above IDLH values for skin absorption. The Level A ensemble is effective against vapors, gases, mists, and even dusts **FIGURE 4-16**.

Ensembles worn as Level A protection must meet the requirements for vapor-protective clothing as outlined in the most current edition of NFPA 1991, *Standard on Vapor-Protective Ensembles for Hazardous Materials Emergencies*.

Recommended PPE of a Level A ensemble includes the following components:

- SCBA or supplied-air respirator (SAR)
- Fully encapsulating vapor-protective chemical-resistant suit
- Inner and outer chemical-resistant gloves
- Chemical-resistant safety boots/shoes (including steel shank and toe)
- Two-way radio

Optional PPE for a Level A ensemble includes the following components:

- Coveralls
- Cooling vest
- Long cotton underwear
- Hard hat
- Disposable gloves and boot covers

Level B

The Level B ensemble consists of chemical-protective clothing, boots, gloves, and SCBA **FIGURE 4-17**. This type of PPE should be used when the type and atmospheric concentration of identified substances require a high level of respiratory protection but less skin protection. To that end, the defining piece of equipment with Level B ensembles is an SCBA or some other type of SAR. Garments and ensembles that are worn for Level B protection should comply with the performance standards outlined in NFPA 1992, *Standard on Liquid Splash-Protective Ensembles and Clothing for Hazardous Materials Emergencies*.

Several types of single-piece and multiple-piece garments on the market can be used to create a Level B

FIGURE 4-17 A Level B ensemble provides a high level of respiratory protection but less skin protection.
© Jones & Bartlett Learning. Photographed by Glen E. Ellman

FIGURE 4-18 A Level C ensemble includes chemical-protective clothing and gloves as well as respiratory protection.
Courtesy of The DuPont Company

ensemble. It is the respiratory protection, however, that distinguishes Level B ensembles from the next (lower) level of protection (Level C).

The types of gloves and boots worn depend on the identified chemical. Wrists and ankles must be properly sealed to prevent splashed liquids from contacting skin.

Recommended PPE worn as part of a Level B ensemble includes the following components:

- SCBA or SAR
- Chemical-resistant clothing
- Inner and outer chemical-resistant gloves
- Chemical-resistant safety boots/shoes
- Two-way radio

Responder Safety Tip

According to the OSHA HAZWOPER regulation, a Level B ensemble is the minimum level of protection to be worn when operating in an unknown environment.

Optional PPE for a Level B ensemble includes the following components:

- Cooling vest
- Coveralls
- Disposable gloves and boot covers
- Face shield
- Long cotton underwear
- Hard hat

Level C

The Level C ensemble consists of standard work clothing plus chemical-protective clothing, chemical-resistant gloves, and a form of respiratory protection. Typically, Level C ensembles are worn with an air-purifying respirator (APR) or a powered air-purifying respirator (PAPR); both are discussed in more detail later in this chapter. The APR could be a half-face (with eye protection) or a full-face mask. A Level C ensemble is appropriate when the type of airborne substance is known, its concentration is measured, the criteria for using APRs are met (see the respiratory protection section of this chapter), and skin and eye exposure is unlikely FIGURE 4-18. The garments selected must meet the performance requirements outlined in NFPA 1992, *Standard on Liquid Splash-Protective Ensembles and Clothing for Hazardous Materials Emergencies*.

Recommended PPE for a Level C ensemble includes the following components:

- Full-face APR
- Chemical-resistant clothing
- Inner and outer chemical-resistant gloves
- Chemical-resistant safety boots/shoes
- Two-way radio

Optional PPE for a Level C ensemble includes the following components:

- Coveralls
- Disposable gloves and boot covers
- Face shield
- Escape mask
- Long cotton underwear
- Hard hat

Level D

The Level D ensemble is the lowest level of protection. This type of ensemble typically comprises coveralls, work shoes, hard hat, gloves, and standard work clothing FIGURE 4-19. It should be used only when the atmosphere contains no known hazard, and when work functions preclude splashes, immersion, or the potential for unexpected inhalation of or contact with hazardous levels of chemicals. A Level D ensemble should be used for nuisance contamination (such as dust) only; it should not be worn on any site where respiratory or skin hazards are known to exist.

Recommended PPE for a Level D ensemble includes the following components:

- Coveralls
- Safety boots/shoes
- Safety glasses or chemical-splash goggles
- Hard hat

Optional PPE for a Level D ensemble includes the following components:

- Gloves
- Escape mask
- Face shield

FIGURE 4-19 The Level D ensemble is primarily a work uniform that includes coveralls and provides minimal protection.

A responder may wear liquid splash–protective clothing over or under structural firefighting clothing in some situations. This multiple-PPE approach provides limited chemical-splash and thermal protection. Those trained to the operational level can wear liquid splash–protective clothing when they are assigned to enter the initial site, protect decontamination personnel, or construct isolation barriers such as dikes, diversions, retention areas, or dams.

Respiratory Protection

NFPA 1994, *Standard on Protective Ensembles for First Responders to CBRN Terrorism Incidents*, was developed to address the performance of protective ensembles and garments (including respiratory protection) specific to weapons of mass destruction. As mentioned earlier, CBRN stands for chemical, biological, radiological, and nuclear. This standard covers three classes of garments (Classes 2, 3, and 4), which differ from the traditional levels of protection listed in the EPA regulation (Levels A, B, C, and D). The CBRN requirements were added to the performance standards of NFPA 1991, so there is no Class 1 garment in NFPA 1994. The main difference between these classifications is that NFPA 1994 covers the performance of the garments and factors in the performance requirements of the respiratory protection. This consideration is critical when it comes to WMD events because some of the chemicals may cause the components of an APR or SCBA to fail, thereby exposing the responder to a highly toxic environment. Essentially, the NFPA standard acknowledges that the entire ensemble is only as good as the individual components. Based on that criterion, NFPA 1994 requires that all components be certified to perform in a CBRN environment. NFPA 1994 typifies a thought process about PPE that should extend beyond the standard—namely, chemical-protective clothing should be thought of as a *system*.

To help clarify the levels of protection, the NFPA 1994 classes are described here:

- *Class 2:* Liquid-splash garment performance with SCBA. (Class 2 standards for PPE are in line with the CBRN requirements for SCBA.)
- *Class 3:* Liquid-splash garment performance with APR. (Class 2 standards for PPE are in line with the CBRN requirements for APR.)
- *Class 4:* Performance requirements for particles and liquid-borne viral protection.

The CBRN performance requirements for SCBA and APR were born out of the terrorist attacks that occurred on September 11, 2001. In response to the growing threat of terrorism, NIOSH set performance guidelines for SCBA and APR relative to anticipated WMD incidents. In addition to meeting the requirements of NFPA 1981, *Standard on Open Circuit Self-Contained Breathing Apparatus (SCBA) for Emergency Services*, any SCBA or APR with a NIOSH CBRN certification must have passed a battery of tests that measured their performance against sarin and sulfur mustard. From a practical standpoint, these tests were intended to ensure that the components of the respiratory protection would stand up to the aggressive nature of these chemicals. To reiterate, NFPA 1994 is intended to serve as an integrated performance guideline, factoring in both the garment and the respiratory protection used.

Physical Capability Requirements

Hazardous materials/WMD response operations put a great deal of physiological and psychological stress on responders. During the incident, personnel may be exposed to both chemical and physical hazards. They may face life-threatening emergencies, such as fire and explosions, or they may develop heat or cold stress while wearing protective clothing or working under extreme temperatures. For these reasons, every emergency response organization should have a comprehensive health and safety management program. The components of a health and safety management system for hazardous materials responders are outlined in the OSHA HAZWOPER regulation. Briefly, a health and safety program should include the following broad elements: medical surveillance including pre-employment screening and periodic medical examinations, treatment plans for acute on-scene illness and injury, thorough recordkeeping of all elements of the program, and a mechanism to periodically review the entire process.

The medical surveillance piece of the overall program is the cornerstone of an effective health and safety management system for responders. The two primary objectives of a medical surveillance program are to determine whether an individual can perform his or her assigned duties, including the use of personal protective clothing and equipment, and to detect any changes in body system functions caused by physical or chemical exposures.

As part of a medical surveillance program, responders should be examined by a physician once a year or biennially based on the physician's recommendations. During this examination, the physician may—among other things—perform a routine exam based on the expected tasks the employee may perform, including wearing PPE; conduct a health questionnaire; and take X-rays and perform a respiratory function test to measure lung capacity and function. The physician may also evaluate an individual's fitness for wearing SCBA and other respiratory protection devices. The specific requirements for any responder who is assigned to any duty where any form of respiratory protection will be used can be found in 29 CFR 1910.134, OSHA's respiratory protection standard. Additional guidance may be found in the OSHA/EPA/NIOSH/USCG (United States Coast Guard) document entitled *Occupational Safety and Health Guidance Manual for Hazardous Waste Site Activities*.

Medical monitoring and support differ in several ways from a medical surveillance program. Medical monitoring is the on-scene evaluation of response personnel who may experience adverse effects because of exposure to heat, cold, stress, or hazardous materials. Such monitoring can quickly identify problems so they can be treated in a timely fashion, thereby preventing severe adverse effects and maintaining the optimal health and safety of on-scene personnel. Medical monitoring, as a support function that takes place on the scene of an emergency, is covered in depth in the chapter, *Implementing the Planned Response*.

RESPONDER TIP

The term *CBRN certified*, when used in reference to SCBA and APR, refers to SCBA and APR that are safe to use during a chemical, biological, radiological, or nuclear incident. CBRN-certified SCBA and APR have undergone rigorous testing to ensure their integrity during such incidents.

An SCBA provides a high level of protection at a hazardous materials/WMD incident. To comply with NFPA 1981, *Standard on Open-Circuit Self-Contained Breathing Apparatus (SCBA) for Emergency Services*, positive-pressure CBRN-certified units must maintain an air flow inside the mask at all times. This is a very important feature when responders are operating in an environment characterized by airborne contamination.

The extra weight and reduced visibility are other factors to consider when choosing to wear an SCBA. As with any piece of PPE, there are as many positive benefits as there are negative points to consider when determining whether SCBA is appropriate. Any responder called upon to wear an SCBA should be fully trained by the AHJ prior to operating in a contaminated environment. All responders should follow manufacturers' recommendations for using, cleaning, filling, and servicing the units they use.

Positive-Pressure Self-Contained Breathing Apparatus

Respiratory protection, which in the fire service is commonly provided by an SCBA, is an important form of respiratory protection at a hazardous materials/WMD incident FIGURE 4-20. Use of positive-pressure SCBA prevents both inhalation and ingestion exposures (two primary routes of exposure), and should be mandatory for fire service personnel. SCBA carries its own air supply, a factor that limits the amount of air and time the user has to complete the job.

FIGURE 4-20 SCBA carries its own air supply, which limits the amount of air and time the user has to complete the job.
Courtesy of Rob Schnepp

Supplied-Air Respirators

Supplied-air respirators (SARs), also referred to as *positive-pressure air-line respirators* (with escape units), use an external air source such as a compressor or a compressed air cylinder. A hose connects the user to the air source and provides air to the face piece. SARs are useful during extended operations such as decontamination, clean-up, and remedial work.

These units are equipped with a small "escape cylinder" of compressed air. Escape cylinders typically provide the user with approximately 5 minutes of breathing air. SARs may be less bulky and weigh less than SCBA, but the length of the air hose may limit movement and there is potential for physical damage or perhaps chemical damage to the hose if it were to come into contact with a released product FIGURE 4-21.

Closed-Circuit SCBA

Some hazardous materials/WMD response teams use a form of respiratory protection referred to as a closed-circuit self-contained breathing apparatus. Also commonly referred to as a "rebreather," this type of unit can be used when long work periods are required. The basic operating principle is different from the SCBA in that exhaled air is scrubbed free of carbon dioxide, supplemented with a small amount of oxygen, and "rebreathed" by the wearer. No exhaled air is released to the outside environment, making this type of unit a closed-circuit system. The earliest rebreathers were developed in the mid-1800s and were used primarily by mine workers.

FIGURE 4-21 A supplied-air respirator is less bulky than an SCBA but is limited by the length and structural integrity of the air hose.
Courtesy of Rob Schnepp

Air-Purifying Respirators

Air-purifying respirators (APRs) are filtering devices or particulate respirators that remove particulates, vapors, and contaminants from the air before it is inhaled. They should be worn only in atmospheres where the type and quantity of the contaminants are known and where sufficient oxygen for breathing is available. APRs should not be used when the atmosphere is IDLH. APRs may be appropriate for operations involving volatile solids and for remedial clean-up and recovery operations where the type and concentration of contaminants are verifiable FIGURE 4-22.

These devices range from full-face piece, dual-cartridge masks, to half-mask, face piece–mounted cartridges with no eye protection. APRs do not have a separate source of air, but rather filter and purify ambient air before it is inhaled. The models used in environments containing hazardous gases or vapors are commonly equipped with an absorbent material that

FIGURE 4-22 Air-purifying respirators can be used only where there is sufficient oxygen in the atmosphere.
Courtesy of Rob Schnepp

soaks up or reacts with the gas. Consequently, cartridge selection is based on the expected contaminants. Particle-removing respirators use a mechanical filter to separate the contaminants from the air. Both types of devices require that the ambient atmosphere contain a minimum of 19.5 percent oxygen.

APRs are easy to wear, but they do have some drawbacks. Because filtering cartridges are specific to expected contaminants, these devices will be ineffective if the contaminant changes suddenly, possibly endangering the lives of responders. The air must also be continually monitored for both the known substance and the ambient oxygen level throughout the incident. For these reasons, APRs should not be employed at hazardous materials/WMD incidents until qualified personnel have tested the ambient atmosphere and determined that the devices can be used safely.

Powered Air-Purifying Respirators

Powered air-purifying respirators (PAPRs) are similar in function to the standard APR described earlier but include a small fan to help circulate air into the mask **FIGURE 4-23**. The fan unit is battery powered and worn around the waist. The fan draws outside air through the filters and into the mask via a low-pressure hose. PAPRs are not considered to be true positive-pressure units like an SCBA because it is possible for the wearer to "outbreathe" the flow of supplied air, thereby creating a negative pressure situation inside the mask, possibly allowing contaminants to enter the facemask due to a poor seal with the face. The main advantages of the PAPR are that it diminishes the work of breathing of the wearer, helps reduce fogging in the mask, and provides a constant flow of cool air across the face.

RESPONDER TIP

The NFPA standards provide guidance on the performance of various types of chemical-protective garments. Performance guidance for responders is provided by NFPA standards and by OSHA regulations.

FIGURE 4-23 PAPR in use during a training exercise.
Courtesy of Rob Schnepp

Decontamination

Even though there should be no intentional contact with the hazardous material involved in a hazardous materials/WMD incident, a procedure or a plan must be established to decontaminate anyone who becomes contaminated **FIGURE 4-24**. According to NFPA 472, contamination is "the process of transferring a hazardous material, or the hazardous component of a weapon of mass destruction (WMD), from its source to people, animals, the environment, or equipment, that can act as a carrier." Decontamination is "the physical and/or chemical process of reducing and preventing the spread of contaminants from people, animals, the environment, or equipment involved at hazardous materials/weapons of mass destruction (WMD) incidents."

Decontamination makes personnel, equipment, and supplies safe by reducing or, in some cases, eliminating the offending substances. Proper decontamination is essential at every hazardous materials/WMD incident so as to ensure the safety of personnel and property. Several types of decontamination are covered in detail in this text. See the chapters *Mission-Specific Competencies: Technical Decontamination* and *Mission-Specific Competencies: Mass Decontamination*. In this section, the focus is on life safety and emergency decontamination.

FIGURE 4-24 There must be a plan in place for decontamination at every hazardous materials/WMD incident.
© Bill Karrow, Dunkirk Observer/AP Photos

Emergency Decontamination

Emergency decontamination is the process of quickly reducing or removing the bulk of contaminants from a victim as rapidly as possible. This procedure is undertaken in potentially life-threatening situations without the formal establishment of a decontamination corridor, a controlled area located within the warm zone where decontamination is performed. A more formal and detailed decontamination process may follow later. Emergency decontamination usually involves removing contaminated clothing and dousing the victim with large quantities of water **FIGURE 4-25**.

If an emergency decon area has not been designated, responders should isolate the exposed victims in a contained area and establish an appropriate location. If possible, try to prevent the runoff from getting into drains, streams, or ponds;

FIGURE 4-25 Emergency decontamination involves the immediate removal of contaminated clothing.
© Jones & Bartlett Learning. Photographed by Glen E. Ellman

instead, divert the stream of water into an area where it can be treated or disposed of later. Emergency decontamination can easily be accomplished from a fire engine. Simply pull a handline, maneuver it into a circle, throw a tarp over the middle of the circle, and pull a booster line or other small hand-line to accomplish the decontamination. This technique is a quick way to handle a conscious victim of a chemical exposure or a responder who may have come into contact with a chemical and requires PPE to be decontaminated.

VIDEO

If adequate decontamination is not performed, the victim should not be allowed into the transport ambulance or the emergency room at the hospital. When possible, it is important to obtain a safety data sheet (SDS) for the substance to which the person was exposed and to make sure the information goes to the hospital with the patient.

Always ensure your own safety first before attempting decontamination of others. Avoid touching contaminated victims and/or entering contaminated environments without the proper level of protection. You will be no help in resolving the incident if you become a victim, too.

Refer back to the scenario at the beginning of the chapter. Would the victim be a candidate for emergency decontamination? If so, how could that process be accomplished at a fixed facility? Most fixed facilities that use chemicals will have emergency safety showers and eyewash stations available to the employees. These sites may be the most easily accessible locations in which responders can perform emergency decontamination. The water is clean and readily available; also, employees of such a facility have typically been trained to use the safety showers FIGURE 4-26.

The chapter, *Implementing the Planned Response*, describes the steps involved in an emergency decontamination.

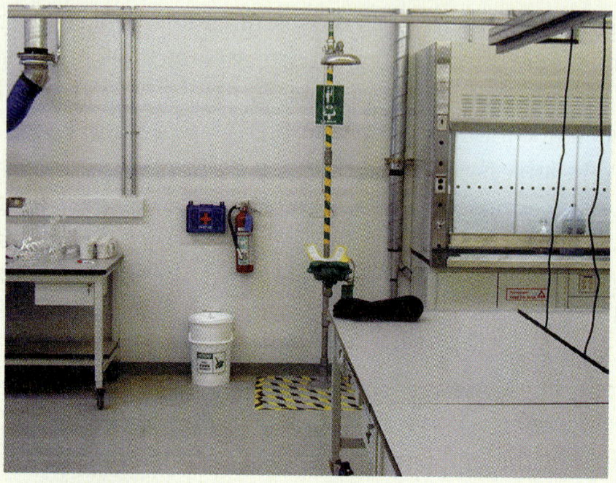

FIGURE 4-26 Indoor safety showers are in rooms that have no drains. Be prepared for a large clean-up process after activating a safety shower.
Courtesy of Rob Schnepp

RESPONDER TIP

An effective emergency decontamination operation depends on an understanding of the contaminant and its chemical and physical properties.

RESPONDER TIP

By removing the appropriate clothing of a contaminated victim, you have significantly reduced the exposure.

Secondary Contamination

Secondary contamination, also known as cross-contamination, is the process of transferring a hazardous material from its source to people, animals, the environment, or equipment, all of which may act as carriers of the contaminant. Secondary contamination occurs when a contaminated person or object comes into direct contact with another person or object. It may happen in several ways:

- A contaminated victim comes into physical contact with another person.
- A bystander or responder comes into contact with a contaminated object from the hot zone (the area immediately around and adjacent to the incident).
- A decontaminated responder reenters the decontamination area and comes into contact with a contaminated person or object.

Because secondary contamination can occur in so many ways, areas at every hazardous materials/WMD incident should be designated as hot, warm, or cold, based on safety considerations and the degree of the hazard found in the area. These areas, known as control zones, should be established, clearly marked, and enforced at hazardous materials/WMD incidents. Establishing control zones is covered in greater detail in the chapter, *Implementing the Planned Response*.

Wrap-Up

Chief Concepts

- The first priority for all responders is to ensure their own safety while operating at the scene.
- Hazardous materials/WMD incident response priorities should be based on the need to protect and/or reduce the threat to life, property, critical systems, and the environment.
- To have some frame of reference for the degree of harm a substance may pose, responders should have a basic understanding of some commonly used toxicological terms and definitions.
- Gather information from detection and monitoring devices, bystanders, reference sources, and environmental conditions to obtain a clear picture of the incident as a whole.
- Exposures can include people, property, structures, or the environment that are subject to influence, damage, or injury as a result of contact with a hazardous material/WMD.
- Immediate protective actions include isolation of the hazard area, denial of entry, evacuation, or sheltering-in-place.
- Tactical control objectives include preventing further injury and controlling or containing the spread of the hazardous material.
- Response objectives should be measurable, flexible, and time sensitive; they should also be based on the chosen strategy.
- Defensive actions include diking and damming, absorbing or adsorbing the hazardous material, stopping the flow remotely from a valve or shutoff, diluting or diverting the material, and suppressing or dispersing vapors.
- The decision to take action at a hazardous materials/WMD incident should be based on the concept of risk versus benefit.
- The type of personal protective equipment required for an incident depends on the material involved, any specific hazards, the physical state of the material, and the tasks to be performed by the operations level responder.
- The selection and use of chemical-protective clothing may have the greatest direct effect on responder health and safety.
- The EPA defines four levels of personal protective equipment: Level A (highest level of protection), Level B, Level C, and Level D (lowest level of protection).
- NFPA 1994, *Standard on Protective Ensembles for First Responders to CBRN Terrorism Incidents*, was developed to address the performance of protective ensembles specific to weapons of mass destruction. It covers three classes of garments: Class 2, Class 3, and Class 4.
- Respiratory protection is so important that it can be viewed as the defining element of personal protective equipment.
- Medical surveillance is the cornerstone of an effective health and safety management system for responders. Before any personal protective equipment is worn, hazardous materials response personnel should be aware of the physiological stress those garments create.
- Even though there should be no intentional contact with the hazardous material involved, a procedure or a plan must be established at every hazardous weapons/WMD incident to decontaminate anyone who accidentally becomes contaminated.

Hot Terms

Air-purifying respirator (APR) A device worn to filter particulates and contaminants from the air before it is inhaled. Selection of the filter cartridge for an APR is based on the expected contaminants.

CBRN Chemical, biological, radiological, and nuclear.

Chemical-resistant materials Clothing (suit fabrics) specifically designed to inhibit or resist the passage of chemicals into and through the material by the processes of penetration, permeation, or degradation.

Closed-circuit self-contained breathing apparatus Self-contained breathing apparatus designed to recycle the user's exhaled air. This system removes carbon dioxide and generates fresh oxygen.

Contamination The process of transferring a hazardous material from its source to people, animals, the environment, or equipment, all of which may act as carriers for the material.

Control zones Areas at a hazardous materials incident that are designated as hot, warm, or cold, based on their safety and the degree of hazard found there.

Decontamination The physical and/or chemical process of reducing and preventing the spread of contaminants from people, animals, the environment, or equipment involved at hazardous materials/weapons of mass destruction incidents.

Decontamination corridor A controlled area within the warm zone where decontamination takes place.

Degradation The physical destruction or decomposition of a clothing material owing to chemical exposure, general use, or ambient conditions (such as storage in sunlight). Materials can also be tested for weight changes, loss of fabric tensile strength, and other properties to measure degradation.

Denial of entry A policy under which, once the perimeter around a release site has been identified and marked out, responders limit access to all but essential personnel.

Emergency decontamination The process of removing the bulk of contaminants off a victim without regard for containment. It is used in potentially life-threatening situations, without the formal establishment of a decontamination corridor.

Evacuation The removal or relocation of those individuals who may be affected by an approaching release of a hazardous material.

High-temperature–protective equipment A type of personal protective equipment that shields the wearer during short-term exposures to high temperatures. Sometimes referred to as a *proximity suit*, this type of equipment allows the properly trained fire fighter to work in extreme fire conditions. It is not designed to protect against hazardous materials or weapons of mass destruction.

Immediately dangerous to life and health (IDLH) The atmospheric concentration of any toxic, corrosive, or asphyxiant substance such that it poses an immediate threat to life or could cause irreversible or delayed adverse health effects.

Isolation of the hazard area Steps taken to identify a perimeter around a contaminated atmosphere. Isolating an area is driven largely by the nature of the released chemicals and the environmental conditions that exist at the time of the release.

Level A ensemble Personal protective equipment that provides protection against vapors, gases, mists, and even dusts. The highest level of protection, Level A requires a totally encapsulating suit that includes a self-contained breathing apparatus.

Level B ensemble Personal protective equipment that is used when the type and atmospheric concentration of substances require a high level of respiratory protection but less skin protection. The kinds of gloves and boots worn depend on the identified chemical.

Level C ensemble Personal protective equipment that is used when the type of airborne substance is known, the concentration is measured, the criteria for using an air-purifying respirator are met, and skin and eye exposure is unlikely. A Level C ensemble consists of standard work clothing with the addition of chemical-protective clothing, chemically resistant gloves, and a form of respiratory protection.

Level D ensemble Personal protective equipment that is used when the atmosphere contains no known hazard, and work functions preclude splashes, immersion, or the potential for unexpected inhalation of or contact with hazardous levels of chemicals. A Level D ensemble is primarily a work uniform that includes coveralls and affords minimal protection.

Liquid splash–protective clothing Clothing designed to protect the wearer from chemical splashes. It does not provide total body protection from gases or vapors and should not be used for incidents involving liquids that emit vapors known to affect or be absorbed through the skin. NFPA 1992 is the performance document pertaining to liquid-splash garments and ensembles.

Penetration The flow or movement of a hazardous chemical through closures such as zippers, seams, porous materials, pinholes, or other imperfections in a material. Liquids are most likely to penetrate a material, but solids (such as asbestos) can also penetrate protective clothing materials.

Permeation The process by which a hazardous chemical moves through a given material on the molecular level. Permeation differs from penetration in that permeation occurs through the material itself rather than through openings in the material.

Permissible exposure limit (PEL) The established standard limit of exposure to a hazardous material. It is based on the maximum time-weighted concentration at which 95 percent of exposed, healthy adults suffer no adverse effects over a 40-hour workweek.

Powered air-purifying respirator (PAPR) A type of air-purifying respirator that uses a battery-powered blower to pass outside air through a filter and then to the mask via a low-pressure hose.

Recommended exposure level (REL) A value established by NIOSH that is comparable to OSHA's permissible exposure limit (PEL) and the threshold limit value/time-weighted average (TLV/TWA). The REL measures the maximum time-weighted concentration of material to which 95 percent of healthy adults can be exposed without suffering any adverse effects over a 40-hour workweek.

Secondary contamination The process by which a contaminant is carried out of the hot zone and contaminates people, animals, the environment, or equipment. Also referred to as *cross-contamination*.

Self-contained breathing apparatus (SCBA) A respirator with independent air supply used by fire fighters to enter toxic or otherwise dangerous atmospheres.

Sheltering-in-place A method of safeguarding people located near or in a hazardous area by keeping them in a safe atmosphere, usually inside structures.

Supplied-air respirator (SAR) A respirator that obtains its air through a hose from a remote source such as a compressor or storage cylinder. A hose connects the user to the air source and provides air to the face piece. SARs are useful during extended operations such as decontamination, clean-up, and remedial work. Also referred to as positive-pressure air-line respirators (with escape units).

Threshold limit value (TLV) The point at which a hazardous material or weapon of mass destruction begins to affect a person.

Threshold limit value/ceiling (TLV/C) The maximum concentration of hazardous material to which a worker should not be exposed, even for an instant.

Threshold limit value/short-term exposure limit (TLV/STEL) The maximum concentration of hazardous material to which a worker can sustain a 15-minute exposure not more than four times daily without experiencing irritation or chronic or irreversible tissue damage. There should be a minimum 1-hour rest period between any exposures to this concentration of the material. The lower the TLV/STEL value, the more toxic the substance.

Threshold limit value/skin The concentration at which direct or airborne contact with a material could result in possible and significant exposure from absorption through the skin, mucous membranes, and eyes.

Threshold limit value/time-weighted average (TLV/TWA) The airborne concentration of a material to which a worker can be exposed for 8 hours a day, 40 hours a week, and not suffer any ill effects.

Vapor-protective clothing Fully encapsulating chemical protective clothing that offers full-body protection from highly contaminated environments and requires air-supplied respiratory protection devices such as self-contained breathing apparatus. NFPA 1991 sets the performance standards for these types of garments, which are commonly referred to as Level A ensembles.

RESPONDER *in action*

It is 9:00 A.M. when your crew is dispatched to a potential chemical spill at a nearby manufacturing facility. Upon your arrival, the incident commander (IC) of the site emergency response team reports that a worker intentionally mixed sodium cyanide and sulfuric acid in a 5–gallon container inside a closed 20" × 20" storage room. The IC gives you a handwritten note from the worker, indicating that the chemical release was an apparent suicide attempt. There is a security camera monitoring the room, and the IC tells you it is clear that the worker is on the floor next to the bucket, not moving. The entire building has been evacuated. The IC urges you to make an immediate rescue. The IC hands you two SDSs—one for sodium cyanide and one for 97 percent sulfuric acid.

1. After reading the SDSs for both chemicals, you realize the mixture of these two chemicals liberates cyanide gas, which is toxic by all routes of entry. Based on this information, which of the following actions would you take?
   - **A.** Take no action that requires entering the storage room. Call for additional resources including law enforcement and a hazardous materials/WMD team.
   - **B.** Don full structural turnout gear without an SCBA and enter the storage area to attempt a rescue.
   - **C.** Instruct a crew member to break out the rear window of the storage room, and then attempt a rescue wearing full turnout gear and SCBA.
   - **D.** Turn on the ventilation system to the entire building.
2. To safely enter the storage room to assess the worker, which level of protection would you choose to wear?
   - **A.** Level A
   - **B.** Level B
   - **C.** Level C
   - **D.** Level D
3. A Level A ensemble offers full-body protection from highly contaminated environments and requires air-supplied respiratory protection such as an SCBA. Which of the NFPA standards spells out the performance requirements for these types of garments, which are more correctly referred to as a vapor-protective ensemble?
   - **A.** NFPA 1994
   - **B.** NFPA 1992
   - **C.** NFPA 1991
   - **D.** NFPA 473
4. Level B ensembles are recommended when the type and atmospheric concentration of a released substance require a high level of respiratory protection but less skin protection. Which of the following items is the defining piece of equipment for a Level B ensemble?
   - **A.** Air-purifying respirator
   - **B.** Tyvek-Saranex chemical-protective garment
   - **C.** Butyl rubber gloves
   - **D.** Self-contained breathing apparatus or supplied-air respirator

Implementing the Planned Response

Greg Henry/ShutterStock, Inc.

Courtesy of Rob Schnepp

© Photos.com

CHAPTER 5

Awareness Level Personnel

Knowledge Objectives

After studying this chapter, you will be able to:

- Describe how to notify the proper authorities and request additional resources. (NFPA 4.4.2, p. 111)
- Describe scene control procedures using control zones. (p. 112–114)
- Describe protective actions, evacuation, and sheltering-in-place. (p. 114–118)

Skills Objectives

This chapter has no skills objectives for Awareness Level personnel.

Operations Level Responders

Knowledge Objectives

After studying this chapter, you will be able to:

- Describe the procedures for requesting additional resources. (NFPA 5.4.3(6), p. 111)
- Describe scene control procedures using control zones. (NFPA 5.1.2.2(3)(a); 5.4.1(1), p. 112–114)
- Describe appropriate locations for incident command posts. (5.4.3(5), p. 123)
- Describe appropriate locations for control zones. (NFPA 5.4.1(2), p. 112–114)
- Describe effective coordinated communication techniques. (NFPA 5.4.1(6), p. 119)
- Describe evidence preservation. (NFPA 5.4.2, p. 119)
- Describe the role of the operations level responder, the incident safety officer, and a hazardous materials branch or group at a hazardous materials incident. (NFPA 5.4.3(1)(4), p. 124–127)
- Describe the role of other agencies that respond to a hazardous materials incident. (NFPA 5.4.3(7), p. 124–127)
- Describe the levels of hazardous materials incidents. (NFPA 5.4.3(2), p. 127)
- Describe the incident command system. (NFPA 5.4.3(3), p. 121–127)
- Describe the importance of the buddy system and backup personnel. (NFPA 5.4.4(1), (2), p. 115–116)
- Describe protective actions during search and rescue, evacuation, and sheltering-in-place. (NFPA 5.4.1(3), p. 114–118)
- Describe the safety precautions to be observed, including safety briefings, as well as physical capability requirements, including those for heat and cold stress, when approaching or working in a hazardous materials environment. (NFPA 5.4.1(6), 5.4.4(3)(4)(5), p. 114–121)
- Describe evaluation and communication of the status of the response. (NFPA 5.1.2.2(4), 5.5.2, 5.5.1, p. 118–119)

Skills Objectives

After studying this chapter, you will be able to:

- Establish scene control procedures. (NFPA 5.4.1, p. 112–114)
- Establish evidence preservation efforts. (NFPA 5.4.2, p. 119)
- Initiate the incident command system. (NFPA 5.4.3, p. 121–123)
- Perform tasks according to the incident action plan. (NFPA 5.1.2.2(2), p. 118–121)
- Perform emergency decontamination. (NFPA 5.1.2.2(2), 5.4.1(4), p. 114)
- Evaluate and communicate progress in accomplishing the response objectives. (NFPA 5.5.1; 5.5.2, p. 118–119)

You Are the Responder

Local law enforcement is requesting a response for a suspicious package leaking a liquid. The package is located in a small storeroom on the ground floor of a three-story office building. While en route, the dispatcher announces this update: "Engine 40, be advised that police officers on scene are reporting several people complaining of eye irritation and nausea. There are also reports of an unusual odor being reported on the second and third floors. A full building evacuation is in progress."

1. Based on the initial reports, how would you go about taking control of the scene?
2. Working with the law enforcement agency on scene is important. Which steps would you take to facilitate a coordinated approach to managing and jointly commanding this incident?
3. Does your agency have the capabilities to communicate via radio with other public safety agencies (various law enforcement agencies, for example)?

Introduction

Scene control is important at all emergencies. At a hazardous materials incident, however, scene control is paramount because it has an influence on both scene security and personnel accountability. Typically, the starting point for implementing any response is sizing-up the situation and taking control of the affected area. Size-up is the rapid mental process of evaluating the critical visual indicators of the incident (what's happening now), processing that information based on your training and experience (what might happen next), and arriving at a conclusion that will serve as the basis to form and implement a plan of action (what you are going to do about the situation). Sometimes that plan includes an aggressive offensive posture—that is, attack the problem. In other cases, a defensive posture is appropriate—that is, isolate the scene, protect exposures, and allow the incident to stabilize on its own. The posture chosen should be driven by prudent decisions based on an accurate size-up and your level of training.

RESPONDER TIP

Think before you take action or give a command to others to take action. Do not put yourself or others at undue risk or become part of the problem. Your safety is the number one priority!

Size-up is always a work in progress. In other words, as the incident progresses, everyone on the scene should be constantly sizing up his or her own piece of the problem. All responders should maintain good situational awareness (SA) and understand the actions they are taking. Remember that responder safety is the number one priority.

RESPONDER TIP

The initial actions taken set the tone for the response and are critical to the overall success of the effort.

The following lists give some examples of initial steps that awareness or operations level responders may take when implementing a response. These actions are based on the acronym SIN: Safety, Isolate, and Notify. Thinking about the basic actions represented by this mnemonic is a good starting point for a hazardous materials response, but remember—these are general guidelines, not absolute actions. Emergencies are seldom black-and-white problems; more typically, they involve many shades of gray that require you to *think*. It is always essential that you understand your job and know how to use the tools at your disposal both effectively and appropriately.

Responder Safety Tip

Situational awareness (SA) refers to the degree of accuracy to which one's perception of the current operating environment mirrors reality. In essence, SA is the act of processing all the information available to you and understanding how it fits together in the big scheme of things.

Safety

- Ensure your own safety—stay upwind, uphill, and out of the problem. Responders must take the time to really pay attention to their surroundings.
- Obtain a briefing from those involved in the incident prior to taking action. (These individuals may include bystanders, law enforcement personnel, emergency medical services [EMS] responders, facility representatives, or other responders.)
- Understand the nature of the problem and the factors influencing the release. This is really where the rubber meets the road in emergency response. If you cannot understand the problem, then you cannot formulate a proper plan to address it.
- Attempt to make a positive identification of the released substance. If possible, obtain a safety data sheet (SDS), shipping papers, or UN ID #, or some other source of reference.

Isolate

- Isolate and deny entry to the scene.
- Recognize that your first priority is to separate the people from the problem—life safety is always the first consideration. If people are involved (exposed or threatened by the release), do not move past this step until you have addressed the life safety issues. This could include removing affected people from the environment, performing decontamination, and rendering medical care.
- Establish a command post in an area where you are protected from the incident and the weather, and where you have access to communications and technical reference materials. With hazardous materials incidents, it is important to establish clear and visible command.
- Determine your response objectives, choose a strategy (offensive or defensive, for example), and begin to formulate an incident action plan. This plan must be carried out as safely as possible and should be well thought out.

Notify

- Decide whether you need to notify anyone else—for example, other specialized responders, law enforcement, or other technical experts. You may also have to notify regulatory agencies such as the local fish and game agency, the state office of emergency services, or county-level agencies such as an air quality control board. Have a current and comprehensive contact list of local, state, and federal resources available, and understand who the key players are in your authority having jurisdiction (AHJ). Large-scale incidents usually draw upon the resources of a large number of agencies.

It is important to follow a decision-making algorithm when facing a hazardous materials incident FIGURE 5-1. Use it as a loose guide for developing an action plan and to focus your thinking.

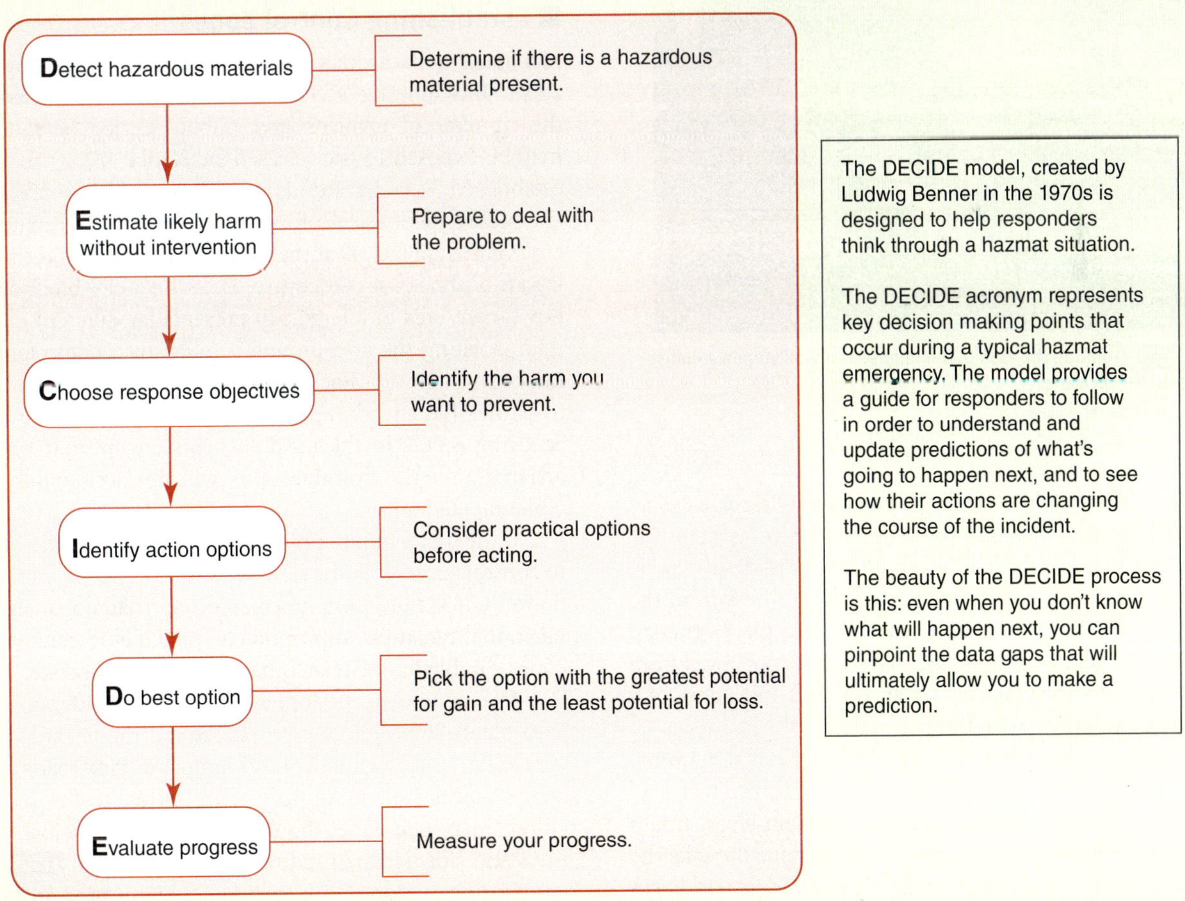

FIGURE 5-1 The DECIDE model. Reproduced with permission from Ludwig Benner.

Data from: the DECIDE model by Ludwig Benner

RESPONDER TIP

Do not let pressure from incoming units push you to make hasty decisions or assignments. Take time to think and formulate a plan of attack.

Response Safety Procedures

As mentioned, isolation of the release area is one of the first measures that awareness level personnel and operations level responders should implement **FIGURE 5-2**. For example, access to the scene can be controlled by stretching banner tape across roads to divert traffic away from a spill or to prevent civilians or other responders from entering or reentering a contaminated building or approaching a rolled-over cargo tank. Law enforcement or site security personnel can be posted in safe areas to prevent bystanders from wandering into a contaminated area. In some instances, scene control actions may be one of your biggest response challenges and could require a coordinated effort among fire, police, EMS, and other agencies. Be generous with your scene control zones, but *take care that you don't mark off more real estate than you can control!*

FIGURE 5-2 Isolating the area of the release is a vital step in gaining control of the incident.

© Jim Ruymen/Reuters/Landov

Responders should consult the *Emergency Response Guidebook (ERG)* for guidance on protective actions or as a starting point for gathering information about a released substance. Other initial actions may include evacuating others from the immediate area, implementing a sheltering-in-place strategy, rendering emergency medical care in a safe location away from the release, and surveying the scene to detect indicators of a hazardous material/WMD release.

These basic protective actions should be taken immediately, while maintaining your own safety. For the first several minutes, not even the first responders themselves should enter the area unless properly protected and trained to take the anticipated actions. During this period, responders can begin to gather necessary scene information and take initial steps to identify the materials involved (size-up). Again, these measures should be completed from a safe distance, away from potential contamination. Any actions taken should follow the predetermined local response plan or your agency's response guidelines.

Another important initial action might consist of identifying and securing any possible ignition sources—especially when the incident involves the release of flammable materials. So as not to create an unintentional ignition source, responders should use only radios and other electrical devices that have been certified as intrinsically safe **FIGURE 5-3**. With intrinsically safe electrical devices, the thermal and electrical energy within the device always remains low enough that a flammable atmosphere could not be ignited. Typically, a portable radio that is certified to be intrinsically safe must be specifically wired and used in conjunction with an intrinsically safe battery.

FIGURE 5-3 All intrinsically safe radios and batteries will be marked by the factory with a specific label denoting them as such.

Courtesy of Rob Schnepp

Establishing Control Zones

Managing a hazardous materials incident by setting control zones and limiting access to the incident site helps reduce the number of civilians and public service personnel who may be exposed to the released substance. Control zones are established at a hazardous materials incident based on the chemical and physical properties of the released material, the environmental factors at the time of the release, and the general layout of the scene. Of course, isolating a city block in a busy downtown area of a large city presents far different challenges than isolating the area around a rolled-over cargo tank on an interstate highway. Each situation is different, requiring you to be flexible and thoughtful about how you secure the area. Securing access to the incident helps ensure that responders arriving after the first-due units will not accidentally enter a contaminated area.

If the incident takes place inside a structure, the best place to control access is at the normal points of ingress and egress—doors. Once the doors are secured so that no unauthorized personnel can enter, appropriately trained emergency response crews can begin to isolate other areas as appropriate.

The same concept applies to outdoor incidents. The goal is to secure logical access points around the hazard. Begin by controlling intersections, on/off ramps, service roads, or other access points (including pedestrian routes) to the scene. Law enforcement personnel should block off streets, close intersections and sidewalks, or redirect traffic as needed **FIGURE 5-4**.

During a long-term incident, highway transportation department or public works department employees may be called upon to set up traffic barriers. Whatever methods or devices are used to restrict access, they should not limit or prevent a rapid withdrawal from the area by personnel working inside the hot zone.

FIGURE 5-4 Police officers might assist by diverting vehicle and foot traffic at a safe distance outside the hazard area.

Courtesy of Rob Schnepp

It is not uncommon to set large control zones at the onset of an incident, only to discover that the zones may have been established too liberally. At the same time, control zones should not be defined too narrowly FIGURE 5-5. As the incident commander (IC) gets more information about the specifics of the chemical or material involved, the control zones may be changed. Ideally, the control zones will be established in the right place, geographically, the first time. This may be accomplished by understanding the state of matter or the nature of the material, visual reconnaissance, or perhaps the control zones will be established based on the results of air monitoring. In all cases, you should be prepared to expand or contract control zones if necessary. Wind shifts are a common reason that control zones are modified during the incident. If there is a prevailing wind pattern or a predicted shift based on time of day in your area, factor that consideration into your decision making when it comes to control zones.

Typically, control zones at hazardous materials incidents are labeled as *hot*, *warm*, or *cold*. You may also hear other terms used, such as *exclusionary zone* (hot zone), *contamination reduction zone* (warm zone), or *outer perimeter* (cold zone). Both sets of terms are correct and common, so make sure you understand the terminology used in your jurisdiction. Be prepared to discover that different jurisdictions may use terminology and setup procedures unlike the ones used in your agency. As long as you understand the concepts behind the actions and remember that safety is the main focus, the act of setting up and naming your zones can remain flexible. Avoid confusion by performing regular and consistent training, and making sure all responders are on the same page with the terminology.

The hot zone is the area immediately surrounding the release, which is also the most contaminated area. All personnel working in the hot zone must wear complete, appropriate protective clothing and equipment. Its boundaries should be set large enough that adverse effects from the released substance will not affect people outside of the hot zone. An incident involving a gaseous substance or a vapor,

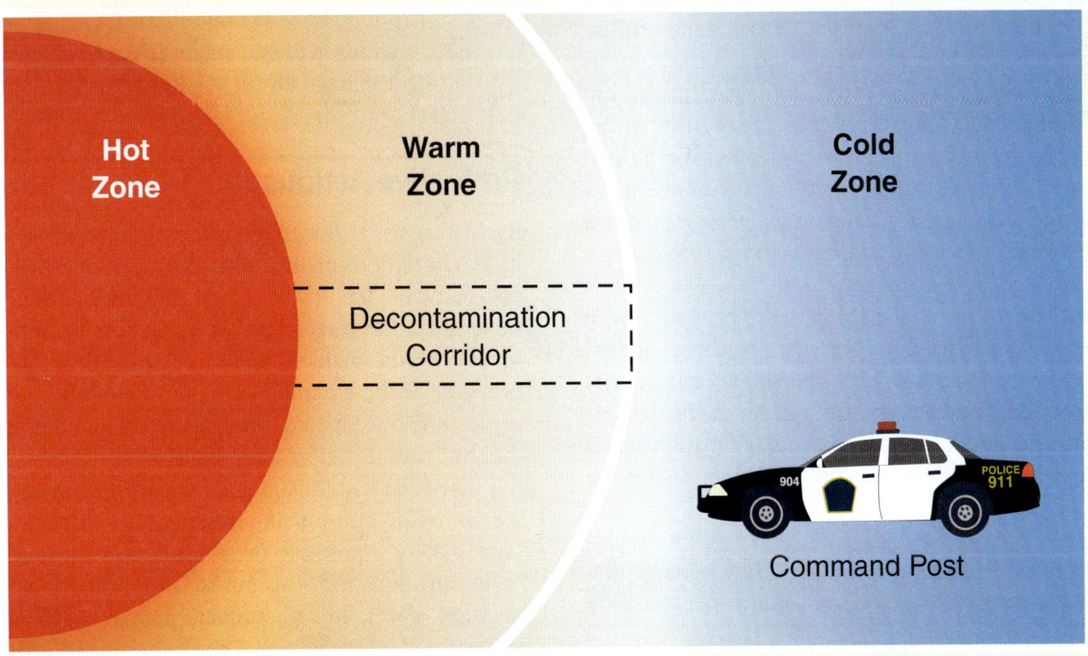

FIGURE 5-5 Control zones spread outward from the hot zone.

© Jones & Bartlett Learning

for example, may require a larger hot zone than one involving a solid or nonvolatile liquid leak. In some cases, atmospheric monitoring, plume modeling, or reference sources such as the *ERG* may prove useful in helping to establish the parameters of a hot zone. Specially trained responders should be tasked with using these tools. Keep in mind that the physical characteristics of the released substance will significantly affect the size and layout of the hot zone. Additionally, all responders entering the hot zone should avoid contact with the product to the greatest extent possible—an important goal that should be clearly understood by those entering the hot zone. Adhering to this policy makes the job of decontamination easier and reduces the risk of secondary contamination.

RESPONDER TIP

It's generally quicker, easier, and safer to *reduce* the size of the control zones than to *enlarge* them.

Personnel accountability is important, so access into the hot zone must be limited to only those persons necessary to control the incident. All personnel and equipment must be decontaminated when they leave the hot zone. This practice ensures that contamination is not inadvertently spread to "clean" areas of the scene.

The warm zone is where personnel and equipment transition into and out of the hot zone. It contains control points for access to the hot zone as well as the decontamination corridor. Only the minimal amount of personnel and equipment necessary to perform decontamination, or support those operating in the hot zone, should be permitted in the warm zone.

Generally, personnel working in the warm zone can use personal protective equipment (PPE) whose protection is rated one level lower than that of the PPE used in the hot zone. For example, if personnel in the hot zone are wearing Level A ensembles, personnel in the warm zone should dress in at least Level B protection. This recommendation is not a hard-and-fast rule, however. You must understand the hazard well enough to make that determination on a case-by-case basis.

Beyond the warm zone is the cold zone. The cold zone is a safe area where personnel do not need to wear any special protective clothing for safe operation. Personnel staging, the command post, EMS providers, and the area for medical monitoring, support, and/or treatment after decontamination are all located in the cold zone. Typically, support personnel for the incident operate in this area.

RESPONDER TIP

The primary functions of warm zone activities are decontamination and providing ready support to personnel operating in the hot zone.

Performing Emergency Decontamination

As discussed in the chapter *Estimating Potential Harm and Planning a Response*, at every hazardous materials incident there must be a plan in place to decontaminate any responder who accidently becomes contaminated. Victims who are removed from a contaminated zone must also be decontaminated by personnel who have donned the appropriate protective gear and who have been trained in the proper methods of victim decontamination. Emergency decontamination is the process of getting the bulk of contaminants off a person as quickly and completely as possible in order to limit or reduce adverse health effects. Emergency decontamination is used in potentially life-threatening situations, without the formal establishment of a decontamination corridor. To perform emergency decontamination, follow the steps in SKILL DRILL 5-1:

1. Ensure that you (the responder) have the appropriate PPE to protect against the contaminant.
2. Stay clear of the product, and avoid physical contact with it if possible.
3. Make an effort to contain runoff by directing the victim out of the hazard zone and into a suitable location for decontamination. This is a recommendation—it is not imperative to capture the runoff when you are in the process of performing emergency decontamination on an exposed person. Getting the contaminant off a person is more important than capturing the runoff.
4. Instruct the victim in removing contaminated clothing. If the victim is capable of removing his or her own clothing, this is a preferred method. This ensures the assisting responder will stay as clean as possible. (STEP 1)
5. Rinse victim with copious amounts of water. Avoid using water that is too warm or too cold; room temperature water is best. (STEP 2)
6. Provide or obtain medical treatment for the victim, and arrange for the victim's transport.

Protective Actions

Evaluating the threat to life is the number one response priority at a hazardous materials incident. The safety of the responders is always taken into account, but the bottom line is this: If there is no life threat—either immediate or anticipated—the severity of the incident is diminished. This is not to say that property or the environment is unimportant; rather, it is simply an acknowledgment that you as a responder must consider the threat to life before anything else.

Life-safety actions include ensuring your own safety and searching for, and possibly rescuing, those persons who were immediately exposed to the substance or those who may now be in harm's way. For example, an IC might be faced with a decision of whether to evacuate people ahead of an advancing vapor cloud or when an explosive device is discovered, or if some other potentially hazardous condition exists. This is not an easy decision, because efforts to relocate people bring up many complications. A risk-based method of decision

SKILL DRILL 5-1 Performing Emergency Decontamination

NFPA 5.1.2.2(2), 5.4.1(4)

VIDEO

1. Confirm that you (the responder) have the appropriate PPE to protect against the contaminant. Stay clear of the product, and avoid physical contact with it if possible. Make an effort to contain runoff by directing the victim out of the hazard zone and into a suitable location for decontamination. It is not imperative to capture the runoff when you are in the process of performing emergency decontamination on an exposed person. Removing the contaminant from the person is more important than capturing the runoff. Instruct the victim in removing contaminated clothing. If the victim is capable of removing his or her own clothing, this is a preferred method. This ensures the assisting responder will stay as clean as possible.

2. Rinse victim with copious amounts of water. Avoid using water that is too warm or too cold; room temperature water is best. Provide or obtain medical treatment for the victim, and arrange for the victim's transport.

making should be employed in such cases. Essentially, the IC should weigh the severity of the threat against the potential effort, impact, and/or resource requirements of moving people from one location to another **FIGURE 5-6**. Evacuating a predominantly ill and elderly population from a care facility, for example, may take a long time, have negative effects on the evacuees, and require many resources. Given these concerns, the IC must decide if the hazard merits such a move or if other methods might be used to protect this type of population.

FIGURE 5-6 The IC must weigh the severity of the threat against the time, personnel, and other logistical challenges required to carry out an evacuation.

The Buddy System and Backup Personnel

It is *not* an accepted practice at a hazardous materials emergency to allow only one responder to don a PPE ensemble and enter a contaminated environment alone. The risk of something going wrong is too great and operating alone should *never* be allowed. To that end, it is an accepted practice to implement the buddy system for those personnel entering contaminated areas. The simplest expression of the buddy system is for no fewer than two responders to enter a contaminated area. This practice should not be deviated from under any circumstance. The use of the buddy system is so important, in fact, that it is required by the Occupational Safety and Health Administration (OSHA) Hazardous Waste Operations and Emergency Response (HAZWOPER) regulation. It is the responsibility of the IC not only to limit the number of responders working in the contaminated area, but also to prevent responders from working alone. More than two responders might enter a contaminated area but never fewer than two. To work alone is to take an undue risk!

On a hazardous materials incident, a backup team—that is, backup personnel, typically wearing the same level of protection as the initial entry team—provides another layer of safety for those entering the contaminated areas. Like the use of the buddy system, the use of backup personnel is required by the OSHA HAZWOPER regulation. The backup team could also be called upon to remove those individuals working in the hot zone if an emergency occurs and the entry crew members are unable to escape on their own. The concept is much like the practice of establishing a rapid intervention team at a structure fire. Backup personnel must be in place, dressed in appropriate PPE (up to the point that respiratory protection can be quickly donned), and ready to spring into action whenever personnel are operating in the hot zone. *A backup team should never be more than a minute or so from being fully dressed and ready for action.*

To reiterate, no one should enter the hot zone alone; at least two team members should suit up and enter together, and at least two additional team members should suit up and stand by to assist those who entered first.

The IC should ensure that backup personnel are prepared for immediate entry into the hot zone **FIGURE 5-7**. Team members should always remain within visual or radio contact of one another and with someone outside of the hot zone.

FIGURE 5-7 All personnel must be briefed before approaching the hazard area or entering the hot zone.

Courtesy of Rob Schnepp

Evacuation

The IC must consider many factors before making decisions concerning evacuation. Some of those factors include considering the nature and duration of the release; the nature of the evacuees, such as their age, underlying health status, and mobility; the ability to support the evacuees with basic services such as food, water, and shelter; and transportation challenges. When considering evacuating a significantly large group of people, it may be helpful to consider the host of "shuns" associated with the task. The "shuns" is a word play on the ending of all of the following terms:

- Contamination (the trigger for the evacuation)
- Communication
- Transportation
- Nutrition
- Sanitation
- Habitation
- Compassion

You may identify more "shuns" that are specific to your AHJ—be creative and complete when thinking about evacuation.

Once the evacuation order is given, fire fighters and law enforcement personnel may be called upon to assist in the physical relocation of residents to a safe area. Their work may include such actions as traveling to homes and informing residents that they must relocate to a temporary shelter.

Before an evacuation order is given, a safe area with suitable facilities should be established. In many cases, schools, fairgrounds, and sports arenas are used as shelters. The evacuation area should be located close enough to the exposure for the evacuation to be practical but far enough away from the incident to be safe. Depending on the time of day and the season, it may take a considerable amount of time to evacuate even a small residential area. Temporary evacuation areas may be needed to shelter residents until evacuation sites or structures indicated in your community's emergency response plan are open and accessible. Security of the evacuation facility is an important consideration and should be accomplished with the assistance of law enforcement. Access should be monitored and controlled to ensure the maximum amount of security and shielding from the media, onlookers, or those interested in taking advantage of the evacuees in any way.

RESPONDER TIP

Evacuation has significant challenges even when the operation is properly planned. Plan carefully and consider all options before proceeding with an evacuation. It is a significant undertaking!

Transportation to temporary evacuation areas must be arranged for all populations when an evacuation is ordered. As part of this effort, the needs of the elderly, handicapped, and special needs persons should be considered. Accommodations for pets should also be considered. It would be a mistake to assume all evacuees are able and/or motivated to move on their own. Initial evacuation distances may be derived from the *ERG* **FIGURE 5-8**. Note that the *ERG* does not give complete detailed information for all conditions that responders might potentially encounter, such as weather extremes, road conditions, or actual conditions encountered at the scene.

GUIDE 161 **RADIOACTIVE MATERIALS (LOW LEVEL RADIATION)** **ERG2008**

POTENTIAL HAZARDS

HEALTH

- Radiation presents minimal risk to transport workers, emergency response personnel and the public during transportation accidents. Packaging durability increases as potential hazard of radioactive content increases.
- Very low levels of contained radioactive materials and low radiation levels outside packages result in low risks to people. Damaged packages may release measurable amounts of radioactive material, but the resulting risks are expected to be low.
- Some radioactive materials cannot be detected by commonly available instruments.
- Packages do not have RADIOACTIVE I, II, or III labels. Some may have EMPTY labels or may have the word "Radioactive" in the package marking.

FIRE OR EXPLOSION

- Some of these materials may burn, but most do not ignite readily.
- Many have cardboard outer packaging; content (physically large or small) can be of many different physical forms.
- Radioactivity does not change flammability or other properties of materials.

PUBLIC SAFETY

- **CALL Emergency Response Telephone Number on Shipping Paper first. If Shipping Paper not available or no answer, refer to appropriate telephone number listed on the inside back cover.**
- **Priorities for rescue, life-saving, first aid, fire control and other hazards are higher than the priority for measuring radiation levels.**
- Radiation Authority must be notified of accident conditions. Radiation Authority is usually responsible for decisions about radiological consequences and closure of emergencies.
- As an immediate precautionary measure, isolate spill or leak area for at least 25 meters (75 feet) in all directions.
- Stay upwind.
- Keep unauthorized personnel away.
- Detain or isolate uninjured persons or equipment suspected to be contaminated; delay decontamination and cleanup until instructions are received from Radiation Authority.

PROTECTIVE CLOTHING

- Positive pressure self-contained breathing apparatus (SCBA) and structural firefighters' protective clothing will provide adequate protection.

EVACUATION

Large Spill

- Consider initial downwind evacuation for at least 100 meters (330 feet).

Fire

- When a large quantity of this material is involved in a major fire, consider an initial evacuation distance of 300 meters (1000 feet) in all directions.

Page 268

FIGURE 5-8 Evacuation distances for small spills or fires are listed in the orange-bordered pages of the ERG.

Courtesy of the U.S. Department of Transportation

In severe weather, evacuation can be challenging at best. Flooded areas, heavy rains, and winds present a danger to the evacuees as well as responders. Even in perfect circumstances, evacuation of residents will be a process that is measured in hours, not minutes. Preplanning for evacuations is key to a safe and successful emergency evacuation.

A good way for the IC to determine the area to be evacuated, including how far to extend actual evacuation distances, is to use detection and monitoring devices to identify areas of airborne contamination. These tools are discussed in the chapter *Mission-Specific Competencies: Air Monitoring and Sampling*. Access to local weather reporting conditions is invaluable in this decision-making process.

Responder Safety Tip

When carrying out an evacuation, make provisions for protecting yourself in case the wind shifts or you unexpectedly come in contact with the hazardous material.

Sheltering-in-Place

As discussed in the chapter *Estimating Potential Harm and Planning a Response*, sheltering-in-place is a method of safeguarding people in a hazardous area by keeping them in an enclosed atmosphere, usually inside structures. Local emergency plans should identify local facilities where vulnerable populations might be found, such as schools, churches, hospitals, nursing homes, and apartment complexes. When residents are sheltered-in-place, they remain indoors with windows and doors closed. All ventilation systems are turned off to prevent outside air from being drawn into the structure.

The toxicity of the hazardous material and the amount of time available to avoid the oncoming threat are major factors in the decision of whether to evacuate or to use a sheltering-in-place strategy. The expected duration of the incident is also a factor in determining whether sheltering-in-place is a viable option. The longer the release is expected to continue, the more time the material has to enter or permeate into protected areas. Short-term events or transient vapor clouds might dictate a sheltering-in-place approach. The IC or Unified Command should conduct plume-modeling activities or use other reliable expertise or methods to determine whether evacuation or sheltering-in-place makes the most sense. In either case, it is generally a time-sensitive decision.

Search and Rescue

Ensuring your safety is the first priority in any emergency response situation, and a hazardous materials incident is no exception. The search for and rescue of people can be more complicated in the setting of a hazardous materials incident. In a structure fire, it is understood that smoke and flame are serious threats to life, and firefighters are properly protected to conduct reasonable operations to perform search and rescue. In a hazardous materials incident, all emergency response personnel (fire, law enforcement, emergency medical services) must first recognize and identify the released substance. This may take some time and is not as straightforward as a typical structure fire. After this is accomplished, the IC should use a risk-based thought process to understand the hazards as they relate to the possibilities of victim survivability and responder safety. Only then should personnel consider search and rescue.

Ultimately, the IC must determine whether entry into a contaminated environment to perform search and rescue is a worthy endeavor. If the released substance is highly toxic, unprotected victims may not have survived the initial exposure. If it is determined that a search can be safely performed, rescue teams wearing proper PPE may then enter the hot zone to look for or retrieve victims. The victims should be removed to the warm zone, where they can be decontaminated and turned over to EMS providers for transport to a medical facility. In nearly all cases, definitive medical care should be provided to victims that have been decontaminated and removed to the cold zone. The chapter *Mission-Specific Competencies: Product Control* offers more information on this topic.

RESPONDER TIP

A written safety plan may be suspended if a rescue is necessary. A plan is still required, but it may take the form of short verbal plan and a rapid safety briefing.

Safety Briefings

Typically, before significant actions are taken at a hazardous materials incident, the IC should ensure that a *written* site safety plan is completed and a *verbal* safety briefing is performed. The purpose of the safety briefing is to inform (at a minimum) all responders of the health hazards that are known or anticipated, the incident objectives, emergency medical procedures, radio frequencies and emergency signals, a description of the site, and the PPE to be worn. Each AHJ should develop templates for site safety plans and verbal safety briefings. The use of a standardized form means that the safety plan can be completed in a timely manner; standardized formats also keep safety briefings on track and provide a logical format that everyone can follow.

There are many ways to conduct a safety briefing. Chiefly, a good safety briefing should be carried out in the manner that its name implies—it should be brief. It should be complete enough to provide the responders on the scene with relevant information, yet not overly detailed with useless or "nice to know" information.

At small incidents where the incident management structure is simple, the IC may be responsible for both putting together a site safety plan and conducting the safety briefing. At larger incidents, where an incident safety officer (also referred to as a safety officer) or hazardous materials safety officer is appointed, it is the responsibility of that officer to participate in the preparation and implementation of the site safety plan. The roles of these officers are discussed in more detail later in this chapter. The depth and scope of the safety briefing should have a direct correlation with the severity and complexity of the incident. For significant incidents, where the released substances are highly toxic or where other significant health hazards are present, the site safety plan and briefing may be comprehensive and documented in writing.

The IC may establish predetermined trigger points, intended to evaluate the status of the planned response, which may lead to withdrawal of responders from the hot zone. Based on a lack of progress toward meeting the incident objectives, the IC may later decide to abandon the current plan of action and withdraw to a safe distance, set a defensive perimeter, and wait for additional resources to arrive—or to allow the hazardous materials incident to run its course. Typically, these decisions are made when the offensive actions are not effective in mitigating the problem, the selected PPE is found to be incorrect or ineffective, the tools and equipment required to solve the problem are ineffective or unavailable, or the problem is simply too complex to handle with the available resources. A reasonable and prudent IC knows when he or she is "outgunned" and does not make a foolhardy decision to risk the health and safety of the personnel. In the event a withdrawal is necessary, the IC should include evacuation signals in the pre-entry briefing.

It is also important, within the context of a safety briefing, to discuss incident communications. *It is common for communication to be disrupted or otherwise hampered during an emergency.* To avoid undue communications complications, the IC may designate command and tactical channels for the incident. Entry teams may be instructed to communicate on a dedicated radio channel to ensure they are not cut off or excluded from making a critical radio transmission by other radio chatter. Radio transmissions are an excellent way to communicate the status of the mitigation efforts and to gauge the success or failures of those efforts **FIGURE 5-9**.

FIGURE 5-9 It is important to use the communication tools provided to you. Pay attention and stay on the right channel!

In the event that things are not going as planned, the entry team may contact the entry team leader, who may have a face-to-face conversation with the safety officer or hazardous materials safety officer (assistant safety officer). That conversation, carried out through the normal chain of command, may result in a change of tactics, modification of the incident objectives, or a complete abandonment of the effort. In any case, by observing the proper chain of command, all parties with management and safety interests at the scene will be included in the appropriate discussions. This approach will also help to ensure coordinated communication to the public.

In some jurisdictions, the safety officer conducts the majority of the safety briefing; in others, the IC or hazardous materials group supervisor conducts the safety briefing in conjunction with the safety officer. Regardless of how it's accomplished, a safety plan should be completed (and approved by the IC), and a safety briefing should be conducted before responders attempt to enter the incident site. Make sure that you are familiar with and follow the standard operating procedures in your jurisdiction.

In today's world, the threat of terrorism or other incidents with criminal intent drives the need for complete and meaningful safety briefings. Briefings at these types of incidents may also include procedures for operating at a crime scene or evidence collection procedures. It is imperative to understand the needs of law enforcement, especially as those needs relate to beginning an investigation or collecting evidence for a possible prosecution. Operating at a crime scene can present a complicated set of circumstances, but a good working relationship between all agencies on the scene (ideally, established before the event) will create better operational efficiency during the response. Refer to the standard operating procedures and policies established by your agency to fully understand your role during a hazardous materials incident with a criminal intent.

A written safety plan should also include information about the heat or cold stress the responders may encounter while engaging in their work. Working in PPE in ideal conditions is challenging enough on its own. When temperature extremes are anticipated, their effects on the responders should be acknowledged and understood by everyone operating at the scene.

Excessive-Heat Disorders

Hazardous materials responders operating in protective clothing should be aware of the signs and symptoms of heat exhaustion, heat stress, dehydration, and heat stroke. If the body is unable to disperse heat because an ensemble of PPE covers it, serious short- and long-term medical issues could result.

Heat exhaustion is a mild form of shock that arises when the circulatory system begins to fail because the body is unable to dissipate excessive heat and becomes overheated. With heat exhaustion, the body's core temperature rises, followed by weakness and sweating. A person suffering from heat exhaustion may become dizzy and have episodes of blurred vision. Other signs and symptoms of heat exhaustion include acute fatigue, headache, and muscle cramps; however, signs and symptoms may vary among individuals. Any individual experiencing heat exhaustion should be removed at once from the heated environment, rehydrated with electrolyte solutions (perhaps by intravenous methods), and kept cool. If not properly treated, heat exhaustion may progress to a potentially fatal condition.

When treating heat-related illnesses, it is important to avoid pouring cold water or otherwise placing the victim in an unusually cold environment. The extreme swing in temperature may have adverse effects on the individual's recovery. Using tepid water for drinking and cooling the skin is the safest approach to take. Keep in mind that rehydration by mouth is much slower than rehydration by intravenous line.

Responder Safety Tip

Most heat-related illnesses are typically preceded by dehydration. It is important to stay hydrated so that you can function at your maximum capacity. As a frame of reference, athletes should consume approximately 500 mL of fluid (water) prior to an event and 200–300 mL at regular intervals. Responders are considered to be occupational athletes—so keep up on your fluids!

Heat stroke is a severe and potentially fatal condition resulting from the failure of the temperature-regulating capacity of the body. It is caused by exposure to the sun or working in high temperatures. Reduction or cessation of sweating is an early symptom. The body temperature can rise to 105°F (40.5°C) or higher and be accompanied by a rapid pulse; hot, red-looking skin; headache; confusion; unconsciousness; and possibly seizures. Heat stroke is a true medical emergency that requires immediate transport to a medical facility.

In an effort to combat heat stress while their personnel are wearing PPE, many response agencies employ some form of cooling technology under the garment. These technologies include, but are not limited to, air, ice, and water-cooled vests, along with phase-change cooling technology. Many studies have been conducted on each form of cooling technology. Each is designed to accomplish the same goal: to reduce the effects of heat stress on the human body. The chapter *Mission-Specific Competencies: Personal Protective Equipment* provides more information on the specific technologies used for cooling.

Cold-Temperature Exposures

Responders at hazardous materials incidents may be exposed to two types of cold temperatures: those caused by the released materials and those caused by the operating environment, including ambient air temperatures or conditions such as rain, snow, or other adverse cold-weather conditions. Hazardous materials such as liquefied gases and cryogenic liquids may expose responders to the same low-temperature hazards as those created by cold-weather environments. Exposure to severe cold for even a short period of time may cause severe injury to body surfaces, especially to the ears, nose, hands, and feet.

Two environmental factors in particular influence the extent of cold injuries: temperature and wind speed. Because still air is a poor heat conductor, responders working in low temperatures with little wind can endure these conditions for longer periods (assuming that their clothing remains dry). However, when low temperatures are combined with significant winds, wind chill occurs. As an example, if the temperature with no wind is 20°F (7°C), it will feel like –5°F (–21°C) when the wind speed is 15 miles per hour.

Regardless of the temperature, anyone will perspire while wearing chemical protective clothing. Wet clothing extracts heat from the body as much as 240 times faster than dry clothing. It may also lead to hypothermia, a condition in which the core body temperature falls below 95°F (35°C). Hypothermia is a true medical emergency.

Responders must be aware of the dangers of frostbite, hypothermia, and impaired ability to work when temperatures are low or when they are working in wet clothing. All personnel should wear layered clothing and be able to warm themselves in heated shelters or vehicles.

The layer of clothing next to the skin, especially the socks, should be kept dry. The combination of cold and wet softens the skin, causing numbness, tingling, and, in some cases, peeling skin.

Responders should carefully schedule their work and rest periods and monitor their physical working conditions. For example, warm or cool shelters should be available where responders may don and doff their protective clothing.

Personal Protective Equipment: Physical Capability Requirements

Medical monitoring is a process of pre-entry and postentry on-scene evaluation of response personnel who may be experiencing adverse effects because they are wearing PPE or because they have been exposed to heat, cold, stress, or hazardous materials. The goal of medical monitoring is to quickly identify medical problems—while responders are on-scene—so they can be treated in a timely fashion, thereby preventing severe adverse effects and maintaining the optimal health and safety of on-scene personnel.

Responders should undergo pre-entry health screening prior to donning any level of chemical protective equipment. (This step may be eliminated in the event that rapid entry is required for rescue or other life-safety reasons.) This screening should include a check of vital signs, including pulse rate, blood pressure, and respiratory rate. Body weight measurements and general health should also be observed. In the event the responder presents with an abnormal reading (determined by the AHJ), that responder may be excluded from wearing PPE or participating in the mitigation phase of the incident.

Once the mission is complete and the responder has been decontaminated and doffed the PPE, a second medical evaluation—similar to the initial evaluation—should be completed. In the event of abnormal findings (again determined by the AHJ), the responder may need to be seen by a physician or transported to an appropriate receiving hospital.

For a more comprehensive set of criteria for medical monitoring/support, consult NFPA 473, *Standard for Competencies for EMS Personnel Responding to Hazardous Materials/Weapons of Mass Destruction Incidents*.

The Incident Command System

In the 1970s, a series of devastating wildland fires occurred in Southern California, requiring the services of numerous local and state resources. That rash of fires illustrated the need for a better way to organize and manage large numbers of agencies and resources called to a major incident. Also as a result of those fires, many of the agencies involved agreed to form a working group, subsequently named FIRESCOPE (Fire Resources of Southern California Organized for Potential Emergencies). The FIRESCOPE group began its work by identifying several problem areas common to almost all major or complex incidents: ineffective communications, span of control challenges, a lack of a common command structure, the inability to track personnel working on the scene, and the inability to effectively coordinate on-scene resources.

RESPONDER TIP

Span of control refers to how many responders can be effectively managed by one supervisor. Typically, an acceptable ratio for emergency operations is one supervisor to five responders. This span of control is flexible and depends on issues such as the criticality of the mission or task being performed, the skill level of the supervisor, and the skill levels of the responders.

Subsequently, the FIRESCOPE consortium developed a standardized yet flexible management system called the incident command system (ICS). Some of the key benefits of using the ICS are summarized here:

- Common terminology
- Consistent organizational structure
- Consistent position titles
- Common incident facilities

The ICS, which was originally designed for managing wildland fires, subsequently evolved into an all-risk management structure suitable for managing resources at all types of fires, natural disasters, technical rescues, and any other type of emergency of virtually any size. When properly used, it provides a strong organizational framework on which to lay the operational goals and objectives. The OSHA HAZWOPER regulation requires that the ICS be implemented in response to all hazardous materials incidents. Now more than ever, it's critical that all responders—regardless of the patch on their uniform or the nature of their job, or even their geographic location—understand the need for this cohesive emergency management tool.

The ICS can be expanded to handle an incident of any size and complexity. Hazardous materials incidents can be extremely complex, such that local, state, and federal responders and agencies may all become involved in many cases of long duration. The basic ICS consists of five functions: command, operations, planning, logistics, and finance/administration FIGURE 5-10.

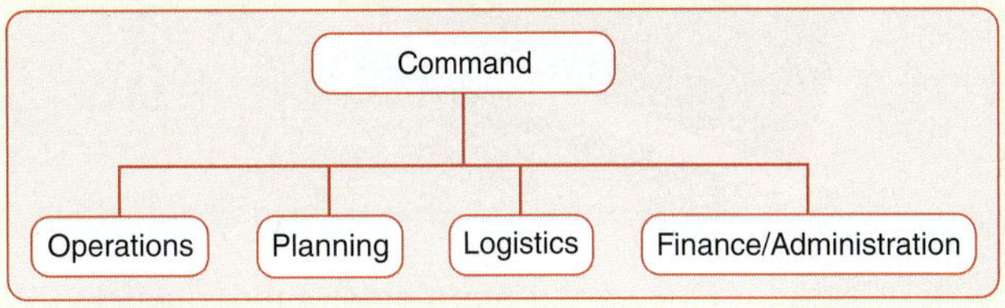

FIGURE 5-10 General staff functions of the ICS.

RESPONDER TIP

The OSHA HAZWOPER regulations mandate the use of the incident command system at hazardous materials incidents.

Command

Command is established when the first unit arrives on the scene and is maintained until the last unit leaves the scene. The concept of incident command is the first main principle of emergency response and the cornerstone of the ICS. Without leadership and a process for organizing and directing personnel and resources, the effort to solve a problem may end up being as chaotic as the problem itself. A certain chain of command must exist from the top to the bottom of the incident organizational structure. Everyone either reports to or is responsible for someone else. Command is directly responsible for at least the following tasks:

- Determining strategy
- Selecting incident tactics
- Creating the action plan
- Developing the ICS organization
- Managing resources
- Coordinating resource activities
- Providing for scene safety
- Releasing information about the incident
- Coordinating with outside agencies

Unified Command

When multiple agencies with overlapping jurisdictions or legal responsibilities are involved in the same incident, a unified command provides several advantages. Using this approach, representatives from various agencies cooperate to share command authority FIGURE 5-11. Oil spills on navigable waterways;

VOICES OF EXPERIENCE

One spring morning at approximately 0900 hours, we responded, along with multiple other agencies, to a neighboring department for an overturned truck tanker that failed to negotiate a turn in the road. At the time of our response, it was unknown what type of chemical we were dealing with.

After I checked in with the incident commander, he assigned me as incident safety officer. Once we were in place, all operational individuals were briefed on the suspected chemical and its properties, along with the response plan. Our first responsibility was to verify the identity of the chemical involved. Next we established the hot, warm, and cold zones. A decontamination corridor was also established, with constant air monitoring.

An entry team, with backup teams in place, identified the damage to the truck tanker and the approximate amount of the chemical spilled. They also conducted diking and absorption activities in containing the spilled chemical.

Through the postbriefing of the entry team, it was determined that the safest and most efficient means to mitigate the situation would be to transfer the chemical to another truck tanker. The incident commander, along with the operations section chief, developed an incident action plan, and the incident was mitigated exactly as planned. Although this incident lasted approximately 15 hours and involved many agencies including private cleanup contractors, we were able to ensure life safety by establishing control zones, delegate tasks through use of the incident command system, and take protective actions such as conducting a safety briefing and utilizing backup teams. Remember to utilize all of your resources when dealing with a potential hazardous materials release.

Jeffry J. Harran
Lake Havasu City Fire Department
Lake Havasu City, Arizona

wildfires that threaten or occur in a combination of state, federal, and local lands; hazardous materials incidents on public roadways or at fixed facilities; or perhaps active shooter incidents represent ideal situations for the establishment of a unified command.

FIGURE 5-11 A unified command involves many agencies directly involved in the decision-making process for a large incident.

Courtesy of Rob Schnepp

Incident Command Post

Regardless of whether there is a single incident command or the incident is run under a unified command, the command function is always located in an incident command post (ICP). The ICP is where the incident commander is located and where coordination, control, and communications are centralized. Command and all direct support staff should be located at the ICP. Ideally, the ICP should be located in an area where it is not threatened by the incident and has the necessary infrastructure (e.g., communications, technology support, bathrooms, meeting space) to support sustained operations if required.

During a hazardous materials incident, the ICP should be established uphill and upwind of the incident, keeping in mind the potential for predicted changes in wind direction based on the time of day.

Responder Safety Tip

It is important to consider the prevailing winds of the area when establishing the location of the incident command post.

Command Staff

The incident commander (IC) is the person in charge of the entire incident and should be qualified to be in the position. The OSHA HAZWOPER regulation [1910.120(q)(6)(v)] states that an on-scene IC, who will assume control of the incident scene beyond the first-responder awareness level, should receive at least 24 hours of training equal to the first-responder operations level and, in addition, have competency in the following areas:

- Know and be able to implement the jurisdiction's incident command system
- Know how to implement the jurisdiction's emergency response plan
- Know and understand the hazards and risks associated with responders working in chemical protective clothing
- Know how to implement the local emergency response plan
- Know about the state emergency response plan and the Federal Regional Response Team
- Know and understand the importance of decontamination procedures

The command staff consists of the safety officer, the liaison officer, and the public information officer FIGURE 5-12. OSHA requires the IC and safety officer positions during a hazardous materials response. These job functions report directly to the IC and are critical to the effective management of a hazardous materials/WMD incident.

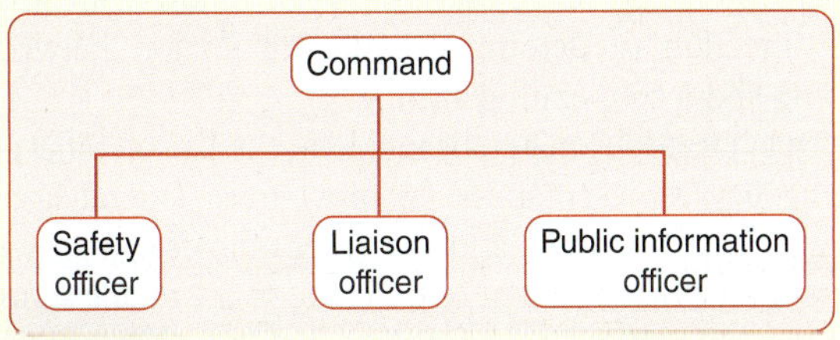

FIGURE 5-12 The command staff members report directly to the incident commander.

© Jones & Bartlett Learning.

Safety Officer

According to the OSHA HAZWOPER regulation [1910.120 (q)(3)(vii)], the role of the safety officer at a hazardous materials incident is clear:

> The individual in charge of the ICS shall designate a safety officer, who is knowledgeable in the operations being implemented at the emergency response site, with specific responsibility to identify and evaluate hazards and to provide direction with respect to the safety of operations for the emergency at hand.

The OSHA HAZWOPER regulation, in 1910.120 (q)(3)(viii), goes on to describe the authority of the safety officer at a hazardous materials incident:

> When activities are judged by the safety officer to be an IDLH [immediate danger to life and health] and/or to involve an imminent danger condition, the safety officer shall have the authority to alter, suspend, or terminate those activities. The safety official shall immediately inform the individual in charge of

the ICS of any actions that need to be taken to correct these hazards at the emergency scene.

A hazardous materials safety officer, sometimes referred to as the assistant safety officer (ASO), is responsible for the hazardous materials team's safety only. When a hazardous materials branch or hazardous materials group has been established, or when the incident requires a dedicated hazardous materials response, the safety officer may appoint a hazardous materials safety officer to the hazardous materials branch or group. This particular hazardous materials safety officer reports directly to the safety officer, who in turn reports directly to the IC.

Liaison Officer

The liaison officer is the point of contact for cooperating and assisting agencies on the scene. The basic distinction between a cooperating agency and an assisting agency relates to the financial stake each has in the management or outcome of the incident. Typically, if an agency has a financial interest/responsibility for the incident, it is considered an assisting agency; all others are considered cooperating agencies. There is much gray area here, however, and you must follow your jurisdiction's direction of determining the distinction between an assisting and a cooperating agency.

On a hazardous materials incident, the liaison officer deals with agency representatives (A reps) from federal agencies, state and local resources, and any other outside agency with an interest in the management or outcome of the incident. Ideally, all of these representatives should have the authority to make decisions on behalf of the agency they represent. It is extraordinarily cumbersome to deal with an agency representative who must constantly go back to a key decision maker within his or her own organization to get approvals. This function is critical when the interests of many jurisdictions are at stake. Liaison officers can find themselves quite busy dealing with questions and providing information to agency representatives during an emergency.

Public Information Officer

The public information officer (PIO) typically functions as a point of contact for the media or any other entity seeking information about the incident FIGURE 5-13. As with all of the other command staff positions, only one person should fill this role. Although many people may work with or for the public information officer, they should serve as assistants. This chain of command is necessary to streamline communications and reduce redundancies in the management system.

The value of the PIO should not be underestimated. Keeping the media (and others) well informed is an important piece of successfully managing almost any type of incident. A wise IC will acknowledge that fact and incorporate good media relations into the incident objectives.

General Staff Functions

When the incident is too large or too complex for just one person (the incident commander) to manage effectively, the

FIGURE 5-13 The public information officer is responsible for gathering and releasing incident information to the media and other appropriate agencies.

Courtesy of Captain David Jackson, Saginaw Township Fire Department

IC may assign other individuals to oversee parts of the incident. Everything that occurs at an emergency incident can be divided among the four major functional components within ICS:

- Operations
- Planning
- Logistics
- Finance/Administration

Operations

The operations section, which is typically led by an operations section chief on larger incidents, carries out the objectives developed by the IC and is responsible for all tactical operations at the incident. The operations section chief directs and manages the resources assigned to his or her section. These resources could include fire units of any type, law enforcement resources, emergency medical units, airborne resources such as helicopters and fixed-wing aircraft, and hazardous materials response resources. This position is usually assigned when complex incidents involve more than 20 single resources or when the IC cannot be involved in all details of the tactical operation. When incident operations require many responders, the operations section can be further divided into groups and divisions.

Groups and Divisions

Organizational units such as groups and divisions are established to aggregate single resources or crews under one supervisor. The primary reason for establishing groups and divisions is to maintain an effective span of control.

A hazardous materials group, led by a hazardous materials group supervisor, is often established when companies and crews are working on the same task or objective, albeit not necessarily in the same location. The term "group" is very specific as it applies to the ICS: A *group* is assembled to relieve span of control issues and is considered to consist of functional assignments that may not be tied to any one geographic location. A division usually refers to companies and crews that are working in the same geographic location. Groups and divisions place several single resources under one supervisor, effectively reducing the IC's span of control. FIGURE 5-14 provides an overview of how divisions and groups are integrated into the overall command structure. The names and functions can be tailored to the specific needs of the incident. The important thing to ensure is that divisions and groups understand their functions and where they fit into the ICS org chart.

Hazardous Materials Branches

If necessary during a hazardous materials incident, a special technical group may be developed under the operations section, known as the hazardous materials branch. The hazardous materials branch consists of some or all of the following positions as needed for the safe control of the incident:

- A hazardous materials group supervisor
- An entry team assigned to enter the designated hot zone
- A decontamination team responsible for reducing and preventing the spread of contaminants from persons and equipment
- A technical reference team that gathers information and reports to the IC and the hazardous materials safety officer

A branch represents a higher level of combined resources than either a division or a group. At a major incident, several different activities may occur in separate geographic locations or involve distinct functions. Span of control might still present a problem, even after the establishment of divisions and groups. In these situations, the IC can establish branches to place a higher-level supervisor (a hazardous materials branch director) in charge of a number of divisions and groups. One incident may require several branches, such as the medical branch, fire branch, or law branch. Keep in mind that each branch would be broken down in a discipline-specific way, based on span of control.

The hazardous materials branch director reports to the operations section chief or directly to the IC if the operations section chief position is not assigned FIGURE 5-15. Responsibilities of the hazardous materials branch director include obtaining a briefing from the IC, staffing the hazardous materials branch functions as required, ensuring that scene control zones are established, ensuring that a site safety plan is developed, maintaining accountability, ensuring that proper PPE is worn, and ensuring that the operational objectives are being met. A hazardous materials group supervisor may be appointed under a hazardous materials branch director to direct and manage positions such as decontamination group supervisor or team leader, entry group supervisor or team leader, or other functions specific to the hazardous materials portions of the

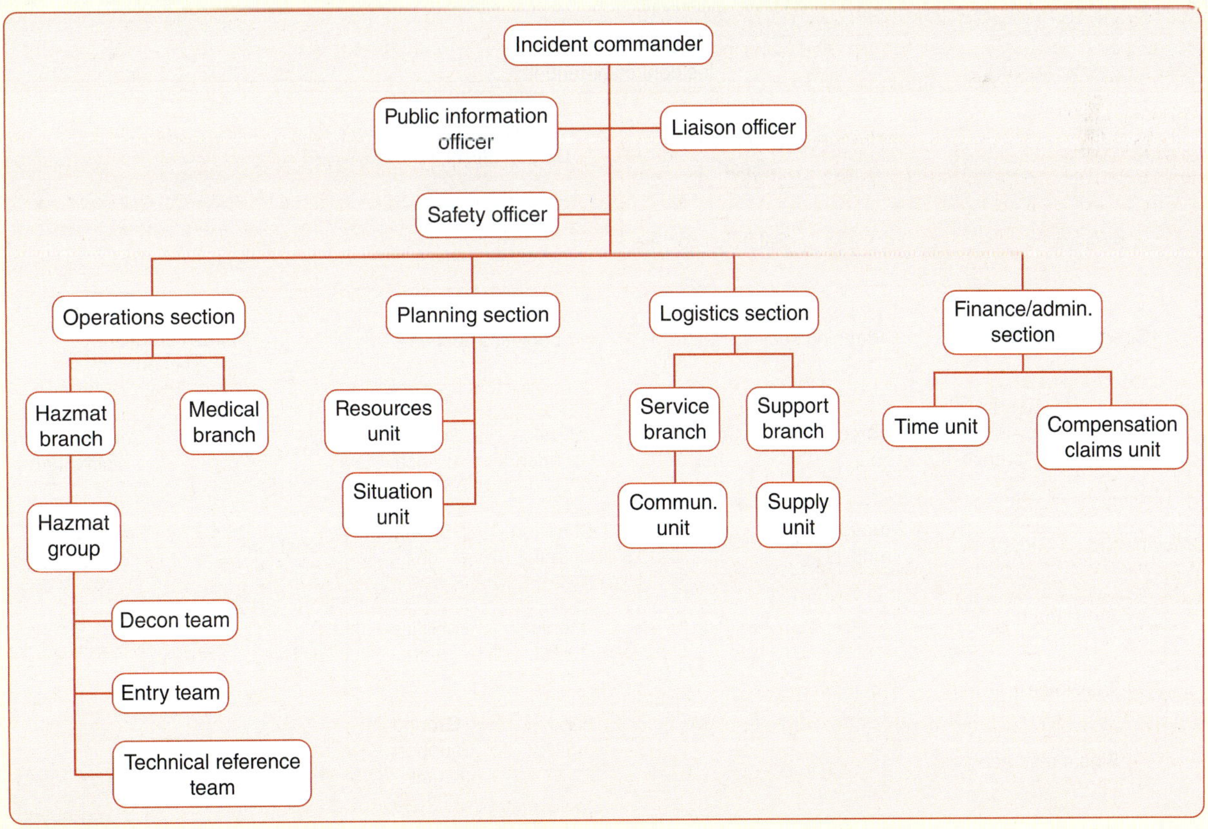

FIGURE 5-14 An overview of the hazardous materials branch.

Modified from: FEMA

response. The management structure and the ICS positions that are ultimately assigned are based on the needs associated with the specific incident.

In some types of incidents, such as a wildland fire or a multistory structure fire, the safety officer may assign an assistant to each division working the fire. (Remember, the command system is intended to be flexible.) Responsibilities of the hazardous materials safety officer include obtaining a briefing from the IC, the safety officer, and/or the hazardous materials branch director; participating in the preparation of the site safety plan; providing or participating in the safety briefing; altering or suspending any activity that poses an imminent threat to the responders; and maintaining accountability for all resources assigned to the hazardous materials mitigation efforts.

RESPONDER TIP

There is only one safety officer (or incident safety officer) for the entire incident, but that position may have many assistants.

Planning

The planning section is responsible for the collection, evaluation, dissemination, and use of information relevant to the incident. The planning section, which is led by a planning section chief, acts as the central point for collecting information on the situation status (sit-stat) of the event, tracking and logging on-scene resources, and disseminating the written incident action plan. On large-scale incidents, the planning section chief facilitates the incident briefings and planning meetings.

Logistics

The logistics section can be viewed as the support side of an incident management structure. This section within the ICS is responsible for providing facilities, services, and materials for the incident. Logistics is headed by a logistics section chief, who is responsible for providing the bulk of the support functions for an incident. This position is generally assigned on long-duration or resource-intensive, complex incidents.

Logistical support includes food, sleeping facilities, transportation needs, sanitation facilities, showers, and all other requests for resources needed to manage the incident. If someone on the incident needs a truckload of sand, for example, the logistics section is the group that makes it happen. When resources are needed, the requesting person follows the appropriate chain of command to route the request to logistics. It is then up to logistics personnel to determine whether they will be able to fill the request. Typically, during a planning meeting at a large incident, the operations section chief, planning

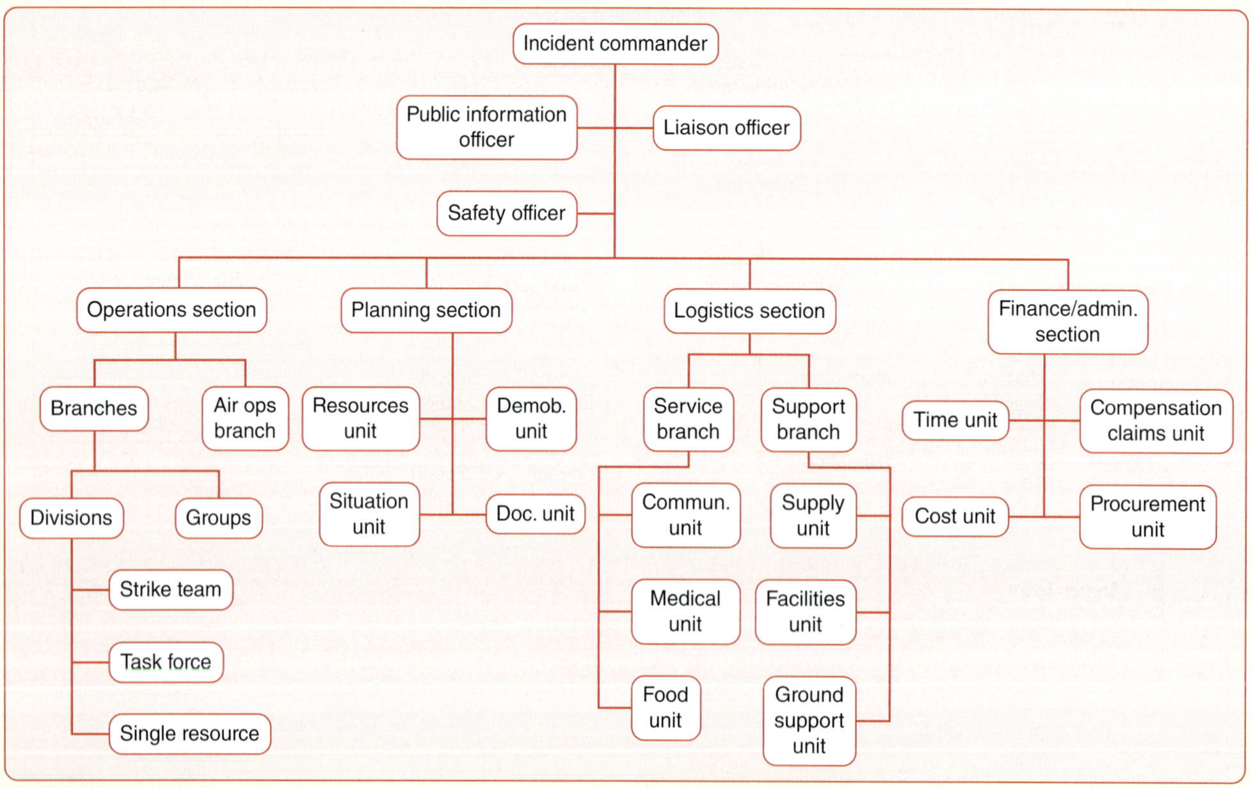

FIGURE 5-15 An example of how to organize a response by function and maintain span of control.

Modified from: FEMA

section chief, safety officer, IC, and a small core of other key management positions meet to discuss the logistical needs of the incident based on the operational objectives. The logistics section then determines whether those needs can be met, and if so, under what time frame.

Finance

The finance/administration section tracks the costs related to the incident, handles procurement issues, records the time that responders are on the incident for billing purposes, and keeps a running cost of the incident. Today, cost is certainly a factor in every response, even when it comes to operational objectives. Large-scale incidents can cost millions of dollars to handle, and it is common to see incident objectives tied to financial constraints. Ideally, an incident should be handled in the most cost-effective way possible, so as to minimize the financial impact to the jurisdiction in which the incident occurs.

Role of the Operations Level Responder

As discussed in the first chapter of this text, according to NFPA 472, operations level responders are those persons who respond to hazardous materials/WMD incidents for the purpose of implementing or supporting actions to protect nearby persons, the environment, or property from the effects of the release; however, these persons may have additional competencies that are specific to their response mission, expected tasks, and equipment and training as determined by the authority having jurisdiction. Clearly, this language identifies operations level responders as integral components of an overall response plan to hazardous materials/WMD incidents.

The scope of the operations level responder definition provides a jurisdiction with much more flexibility in its operational procedures than ever before, especially where mission-specific competencies are concerned. An AHJ may, for example, choose to train and use operations level responders strictly as support personnel for a group of hazardous materials technicians. Alternatively, the AHJ could provide its personnel with the training to satisfy operations level requirements, including both core competencies and all mission-specific competencies, thereby ensuring that its responders can handle a wide variety of hazardous materials/WMD incidents.

As a responder, you should be familiar with all emergency response plans for hazardous materials/WMD incidents that may occur within your jurisdiction. If none exists, as a responder you should default to the scope of practice established by your training when you are called into service, regardless of the nature of the incident.

It is common for a jurisdiction to predetermine response levels and response configurations to anticipated hazards. To that end, agencies use a wide variety of ways to identify and denote the severity of certain types of incidents. Establishing these thresholds takes the guesswork out of determining the type and amount of resources required to handle a particular problem as well as the levels of training required for the responding personnel. TABLE 5-1 illustrates how an agency might choose to denote the different levels of response. For obvious reasons, the role of the operations level responder will differ depending on the severity of the incident.

Table 5-1 Hazardous Materials Incident Levels

Level I: Lowest Level of Threat	Level II: Medium Level of Threat	Level III: Highest Level of Threat
A small amount of a low-toxicity/low-hazard substance is involved.	An organized hazardous materials team is needed.	Highly toxic chemicals are involved.
The incident can usually be handled by a single resource such as an engine company.	Additional chemical protective clothing will be required.	Mitigation efforts may require multiple jurisdictions.
Appropriate level of protection includes turnout gear and SCBA.	Civilian evacuations may be required.	Large-scale evacuations may be needed.
Example: A small gasoline spill from a motor vehicle accident.	Decontamination may need to be performed.	Federal agencies will be called in.
	Example: A tanker carrying sulfuric acid has overturned in a tunnel and is leaking onto the freeway.	Example: A ship in a highly populated harbor catches fire and begins to release chlorine vapors from its cargo area.

Wrap-Up

Chief Concepts

- The acronym SIN—which stands for Safety, Isolate, and Notify—is a good place to start when undertaking a hazardous materials response.
- Awareness level responders should ensure scene safety by setting up a barrier around the scene to keep out bystanders and those who are not properly trained or outfitted to engage in the response.
- The three levels of control zones around a hazardous materials incident are the cold zone, the warm zone, and the hot zone. The hot zone is the most contaminated; the cold zone is a safe area.
- At the scene of a hazardous materials incident, OSHA requires the use of the buddy system.
- A risk-based method of decision making should be employed before people are told to engage in sheltering-in-place or to evacuate from the scene of a hazardous materials incident.
- A safety briefing informs all responders about the health hazards that are known or anticipated at the hazardous materials/WMD incident.
- It is essential for responders to be aware of the signs and symptoms of excessive-heat exposures such as heat exhaustion, heat stroke, heat stress, and dehydration. Become familiar with the cooling technologies used by your AHJ and your own physical capabilities and requirements.
- Two types of cold-temperature exposures are possible: those caused by the release of a material and those caused by the operating environment.
- The incident command system (ICS) has many benefits, including the use of common terminology, consistent organizational structure and position titles, and common incident facilities.
- The ICS can be expanded to handle an incident of any size and complexity.
- The ICS incorporates five main functions: command, operations, planning, logistics, and finance/administration.
- The incident commander (IC) is the person in charge of the incident site; he or she is responsible for all decisions relating to the management of the incident.
- The command staff consists of the safety officer, the liaison officer, and the public information officer.
- Groups and divisions are established to aggregate single resources and/or crews under one supervisor.
- A hazardous materials branch consists of some or all of the following staff, as needed: a hazardous materials safety officer (assistant safety officer), an entry team, a decontamination team, a backup entry team, and a technical reference team.
- The operations level responder responds to hazardous materials incidents for the purpose of implementing or supporting actions to protect nearby persons, the environment, or property from the effects of the release.

Hot Terms

Backup team Individuals who function as a stand-by rescue crew or relief for those entering the hot zone (entry team). Also referred to as backup personnel.

Buddy system A system in which two responders always work as a team for safety purposes.

Cold zone A safe area at a hazardous materials incident for those personnel involved in the operations. The incident commander, the command post, EMS providers, and other support functions necessary to control the incident should be located in the cold zone. Also referred to as the clean zone or the support zone.

Command staff Staff positions that assume responsibility for key activities in the incident command system. Individuals at this level report directly to the incident commander. Command staff members include the safety officer, public information officer, and liaison officer.

Control zones Areas at a hazardous materials incident that are designated as hot, warm, or cold, based on safety issues and the degree of hazard found there.

Decontamination team The team responsible for reducing and preventing the spread of contaminants from persons and equipment used at a hazardous materials incident. Members of this team establish the decontamination corridor and conduct all phases of decontamination.

Division An organizational level within the incident command system that divides an incident in one location into geographic areas of operational responsibility.

Emergency decontamination The process of removing the bulk of contaminants off a victim without regard for containment. It is used in potentially life-threatening situations, without the formal establishment of a decontamination corridor.

Entry team A team of fully qualified and equipped responders who are assigned to enter the designated hot zone.

Finance/administration section The command-level section of the incident command system responsible for all costs and financial aspects of the incident, as well as any legal issues that arise.

Hazardous materials branch A unit consisting of some or all of the following positions as needed for the safe control of a hazardous materials incident: a second safety officer known as the hazardous materials safety officer (or assistant safety officer), who reports directly to the incident safety officer; a hazardous materials group supervisor; the entry team; the decontamination team; and the technical reference team.

Hazardous materials group A group often established when companies and crews are working on the same task or objective, albeit not necessarily in the same location. A group is specific because it applies to the incident command system and is assembled to relieve span of control issues.

Hazardous materials safety officer A second safety officer dedicated to the safety needs of the hazardous materials branch. Also referred to as the assistant safety officer.

Heat exhaustion A mild form of shock that occurs when the circulatory system begins to fail as a result of the body's inadequate effort to give off excessive heat.

Heat stroke A severe, sometimes fatal condition resulting from the failure of the body's temperature-regulating capacity. Reduction or cessation of sweating is an early symptom; body temperature of 105°F (40.5°C) or higher, rapid pulse, hot and dry skin, headache, confusion, unconsciousness, and convulsions may occur as well.

Hot zone The area immediately surrounding a hazardous materials spill/incident site that is directly dangerous to life and health. All personnel working in the hot zone must wear complete, appropriate protective clothing and equipment.

Incident commander (IC) A level of training intended for those assuming command of a hazardous materials incident beyond the operations level. Individuals trained as incident commanders should have at least operations level training and additional training specific to commanding a hazardous materials incident.

Incident command post (ICP) The location in the cold zone where the incident command is located; it is where the command, coordination, control, and communications functions are centralized.

Incident command system (ICS) The combination of facilities, equipment, personnel, procedures, and communications under a standard organizational structure organized so as to manage assigned resources and effectively accomplish stated objectives for an incident.

Incident safety officer The position within the incident command system responsible for identifying and evaluating hazardous or unsafe conditions at the scene of an incident. Safety officers have the authority to stop any activity that is deemed unsafe. Also referred to as the safety officer.

Liaison officer The position within the incident command system that establishes a point of contact with outside agency representatives.

Logistics section The section within the incident command system responsible for providing facilities, services, and materials for the incident.

Logistics section chief The general staff position responsible for directing the logistics function. It is generally assigned on complex, resource-intensive, or long duration incidents.

Operations section The section within the incident command system responsible for all tactical operations at the incident. Its personnel carry out the objectives developed by the incident commander.

Operations section chief The general staff position responsible for managing all operations activities. It is usually assigned when complex incidents involve more than 20 single resources or when command staff cannot be involved in all details of the tactical operation.

Planning section The section within the incident command system responsible for the collection, evaluation, and dissemination of tactical information related to the incident and for preparation and documentation of incident management plans.

Planning section chief The general staff position responsible for planning functions and for tracking and logging resources. It is assigned when command staff members need assistance in managing information.

Public information officer (PIO) The position within the incident command system responsible for providing information about the incident. Functions as a point of contact for the media.

Sheltering-in-place A method of safeguarding people in a hazardous area by keeping them in a safe atmosphere, usually inside structures.

Size-up The rapid mental process of evaluating the critical visual indicators of the incident, processing that information based on training and experience, and arriving at a conclusion that will serve as the basis to form and implement a plan of action.

Technical reference team A team of responders who serve as an information-gathering unit and referral point for both the incident commander and the hazardous materials safety officer (assistant safety officer).

Unified command An incident command system option that allows representatives from multiple jurisdictions and agencies to share command authority and responsibility, thereby working together as a "joint" incident command team.

Warm zone The area located between the hot zone and the cold zone at an incident. Personal protective equipment is required for all personnel in this area. The decontamination corridor is located in the warm zone.

RESPONDER in action

It is 4:00 A.M. when your engine is dispatched to an old manufacturing plant for a report of a 2000-gallon above ground storage tank slowly leaking. The on-scene security guard tells you the storage tank has been there for years and he thinks it might contain polychlorinated biphenyls (PCBs). Approximately 50 gallons of liquid has spilled onto the asphalt and is entering a storm drain. You and your crew must establish control zones to limit exposure to the chemical.

1. Which of the following most accurately describes the warm zone?
   - **A.** The area immediately around and adjacent to the incident.
   - **B.** The area located between the hot zone and the cold zone. The decontamination corridor is located in the warm zone.
   - **C.** An area where personnel do not need to wear any special protective clothing for safe operation.
   - **D.** The area where the incident command post is located.
2. After doing some research on PCBs, you learn the following:

   PCB is a toxic chemical that was banned in the United States in the late 1970s but can still be found in some older electrical devices. This chemical is very harmful to the environment; it can also cause skin irritation, liver damage, nausea, dizziness, and eye irritation. It is a thick, oily liquid with a low vapor pressure at ambient atmospheric temperature.

   The leak is occurring approximately 200 yards away from an occupied apartment complex. Which of the following actions would be appropriate for this incident?
   - **A.** Evacuate the apartment complex.
   - **B.** Take no action, because this spill poses no threat of airborne contamination to the apartment complex and the possibility of the material being PCB is very remote.
   - **C.** Notify local law enforcement to shelter-in-place all occupants of the apartment complex.
   - **D.** Instruct the public information officer to make an announcement of the spill during the evening news broadcast.

3. The IC decides that the entry team will wear a level of protection that consists of a chemical-resistant suit, boots, gloves, and SCBA. Which of the following classifications best describes this level of protection?
   - **A.** Level A
   - **B.** Level B
   - **C.** Level C
   - **D.** Level D

4. After working in PPE for 30 minutes and going through decontamination, one of the responders complains of dizziness and nausea. He is sweating and tells you he feels weak. Which of the following medical conditions most accurately describes his condition?
   - **A.** Heat stroke
   - **B.** Hypothermia
   - **C.** Hypocalcemia, due to the PCB exposure
   - **D.** Heat exhaustion

Terrorism

Greg Henry/ShutterStock, Inc.

© Anthony Correia/ShutterStock, Inc.

CHAPTER 6

Awareness Level Personnel and Operations Level Responders

Knowledge Objectives

After studying this chapter, you will be able to:

- Describe the threat posed by terrorism. (p. 134)
- Understand the definition of terrorism from a broad perspective. (p. 134)
- Describe various types of potential terrorist targets. (p. 135)
- Understand the dangers posed by explosive devices and secondary explosive devices. (p. 138–140)
- Define weapons of mass destruction. (p. 140)
- Understand the basic differences and indicators of chemical, biological, and radiological threats. (p. 140–149)
- Describe operations considerations at a terrorism event, including initial actions, interagency coordination, decontamination, mass casualties, and triage. (p. 149–151)

Skills Objectives

This chapter has no skills objectives.

You Are the Responder

Your engine company has been dispatched to a local government office building for a reported explosion. When you arrive on the scene you do not observe any indications of fire or structural damage, but people are running from the building. Many of them complain of eye irritation and difficulty breathing. One person tells you that she heard a loud popping sound from a nearby office and that some employees were showered with an unknown liquid substance. You quickly learn that there are still people in the office. Some are unconscious; others appear to be having seizures.

1. Is the on-scene information provided at this incident typical for most fires or explosions?
2. Which characteristics of this situation would cause you concern?
3. What is your primary concern as you enter the office building?

Introduction

Reflecting the rising threat posed by terrorism, NFPA 472, *Standard for Competence of Responders to Hazardous Materials/Weapons of Mass Destruction Incidents*, outlines the following operational philosophy:

> Emergency response operations to a terrorism or criminal scenario using hazardous materials are based on the basic operational concepts of hazardous materials response. In fundamental terms, responders cannot safely and effectively respond to a terrorism or criminal scenario involving hazardous materials/weapons of mass destruction (WMD) if they do not first understand basic hazardous materials response.

This chapter provides an overview of terrorism and profiles some of the substances/agents that emergency personnel may encounter at these incidents. Readers seeking additional information on agent-specific information and tasks—such as those associated with incidents involving biological, chemical, and radiological agents, for example—should consult the annex sections of NFPA 472. The annex information is nonmandatory and is not considered to be part of the requirements of NFPA 472 but is designed to provide additional information over and above the core competencies and certain mission-specific competencies.

Terrorism, as defined in the U.S. Code of Federal Regulations, can be described as "the unlawful use of force and violence against persons or property to intimidate or coerce a government, the civilian population, or any segment thereof, in furtherance of political or social objectives." This broad definition encompasses a wide range of criminal acts committed by groups or individuals for a variety of reasons.

At its core, the goal of terrorism is to produce feelings of fear in a population or group. Terrorism is not limited to incidents involving grandiose or complex methods or weapons, however. In fact, almost anything can be used as a tool by terrorists. For example, an MC 306 type cargo trailer carrying gasoline could be intentionally wrecked and burned in an attempt to destroy property or injure people in the name of political cause. Such an event could be viewed as having criminal intent or viewed as an act of terrorism. Yet, if an identical gasoline tanker crashed and caught fire due to driver error, the response would be deemed "an accident." The challenge lies in determining intent, and in the first few minutes of the event, the distinction between criminal intent and an accident may not be clear. It is in those first critical moments that responders must rely on their training and standard operating procedures to gain an understanding of the circumstances, and safely and efficiently handle the problem—whether it was intentionally caused or not.

The Federal Bureau of Investigation (FBI) classifies terrorism as either domestic or international in nature. *Domestic terrorism* refers to acts that are committed within the United States by individuals or groups that operate entirely within the United States and are not influenced by any foreign interests. *International terrorism* includes any acts that transcend national boundaries.

On September 11, 2001, the largest terrorist events in the history of the United States occurred—the attacks on the World Trade Center, the Pentagon, and United Airlines Flight 93, the airplane that crashed in Shanksville, Pennsylvania. Collectively, these events resulted in almost 3000 deaths **FIGURE 6-1**.

Although acts of terrorism can occur at any time, these types of incidents have yet to become commonplace in the United States.

Responding to Terrorist Incidents

The roles of all public safety responders in handling terrorist events involve many of the same functions that responders perform on a day-to-day basis. Regardless of the intent behind the incident, responders are required to make risk-based decisions on tactics and strategy, personal protective equipment (PPE),

FIGURE 6-1 The September 11 attacks on the World Trade Center and the Pentagon accounted for the majority of the deaths caused by terrorists in 2001.
© Todd Hollis/AP Photos

victim rescue and/or evacuation, decontamination, and information obtained during detection and monitoring activities—the entire gamut of actions required to solve the problem.

What is different during an incident with criminal intent is the landscape upon which that incident is handled and the interagency cooperation that must occur. An incident with criminal intent, causing widespread destruction and/or numerous casualties, will require the involvement of many local, state, and federal law enforcement agencies; emergency management agencies; allied health agencies; and perhaps the military. It is critical that all of these agencies train together and work together in a coordinated and cooperative manner. A mass casualty incident involving a WMD could quickly overwhelm your agency, neighboring agencies, and the local healthcare system, in addition to quickly becoming national and international news. To that end, it is important to understand which assets and capabilities of those regional agencies may be available to assist with a large-scale incident.

Potential Targets and Tactics

Terrorists are motivated by a cause and choose targets they believe will help them advance or highlight their political or social agenda. Given this goal, terrorists tend to choose symbolic targets, such as places of worship, embassies, monuments, prominent government buildings or other significant landmarks.

Terrorists choose a method of attack they think will make the desired statement or achieve the maximum results. Explosive devices have remained popular weapons for terrorists, however, and have been used in thousands of terrorist attacks. In recent years, the number of suicide bombings has increased significantly, emphasizing the high threat level posed by explosives.

In the 1980s, many terrorist incidents involved the taking of hostages on hijacked aircraft or cruise ships. Sometimes only a few people were held; at other times, hundreds of people were taken hostage. Diplomats, journalists, and athletes were targeted in several incidents, and the terrorists often offered to release their hostages in exchange for the release of imprisoned individuals allied with the terrorist cause. More recently, terrorist actions worldwide have killed or injured thousands of lives without giving authorities an opportunity to negotiate for the victims' safety. Terrorism can occur in any community and in connection with many different issues. Many causes inspire passion in their supporters, who may range from members of peaceful, nonviolent organizations to advocates of fanatical fringe groups. For example, an environmental group upset over construction or urban encroachment into a sensitive wetland area may attack or disable earthmoving equipment. A small retailer in an upscale suburban community could become the target of an animal rights group that objects to the sale of fur coats. An antiabortion group might plant a bomb at a local community health clinic.

Law enforcement agencies routinely gather intelligence about terrorist groups, threats, and potential targets, and may have some inclination of a pending attack. The fire service, in conjunction with law enforcement agencies, is increasingly engaging in intelligence sharing so that preincident plans can be developed for possible targets and scenarios. Even if no specific threats have been made, certain types of occupancies, landmarks or critical infrastructure (dams, water supplies, power grids, etc.) should be viewed as potential targets; preincident planning for those locations should take into account the possibility of a terrorist attack (or damage from a weather event or unplanned failure).

News reports about terrorist incidents abroad can help responders stay current with trends in terrorist tactics. These accounts can provide useful information about situations that could occur in your jurisdiction in the future. A variety of Internet resources can also help responders to keep abreast of current threats. Periodic nonclassified bulletins and reports released through governmental response agencies can provide a variety of useful information. Responders should seek out these informational conduits within the authority having jurisdiction (AHJ) and subscribe to those outlets that might be the most relevant.

Classification of Terrorist Targets

Potential targets for terrorists can be classified into three broad categories: infrastructure targets, symbolic targets, and civilian targets.

Infrastructure Targets

Terrorists might select bridges, tunnels, subways, or hospitals as targets in an attempt to disrupt transportation and inflict a large number of casualties FIGURE 6-2. They might also plan to attack the public water supply or try to disable the electrical power distribution system, telephones, or computer based

© Arthur S. Aubry/Photodisc/Getty Images

© Steve Allen/Brand X Pictures/Alamy Images

© AbleStock

© Jones and Bartlett Learning. Photographed by Christine McKeen

FIGURE 6-2 Subways, airports, bridges, and hospitals are all vulnerable to attack by terrorists who seek to interrupt a country's infrastructure.

targets. Disruption of a community's 911-system or public safety radio network would have a direct impact on emergency response agencies.

Symbolic Targets

National monuments such as the Lincoln Memorial, Washington Monument, or Mount Rushmore could be targets for groups who want to attack symbols of national pride FIGURE 6-3. Groups promoting revolution within their own countries or protesting their international policies might attack foreign embassies and institutions. Religious institutions are potential targets of hate groups. By targeting these symbols, terrorist groups seek to make people aware of their cause and to create a sense of fear in the civilian population.

Civilian Targets

Civilian targets include places where large numbers of people gather, such as shopping malls, schools, or sports stadiums. In incidents involving these sites, the goal of terrorists is to indiscriminately kill or injure large numbers of people, thereby creating fear among society at large FIGURE 6-4.

Terrorism can also be classified according to the type of target—for example, ecoterrorism, agroterrorism, and cyberterrorism.

Ecoterrorism

Ecoterrorism refers to illegal acts committed by groups supporting environmental or environmental related causes. Examples include spiking trees to sabotage logging operations, bombing a university research laboratory conducting animal research, or setting fire to a store that sells fur coats. Construction projects extending into wilderness areas are also often targets of ecoterrorists.

Agroterrorism

Agroterrorism includes the use of chemical or biological agents to attack the agricultural industry or the food supply FIGURE 6-5. The deliberate introduction of an animal-targeted disease, such as foot-and-mouth disease within the livestock population, could result in major losses to the food industry and produce fear among members of the general population. Contaminating crops or rural water supplies may also be methods of carrying out agroterrorism.

FIGURE 6-3 Terrorists might attempt to destroy visible national icons.

© James P. Blair/Photodisc/Getty Images

FIGURE 6-4 By attacking civilian targets such as a crowded stadium, terrorists might make citizens feel vulnerable in their everyday lives.

© Photodisc/Getty Images

FIGURE 6-5 Agroterrorism affects the food supply and the agricultural industry.

© Larry Rana/USDA

RESPONDER TIP

Potential Terrorist Targets

Infrastructure Targets

Bridges and tunnels
Emergency facilities
Hospitals
Oil refineries
Pipelines
Power plants
Railroads
Telecommunications systems
Water reservoirs and treatment plants
Data centers

Symbolic Targets

Embassies
Government buildings
Military bases
National monuments
Places of worship

Civilian Targets

Arenas and stadiums
Airports and railroad stations
Mass-transit systems
Schools and universities
Shopping malls
Theme parks

Ecoterrorism Targets

Controversial development projects
Environmentally sensitive areas
Research facilities

Agroterrorism Targets

Crops
Feed storage
Grain elevators
Livestock and poultry

Cyberterrorism Targets

Banking and finance systems
Cloud-based data storage services
Judicial system electronic record-keeping
Government data storage systems or secured access sites
Law enforcement communications systems
Military communications systems

Cyberterrorism

Individuals or terrorists groups could engage in cyberterrorism or carry out cyber attacks by electronically compromising government, industrial, or private computer systems. Recently many successful attempts have been made to access credit card data, personal information, or infect systems with a virus. Daily, government computer systems and other critical networks are probed or attacked by hackers searching for ways around firewalls or other security measures. The move to cloud-based data storage and companies that provide software as a service (SaaS) have concentrated the risk of compromise/ disablement to higher levels than ever before. An unimaginable amount of information is stored in data centers around the globe, and billions of electronic transactions are carried out each day. Over the last several years, technology has provided many electronic solutions to ease daily life. The potential downside is our collective reliance on electronic solutions, and the greater risk of loss of compromise by those with bad intentions. Think about the reliance of your AHJ on the various aspects of technology such as mobile data terminals in fire apparatus and police cars; digital communication systems and dispatching; links between detection and monitoring devices and onboard computers; GPS enabled location devices; and all the other ways electronic devices and systems are part of your response framework. Now imagine those devices and systems are nonfunctional or, in some cases, used against you to supply criminals with information.

Other Methods

Terrorists can turn the most ordinary objects into powerful weapons. For example, we tend to think of gasoline tankers and commercial airliners as valuable elements of transportation and commerce, but in the hands of determined terrorists, these devices can turn into deadly weapons. As mentioned earlier, designating an incident as an act of terrorism reflects more the intent of the attacker than the use of a certain device. For example, demolition companies use explosives to bring down unneeded structures quickly and safely; terrorists could use the same explosives to bring down an occupied building, deliberately killing many people. Again, it is the intent that distinguishes between an accident and a terrorist event.

Although bombings are the most frequent terrorist acts, responders also must be aware of the threats posed by other potential weapons. Shooting into a crowd at a shopping mall or a train station with an automatic weapon could cause devastating carnage. The release of a biological agent into a subway system could make numerous people become ill and die, simultaneously creating a public panic response that could overload the local fire department, EMS, law enforcement, and hospitals. A computer virus that attacks the banking industry could cause tremendous economic losses. Given the breadth of these threats, planning should consider the full range of possibilities.

Explosives and Incendiary Devices

Each year, thousands of pounds of explosives are stolen from construction sites, mines, and military facilities FIGURE 6-6. How much of this material makes its way to criminals and terrorists is not known. It is true, however, that explosives are the most common and prevalent weapon used to cause intentional harm.

An improvised explosive device (IED) is any explosive device that is fabricated from readily available materials. An IED could be contained in almost any type of package from a letter bomb to a briefcase or backpack or even a truckload of explosives. The Unabomber, for example, constructed at least 16 bombs that were delivered in small packages through the U.S. Postal Service. By contrast, the bombings of the World Trade Center in 1993 and the Alfred E. Murrah Federal Building in 1995 both involved delivery vehicles loaded with ammonium nitrate fertilizer and fuel oil (ANFO) and detonated by a simple timer FIGURE 6-7.

Pipe Bombs

The most common IED is the pipe bomb. A pipe bomb is simply a length of pipe filled with an explosive substance and rigged with some type of detonator FIGURE 6-8. Most pipe bombs are simple devices that contain black powder or smokeless powder and are ignited by a hobby fuse. More-sophisticated pipe bombs may contain a variety of chemicals and incorporate electronic timers, mercury switches, vibration switches, photocells, or remote control detonators as triggers.

Pipe bombs are sometimes packed with nails or other objects that act as shrapnel; their expulsion is intended to inflict as much injury as possible on anyone in the vicinity. A chemical or biological agent or radiological material could be added to a pipe bomb to create a much more complicated and

FIGURE 6-6 Every year, thousands of pounds of explosives are stolen from their rightful owners.
Courtesy of Dennis Krebs

FIGURE 6-7 The Alfred E. Murrah Federal Building in Oklahoma City was destroyed by a truck bomb in 1995.
Courtesy of FEMA

FIGURE 6-8 Pipe bombs come in many shapes and sizes.
Courtesy of Captain David Jackson, Saginaw Township Fire Department

dangerous incident. Experts can only speculate at the number of casualties that could result from the use of such a weapon.

Secondary Devices

Emergency responders must realize that terrorists may have placed a secondary device in the area where an initial event has occurred. These devices are intended to explode some time after the initial device detonates. As discussed in the chapter *Estimating Potential Harm and Planning a Response*, secondary devices are designed to kill or injure emergency responders, law enforcement personnel, spectators, bystanders or news reporters. Terrorists have used this tactic to attack investigators and/or emergency responders, or simply to increase the levels of fear and chaos following an attack. Remember the EVADE mnemonic from NFPA 473, *Standard for Competencies for EMS Personnel Responding to Hazardous Materials/Weapons of Mass Destruction Incidents*, when considering the presence of a secondary device:

- **E**valuate the scene for likely areas where secondary devices may be placed.
- **V**isually scan operating areas for a secondary device before providing patient care.
- **A**void touching or moving anything that may conceal an explosive device.
- **D**esignate and enforce scene control zones.
- **E**vacuate victims, other responders, and nonessential personnel as quickly and safely as possible.

The use of secondary devices is a common tactic in incidents abroad and has occurred at a few incidents in North America. For example, in 1998, a bomb exploded outside a Georgia abortion clinic. Approximately an hour later, a second explosion at the same site injured seven people, including two emergency responders. A similar secondary device was discovered approximately 1 month later at the scene of a bombing in a nearby community, although responders were able to disable this device before it detonated.

Potentially Explosive Devices

Responders, fire department, and law enforcement units responding to an incident that involves a potentially explosive device—one that has not yet exploded—should remove all civilians from the area and establish a perimeter at a safe distance. At no time should any emergency responder handle a potential explosive device unless he or she has received specialized training. Trained explosive ordnance disposal (EOD) personnel should assess the device and render it inoperative FIGURE 6-9.

While waiting for the properly trained personnel to arrive, responders should establish a command post and a staging area FIGURE 6-10. Because secondary devices may be present, the staging area should be located at a safe distance from the incident site. Follow your standard operating procedures to determine where the command post should be established. Nevertheless, responders must make informed decisions as to appropriate locations for staging areas and evacuation actions. In this type of incident, a joint command structure (commonly referred to as a unified command) is usually established to coordinate the activities of all agencies participating in the response.

If the bomb tech decides to disarm the device, an emergency action plan must be developed in case an accidental detonation occurs FIGURE 6-11. The incident commander (IC) should work in concert with law enforcement/bomb squad to determine a safe perimeter where responders will be staged.

In some cases, a forward staging area may be established. A rapid intervention team then stands by to provide immediate assistance to the bomb disposal team if something goes wrong. Other fire fighters and emergency medical personnel would remain in a remote staging area in this scenario.

FIGURE 6-9 EOD personnel are trained and have the necessary tools and equipment to evaluate and disable or disrupt a device or suspected device.
Courtesy of Rob Schnepp

FIGURE 6-10 The first responders to arrive at the scene of an explosion should establish a command post in a safe location and begin the process of setting up a unified command with law enforcement.

Courtesy of Captain David Jackson, Saginaw Township Fire Department

FIGURE 6-11 EOD personnel disabling a device during a training exercise.

Courtesy of Rob Schnepp

Actions Following an Explosion

As a first responder, your first priority should be to ensure your own safety, followed by ensuring the safety of the scene. During the initial stages of an incident, you will not know whether the event was caused by an intentional act or resulted from accidental circumstances. In any incident following an explosion, follow departmental procedures to ensure the safety of rescuers, victims, and bystanders. As mentioned earlier in the chapter, it is important to consider the possibility of the presence of a secondary device. Quickly survey the area for any suspicious bags, packages, or other items.

It is also possible that chemical, biological, or radiological agents may be involved in a terrorist bombing. For this reason, qualified personnel with appropriate monitoring instruments should be assigned to check the area for potential contaminants and/or radiation. Likewise, the stability of any building involved in the explosion must be evaluated before anyone is permitted to enter. Entering an unstable area without proper training and equipment may complicate rescue and recovery efforts.

Working with Other Agencies

Joint training with local, state, and federal agencies charged with handling incidents involving explosive devices should occur on a routine basis. Among these agencies are local and state police; the FBI; the Bureau of Alcohol, Tobacco, Firearms and Explosives; and military EOD units. In some jurisdictions, the fire department is responsible for handling explosive devices. In other communities, specially trained law enforcement personnel perform this function.

Agents

The greatest threat posed by terrorists is the use of weapons of mass destruction (WMD) agents. These weapons include chemical, biological, and radiological agents, as well as conventional weapons and explosives. The annex information found in NFPA 472 provides additional, specific information about the various agents that may be used in a terrorist attack. Take extra care and be observant when responding to incidents where large numbers of victims are found, or where an explosion or shooting has occurred at a mass-gathering event such as a sporting or political event. The following sections are intended to provide you with an overview of some of the agents that might be used with criminal intent.

■ Chemical Agents

Chemical agents have the potential to kill or injure large numbers of people. Many chemical agents are readily available in this country because they are widely used and produced in the United States and found in a broad range of industrial and commercial processes. For instance, chlorine is used for many purposes, including purification of drinking water and swimming pools. These chemicals are stored and transported in various-size containers, ranging from small pressurized cylinders to railroad tank cars. These same chemicals, if intentionally released, could cause injury or death to many people. Mitigating an accidental release of chlorine gas, however, requires the same safety precautions as mitigating an intentional release of chlorine gas.

Chemical agents are not new. In fact, they have been used as weapons for well over 100 years. Phosgene, chlorine, and mustard agents were used in World War I, resulting in thousands of battlefield deaths and permanent injuries. Chemical weapons were also used during the Iran–Iraq War (1980 to 1988). During that time period, the Iraqi government used these types of weapons against the minority Kurdish people in its own country. Although international agreements prohibit the use of chemical and biological agents on the battlefield, great concern remains that terrorists could use chemical weapons.

In 1995, the religious cult Aum Shinrikyo released a chemical nerve agent, sarin, into the Tokyo subway system. Although the attack resulted in 12 deaths and several thousand injuries, many experts suggest that this attack was ineffective because relatively few casualties occurred even though thousands of subway riders were potentially exposed to the chemical agent. Even so, this attack achieved one of the major goals of terrorism: It instilled fear in a large population.

VOICES OF EXPERIENCE

Knowing the resources available in your area and following your standard operating procedures (SOPs) can be valuable to your response and the overall outcome of an incident involving weapons of mass destruction (WMD).

Our hazardous materials response team was dispatched to a neighboring city's police department to check an unknown white powder leaking from an envelope. A citizen had received the envelope in the U.S. mail and, concerned that the powder might be a WMD substance, called the police department.

As acting captain that day, upon arrival at the police department I established a hot zone and gathered the necessary information related to the incident. The police officer told us he had visited the citizen's home, placed the envelope in a Ziploc bag, and stored it in the trunk of his police cruiser. Because the incident involved the U.S. mail, the first thing I did was request a postal service inspector to respond to the scene. After a few minutes, I received a call from a postal inspector who was also a hazardous materials technician. I relayed the details of this incident and told him that we had a Hazmat ID (a portable solids and liquids identifier that uses Fourier-transform infrared spectroscopy to identify unknown chemicals), as well as other capabilities to identify biological agents.

The inspector said that he was responding to the scene, he also had a Hazmat ID, and he would handle the identification and the investigation. We did not have to touch the envelope. The postal inspector arrived 45 minutes later and took possession of the envelope directly from the trunk of the police cruiser and began sample testing the unknown substance, using his Hazmat ID.

The substance turned out to be ground plastic made to look like snow on a Christmas promotional mailer. The envelope was written as if it was coming from a person, not a business.

As hazardous materials technicians, we were trained in public safety sampling techniques and procedures training. At this incident, we decided to reach out and make the proper notifications per our SOPs to the proper authorities. The message this story conveys is to ensure you have a WMD policy and procedure, know the resources available to you, and make the proper notifications to the appropriate state and federal agencies. This can help preserve the evidence and maintains the proper procedures for handling and sampling unknown substances, particularly if the incident is determined to be a criminal case and therefore the responsibility of maintaining and preserving the evidence is out of your hands.

Hakim Benmoussa
Hazardous Materials Response Team
City of Goodyear Fire Department, Arizona

The basic instructions for making chemical weapons can be found on the Internet, in books, and in other print or online sources. Many of the chemicals routinely used in legitimate industrial processes are relatively easy to acquire by theft and/or purchase. For example, as mentioned earlier, chlorine gas is used in swimming pools, at water treatment facilities, and in industry FIGURE 6-12. Despite its legitimate uses, chlorine is classified as a pulmonary agent (a choking agent) because its inhalation causes severe pulmonary damage.

Responder Safety Tip

Preincident planning at infrastructure, symbolic, and civilian targets should take into account the possibility of a terrorist attack.

Chemical weapons can be disseminated in several ways. For example, intentionally releasing chlorine gas inside a building or a crowded gathering place could cause a large number of injuries and deaths. To ensure broader distribution of a chemical agent, however, the agent might be added to an explosive device, or mechanically dispersed. Crop-dusting aircraft, truck-mounted spraying units, or machine/hand-operated pump tanks could all potentially be used to disperse an agent. FIGURE 6-13.

The extent of dissemination of a toxic gas or suspended liquid particles depends on wind direction, wind speed, air temperature, and humidity at the time of the material's release. Because these factors can change quickly, it is difficult to predict the exact direction that might be taken by a chemical release cloud. Hazardous materials teams use computer models to predict the pathway of a toxic cloud. They also have the training and equipment to safely handle these situations.

FIGURE 6-12 Although chlorine is regularly used in swimming pools and water treatment facilities, it is classified as a pulmonary agent that can be used in a terrorist attack.

FIGURE 6-13 Crop-dusting equipment could be used to distribute chemical agents.

Nerve Agents

Nerve agents are toxic chemical agents that attack the central nervous system. These weapons were first developed in Germany before World War II. Nerve agents are similar to some pesticides (organophosphates) but are much more toxic—in some cases, 100 to 1000 times more toxic than similar pesticides. Exposure to these substances can result in injury or death within minutes.

RESPONDER TIP

When pure, nerve agents are colorless and, in most cases, take the form of mobile liquids. In an impure state, nerve agents may appear as yellowish to brown liquids. Some nerve agents may have a faint fruity odor.

In their normal states, most nerve agents are liquids FIGURE 6-14. To be an effective weapon, the liquid must either be dispersed in aerosol form or be broken down into fine droplets so that it can be inhaled or absorbed through the skin.

Pouring a liquid nerve agent, such as V agent (VX), onto the floor of a crowded building may not affect large numbers of people because at ambient temperature, it is not highly vaporous. A more effective method to disperse it would be aerosolize it in some fashion and distribute it across the widest populated area possible. In such a scenario, the effectiveness of the agent

FIGURE 6-14 In their normal states, nerve agents are liquids. They must be dispersed in aerosol form if they are to be inhaled or absorbed by the skin.

© AbleStock

Table 6-1 Common Nerve Agents

Nerve Agent	Route of Exposure	Characteristics
Tabun (GA)	Skin contact Inhalation	Nonpersistent
Soman (GD)	Skin contact Inhalation	Nonpersistent
Sarin (GB)	Skin contact Inhalation	Nonpersistent
V-agent (VX)	Skin contact	Persistent

© Jones & Bartlett Learning

would depend on how long it stayed in the air and how widely it became dispersed throughout the building. VX is considered to be persistent because it has a low vapor pressure and will not evaporate quickly. Sarin, on the other hand, is considered nonpersistent because it evaporates at about the same rate as water. Common nerve agents, their method of contamination, and specific characteristics are listed in TABLE 6-1.

When a person is exposed to a nerve agent, symptoms of that exposure will become evident within minutes. The symptoms may include pinpoint pupils, runny nose, drooling, difficulty breathing, tearing, twitching, diarrhea, convulsions or seizures, and loss of consciousness FIGURE 6-15. The same symptoms are seen in individuals who have been exposed to pesticides.

Several mnemonics can help you remember the symptoms of a nerve agent exposure. The mnemonic used most often in the emergency response community is SLUDGE TABLE 6-2.

Keep in mind that such mnemonics represent a very basic and limited way to identify a nerve agent exposure. From a medical standpoint, an exposure of this type involves both the sympathetic and parasympathetic nervous systems—each having its own unique set of signs and symptoms. Advanced life support providers (paramedics, for example) should have in-depth knowledge when it comes to recognizing the signs and symptoms of a nerve agent exposure. Additionally, many internet and print resources are available for further study on the subject.

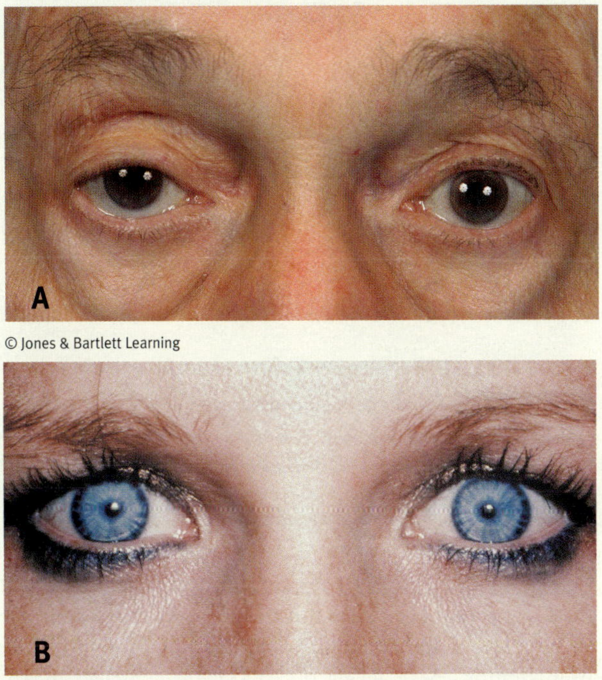

© Jones & Bartlett Learning

FIGURE 6-15 Pupil responses. A. Normal. B. Pinpoint.

© Jones & Bartlett Learning

Table 6-2 Symptoms of Nerve Agent Exposure

S: salivation (drooling)
L: lacrimation (tearing)
U: urination
D: defecation
G: gastric upset (upset stomach, vomiting)
E: emesis (vomiting)

© Jones & Bartlett Learning

In terms of medical treatment for a nerve agent exposure, the most common field level treatment is the DuoDote™, which contains 2.1 mg of atropine and 600 mg of 2-PAM, delivered as a single dose through one needle FIGURE 6-16. These kits have been provided to many fire departments' hazardous materials teams. In addition, many law enforcement agencies and EMS

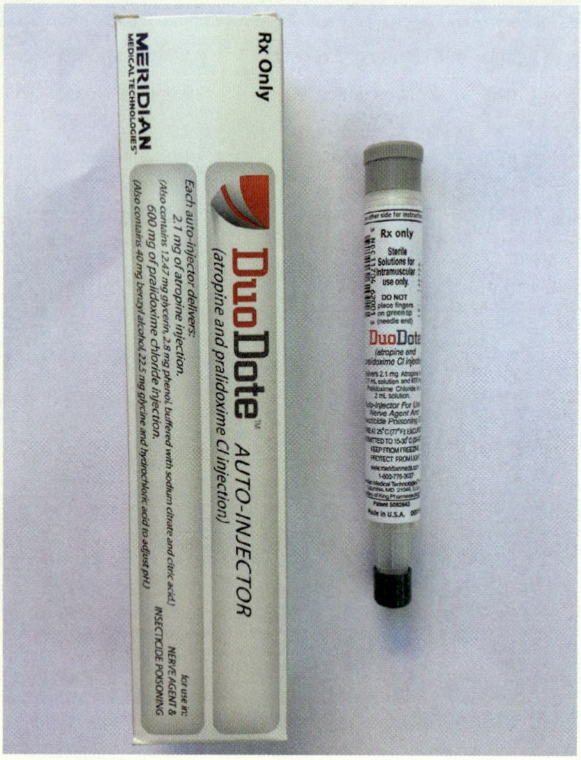

FIGURE 6-16 A DuoDote™ auto-injector.

Courtesy of Rob Schnepp

units around the United States carry these antidotes on their response vehicles.

The antidote medications can be quickly injected into a person who has been contaminated by a nerve agent. Peak atropine levels are reached approximately 5 minutes after administration; peak levels of 2-PAM are achieved in 15 to 20 minutes. For the emergency responder, this delay means that you should not expect to administer this antidote to a person exhibiting serious signs and symptoms of a nerve agent exposure and have the patient recover right away, or even at all. Auto-injectors are not an immediate cure-all. In fact, they may be ineffective if the victim has suffered a significant exposure to a nerve agent or pesticide.

RESPONDER TIP

The DuoDote™ is intended for patients between the ages of 18 and 55 years of age.

Each drug in the auto-injector has a specific target and acts independently to reverse the effects of a nerve agent exposure. Atropine, for example, may reverse muscarinic effects such as runny nose, salivation, sweating, bronchoconstriction, bronchial secretions, nausea, vomiting, and diarrhea; essentially, atropine is intended to deal with the SLUDGE effects of a nerve agent exposure. Atropine dosing is guided by the patient's clinical presentation and should be given until secretions are dry or drying and ventilation becomes less labored.

2-PAM, by contrast, does not reverse muscarinic effects on glands and smooth muscles. Instead, this medication's main goal is to decrease muscle twitching, improve muscle strength, and allow the patient to breathe better, but it has little effect on the muscarinic effects described earlier.

When it comes to treating victims of nerve agent exposure, emergency responders should understand that all nerve agents "age" once they are absorbed by the body. Thus, after a certain period of time (which varies depending on the agent), the administration of the DuoDote™ kit may be largely ineffective. Soman, for example, has an aging time of approximately 2 minutes. Sarin's aging time is approximately 3 to 4 hours. The other nerve agents have longer aging times. Follow local protocols before administering any antidote.

Blister Agents

Two chemical agents are generally classified as blister agents, so called because contact with these chemicals causes the skin to blister.

- Sulfur mustard (H) is a clear, yellow, or amber oily liquid with a faint sweet odor of mustard or garlic. It vaporizes slowly in cool temperatures and may be dispersed in an aerosol form.
- Lewisite (L) is an oily, colorless-to-dark-brown liquid with an odor of geraniums.

Blister agents (also known as vesicants) produce harmful and painful burns and blisters with even minimal exposure to the skin FIGURE 6-17. The major difference between sulfur mustard and lewisite is that lewisite causes pain immediately upon contact with the skin, whereas the signs and symptoms from exposure to sulfur mustard may not appear for several hours. The patient will complain of burning at the site of the exposure, the skin will redden, and blisters will appear. The eyes may also itch, burn, and turn red. Both of these vesicants can cause significant—and possibly fatal—respiratory damage.

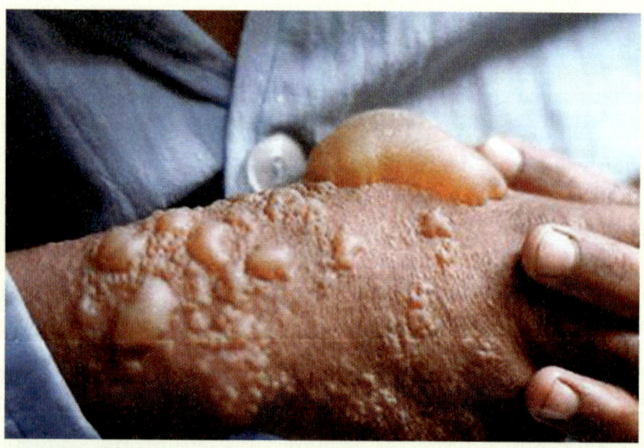

FIGURE 6-17 Typical effects of blister agents.

Courtesy of Dr. Saeed Keshavarz/RCCI (Research Center of Chemical Injuries)/IRAN

Pulmonary Agents

Pulmonary agents (choking agents) cause severe damage to the lungs and lead to asphyxia. Two chemicals that might be used as pulmonary agents in terrorist attacks are phosgene and chlorine. Both of these agents were used extensively as weapons in World War I, and both have several industrial uses.

Phosgene and chlorine gas are both heavier than air, so they tend to settle in low areas when released. Thus subways, basements, trenches, and sewers are prime areas for these agents to accumulate.

Exposure to high concentrations of phosgene or chlorine will immediately irritate the eyes, nose, and upper airway. Within hours, the exposed individual will begin to develop pulmonary edema (fluid in the lungs). Individuals who are exposed to lower concentrations may not exhibit any initial symptoms but can still experience respiratory damage.

Although neither agent is absorbed through the skin, both phosgene and chlorine can cause skin burns on contact. Decontamination should consist of removing exposed individuals from the area and flushing the skin with lots of water.

Blood Agents

Blood agents, such as cyanide, interfere with the utilization of oxygen by the cells of the body. Cyanide compounds are highly toxic, and depending on the airborne concentration, can cause death within minutes of exposure. The most commonly encountered cyanide compounds are hydrogen cyanide and cyanogen chloride, both of which have legitimate uses in industry.

Cyanide can be either inhaled or ingested. Historically, the gaseous form was used as a means of carrying out the death penalty (hence the "gas chamber"). Liquid cyanide mixed with fruit punch was used in the mass suicide of 913 members of a religious cult in Guyana in 1978.

Indicators of Chemical Agents

As discussed in the chapter *Recognizing and Identifying the Hazards*, indicators of possible criminal or terrorist activity involving chemical agents may vary depending on the nature of the illicit laboratory or production process. Sometimes protective equipment such as rubber gloves, chemical suits, and respirators may be evident. Chemical containers of various shapes and sizes may also be present. Glass containers are very prevalent as are portable heaters. It is important to be observant at any scene, as there is no "typical" profile for a laboratory or explosive making operation.

The chemicals themselves may provide unexplained odors that are out of character for the surroundings, and residual chemicals (liquid, powder, or gas form) may be found in the area. Additionally, the presence of dead or dying vegetation and dead animals in the immediate vicinity may indicate the presence of chemical weapons. Chemistry books or other reference materials may be seen, as well as materials that are used to manufacture chemical weapons (such as scales, thermometers, or torches).

Protection from Chemical Agents

The most common and effective dispersal method for chemical agents is releasing the agent into the air. Agents released outdoors will follow wind currents; agents released inside structures will be dispersed through a building's air circulation system. Some chemical agents have distinctive odors. Hence, if any unusual odor is noted at an emergency scene, emergency responders must either leave the area immediately or don PPE, including self-contained breathing apparatus (SCBA) **FIGURE 6-18**.

It is never sufficient to rely on odor to determine the presence of a chemical agent. Instead, definitive detection and identification of a chemical agent require the activation of a well-trained hazardous materials team. The mission specific chapter on PPE provides more detail on the types of PPE available and some guidance on selecting the proper garment for the mission.

Biological Agents

Biological agents are organisms that cause disease and attack the body. They include bacteria, viruses, and toxins. Some of these organisms, such as anthrax, can live in the ground for years; others are rendered harmless after being exposed to sunlight for only a short period of time. The effects of a biological agent depend on the specific organism, the dose, and the route of entry. Most experts believe that a biological weapon

FIGURE 6-18 If an unusual odor is reported at the scene, responders must don appropriate PPE including SCBA.

would probably be spread by a device capable of widespread dispersion.

Some of the diseases caused by biological agents, such as smallpox and pneumonic plague, are contagious and can be passed from person to person. Doctors are concerned about the use of contagious diseases as weapons, because the resulting epidemic could overwhelm the healthcare system. Experts have different opinions about how difficult it would be to infect large numbers of people with one of these naturally occurring organisms. Because of their incubation period, people would not begin to show signs of being infected until 2 to 17 days after exposure to these organisms.

RESPONDER TIP

Bacteria

Single-cell microscopic organisms with a nucleus, and cell wall

May form a spore

Examples: anthrax, plague, tularemia

Virus

Submicroscopic agent; protein coated with DNA or RNA

Requires a host to live and reproduce

Examples: smallpox, viral hemorrhagic fever (VHF)

Anthrax

Anthrax is an infectious disease caused by the bacterium *Bacillus anthracis*. These bacteria are typically found around farm animals such as cows and sheep. For use as a weapon, the bacteria must be cultured to develop anthrax spores. The spores, weaponized into an ultra fine powder, can then be dispersed in a variety of ways FIGURE 6-19. Approximately 8000 to 10,000 spores are typically required to cause an anthrax infection. Spores infecting the skin cause cutaneous anthrax, ingested spores cause gastrointestinal anthrax, and inhaled spores cause inhalational anthrax. Anthrax has an incubation period of 2 to 6 days. The disease can be successfully treated with a variety of antibiotics if it is diagnosed early enough.

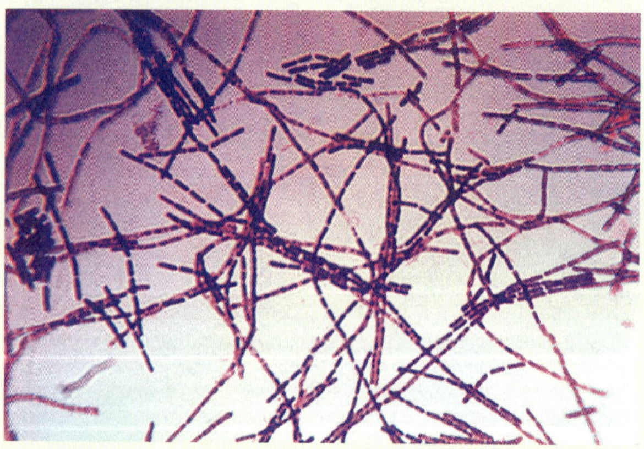

FIGURE 6-19 Anthrax spores can be dispersed in a variety of ways.

Courtesy of CDC

The threat posed by anthrax-related terrorism is quite real. In 2001, four letters containing anthrax were mailed to locations in New York City; Boca Raton, Florida; and Washington, D.C. Five people died after being exposed to the contents of these letters, including two postal workers who were exposed as the letters passed through postal sorting centers. Several major government buildings had to be shut down for months to be decontaminated. These incidents followed shortly after the terrorist attacks of September 11, 2001, and they caused tremendous public concern. In the wake of these incidents, emergency personnel had to respond to thousands of incidents involving suspicious packages and citizens who believed that they might have been exposed to anthrax.

Today, presumptive field tests are available that hazardous materials teams can use to determine whether the threat of anthrax or other biologic agents is legitimate.

The gold standard to positively identify anthrax (and other biologic agents), however, is not a field test. Anthrax must be cultured in a lab, by qualified microbiologists, to be positively identified. Your agency should have established procedures for handling suspicious powders and getting samples to a qualified laboratory. Your local FBI office can provide guidance on the different types of labs available through the Laboratory Response Network, and how to work with law enforcement to get a sample to the appropriate lab.

Plague

Plague is caused by *Yersinia pestis*, a bacterium that is commonly found on rodents. These bacteria are most often transmitted to humans by fleas that feed on infected animals and then bite humans.

The two main forms of plague are bubonic and pneumonic. Individuals who are bitten by fleas generally develop bubonic plague, which attacks the lymph nodes FIGURE 6-20. Pneumonic plague can be contracted by inhaling the bacterium.

Yersinia pestis can survive for weeks in water, moist soil, or grains. These bacteria might also be cultured for distribution as

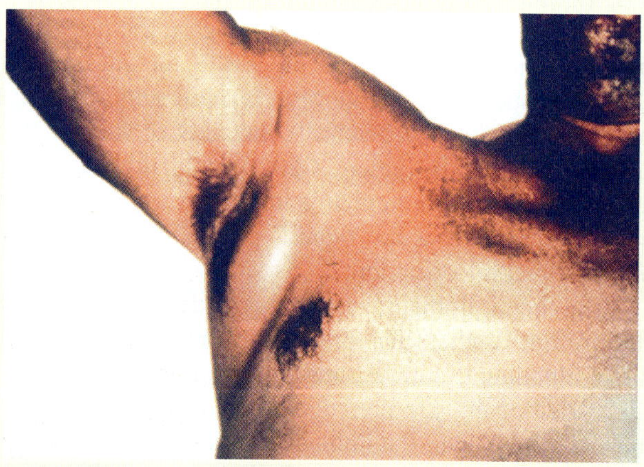

FIGURE 6-20 A bubo—one of the symptoms of the plague—consists of a swollen, painful lymph node.

Courtesy of CDC

a weapon in aerosol form. Inhalation of the aerosol form would put the target population at risk for pneumonic plague.

The incubation period for the plague ranges from 2 to 6 days. This disease can be treated with antibiotics.

Smallpox

Smallpox is a highly infectious and often fatal disease caused by *Variola*, a virus; it kills approximately 30 percent of all persons who become infected with this pathogen. Smallpox first presents with small red spots or as a rash in the mouth. The rash then progresses to the face, followed by the arms and legs, and then farther outward to the hands and feet. Smallpox lesions are unique in that all lesions appear to be in the same stage of development at the same time. In contrast, in chickenpox, the lesions are in different stages of development across the body. Additionally, smallpox lesions can be found on the palms of the hands and the soles of the feet, whereas chickenpox lesions are seldom found on the palms and/or soles of the feet.

Although smallpox was once routinely encountered throughout the world, by 1980 it had been successfully eradicated as a public health threat through the use of an extremely effective vaccine. Officially, two countries (the United States and Russia) have maintained cultures of the disease for research purposes. It is possible, however, that international terrorist groups may have acquired the virus.

The smallpox virus could potentially be dispersed over a wide area in an aerosol form; however, widespread broadcasting of this agent may not be necessary to cause a devastating outbreak. Infecting a small number of people could lead to a rapid spread of the disease throughout a targeted population, given smallpox's highly contagious nature: The disease is easily spread by direct contact, droplet, and airborne transmission. Patients are considered highly infectious and should be quarantined until the last scab has fallen off **FIGURE 6-21**. The incubation period for smallpox is between 4 and 17 days (average = 12 days).

Currently there are millions of people who have never been vaccinated for the disease, and millions more have reduced immunity because decades have passed since their last immunization.

Indicators of Biological Agents

As discussed in the chapter *Recognizing and Identifying the Hazards*, indicators of the presence of biological agents are somewhat similar to the indicators of chemical agents. Additionally, microbiology or biology textbooks or reference manuals may be present. Containers used to transport biological agents may be observed, including metal cylindrical cans or plastic boxes or bags **FIGURE 6-22**. Personal protective equipment such as respirators, chemical or biological suits, and latex gloves may be present. Excessive amounts of antibiotics may be present as a means to protect personnel working with the agents. Other potential signs of biological agents may include abandoned spray devices and unscheduled or unusual sprays being disseminated (especially outdoors at night). Finally, personnel working in the area of the lab may exhibit symptoms consistent with the biological weapons with which they are working.

Protection from Biological Agents

It is unlikely that fire fighters or other emergency responders would immediately recognize that a biological weapon had been released in their communities, in large part because of these agents' incubation period—that is, the lag between the actual infection and the appearance of symptoms. The signs and symptoms of a biological agent attack would typically manifest over a period of days after the exposure. If left untreated, many of these diseases will cause death in a large proportion of those infected.

The Centers for Disease Control and Prevention (CDC) and/or area hospitals would typically be the first to recognize the situation, after identifying significant numbers of people arriving at their facilities with similar symptoms. Multiple medical responses with patients exhibiting similar symptoms could provide a clue about the use of a biological weapon, especially if the location is considered to be a potential target.

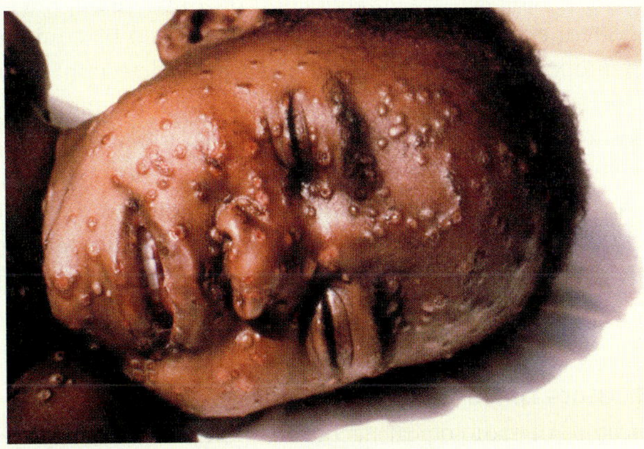

FIGURE 6-21 Smallpox is a highly contagious disease with a mortality rate of approximately 30 percent.

Courtesy of CDC

FIGURE 6-22 Biological agents may be evident in production or containment equipment, such as Petri dishes, vented hoods, Bunsen burners, pipettes, microscopes, and incubators.

Courtesy of Rob Schnepp

At a minimum, fire fighters or emergency responders treating people who may have been exposed to a biological agent should follow the recommendations of their AHJ regarding universal precautions (protective measures for use in dealing with objects, blood, body fluids, or other potential exposure risks of communicable diseases). These measures include wearing gloves, full-face or half-face respirators with N100 or P100 filtration, eye protection, and surgical-type gowns when treating patients.

Emergency responders who exhibit flu-like symptoms (keep in mind incubation periods for biological agents) after an incident should seek medical care immediately. It is important that responders report any possible exposure to department officials. Postincident actions may include logging those responders who may have made contact with exposed persons, follow-up medical screening, testing, or vaccinations.

Radiological Agents

A third type of agent that may potentially be used by terrorists is radiation. The most probable scenarios involve the use of a variety of approaches to disperse radiological agents and contaminate an area with radioactive materials. This threat is very different from that posed by a nuclear detonation.

Radiation

Radioactive materials release energy in the form of electromagnetic waves or energy particles. The resulting radiation cannot be detected by the normal senses of smell and taste. Several instruments have been developed that can detect the presence of radiation and measure dose rates. All responders should become familiar with the radiation detectors used by their departments. Do not enter any place that may contain radiation unless you have been trained in measuring radiation, have a working radiation detector, and are properly protected.

Three types of radiation exist: alpha particles, beta particles, and gamma rays.

- Alpha particles quickly lose their energy and, therefore, can travel only 1 to 2 inches from their source. Clothing or a sheet of paper can stop this type of energy. If ingested or inhaled, alpha particles can damage a number of internal organs.
- Beta particles are more powerful, capable of traveling 10 to 15 feet. Heavier materials such as metal, plastic, and glass can stop this type of energy. Beta radiation can be harmful to both the skin and eyes, and ingestion or inhalation will damage internal organs.
- Gamma radiation can travel significant distances, penetrate most materials, and pass through the body. Gamma radiation is the most destructive to the human body. The only materials with sufficient mass to stop gamma radiation are concrete, earth, and dense metals such as lead.

Effects of Radiation

Symptoms of low-level radioactive exposure might include nausea and vomiting. Exposure to high levels of radiation can also cause vomiting and digestive system damage within a short time. Other symptoms of high-level radiation exposure include bone marrow destruction, nerve system damage, and radioactive skin burns. An extreme exposure may cause death rapidly; more typically, it takes time before the signs and symptoms of radiation poisoning become obvious. As the effects progress over the years, a prolonged spiral into death caused by leukemia or carcinoma may occur.

A person contaminated by radioactive material will have particulates on the exposed skin and clothing. For this reason, decontamination procedures are required to remove the radioactive materials. Decontamination must be complete and thorough because the medical condition of the contaminated person could deteriorate quickly if this procedure is inadequate. Additionally, the contamination could quickly spread to other responders, ambulances, hospitals, and other locations where contaminated individuals or items are transported.

Radioactive contamination poses a risk to both rescuer and victim. Anyone who comes in contact with a contaminated person must be protected by appropriate PPE, depending on the particular contaminant; however, the only way to know whether radiation is present on individuals' bodies or clothing is to use radiation detection devices.

A person can also be injured by "exposure" to radiation without being contaminated by the radioactive material itself. Radiation exposures occur from certain beta and/or gamma emitters absent skin contamination. Someone who has been exposed to radiation, but who has not been contaminated by radioactive materials, requires no decontamination as the damage is internal to the body.

There are three ways to limit exposure to radioactivity:

- Keep the *time* of the exposure as short as possible.
- Stay as far away (*distance*) from the source of the radiation as possible.
- Use *shielding* to limit the amount of radiation absorbed by the body.

In addition, do not enter any area that might contain radiation or radioactive contamination until trained personnel have assessed the radiation level using approved detection devices. If radioactive contamination or a radiation source is suspected, everyone who enters the area should be equipped with proper PPE; an appropriate radiation detection device and personal dosimeter to measure the amount of radioactive exposure **FIGURE 6-23** and **FIGURE 6-24**.

The chapter *Hazardous Materials: Properties and Effects* provides a more thorough review of radiation and its potential health effects; the chapter *Mission-Specific Competencies: Air Monitoring and Sampling* provides an overview of radiation detection devices.

Indicators of Radiological Agents

Indicators of radiological agents may include production or containment equipment, such as lead or stainless steel containers (with nuclear or radiological labels), and equipment that may be used to detonate the radioactive source, such as containers (e.g., pipes), caps, fuses, gunpowder, timers, wire, and

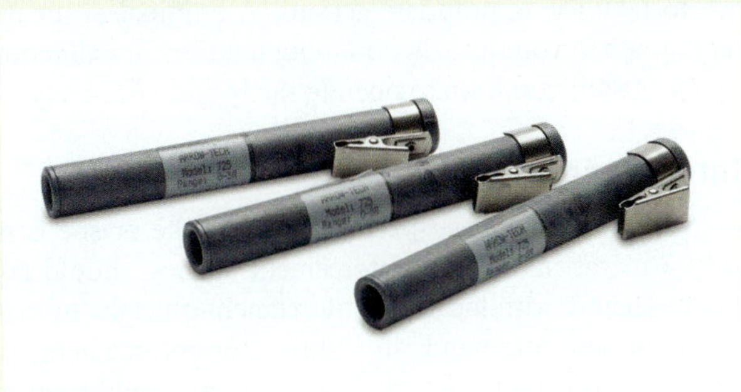

FIGURE 6-23 A personal dosimeter measures the amount of radioactive exposure received by an individual.
Courtesy of Fluke Biomedical

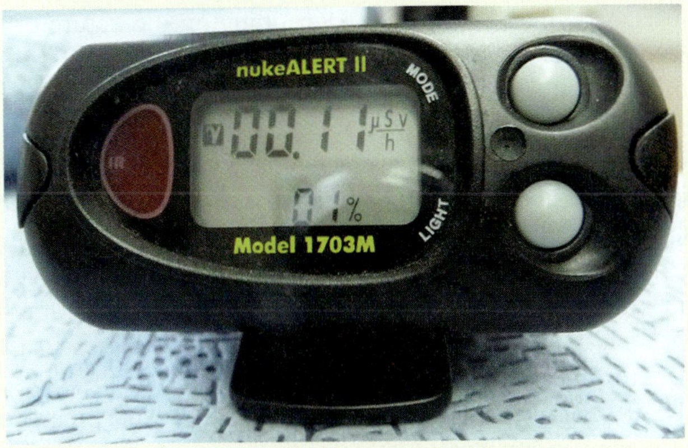

FIGURE 6-24 Only properly trained and equipped personnel, each of whom must carry an approved radiation monitor to measure the amount of radiation present, should make rescue attempts.
Courtesy of Rob Schnepp

detonators. Radiation monitoring equipment such as Geiger counters or radiation pagers may be evident as well. Personnel working around the radiological agents may exhibit exposure symptoms such as burns or experience difficulty breathing. It is unlikely you will encounter sophisticated radiation handling equipment outside of a laboratory, but the possibility does exist.

The Dirty Bomb

In recent years, the radiation dispersal device (RDD) or "dirty bomb" has emerged as a source of serious concern in terms of terrorism. Although not considered a weapon of mass destruction, an RDD has been described by NFPA as "any device that causes the purposeful dissemination of radioactive material across an area without a nuclear detonation." Packing radioactive material around a conventional explosive device could contaminate a wide area, with the size of the affected area ultimately depending on the amount of radioactive material and the power of the explosive device. There are only a select number of radioactive sources that can be used effectively in an RDD. It is possible for a criminal to construct a nonexplosive RDD, disseminating radioactive material via pressurized sprayers or air handling systems in buildings. To limit this threat, radioactive materials, even in small amounts, are kept secure and protected. Such materials are widely used in industry and health care and a criminal could potentially construct a dirty bomb with just a small quantity of stolen radioactive material.

Other types of radiological devices that may be used as weapons include improvised nuclear devices (IND) and radiation exposure devices (RED):

- *Improvised nuclear device (INDs):* A weapon fabricated from fissile material (highly enriched uranium or plutonium) capable of producing a nuclear explosion. A generally accepted successful yield in the 10- to 20-kiloton range—the equivalent to 10,000–20,000 tons of dynamite—in addition to high levels of radiation, would make this a very lethal device.
- *Radiation exposure device (REDs):* Radioactive material in a sealed container located where persons nearby would receive a direct exposure. Although not considered a device that would cause widespread death or sickness, it could constantly impact a steady stream of persons over a given period of time. If the container failed, then contamination might occur because the material could come into contact with victims.

Operations

Although responders must ensure their own safety at every incident, a terrorist incident may carry an extra dimension of risk. Because the terrorist's objective is to cause as much harm as possible, emergency responders are just as likely to be targets as are ordinary civilians.

In most cases, the first responder emergency units will not be dispatched for a known WMD or terrorist incident. Rather, the initial dispatch report might cite an explosion, a possible hazardous materials incident, a single person with difficulty breathing, or multiple victims with similar symptoms. Emergency responders will usually not know that a terrorist incident has occurred until personnel on the scene begin to piece together information gained from their own observations and from interviews with witnesses.

If appropriate precautions are not taken, the initial responders may find themselves in the middle of a dangerous situation before they realize it. For this reason, initial responders should take note of any factors that suggest the possibility of a terrorist incident and immediately implement appropriate procedures. The possibility of a terrorist incident should be considered when responding to any location that has been identified as a potential terrorist target. It could be difficult to determine the true nature of the situation until a scene size-up is conducted.

Initial Actions

Responders should approach a known or potential terrorist incident just as they would a hazardous materials incident. If possible, apparatus and personnel should approach the scene from a position that is uphill and upwind. Emergency responders should don PPE, including SCBA. Later arriving units should be staged an appropriate distance away from the incident.

The first units to arrive should establish an outer perimeter to control access to and from the scene. They should deny access to all persons except emergency responders, and they should prevent potentially contaminated individuals from leaving the area before they have been decontaminated. The perimeter must completely surround the affected area, with the goal being to keep people who were not initially involved from becoming additional victims. This operation sounds quite orderly while you are reading this text, but rest assured that a real-world scene would be chaotic. There will be widespread panic and you will probably be overwhelmed. It will take some time and many more responders before you get a handle on the scene. Expect this kind of tumultuous environment if the incident is significant.

Incident command should be established in a safe location, which could be as far as 3000 feet away from the actual incident scene. The incident command post must be set up outside the area of possible contamination and beyond the distance where a secondary device may be planted. The initial task should be to determine the nature of the situation, the types of hazards that could be encountered, and the magnitude of the problems that must be faced.

An initial reconnaissance (recon) team should be sent out to quickly examine the involved area and to determine how many people are involved. Proper use of PPE, including SCBA, is essential for the recon team, and the initial survey must be conducted very cautiously, albeit as rapidly as possible. The possibility that chemical, biological, or radiological agents are involved cannot be ruled out until qualified personnel with appropriate instruments and detection devices have surveyed the area. Responders should begin their reconnaissance mission from a safe distance, working inward toward the scene, while taking care not to touch any liquids or solids or to walk through pools of liquid. Emergency responders who become contaminated must not leave the area until they have been decontaminated.

A process of elimination may be required to determine the nature of the situation. Occupants and witnesses should be asked if they observed any unusual packages or detected any strange odors, mists, or sprays. The presence of a large number of casualties with no outward signs of trauma could indicate a possible chemical agent exposure. In such a case, victims' symptoms might include trouble breathing, skin irritations, or seizures.

A visible vapor cloud would be another strong indicator of a chemical release. Such a release could have occurred as the result of either an accident or a terrorist attack. The presence of dead or dying animals, insects, or plant life might also point to a chemical agent release as the culprit.

When approaching the scene of an explosion, responders should consider the possibility of a terrorist bombing incident and remain vigilant for secondary explosive devices. They should note any suspicious packages and notify the IC immediately. Responders who have not been specially trained should never approach a suspicious object. Instead, EOD personnel should examine any suspicious articles and disable them.

The guidelines presented in this section are general recommendations, of course. You must also rely on your agency's standard operating procedures, your training, and your experience to take the appropriate actions. A criminal event may never happen in your jurisdiction—but if it does, it will require every bit of skill you have to operate safely and effectively.

Interagency Coordination

If a terrorist incident is suspected, the IC, if he or she is not already a member of a law enforcement agency, should consult immediately with local law enforcement officials. In many cases, a unified command, including law enforcement, fire department, and emergency medical services, should be established. If there are casualties, or if a mass casualty situation is evident, the IC should notify area hospitals and activate the mass casualty incident (MCI) medical plan (if one exists). Typically, local hospitals will begin to conduct an open-bed count and communicate with other hospitals about the availability of specialized medical services. Depending on the nature of the event, specialists like tactical law enforcement (SWAT teams) and bomb techs would be involved in the mitigation of the event FIGURE 6-25.

State emergency management officials should be notified as soon as possible. This will help ensure a quick response by both state and federal resources to a major incident. Regional emergency operation centers (REOCs) and state-level emergency operation centers (EOCs) may be established, depending on the severity of the incident. Large-scale search-and-rescue incidents could require the response of Urban Search and Rescue (USAR) task forces activated through the Federal Emergency Management Agency (FEMA). Medical response teams, such as disaster medical assistance teams (DMATs), may be needed for incidents involving large numbers of people. The CDC Strategic National Stockpile (SNS) may be requested when large caches of life-saving pharmaceuticals such as antidotes and medical supplies are needed.

An EOC can help to coordinate the actions of all involved agencies in a large-scale incident, particularly if terrorism is involved. The EOC is usually set up in a predetermined remote location and is staffed by experienced command and staff

FIGURE 6-25 Tactical law enforcement teams may be required to render the area safe for fire and EMS responders to mitigate additional problems and render patient care.

Courtesy of Rob Schnepp

personnel FIGURE 6-26. The IC, who remains at the scene of the incident, should provide detailed situation reports to the EOC and request additional resources as needed.

Responders must remember that a terrorist incident is also a crime scene. To avoid destroying important evidence that could lead to a conviction of those responsible for perpetrating the attack, responders should not disturb the scene any more than is necessary. Where possible, law enforcement personnel should be consulted prior to overhaul and before the removal of any material from the scene. Responders should also realize that one or more terrorists could be among the injured. Be alert for threatening behavior, and make note of anyone who seems determined to leave the scene. The chapter *Mission-Specific Competencies: Evidence Preservation and Sampling* provides more information on this topic.

FIGURE 6-26 An EOC is set up in a predetermined location for large-scale incidents.

© Jones & Bartlett Learning. Courtesy of MIEMSS

Decontamination

Everyone who is exposed to chemical, biological, or radiological agents must be decontaminated to remove or neutralize any chemicals or substances on their bodies or clothing. Decontamination should occur as soon as possible to prevent further absorption of a contaminant and to reduce the possibility of spreading the contamination. Likewise, equipment must be decontaminated before it leaves the scene. The chapters *Mission-Specific Competencies: Technical Decontamination* and *Mission-Specific Competencies: Mass Decontamination* explain decontamination procedures in detail.

The perimeter established for the incident response must fully surround the area of known or suspected contamination. Every effort must be made to avoid contaminating any additional areas, particularly hospitals and medical facilities. Qualified personnel must monitor the perimeter with instruments or detection devices to ensure that contaminants are not spread.

Standard decontamination procedures usually involve processing of individuals through a series of stations. At each station, clothing and protective equipment are removed and the individual is cleaned. Although some contaminants require only soap and water for their removal, elimination of chemical and biological agents may require the use of special neutralizing solutions. Because of the chemical reaction that takes place during neutralization, this may not be the preferred method of decontamination.

A terrorist incident that results in a large number of casualties may require a somewhat different decontamination process. Special procedures for mass decontamination have been devised for incidents involving large numbers of people. These procedures range from the use of master stream devices from engine companies and aerial apparatus to create high-volume, low pressure showers, to complete systems that include specially designed tents with preplumbed sprayers and conveyors for nonambulatory patients. In either case, this approach allows large numbers of people to be decontaminated rapidly FIGURE 6-27.

FIGURE 6-27 Mass-decontamination procedures may be required to handle large groups of contaminated victims.

Courtesy of Rob Schnepp

Mass Casualties

A terrorist or WMD incident may result in a large number of casualties—hundreds or thousands of injuries in some scenarios. Special mass casualty plans are essential to manage this type of situation, which would quickly overwhelm the normal capabilities of most emergency response systems. Mass casualty plans typically require resources from multiple agencies to handle large numbers of patients efficiently. A terrorist incident, however, might potentially involve several additional complications and considerations, including the possibility of contamination by chemical, biological, or radiological agents. The mass casualty plan must be expanded to address these problems.

If contamination is suspected, the plan must ensure that it does not spread beyond a defined perimeter. In some cases, patients may need to be decontaminated as they are moved to the triage and treatment areas so as to keep these areas free of contaminants FIGURE 6-28. In other cases, the triage and treatment areas may be considered contaminated zones. In this scenario, patients would be decontaminated before they are transported from the scene.

FIGURE 6-28 Patients should be decontaminated before they are placed in triage and treatment areas.

During the early stages of an incident, it may be difficult to determine which agent was used and, therefore, which treatment and decontamination procedures are appropriate. In such a case, it may be necessary to quarantine exposed individuals until further analysis is done. Field categorization of the substance may involve performing a series of chemical tests on a sample of the agent.

Do not withhold decontamination when life safety is paramount: Default to a water or soap-and-water decontamination so that medical treatment can be rendered promptly. The specific actions to be taken will always be dictated by your size-up of the situation and the most prudent course of action based on the nature of the contaminant and the circumstances under which it was released.

Wrap-Up

Chief Concepts

- Terrorism is the unlawful use of violence or threats of violence to intimidate or coerce a government, the civilian population, or any segment thereof, to further political or social objectives. This broad definition encompasses a wide range of acts committed by different groups for different purposes.
- The goal of terrorism is to produce feelings of fear in a population or a group.
- Terrorism can occur in any community, so it is essential that responders be aware of all potential targets in their area.
- Terrorists can and do turn ordinary objects into weapons.
- Secondary devices are intended to explode sometime after the initial device detonates.
- Weapons of mass destruction include chemical, biological, and radiological agents, as well as conventional weapons and explosives.
- As part of the response to a potential terrorism incident, it is important to be able to identify which type of agent is involved.
- When dealing with a potential terrorist-related incident, responders should establish a staging area at a safe distance from the scene and follow the direction of the incident commander.
- Interagency coordination is an important part of responding to a terrorist event.

Hot Terms

Agroterrorism The intentional act of using chemical or biological agents against the agricultural industry or food supply.

Alpha particles A type of radiation that quickly loses energy and can travel only 1 to 2 inches from its source. Clothing or a sheet of paper can stop this type of energy. Alpha particles are not dangerous to plants, animals, or people unless the alpha-emitting substance has entered the body.

Ammonium nitrate fertilizer and fuel oil (ANFO) An explosive made of commonly available materials.

Anthrax An infectious disease spread by the bacterium *Bacillus anthracis*; typically found around farms, infecting livestock.

Beta particles A type of radiation that is capable of traveling 10 to 15 feet from its source. Heavier materials, such as metal, plastic, and glass, can stop this type of energy.

Biological agents Disease-causing bacteria, viruses, and other agents that attack the human body.

Blister agents A chemical that causes the skin to blister. Also known as a vesicant.

Blood agents Chemicals that interfere with the utilization of oxygen by the cells of the body. Cyanide is an example of a blood agent.

Chlorine A yellowish gas that is approximately 2.5 times heavier than air and slightly water soluble. Chlorine has many industrial uses. It damages the lungs when inhaled; it is a choking agent.

Choking agent A chemical designed to inhibit breathing and typically intended to incapacitate rather than kill its victims.

Cyanide A highly toxic chemical agent that prevents cells from using oxygen.

Cyberterrorism The intentional act of electronically attacking government or private computer systems.

Ecoterrorism Terrorism directed against causes that radical environmentalists think would damage the earth or its creatures.

Explosive ordnance disposal (EOD) personnel Personnel trained to detect, identify, evaluate, render safe, recover, and dispose of unexploded explosive devices.

Forward staging area A strategically placed area, close to the incident site, where personnel and equipment can be held in readiness for rapid response to an emergency event.

Gamma radiation A type of radiation that can travel significant distances, penetrating most materials and passing through the body. Gamma rays are the most destructive type of radiation to the human body.

Improvised explosive device (IED) An explosive or incendiary device that is fabricated in an improvised manner.

Incubation period The time period between the initial infection by an organism and the development of symptoms by a victim.

Lewisite A blister-forming agent that is an oily, colorless-to-dark-brown liquid with an odor of geraniums.

Mass decontamination The physical process of reducing or removing surface contaminants from large numbers of victims in potentially life-threatening situations in the fastest time possible.

Muscarinic effects Effects such as runny nose, salivation, sweating, bronchoconstriction, bronchial secretions, nausea, vomiting, and diarrhea.

Nerve agent A toxic substance that attacks the central nervous system in humans.

Personal dosimeter A device that measures the amount of radioactive exposure incurred by an individual.

Phosgene A chemical agent that causes severe pulmonary damage.

Pipe bomb A device created by filling a section of pipe with an explosive material.

Plague An infectious disease caused by the bacterium *Yersinia pestis*, which is commonly found on rodents.

Pulmonary agent Chemicals that cause severe damage to the lungs and lead to asphyxia. Chlorine and phosgene are examples. Pulmonary agents may also be referred to as choking agents.

Radiation dispersal device (RDD) Any device that causes the purposeful dissemination of radioactive material without a nuclear detonation; a dirty bomb.

Radiological agents Materials that emit radioactivity (the spontaneous decay or disintegration of an unstable atomic nucleus accompanied by the emission of radiation).

Sarin A liquid nerve agent that is primarily a vapor hazard.

Secondary device An explosive or incendiary device designed to harm emergency responders who have responded to an initial event.

Smallpox A highly infectious disease caused by the *Variola* virus.

Soman A nerve gas that is both a contact and a vapor hazard; it has the odor of camphor.

Sulfur mustard A clear, yellow, or amber oily liquid with a faint sweet odor of mustard or garlic that may be dispersed in an aerosol form. It causes blistering of exposed skin.

Tabun A nerve agent that disables the chemical connections between nerves and targets organs.

Terrorism As defined by the Code of Federal Regulations, the unlawful use of force and violence against persons or property to intimidate or coerce a government, the civilian population, or any segment thereof, in furtherance of political or social objectives.

Triage The process of sorting victims based on the severity of their injuries and medical needs to establish treatment and transportation priorities.

Universal precautions Procedures for infection control that treat blood and certain bodily fluids as capable of transmitting bloodborne diseases.

V agent (VX) A nerve agent, principally a contact hazard; an oily liquid that can persist for several weeks.

Weapons of mass destruction (WMD) Weapons whose use is intended to cause mass casualties, damage, and chaos. The NFPA includes the use of WMD in its definition of *hazardous materials*.

RESPONDER *in action*

During the afternoon rush hour, an engine company and a basic life support ambulance respond to a report of an unconscious person at a bus station. En route, the dispatcher advises you that there are now several patients feeling ill at this location.

1. The bus station is a confusing scene, with many passengers entering and exiting the structure. You see several individuals sitting on the sidewalk close to the entrance. They look disoriented and confused. What is your first priority?

A. Without chemical protective gear, enter the bus station to search for additional victims.

B. Immediately transport all patients to the closest hospital.

C. Establish a perimeter around the scene and prevent anyone from entering the bus station. Notify local law enforcement because you suspect an event with criminal intent.

D. Set up a fan to blow fresh air into the bus entrance.

2. A witness tells you that several individuals came out of the bus station complaining of burning eyes, a choking sensation, nausea, and dizziness. The witness does not have any information about the situation inside the station. Which actions should be taken now?

A. Establish an incident command post and request additional units.

B. Advise the bus authority to prevent the discharge of any additional passengers at this station.

C. Move patients away from the bus entrance and establish a treatment area.

D. All of the above.

3. As hazardous materials teams arrive, one of the first assignments is to enter the bus station to rescue anyone who is still inside and determine if there are any additional patients. The crew that is assigned to enter the station should wear:
   - **A.** Latex gloves.
   - **B.** Firefighting PPE and SCBA.
   - **C.** Level B hazardous materials ensembles.
   - **D.** Level A hazardous materials ensembles.

4. Two additional patients who are coughing and have extremely red, watery eyes exit from the station. These individuals report that they observed a canister on the floor of the station leaking a clear liquid. The incident commander should issue the following order:
   - **A.** Order an engine company to begin decontaminating the patients.
   - **B.** Order ambulances to transport all of the patients to a nearby hospital immediately.
   - **C.** Send the crew back into the station to retrieve the canister and bring it outside for close examination.
   - **D.** Assign a hazardous materials team to attempt to identify the substance.

Fire Smoke: Responder Health and Safety

Greg Henry/ShutterStock, Inc.

Courtesy of Rob Schnepp

CHAPTER 7

Awareness Level Personnel and Operations Level Responders

Knowledge Objectives

After studying this chapter, you will be able to:

- Discuss the hazards of fire smoke. (p. 160–164)
- Identify different classifications of fire gases generated during the typical combustion process encountered in various types of fires. (p. 162–167)
- Discuss the need for wearing respiratory protection during and after active firefighting. (p. 158–160)
- Discuss the various technologies available for fire-ground detection and monitoring. (p. 168–170)
- Discuss general fire-ground monitoring principles and practices. (p. 170–171)
- Understand the need for prompt recognition and treatment of smoke inhalation. (p. 162–165)
- Identify NFPA 473, Chapter 6.4, *Mission-Specific Competencies Advanced Life Support (ALS) Responder Assigned to Treatment of Smoke Inhalation Victim*, as a reference source for treating smoke-related illness and injury. (p. 165)
- Identify appropriate clinical interventions for treating smoke inhalation. (p. 165)

Skill Objectives

This chapter has no skills objectives.

You Are the Responder

Jones & Bartlett Learning. Photographed by William Larkin

Just before lunch, your engine company responds to a working fire in a single-family wood frame dwelling. The fire started in a bedroom and extended into the adjacent hallway. Your apparatus is first in and your crew extends the initial attack line, makes access to the seat of the fire, and quickly knocks it down. The fire is declared under control and the crew transitions to overhaul. A few minutes later, your officer directs the crew to exit the building. After a short rehab period, your crew reenters to assist with overhaul, wearing bunker pants, jacket, helmet, gloves, and eye protection. Your engine company remains on scene for another hour, mopping up and eventually turning the building back over to the homeowner. The crew returns to quarters and places the engine back in service. You have a headache, mild nausea, and a feeling of fatigue. It doesn't resolve over the remainder of the shift, and you go home the next morning still not feeling 100%.

1. Do you believe the ill feeling you have is related to the smoke exposure? If so, what do you think you were exposed to?
2. Should air monitoring be conducted at the scene prior to reentering the structure for overhaul? At what point would you deem the environment safe to operate without SCBA?
3. If you began feeling worse either at the firehouse or later at home, and sought medical attention at a local hospital, do you think you would receive medical treatment appropriate for a smoke exposure?

Introduction

Research conducted over the years has proven beyond question that breathing smoke is bad for you. There may be debate about how much of a certain fire gas is present at a given fire scene or what combustion byproducts are identified in certain studies, but the fact remains that every time something burns, the combustion process liberates fire gases (many of which are toxic) and particulates.

When asked about fire fatalities, fire fighters typically observe that smoke kills people before the flames ever get to them, and fire death statistics prove likewise. Fire fighters will also acknowledge that after working at a fire and breathing smoke during the firefight, or perhaps during the overhaul phase, they don't feel well in the hours and sometimes days following a fire. Headache, nausea, dizziness, and fatigue are common signs and symptoms that can be traced back to breathing smoke from all types of fires.

U.S. Fire Administration fire death statistics for the United States show that fire-related mortality is declining overall, but the 2010 fire death data report lists approximately 3445 fire-related deaths. This places the United States within the top 15 countries in the world for per capita fire death. Many thousands of fire fighters and civilians also suffer from smoke-related illnesses (such as the one described in the opening scenario).

After the fire is knocked down, it is not uncommon to see fire fighters remove their self-contained breathing apparatus (SCBA) and perform overhaul without respiratory protection, breathing smoke at the point in time when smoldering debris is producing volumes of particulates and a variety of gases. It is becoming more common for fire agencies to mandate the use of SCBA for longer periods of time on the fire ground—during and after the active firefighting phase—but there is a long way to go to change the perception of some fire fighters that breathing smoke is just part of the job. Like a scuba diver, a fire fighter advancing a hose line during an interior attack at a working structure fire is completely enveloped by the operating environment—the smoke—and if a supply of breathable air is lost or interrupted, the environment invades the body, causing harm **FIGURE 7-1**.

To underscore this point, consider the case of Captain James Carter from Springdale, Arkansas, as told in the following *Voices of Experience*.

FIGURE 7-1 Think of yourself as a "smoke diver" when you are working at a structure fire.

Courtesy of Rob Schnepp

VOICES OF EXPERIENCE

On January 13, 2012, I was involved in an accident on a fire ground that changed my family and me forever. I am a captain on an engine company with more than 20 years of fire service experience. My engine company was dispatched to a residential structure fire with victims. On our arrival, we found heavy smoke showing from a single-family dwelling. The fire load in this residence was extreme and was well involved.

The postfire interior conditions.

During interior attack operations, my face mask was knocked off. It was off for only a few seconds—time for approximately four breaths. I was on a hose line under a heavy workload, breathing hard (full tidal volume). The pain was instant—a burning sensation in my entire respiratory tract, and I was unable to get a good breath. I immediately exited the structure. Once outside I removed my mask, and it was difficult to breathe. Multiple times, fire department personnel checked on me to see if I was ok. I told them, "I just need to catch my breath." I continued fire-ground operations after a short while, ignoring the symptoms I was having. At this point, I did not know how seriously I was injured. My memory starts to fade on events that happened at the fire, 10–15 minutes into the incident. To this day, I can't remember all the events on the fire ground. I have read radio logs and reports, but it is not in my memory.

Another engine company relieved my company, and we returned to quarters. I still was having difficulty breathing with an intense headache. My conditions did not improve, and I requested medical treatment. I was taken to the local emergency room for evaluation. I was treated at that facility and released. Within 1 to 2 days, my condition worsened again with respiratory distress and headaches. For the next 2 1/2 weeks, I spent approximately 1 week's time in and out of the hospital. I was off of work on supplemental oxygen for 2 months and light duty for 9 months. As I write this, it has been 18 months, and I am still being treated for neurological and respiratory problems.

My lesson to you is to realize the dangers of smoke and smoke inhalation. Take time to learn the contents of the smoke that surrounds you in a fire, and know the effects. Take any exposure as seriously as if your life depends on it. Stay safe and train like your family would want you to so that you can go home at the end of your shift.

Captain James Carter
Springdale Fire Department
Springdale, Arizona

The goal of this chapter is to provide an overview of the hazards of fire smoke and highlight a linkage between smoke exposure and the resulting acute and chronic health effects. The anticipated result is a modification of the fire fighter's attitude toward smoke and an increased interest in better managing repeated dermal and inhalation exposures. This chapter is aimed mostly toward rank and file fire fighters, but other groups of emergency responders such as law enforcement, emergency medical service providers, fire investigators, and fire training instructors will find the information useful.

Simply put, if you go into situations where smoke is present you should be informed about the potential health hazards. Envision smoke as reaching out and touching nearly every aspect of the life of anyone who routinely works in or around it. **FIGURE 7-2** represents some of the places and situations in which people interact with particulates and gaseous toxins.

Ideally, the authority having jurisdiction (AHJ) should establish reasonable and safe work practices, but at the most basic level, it is your responsibility to be informed about the hazards of breathing smoke and the need to take all precautions for reducing your repeated exposure. Keeping bunker gear clean is a good example, as is taking a shower after working at a fire scene and ensuring crews are wearing SCBA during active fireground operations, even when engaged in exterior operations or working on a roof. Think about how overhaul is carried out in your organization and whether or not you are doing everything possible to reduce your dermal and inhalation exposure.

Some questions to ask about your fire scene work practices include:

- Do crews wear SCBA until the building is clear of visible particulates (smoke)?
- If you are doing postfire detection and monitoring, have you established action levels (keeping face mask on) for the gases you are monitoring?
- Are personnel trained to understand the benefits and limitations of detection and monitoring at the fire scene, and the operational parameters of the technology used to perform the detection and monitoring?
- Are rapid intervention crews located so close to the building that they are breathing smoke prior to being called into service? Is there a better place for those crews to set up and still be available if the need arises?
- Is the command post located in drift smoke?
- Does your prehospital care system have a treatment protocol for smoke inhalation?
- Is the local receiving hospital ready to treat a civilian or fire fighter in the event of a significant smoke-related illness (like the one suffered by Captain Carter)?

Think through the potential pathways of a smoke exposure, and identify ways you could reduce or address the places and situations whereby you could be exposed.

Overview of Fire Smoke

Even though smoke is the constant companion of the fire fighter, there is not always a thorough understanding of the link between what's in smoke and the reason people get sick or die when they breathe it, or why fire fighters contract heart disease, neurological dysfunction, cancer, or some other illness

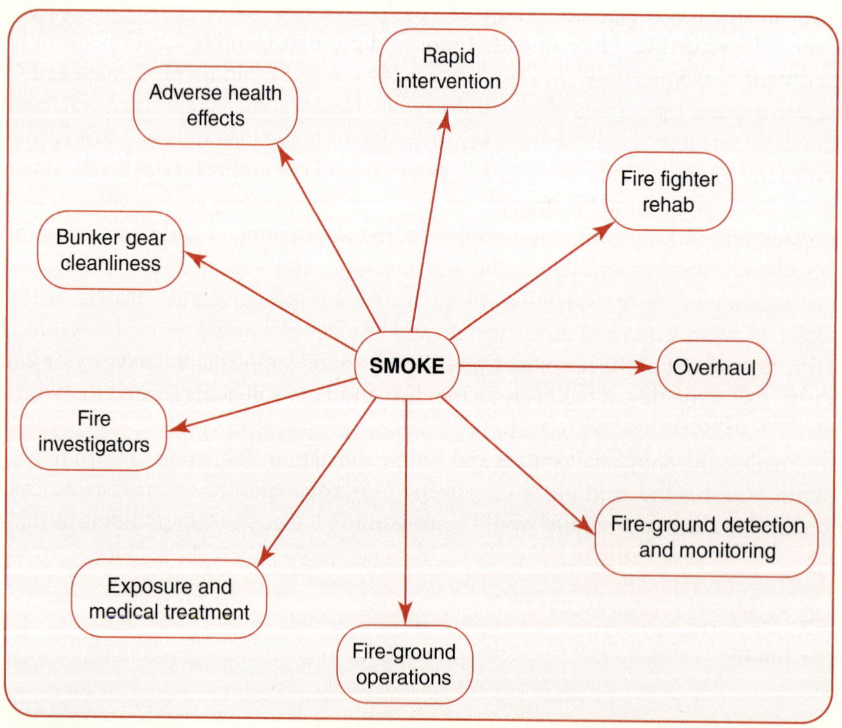

FIGURE 7-2 Think about all the ways that smoke exposures can reach out and affect a fire fighter.

related to a single or repeated exposure to smoke. Cancer is prevalent in the fire service, more so than most any other profession, and this should be no surprise once the byproducts of combustion are understood. Known human carcinogens typically are liberated as fire gases and particulates. (This will be discussed later in this chapter.)

Most people identify carbon monoxide as the main harmful component of fire smoke but struggle to list more than a handful of substances beyond that. Less often acknowledged are compounds such as ammonia, hydrogen chloride, sulfur dioxide, hydrogen sulfide, hydrogen cyanide, carbon dioxide, the oxides of nitrogen, formaldehyde, acrolein, polycyclic aromatic hydrocarbons, and soot FIGURE 7-3.

Smoke production depends on several factors, including the chemical make-up of the burning material, the temperature of the combustion process, and the influence of ventilation (oxygenation). Make no mistake that fire (combustion) is a complex process and the smoke produced is an intricate collection of particulates, superheated air, and gaseous chemical compounds.

RESPONDER TIP

Smoke is a collection of gaseous products from burning materials—especially organic materials such as plastic, nylon, rubber products, wood, and paper—made visible by the presence of small particles of incomplete combustion (carbon).

The extensive use of synthetic manufacturing and construction materials (e.g., plastics, nylon, rubber products, fire retardants, laminates, and foams such as Styrofoam and polystyrene) has a significant effect on fire behavior and smoke production. Synthetic substances ignite and burn fast, causing rapidly developing fires and toxic smoke and making structural firefighting more dangerous than ever before.

Polyurethane foam is the most predominant substance in a typical mattress and is made up of many different chemicals including polyol (an organic alcohol molecule and the majority of the polyurethane compound), toluene diisocyanate (TDI), methylene chloride, and ammonia-based catalysts. When polyurethane foam is exposed to heat, the parent substances break down and bond with each other, creating many new compounds. Some of those compounds are irritants, such as hydrogen chloride and ammonia, causing eye irritation or airway problems in smoke exposures. Other compounds, like carbon monoxide and cyanide compounds, are acutely toxic when inhaled.

The thermal decomposition of polyurethane foam in the mattress fire scenario is broadly representative of the way gases are liberated by a working fire. (Keep in mind this is a very limited representation of the process; many more substances are also liberated.) Certainly, each substance is present in varying levels depending on the material(s) involved, the heat of combustion, and the available oxygen (ventilation of the fire). Aside from the toxic gases produced, it is equally important to recognize the visible part of combustion (aside from the flames). These are carbon particles of varying sizes, ranging from large embers to particles you can't see with the naked eye. This collection of particulates—what we traditionally called smoke, is actually soot. Soot is important to acknowledge because it is a known human carcinogen, just like benzene and formaldehyde—also byproducts of combustion. With that in mind, you should take a moment and think about your perception of smoke and revise it to this—the combustion process liberates things you can see (particulates) and things you can't see (gases). TABLE 7-1 breaks "smoke" down into those two main categories to make it easier to understand and separate the hazards. The matrix isn't inclusive of all substances that could fall into the four boxes, but it does illustrate the point that there are many properties in smoke that cause acute and chronic health effects. It also shows that smoke is not only one *thing*; rather, it is a dynamic multifaceted mixture of gases and particulates that changes from minute to minute at any fire.

Another group of compounds commonly found in fire smoke is polycyclic aromatic hydrocarbons (PAH), classified as probable or possible human carcinogens. There are well over 100 PAHs identified and categorized by various regulatory agencies. This group of substances occurs naturally in materials such as coal or crude oil. These substances also are generated during the combustion of organic materials, and can be found in vehicle exhaust, tobacco smoke, and the smoke generated from structure fires, vehicle fires, wildland fires, or any other type of fire. PAHs can exist as a particle or a gas. Chances are that you have been exposed to PAHs throughout your life from a common and perhaps surprising activity—grilling food FIGURE 7-4.

When PAHs are generated during a structure fire they may bind with the soot, resulting in dermal and inhalation exposures. PAHs are believed to be immunosuppressant, perhaps contributing to the mechanism by which PAHs are suspected

FIGURE 7-3 A different way to look at fire smoke: If this cargo trailer were carrying smoke, this is how it might be placarded. If these placards were on the next house fire you went to, would you think differently about the potential health hazards?

Courtesy of Rob Schnepp

Table 7-1 Smoke Matrix

	Harm You Now	Harm You Later
What You See		Soot, Particulates
What you Don't See	CO, HCN, NOx, SO_2, HCL, H_2S	Aldehydes, Benzene, etc.

FIGURE 7-4 The thermal decomposition of muscle meats such as chicken, beef, and pork generates PAHs and heterocyclic amines (HCAs) when cooked over an open flame.

Courtesy of Rob Schnepp

to cause cancer. Some examples of polycyclic aromatic hydrocarbons are:

- Anthracene
- Benzopyrene
- Methylchrysene
- Phenanthrene
- Pyrene

Not only do responders have to be concerned about acute exposure to PAHs and other materials, it is also important to consider this group of chemicals as contaminants to fire fighter bunker gear. A 2013 study, *Evaluation of Dermal Exposure to Polycyclic Aromatic Hydrocarbons in Fire Fighters*, released by the Health Hazard Evaluation Program of the U.S. Department of Health and Human Services, Centers for Disease Control and Prevention, National Institute for Occupational Safety and Health, discusses PAHs and other substances in terms of fire fighter dermal exposure during firefighting, overhaul, and postfire activities. In short, the study illustrates that it is possible to absorb PAHs and other particulates through your skin and that dirty bunker gear is unhealthy and is likely to contribute to sustained exposure to the fire, long after you've left the scene. NFPA 1851, *Standard on Selection, Care, and Maintenance of Protective Ensembles for Structural Fire Fighting and Proximity Fire Fighting*, offers guidance on bunker gear cleaning and maintenance.

Carbon Monoxide and Hydrogen Cyanide: Silent Killers

Smoke is one of the first observable signs of a working fire. Fire fighters note the volume, color, and force as smoke is exiting a fire building—all good indicators of what the fire is doing inside—and use that information to implement appropriate fire-ground tactics. Fire fighters may aggressively enter smoky buildings to search for victims but rarely perform a conscious evaluation of the toxic substances lurking in the smoke. This could be a significant oversight in terms of treating victims, according to studies performed in Paris, France, and Dallas County, Texas. These studies focused on carbon monoxide (CO) and hydrogen cyanide (HCN) specifically because they are acutely toxic, present to some degree in nearly all fires, and have clinical interventions available to reverse the adverse health effects of the exposure. Clearly many toxic substances are generated during a typical structure fire, and smoke inhalation is a complicated illness. It is also clear that successful medical treatment isn't accomplished by a single drug or single action, and that smoke inhalation victims are quite sick.

RESPONDER TIP

The discussion of CO and HCN isn't framed in the light that those two substances make smoke toxic. Rather, the important point is that they are toxic substances that are common byproducts of combustion (present at some level at most fires) and treatable with clinical interventions.

These studies were done more than 20 years ago, which underscores the fact that the presence of HCN and CO in smoke isn't a new revelation; however, the fire service and other disciplines have recently been connecting the dots to understand that smoke is a bigger issue than previously thought. In short, these studies identify and evaluate the impact of hydrogen cyanide and carbon monoxide on smoke inhalation patients. *Again, there are many other toxins found in smoke—the goal here is to highlight two fire gases found at nearly any kind of fire.*

The Paris study was designed to prospectively assess the role of cyanide in smoke-related morbidity and mortality. Blood samples were drawn from survivors as well as fatalities (at the time of exposure), and cyanide and carbon monoxide levels were measured. In several deaths, cyanide levels were in the lethal range whereas carbon monoxide levels were in nontoxic concentrations. This suggests cyanide toxicity as the primary cause of death. The study also revealed another bit of interesting information: death occurred in victims with cyanide *and* carbon monoxide levels in the nontoxic range, perhaps revealing a relationship between carbon monoxide and cyanide in smoke inhalation patients. The following summarizes the results of the Paris study:

- Cyanide and carbon monoxide were both important determinants of smoke inhalation–associated morbidity and mortality.
- Cyanide concentrations were directly related to the probability of death.
- Cyanide poisoning may be more predominant than carbon monoxide poisoning as a cause of death in certain fire victims.
- Cyanide and carbon monoxide may potentiate the harmful effects of one another.

The Dallas County study measured blood cyanide levels in victims after exposure to fire smoke, and in many respects echoed the findings of the Paris study. TABLE 7-2 shows a summary of the findings. In Dallas County over a 2-year period, blood samples were collected from a total of 187 smoke inhalation patients, within 8 hours of exposure. There were 144 viable patients at the University of Texas Health Sciences emergency department; 43 victims were dead on arrival at the Dallas County Medical Examiner's office.

Table 7-2 Cyanide Levels After Exposure to Fire Smoke

Patient #	Age, y	% Total Body Surface Area Burn	Blood Cyanide, mg/L	HbCO, %	Outcome
1	47	98%	1.20	18.6%	Died
2	80	3%	1.60	22.6%	Died
3	29	4%	1.40	6.0%	Died
4	22	55%	5.20	35.6%	Lived
5	30	63%	1.40	5.0%	Lived
6	58	32%	2.60	10.9%	Died
7	32	5%	6.00	32.0%	Lived
8	50	25%	2.20	17.2%	Lived
9	4	0%	11.50	22.4%	Died
10	36	40%	5.70	3.8%	Died
11	19	90%	1.20	40.0%	Died
12	30	76%	2.72	37.0%	Died

Reproduced from: Silverman SH et al. J Trauma 1988; 28:171–176.

Of the 144 living patients that reached the emergency room, 12 had blood cyanide concentrations exceeding 1 mg/L (see Table 7-2). Of these 12 patients, 8 eventually died. *None had blood carboxyhemoglobin (COHb) concentrations suggesting carbon monoxide as the cause of death (i.e., ≥ 50 percent)*. Although some of the patients had extensive burns that may have contributed to their death, three of them (patients 2, 3, and 9) had ≤ 4 percent total body surface area burns. According to the study, blood cyanide levels greater than 1 mg/L had a significant impact on patient outcome. More importantly, the study found that elevated cyanide levels were pervasive in smoke inhalation victims, and cyanide concentrations were directly related to the probability of death.

Lastly, both studies dispel a long-held belief in the fire service—that carbon monoxide is the predominant killer in fire smoke. In fact, it appears that cyanide plays a role in smoke-related death and injury, perhaps more often than we think. Therefore, any victim(s) exposed to significant amounts of smoke or rescued from a closed space structure fire may be suffering from cyanide toxicity. However, remember that there are many properties in smoke that are capable of causing acute illness and leaving an indelible impression on your body.

Morbidity: A disease or the incidence of disease within a population.
Mortality: The incidence of death in a population.

Depending on the dose, hydrogen cyanide has the ability to incapacitate a victim, preventing escape from the fire environment and thereby increasing the exposure to more cyanide, carbon monoxide, and other toxic byproducts of combustion. Although this theory is currently unsupported with human data related to smoke exposures, there is information to substantiate the "knock down" potential of cyanide. In the mid-1980s, studies were conducted on monkeys exposed to the fumes of heated polyacrylonitrile. (When this substance is broken down by pyrolysis, cyanide is liberated.) Cyanide-exposed monkeys first hyperventilated and then rapidly lost consciousness at a dose-dependent concentration. A concentration of 200 parts per million (ppm) was associated with rapid incapacitation, but not with elevated blood cyanide concentrations measured hours after exposure. The direct correlation to human data is unknown at present but could be interpreted in the following way: Hydrogen cyanide could be partly responsible for rendering fire fighters and civilians incapable of self-rescue when exposed to smoke.

RESPONDER TIP

Cyanide exposures can be fatal in the presence of normal oxygen levels.

To appreciate cyanide's mechanism of action, it is first necessary to understand the process of oxygen transportation and use in the body and the basic idea of aerobic metabolism. To simplify the concept, imagine the circulatory system as a very efficient transit system, full of "buses" (red blood cells) carrying passengers (oxygen) to and from a multitude of bus stops (the cells). The circulatory system, similar to a network of streets, is loaded with red blood cells (RBCs)—hemoglobin buses—each carrying four oxygen passengers FIGURE 7-5.

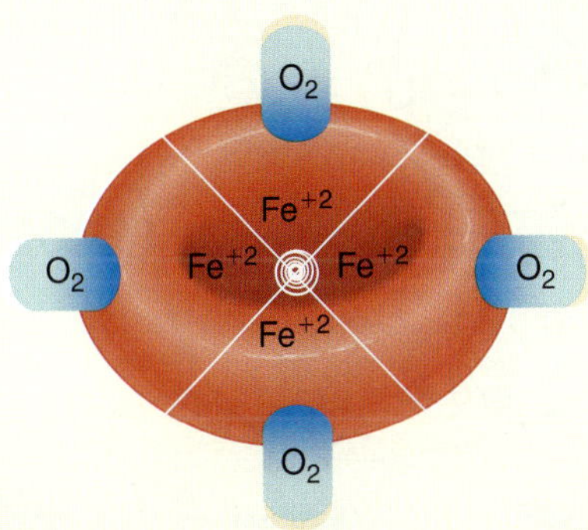

FIGURE 7-5 The circulatory system is loaded with red blood cells (RBCs), each of which carries four oxygen "passengers."

During normal cellular respiration, the bus system transports oxygen passengers to the bus stops (cells). At the appropriate stop, four oxygen molecules get off and move through an electron chain, ultimately combining with the final electron acceptor—cytochrome oxidase (an enzyme)—before entering the mitochondria of each cell.

Mitochondria are responsible for converting nutrients into energy-yielding molecules of adenosine triphosphate (ATP) to fuel the cell's activities FIGURE 7-6. ATP production is highly dependent on oxygen (and glucose); without it, normal aerobic metabolism is impossible. If this process is seriously compromised, death is imminent.

Aerobic metabolism: The creation of energy through the breakdown of nutrients in the presence of oxygen. The byproducts are carbon dioxide and water, which the body disposes of by breathing and sweating.
Anaerobic metabolism: The creation of energy through the breakdown of glucose.
Without oxygen, the metabolic process results in the production of lactic acid.

Cyanide compounds, once absorbed in the body, "poison" the cytochrome oxidase, barring oxygen from entering the mitochondria and effectively shutting down the process of aerobic metabolism. In short, the buses may be transporting some amount of oxygen passengers, but when they get off the bus, they find their ultimate destination locked. Without oxygen, the cells switch to anaerobic metabolism, producing toxic byproducts such as lactic acid, ultimately destroying the cell. Therefore, cyanide toxicity is not about the amount of oxygen available to the body; rather, it's about the inability of the body to *use* oxygen for aerobic (life-sustaining) metabolism. Consequently, an elevated lactic acid level is a key indicator of cyanide toxicity.

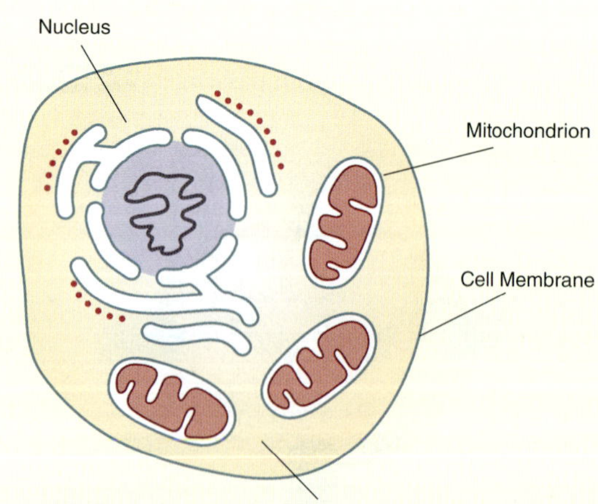

FIGURE 7-6 Mitochondria are responsible for converting nutrients into energy-yielding molecules of adenosine triphosphate (ATP) to fuel the cell's activities.

RESPONDER TIP

The signs and symptoms of acute cyanide toxicity are similar to those of CO poisoning and mimic the nonspecific signs and symptoms of oxygen deprivation including headache, dizziness, stupor, anxiety, rapid breathing, and increased heart rate. In extreme cases of cyanide poisoning, patients may present with seizures, a significant altered level of consciousness including coma, severe respiratory depression or respiratory arrest, and complete cardiovascular collapse.

According to the studies done in Paris and Dallas County, there is a possible deleterious relationship between carbon monoxide and cyanide in smoke inhalation victims. This relationship could be attributed to the inability of the cells to use oxygen (due to cyanide), coupled with the adverse impact of CO on the RBCs. Carbon monoxide binds in place of oxygen on the RBC, excluding oxygen from riding on the hemoglobin bus. (The oxygen-carrying capacity of the RBC becomes limited or nonexistent, thereby reducing the amount of oxygen transported to the cells.)

Carbon monoxide is one of the most common industrial hazards. It is colorless and odorless and produced during incomplete combustion. As mentioned, CO affects the oxygen-carrying capacity of the red blood cell, causing hypoxia. Consequently, the signs and symptoms of exposure will be similar to those caused by cyanide poisoning and consistent with hypoxia—headache, nausea, vomiting, disorientation, and if the exposure is extreme, seizures, coma, and death. FIGURE 7-7 represents the signs and symptoms of a CO exposure. As the arrow indicates, the severity of health effects is in relation to the severity of the exposure. In summary, you can look at CO and HCN exposures in this way: carbon monoxide reduces the amount of oxygen carried to the cells; cyanide renders the cells incapable of using whatever oxygen is present.

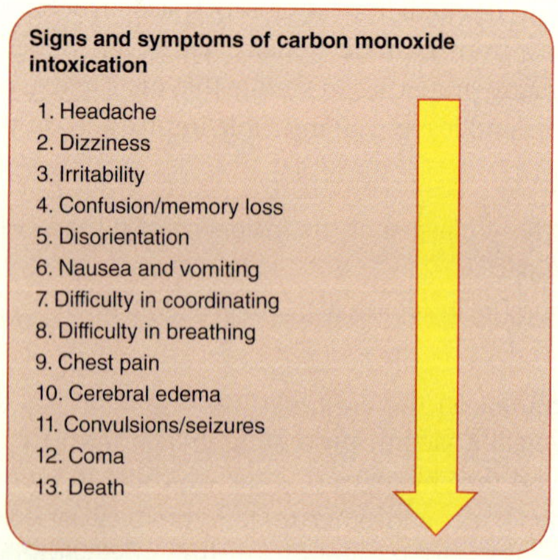

FIGURE 7-7 As the level of COHb rises, the severity of the signs and symptoms increases. The downward arrow indicates the progression from the most common symptom of CO poisoning, headache, to the most extreme, seizures, coma, and death.

Smoke Inhalation Treatment

Smoke inhalation is one of the most complex and challenging patient presentations faced by medical care providers. Patient outcomes vary greatly, influenced by such factors as the extent and duration of the smoke exposure, amount and nature of toxicants in the smoke, degree of thermal burns to the skin and lungs, quantity/size of inhaled particulates (soot), patient's age, and underlying medical conditions. Nationwide, there are few established protocols for treating smoke inhalation, leaving paramedics and other prehospital care providers with limited tools and/or training to properly care for smoke inhalation victims. In most cases, treating smoke inhalation outside the hospital boils down to supportive care: monitoring vital signs, providing high-flow oxygen, establishing intravenous (IV) lines, performing advanced airway management techniques such as endotracheal intubation, cardiac monitoring, and rapid transport.

Referring to the studies done in Paris and Dallas County, however, it is evident that supportive care alone will not correct the underlying cause of death in smoke inhalation patients—the adverse effects of hydrogen cyanide and carbon monoxide on the body's oxygen transportation and utilization system, causing asphyxia at the cellular level. The bottom line is this: until the underlying cause of asphyxia is reversed at the cellular level, normal oxygenation is not possible. This requires a clinical intervention—administering an antidote—to restore the body's ability to use oxygen.

RESPONDER TIP

Hydrogen cyanide poisoning should be suspected in smoke inhalation patients with significant hypotension, soot in the nose or mouth, and/or altered level of consciousness.

Oxygen is the natural antidote for CO poisoning, and in all cases of smoke inhalation high-flow oxygen should be administered. For CO poisoning, which primarily targets the heart and brain, the administration of high-flow oxygen reduces the half-life of CO in the body to around 1 hour. If hyperbaric therapy is indicated, the amount of CO in the body can be reduced by one half in approximately 30 minutes. These are approximations, and patient response may vary from case to case. The point is that oxygen is good for CO inhalation victims, whether from fire smoke or some other source.

When it comes to treating the HCN portion of smoke inhalation, a different kind of antidote is indicated along with oxygen administration. Currently, in the United States there are two types of FDA-approved interventions for treating cyanide poisoning. The Lilly kit (also called the Taylor or Pasadena kit) was the original and only approved antidote for many years. The contents of the kit— referred to generically as the cyanide antidote kit (CAK)—are still available, but configured differently, and sold separately by different manufacturers. The CAK contains amyl nitrite, sodium nitrite, and sodium thiosulfate. (Amyl nitrite is administered as an inhalant; sodium nitrite and sodium thiosulfate are given intravenously.) Another antidote called Nithiodote contains sodium nitrite and sodium thiosulfate. In either case, the two main components in a CAK are nitrites and sodium thiosulfate. The nitrites are given to convert hemoglobin in the RBC to methemoglobin. Methemoglobin attracts cyanide away from the cytochrome oxidase, restoring the cell's ability to take in oxygen and continue the process of aerobic metabolism. Thiosulfate is then administered to chemically bond with cyanide, rendering it less harmful to the body. Although methemoglobin does draw cyanide away from the cytochrome oxidase, it further reduces the oxygen-carrying capacity of the RBC (already compromised by the presence of CO)—a bad trade-off in smoke inhalation patients. Additionally, nitrites may cause a precipitous drop in blood pressure, exacerbating the hypotension commonly found with smoke inhalation exposures. Because of these adverse effects, most experts agree that administering the current cyanide antidote kit is a risky proposition for smoke inhalation patients, especially in a prehospital setting.

Another type of cyanide antidote, hydroxocobalamin, is also approved for use in the United States. This drug (a precursor to vitamin B_{12}) is a relatively benign substance, with minimal side effects, making it well suited for use in the prehospital setting.

Hydroxocobalamin has no adverse effect on the oxygen-carrying capacity of the RBCs and no negative effect on the patient's blood pressure—significant benefits when treating victims of smoke inhalation. Surprisingly, the mechanism of action is simple: Hydroxocobalamin binds to cyanide, forming vitamin B_{12} (cyanocobalamin), a nontoxic compound ultimately excreted in the urine.

Hydroxocobalamin is the antidote in the Cyanokit. The Cyanokit can be administered to a smoke inhalation patient without first verifying the presence of cyanide in the body, with little fear of making the patient worse. Numerous fire departments and emergency medical services carry and routinely administer cyanide antidotes to smoke inhalation with success.

It is important that you understand how smoke inhalation victims are treated in your AHJ and if a cyanide antidote is carried on advanced life support transport units or stocked in the local hospitals. It is also important that you understand that smoke inhalation is an illness just like congestive heart failure, asthma, or any other medical condition you may encounter in the field, and that aggressive clinical intervention is key to the possibility of a good outcome. As a reference source, consult NFPA 473, Chapter 6.4, *Mission-Specific Competencies Advanced Life Support (ALS) Responder Assigned to Treatment of Smoke Inhalation Victim*. This companion document to NFPA 472, *Standard for Competence of Responders to Hazardous Materials/Weapons of Mass Destruction Incidents*, provides background on treating smoke inhalation patients.

Postfire Detection and Monitoring

Although gas detection and atmospheric monitoring is common in hazardous materials response, the typical line fire fighter may be unfamiliar with gas detection devices, their methods of use, and procedures for detecting select gases on the fire scene. It is important to note that (at the time of this writing) there is no concrete best practice when it comes to detection and monitoring in the fire environment, specifically during overhaul. To that end, many fire agencies are investing in technologies for detecting toxic gases at the fire scene without a clear understanding of the mission, the limitations of the devices, or an understanding of what it means to check to see if the building is clear and, more commonly, when it is safe to remove SCBA.

Generally, fire fighters are wearing SCBA during active interior firefighting, but there are still places along the edges of the fire scene, and times during the progression of the incident, where smoke exposures occur. The most common and repeated time and place is during overhaul. The fire service at large is getting better about wearing SCBA during overhaul, but it's still common to see this task being done with no respiratory protection at all. Pump operators working at the panel during a fire, shrouded in drift smoke, are also common victims of exposure, as is a chief officer commanding the fire, standing in the haze at a poorly located command vehicle. Personnel operating exterior lines for extended periods of time are also breathing fire gases and particulates, as are rapid intervention crews setting up in the front yard of a single-family dwelling fire. Think about all the places on the fire ground, outside of the interior attack crew, where fire personnel are repeatedly exposed to smoke.

Responder Safety Tip

Wearing SCBA means the cylinder *and* the mask!

Courtesy of Rob Schnepp

Consequently, the increased appetite for education regarding fire gas toxicity and fire fighter safety, coupled with an interest in postfire atmospheric monitoring, adds an entirely new category for detection, outside of the traditional paradigm of detection and monitoring. Over time, a new normal may exist in the fire service based on understanding the benefits and limitations of gas detection in the hostile environment of the fire scene.

To begin the discussion in the right context, consider this before embarking on your quest to perform postfire detection and monitoring: Atmospheric monitoring has a place on the fire scene. The technology and devices selected should be user-friendly, durable, cost effective, and easy to maintain, and the benefits and limitations of any instrumentation should be understood. Because there is no single device that will identify all possible toxins on the scene, it is important to know that targeting certain gases (CO and HCN as an example) may be broadly representative of the airborne environment but not an exact indicator of the presence, absence, or concentration of any other gas or particulate in the air. Wearing SCBA is still the gold standard of respiratory protection and is the best way to reduce the possibility of inhalation exposures.

Until recently, there have been three primary uses for detection devices outside of the traditional fire-based hazardous materials response:

- Rescue response including confined space
- Building collapse and trench rescue
- CO detector responses

The intent of this section is to provide some general information about fire scene gas detection, offer descriptions of other types of detection equipment available that could be used at the fire scene, and provide an overview of electrochemical sensor technology (commonly used in fire-ground detection and monitoring). As you will see, the focus is on detecting gases, not particulates. This is important to understand, because the visible part of smoke—the particulates—is hazardous as well, and often is overlooked in favor of evaluating the environment for gases like CO and HCN.

Why Use Detection Devices at the Fire Scene?

As stated throughout this chapter, fire smoke is a collection of fire gases and particulates. Most of them can be detected and measured by one type of technology or another, but no single device can detect them all; this is important to understand. Therefore, an initial decision must be made about what substance(s) you want to detect and/or monitor, and at what concentrations those substances pose a risk of exposure. To that end, it is vital to understand and apply some basic toxicological terms and definitions. The National Institute for Occupational Safety and Health (NIOSH) establishes safe levels for chemical exposures in the workplace; however, the values are for average worker exposure and can be only *estimated* for the rigors of firefighting. Elevated respiratory rate, blood pressure, and heart rate; increased skin temperature resulting in increased permeability; and dehydration are only some of the factors that would add to the intake and effects of these toxic gases on responders. For the purpose of this section, the more conservative NIOSH levels will be used to guide suggested operations and assist in evaluating airborne contamination levels. The exposure levels are as follows:

- Immediately dangerous to life and health (IDLH)
- Short term exposure limits (STEL)
- Recommended exposure limits (REL)

In short, levels at or above IDLH require use of breathing apparatus, or withdrawal from the area if no SCBA is worn. The definition of IDLH as found in OSHA 29 CFR 1910.120 means an atmospheric concentration of any toxic, corrosive,

Table 7-3 The Values for the Toxic Gases Commonly Present in Fire Smoke

Gas	REL	STEL	IDLH	Density
Ammonia (NH_3)	25 ppm	35 ppm	300 ppm	Lighter than air
Carbon dioxide (CO_2)	5000 ppm (0.5% vol.)	30,000 ppm (3% vol.)	40,000 ppm (4% vol.)	Heavier than air
Carbon monoxide (CO)	35 ppm – no higher than 200 ppm allowed (ceiling)	NR	1200 ppm	Lighter than air
Hydrogen chloride (HCl)	No higher than 5 ppm (ceiling)	NR	50 ppm	Heavier than air
Hydrogen cyanide (HCN)	4.7 ppm (15 minutes only)	4.7 ppm – 1 time only	50 ppm	Lighter than air
Hydrogen sulfide (H_2S)	10 ppm (10 minutes only)	10 ppm – 1 time for 10 minutes	100 ppm	Heavier than air
Oxides of nitrogen (NO_x) (NO_2) (NO)	NO_2 – 3 ppm; NO – 25 ppm	NO_2 – 5 ppm; NO – NR	NO_2 – 20 ppm; NO – 100 ppm	Heavier than air
Sulfur dioxide (SO_2)	2 ppm	5 ppm	100 ppm	Heavier than air

or asphyxiant substance that poses an immediate threat to life or would interfere with an individual's ability to escape from a dangerous atmosphere is considered IDLH.

A STEL represents a 15-minute exposure no more than four times a day, and a REL is for a 10-hour exposure. The Occupational Safety and Health Administration (OSHA) also has set comparable levels and, in many cases, uses the NIOSH values. OSHA uses permissible exposure limit (PEL) instead of REL; the PEL is an average calculated over an 8-hour period. TABLE 7-3 lists the values for the toxic gases commonly present in fire smoke.

A number of technologies are available to detect fire gases, but the most common method involves the use of single-gas or multi-gas detection devices. (There is more detail on this technology later in the chapter.) When monitoring for fire gases, responders can reference the levels listed in Table 7-3 to determine if a reading is above or below levels set by NIOSH or OSHA (for the listed set of substances). As an example, the IDLH for CO, likely the most common and prevalent toxic fire gas, is 1200 ppm. Therefore, if responders obtain meter readings that indicate levels at or above 1200 ppm, the environment or situation should be considered immediately dangerous to life and health (IDLH).

The use of the term "immediately dangerous" imparts a sense of high hazard, and is usually interpreted as such, meaning that personnel operating in the environment could suffer serious adverse health effects if not properly protected. Airborne contamination at this level is usually understood to be a high hazard environment. It's a safe estimation to say that during the active phase of combustion and firefighting, and for some period of time postextinguishment, some gas or the level of aggregate particulates will be above IDLH, therefore making the environment at large above IDLH, requiring the use of SCBA. What can be more difficult to navigate are the situations where the level of a specific gas that you may be looking for, again using CO as an example, is below the IDLH. Consider this question: If CO is below the REL, does that mean any or all other gases are below REL? The answer is unknown unless a host of other gases are being evaluated in the same place at the same time. CO readings, then, are more of a rough estimation of the total airborne environment.

Table 7-4 Common Interfering Gases to CO and H_2S

Other substances picked up by a CO sensor		
Acetylene	Hydrogen	Methyl Sulfide
Ammonia	Hydrogen cyanide	Nitric oxide
Carbon disulfide	Hydrogen sulfide	Nitrogen dioxide
Dimethyl Sulfide	Isobutylene	Phosphine
Ethyl alcohol	Isopropyl alcohol	Propane
Ethyl Sulfide	Mercaptans	Sulfur dioxide
Ethylene	Methyl alcohol	Turpentine

The REL is the most conservative level to acknowledge, and therefore a safer end point for detection and monitoring, so it will be used for examples herein. Remember, the REL is an average over a 10-hour period.

RESPONDER TIP

To determine the REL on an emergency scene would involve calculations of the dose of the toxic gas over time. Considering the dynamic nature of fire scenes and that time isn't typically available for emergency responders, this isn't a real possibility. The easiest and safest route is to use the REL level itself as the safe point.

For example, the REL for CO is 35 ppm. Any level above 35 ppm is above the regulated baseline value, but any amount detected below 35 ppm is not necessarily safe—it's below the established regulatory limit for a given timeframe. What isn't considered in this example are the other substances present—the other nasty things in smoke—because if you are monitoring CO on the scene, you are primarily looking for that gas, but there could be others that the sensor will "see." (TABLE 7-4 lists other gases that would be picked up by a CO sensor.) Therefore, it isn't possible to draw a straight-line correlation between CO and/or HCN, and the presence of any other gas or particulate that might be present at the scene. Detection of a single gas represents a snapshot of that single gas, at that point in time, at that location. It does not mean the entire building is safe to operate in without SCBA, or that the entire scene or even a specific location is entirely clear.

With that in mind, it is important to understand the limitations of postfire detection and monitoring when a single gas is being targeted. A common misconception is this: if that single gas is found to be below the REL, the entire airborne environment is safe. The shortcomings of this approach should be clear in that CO levels may be below REL, but something else generated by the combustion process, especially the particulates, may not be in a safe range. Therefore, the philosophy of using the REL of a single gas to determine whether a room, portion of a room, or entire building is safe may be subjective. Again, it's better than nothing, but likely not definitive. When the entire amount of materials involved in the combustion process are cooled down past their ignition temperature, and therefore no longer thermally decomposing and off-gassing, the environment is likely safe to occupy without respiratory protection. *CO levels may help ascertain that, because CO is always present during the combustion process.*

Because one gas may not be an indicator of the level of airborne contamination across the entire scene, the next logical step is to widen the focus and evaluate the fire environment for multiple gases. A common approach is a multi-gas meter configured with a variety of sensors—commonly CO and hydrogen sulfide (H_2S) along with oxygen and lower explosive limit (LEL) sensors.

Certainly, looking for more than a single gas is better because it casts a wider detection and monitoring net, but the practice is still limited in terms of evaluating the particulates and gases other than CO and HCN. Choosing to detect these two gases at the scene is beneficial in that it puts an atmospheric monitoring "stake in the ground" by identifying a couple of nasty players in smoke and drives responders to understand the correlation of feeling bad after breathing smoke and the identification of CO and HCN. Again, looking for these two gases is better than looking for only one, but it is still limited in the view it gives about other substances that might be present.

Another method of postfire detection and monitoring is to use more than one technology and look for multiple gases. Although this is interesting and provides useful information, it still does not address airborne contamination in total (mostly regarding the particulates) and should ultimately lead you to the most conservative conclusion—the safest way to work in smoke is to keep a breathing apparatus on when you are near or in smoke. At the end of the day, each agency must determine how to balance the scales of the impact of wearing SCBA for longer periods of time at the scene versus the attempt to determine if and when the postfire environment is safe enough to work without it.

Common Fire Scene Detection and Monitoring Technologies

This section will describe a few common technologies that could be used for detection and monitoring at the fire scene. The list is not definitive in that there are many other instruments/technologies available. For a more in-depth discussion of a particular technology, refer to the chapter *Mission-Specific Competencies: Air Monitoring and Sampling.*

Electrochemical Sensors

One of the most common technologies used in postfire detection and monitoring is electrochemical sensors. Broadly speaking, these sensors are filled with a chemical reagent that reacts with a target gas and results in a meter reading **FIGURE 7-8**.

For example, a carbon monoxide sensor is filled with a jelly-like substance containing sulfuric acid and has two electrical poles within the sensor. When carbon monoxide enters the sensor, either passively or drawn into the device with a pump, there is a chemical reaction that changes the electrical balance in the sensor. The poles detect the change, which results in a reading seen on a screen. If the amount of gas present is above a preset level, an alarm will sound.

Typical electrochemical sensors used in postfire detection and monitoring include oxygen (O_2), hydrogen cyanide, carbon monoxide, hydrogen sulfide, ammonia, and chlorine. These sensors are also commonly found in single-gas devices. There are a few important factors to understand about electrochemical sensors, specifically those used to detect O_2, CO, H_2S, and HCN:

- Toxic sensors react to other gases, so you cannot be sure whether you are reading a level of the intended gas or an interfering gas. For example, if you are reading levels of both CO and H_2S there is probably an acidic gas present that is causing the reaction and resulting in the readings displayed. Most of the interfering gases are

FIGURE 7-8 Electrochemical sensors ready to be installed and used.

Courtesy of Rob Schnepp

also toxic, so when readings are found on the detection devices fire fighters should be aware that there are toxic gases present. Some of these gases are also flammable, which means you may also get a reading in your flammable gas sensor. There are ways to determine which gases are present through the use of detection devices your hazardous materials team carries. Nonetheless, the sensor is alerting to the presence of a gas.

- Electrochemical sensors can be easily overwhelmed and will max out with regard to their readings. For example, most HCN sensors will max out at 50 ppm (IDLH) and will not tell you if you are in levels higher than the maximum. This is largely irrelevant; however, once the readings pass the REL for HCN (4.7 ppm), SCBA should be worn. The same holds true for CO sensors. They usually have a maximum reading of 500 ppm. Remember that CO has an IDLH value of 1200 ppm. High exposures to a gas that causes a reaction will result in the sensor failing sooner than its intended life.
- Electrochemical sensors fail to the 0 (zero) point. With O_2 that's not necessarily a problem because 0% oxygen is dangerous and you would not enter a potentially hazardous environment. But when a CO or HCN sensor fails and reads 0 it may not indicate it has failed, and that can lead responders to believe a toxic gas is not present—a dangerous situation. Proper care and maintenance will help prevent this from occurring.
- Responders should ensure their devices are calibrated according to manufacturers' recommendations and should at least be bump tested prior to use. (For a review of calibration and bump testing, refer to the chapter *Mission-Specific Competencies: Air Monitoring and Sampling*.) Some instrument manufacturers recommend that the devices be calibrated before each use, which is not practical in an emergency response environment, but it should be bump tested at a minimum.
- Responders must be mindful of the reaction time of electrochemical sensors because they can take as little as 20 seconds or as much as 200 seconds to react based on the type of sensor.

Photoionization (PID) Sensor

Another type of sensor is a photoionization (PID) sensor. A PID is available as a stand-alone unit or may be incorporated into a multi-gas meter FIGURE 7-9. In either case, the PID sensor uses ultraviolet light to ionize the gases that move through the sensor. When gases are ionized, sensors are able to determine the ionization of the molecules. The PID sensor generally detects common materials such as benzene, acetone, toluene, ammonia, ethanol, butane, and many others. It is primarily a sensor that detects organic materials (materials that contain carbon as part of the molecular structure). Ammonia is the most common inorganic chemical detected by a PID. The PID does not identify the material that is present; it only alerts the user of its presence. A PID in a smoke situation will detect parts of the toxic soup that is in the air and alert you that something is present, but it will not identify which toxic gas is in the air. An important thing to remember is that PIDs won't detect the presence of CO or HCN or compounds such as natural gas.

FIGURE 7-9 A photoionization (PID) sensor in service during a session of fire investigation research.

Courtesy of Rob Schnepp

Responders should be aware of differences in the reaction time of the various detection devices: electrochemical sensors react within 20–200 seconds, LEL sensors react in 7–10 seconds, and the PID reacts within 1–2 seconds.

Another consideration with any electronic detection device is the dirty nature of the fire environment. There is a lot of particulate matter in the air, and many detection devices have an internal pump that is drawing the gas into the device for analysis, so any particulates in the air will be drawn in as well. To combat this problem, most devices have a protective filter, which keeps out particulate material and will offer some limited protection against liquids FIGURE 7-10.

FIGURE 7-10 The filter on the top is clean; the one on the bottom was used during live burns at a training session. The bottom filter was so contaminated the device could not be fresh-air calibrated.

Courtesy of Rob Schnepp

References

1. Baud FJ, Barriot P, Toffis V, et al. "Elevated blood cyanide concentrations in victims of smoke inhalation." *New England Journal of Medicine*. 1991, 325: 1761–1776.
2. Silverman SH, Purdue GF, Hunt JL, et al. "Cyanide toxicity in burned patients." *Journal of Trauma*. 1988, 28(2): 171–176.
3. Birky MM, Clarke FB. "Inhalation of toxic products from fires." *Bulletin of the New York Academy of Medicine*. 1981, 57(10): 997–1013.
4. Teige B, et al. "Carboxyhemoglobin concentrations in fire victims and in cases of fatal carbon monoxide poisonings." *Zeitschrift fur Rechtsmedizin*. 1977, 80(1): 17–21.
5. Norris JC, et al. "Synergistic lethality induced by the combination of carbon monoxide and cyanide." *Toxicology*. 1986, 40: 121–129.
6. Purser DA, Grimshaw P, Berril KR. "Intoxication by cyanide in fires: A study in monkeys using polyacrylonitrile." *Archive of Environmental Health*. 1984, 39: 394–400.
7. Fortin J-L. "Use of hydroxocobalamin in fire victims by 'The Brigade De Sapeurs Pompiers De Paris' from 1998-2003." Presentation at the Second World Congress on Chemical, Biological and Radiological Terrorism, September 2003.
8. Hawley C. *Hazardous Materials Air Monitoring and Detection Devices*. 2nd ed. Clifton Park, NY: Delmar Cengage Learning, 2007.
9. RAE Systems Application Note AP-222 Rev. 1, *Carbon Monoxide and Natural Gas Detection for Fire Department Engine Companies*, San Jose, CA.
10. RAE Systems Technical Note TN-106 (Revised 8/12), *Correction Factors, Ionization Energies, and Calibration*, San Jose, CA.
11. RAE Systems Application Note AP-206 Rev. 4, *Guide to Atmospheric Testing in Confined Spaces*, San Jose, CA.
12. Hall JR. *Characteristics of Home Fire Victims*. NFPA Fire Analysis and Research Division, Quincy, MA, July 2006.
13. Geneva Association. "World Fire Statistics." *Geneva Association Information Newsletter*, 2012, 28. Note: Belgium was excluded from this review because its 2007–2009 death rates were unavailable.

Wrap-Up

Chief Concepts

- U.S. Fire Administration fire death statistics for the United States report approximately 3000 fire-related deaths annually. That places the United States in the top 15 countries in the world for per capita fire deaths.
- Like a scuba diver, a fire fighter initiating an interior attack at a working structure fire is completely enveloped by the operating environment. If a supply of breathable air is lost or interrupted, the environment invades the body, causing harm.
- The AHJ should establish reasonable and safe work practices for those responders working in, and possibly exposed to, fire smoke.
- Think about all the places on the fire ground, outside of the interior of the structure, where fire personnel are repeatedly exposed to smoke.
- Cancer is prevalent in the fire service, more so than nearly any other profession. This should be no surprise once the byproducts of combustion are understood. A number of known human carcinogens are liberated as fire gases and particulates.
- Smoke is a collection of gaseous products of burning materials—especially organic materials—made visible by the presence of small particles of incomplete combustion.
- Soot is a known human carcinogen, just like benzene and formaldehyde—also common byproducts of combustion.
- PAHs are believed to be immunosuppressant, perhaps contributing to the mechanism by which PAHs are suspected to cause cancer.
- NFPA 1851, *Standard on Selection, Care, and Maintenance of Protective Ensembles for Structural Fire Fighting and Proximity Fire Fighting*, offers guidance on bunker gear cleaning and maintenance.
- Carbon monoxide is one of the most common industrial hazards. It is colorless and odorless and produced during incomplete combustion.
- Cyanide compounds, once absorbed into the body, "poison" the cytochrome oxidase, barring oxygen from entering the mitochondria, effectively shutting down the process of aerobic metabolism.
- Carbon monoxide reduces the amount of oxygen carried to the cells; cyanide renders the cells incapable of using whatever oxygen is present.
- Oxygen is the natural antidote for CO poisoning. In all cases of smoke inhalation, high-flow oxygen should be administered.
- Hydrogen cyanide poisoning should be suspected in smoke inhalation patients with significant hypotension, soot in the nose or mouth, and/or an altered level of consciousness.
- It is important to understand that smoke inhalation is an illness just like congestive heart failure, asthma, or any other medical condition you may encounter in the field, and that aggressive clinical intervention is key to the possibility of a good outcome.
- NFPA 473, Chapter 6.4, *Mission-Specific Competencies Advanced Life Support (ALS) Responder Assigned to Treatment of Smoke Inhalation Victim* is a companion document to NFPA 472 that provides background on treating smoke inhalation patients.
- Atmospheric monitoring has a place on the fire scene. The technology and devices selected should be user-friendly, durable, cost effective, and easy to maintain, and the benefits and limitations of any instrumentation should be understood. Because no single device will identify all possible toxins on the scene, it is important to know that targeting certain gases (CO and HCN as an example) may be broadly representative of the airborne environment, but not an exact indicator of the presence, absence, or concentration of any other gas or particulate in the air. Wearing SCBA is still the gold standard of respiratory protection and is the best way to reduce the possibility of inhalation exposures.
- It is important to understand the limitations of post-fire detection and monitoring when a single gas is being targeted. A common misconception is this: if that single gas is found to be below the REL, the entire airborne environment is safe.
- Regardless of the technology selected, responders should develop a strategy for the use of detection devices. There are no hard and fast rules for post-fire detection and monitoring, so the AHJ should develop some broad operational guidance.
- Detection and monitoring at the fire scene should be a highly mobile and continuous process.

Hot Terms

aerobic metabolism The creation of energy through the breakdown of nutrients in the presence of oxygen. The byproducts are carbon dioxide and water, which the body disposes of by breathing and sweating.

anaerobic metabolism The creation of energy through the breakdown of glucose. Without oxygen, this metabolic process results in the production of lactic acid.

colorimetric tubes These are designed to determine the levels of a known gas or vapor; they can also be used to help determine what unidentified airborne substances may be present.

cytochrome oxidase Found in the mitochondria, this is important in cell respiration as an agent of electron transfer from certain cytochrome molecules to oxygen molecules.

hypoxia A deficiency of oxygen supply to the tissues in the human body.

methemoglobin Formed by a change in the iron atom in hemoglobin from the ferrous (+2) to the ferric state (+3). It's normal to have a small amount of methemoglobin in the blood, but substances such as nitrites and nitrile-containing substances convert a larger proportion of hemoglobin into methemoglobin, which does not function as an oxygen carrier.

mitochondria Responsible for converting nutrients into energy, yielding molecules of adenosine triphosphate (ATP) to fuel the cell's activities.

photoionization (PID) sensor A sensor that uses ultraviolet light to ionize the gases that move through the sensor; available as a stand-alone unit, or may be incorporated into a multi-gas meter.

polycyclic aromatic hydrocarbons (PAH) A group of substances that occurs naturally in materials such as coal or crude oil. These substances also are generated during the combustion of organic materials and can be found in vehicle exhaust, tobacco smoke, and the smoke generated from structure fires, vehicle fires, wildland fires, or any other type of fire. PAHs can exist as a particle or gas.

pyrolysis A chemical change brought about by the action of heat.

red blood cell An oxygen-carrying cell found in mammals. Red blood cells contain hemoglobin.

smoke A collection of gaseous products of burning materials—especially organic materials—made visible by the presence of small particles of incomplete combustion (carbon).

RESPONDER *in action*

You respond to a working fire with a report of victims trapped. Upon arrival, the homeowner states that all occupants are out of the building. You are directed to a male victim sitting on the ground next to a parked car. The victim is alert and responding to your questions but appears to be disoriented. You suspect the victim is suffering from smoke inhalation.

1. What is the definitive way to confirm cyanide poisoning in the field?

A. Confirm a blood oxygenation level of less than 92%.

B. Use a transcutaneous or in-line cyanide oximeter (cyanide oximeter).

C. Confirm carbon monoxide poisoning—where there's carbon monoxide poisoning, there's cyanide poisoning.

D. There is no detection method for confirming cyanide poisoning in the field.

2. Which of the following organs quickly suffer from the oxygen-deprivation effects of cyanide poisoning?

A. Lungs and kidneys

B. Lungs and heart

C. Heart and brain

D. Heart and kidneys

3. In the face of smoke inhalation, which of the following is an acceptable antidote to administer if you suspect HCN poisoning?

A. Sodium hydroxide

B. Amyl nitrite

C. Methylene blue

D. Hydroxocobalamin

Mission-Specific Competencies: Personal Protective Equipment

Greg Henry/ShutterStock, Inc.

Courtesy of Rob Schnepp

CHAPTER 8

Operations Level Responders Assigned Mission-Specific Responsibilities

This chapter shall address competencies found in NFPA 472 for the operations level responders assigned mission-specific responsibilities at hazardous materials/WMD incidents by the authority having jurisdiction beyond the core competencies at the operations level.

Knowledge Objectives

After studying this chapter, you will be able to:

- Describe types of personal protective equipment (PPE) used for hazardous materials incidents. (NFPA 6.2.3.1(1)(2), p. 179–183)
- Describe the capabilities of the PPE provided by the authority having jurisdiction (AHJ) so as to perform any mission-specific tasks assigned. (NFPA 6.2.3.1, p. 178–183)
- Define degradation, penetration, and permeation and their effects on the selection of PPE. (NFPA 6.2.3.1(3)(a)(b), p. 181)
- Explain the advantages and disadvantages of vapor-protective and splash-protective clothing. (NFPA 6.2.3.1(3)(c), p. 183)
- Describe common heat exchange units used for cooling personnel operating in personal protective equipment. (NFPA 6.2.3.1(3)(d), p. 197–199)
- Describe how to don, work in, and doff the PPE provided by the AHJ. (NFPA 6.2.4.1, p. 186–195)
- Describe safety and emergency procedures for personnel wearing PPE. (NFPA 6.2.4.1(1)(2), p. 183–185, 193–199)
- Describe maintenance, testing, inspection, storage and documentation procedures for personal protective equipment. (NFPA 6.2.4.1(5), p. 178–181)
- Terminate the incident by completing the reports and documentation pertaining to PPE. (NFPA 6.2.5.1, p. 200)

Skills Objectives

After studying this chapter, you will be able to:

- Demonstrate the ability to properly don and doff a Level A ensemble. (NFPA 6.2.4.1(3), p. 185–187)
- Demonstrate the ability to properly don and doff a Level B nonencapsulating chemical-protective clothing ensemble. (NFPA 6.2.4.1(3),p. 188–191)
- Demonstrate the ability to properly don and doff a Level C chemical-protective clothing ensemble. (NFPA 6.2.4.1(3), p. 190–193)
- Demonstrate the ability to properly don and doff a Level D chemical-protective clothing ensemble. (NFPA 6.2.4.1(3), p. 194–195)

Note: For specific information regarding technical decontamination of personal protective equipment, see the chapter *Mission Specific Competencies: Technical Decontamination*.

You Are the Responder

Your engine company arrives on the scene of a vehicle accident involving a small passenger vehicle and a tanker truck carrying 20 tons of anhydrous ammonia. The tanker rolled over on its side as a result of the accident and slid down the highway for approximately 100 feet. There are no injuries to the three victims in the passenger vehicle, but the driver of the tanker truck is pinned inside the cab. You notice the smell of ammonia in the air but see no visible signs of a product release. The regional hazardous materials team, fully staffed with technician level responders, is also on scene. Your company officer confers with the hazardous materials team and then directs you and another fire fighter to don your SCBA and full turnout gear and evaluate the driver for injuries.

1. Would full structural fire fighters' turnout gear and SCBA offer adequate protection in this situation?
2. Based on your level of training—operations level core competencies and all of the mission-specific competencies—would you be qualified to perform this task?
3. Describe the steps you would need to take, including obtaining information about ammonia, to complete your assignment.

Introduction

Emergency responders should be familiar with the policies and procedures of the authority having jurisdiction (AHJ) to ensure a consistent approach to selecting the proper personal protective equipment (PPE) for an expected task or set of tasks. Additionally, all responders charged with responding to hazardous materials/weapons of mass destruction (WMD) incidents should be proficient with local procedures for technical decontamination as well as the manufacturers' guidelines for maintenance, testing, inspection, storage, and documentation procedures for the PPE provided by the AHJ. Refer to the chapter *Estimating Potential Harm and Planning a Response* for the specifics of the NFPA standards on protective clothing and specific information on the Occupational Safety and Health Administration (OSHA)/Environmental Protection Agency (EPA) levels of protection.

Much of the chemical-protective equipment on the market today is intended for a single use (i.e., it is disposable), and usually discarded along with the other hazardous waste generated by the incident. As a consequence of this intention, single-use PPE is decontaminated to the point that it is safe for the responder to remove but not so extensively that the garment is completely free of contamination. Single-use PPE is generally less expensive than reusable gear, and it needs to be restocked and/or replenished after the incident. Before the use of any PPE, it should undergo a thorough visual inspection to ensure that piece of equipment is absolutely response ready.

Reusable garments are required to be tested at regular intervals and after each use. Level A suits, for example, are required to be pressure tested—usually upon receipt from the manufacturer, after each use, and annually **FIGURE 8-1**. Individual manufacturers will have well-defined procedures for this activity. Prior to purchasing any type of PPE, the AHJ should understand the maintenance and upkeep requirements. PPE is not intended to be purchased and completely left alone until it is needed. There must be a level of care and attention devoted to the barrier between you and the released substance. Also, keep in mind that single-use and reusable garments have a shelf life. This time frame should be noted and adhered to. Don't use PPE that is beyond its shelf life—you may not be able to depend on it!

Chemical-protective garments are tested in accordance with the manufacturer's recommendation or following the method outlined in OSHA's Hazardous Waste Operations and Emergency Response (HAZWOPER) standard (29 CFR 1910.120, Appendix A). Generally speaking, a pressure test is accomplished by using a specially designed kit to pump a certain amount of air into the suit and leaving it pressurized for a specified period of time. At that point, if the garment has lost more than a certain percentage (usually 20 percent)

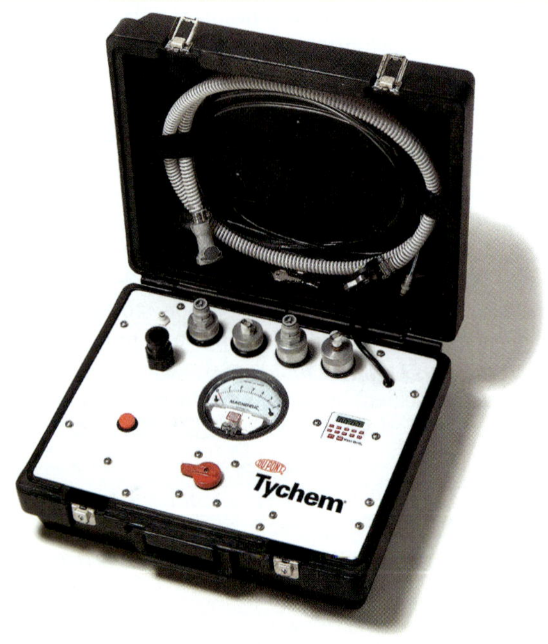

FIGURE 8-1 Level A suit testing kit.

Courtesy of the DuPont Company

of pressure, it is assumed the suit has a leak. Often, leaks are located by gently spraying or brushing the inflated suit with a solution of soapy water. Small bubbles begin to form in the area of the leak, alerting you to its presence and location. Any garment with a leak should be removed from service until the defect is identified and repaired in accordance with the manufacturer's specification.

Chemical-protective equipment should be stored in a cool, dry place that is not subject to significant temperature extremes and/or high levels of humidity. Furthermore, the equipment should be kept in a clean location, away from direct sunlight, and should be inspected at regular intervals based on the manufacturer's recommendations. If repairs are required, consult the manufacturer prior to performing any work—there is a risk that the garment will not perform as expected if it has been modified or repaired incorrectly. Again, individual manufacturers have well-defined procedures for this activity; prior to purchasing any PPE, the AHJ should understand what is required in terms of maintenance and upkeep.

As the title of this chapter implies, responders must correlate the mission they are expected to perform with the anticipated hazards. For example, in the ammonia scenario described earlier, the responders should understand that ammonia presents a significant health hazard because it is corrosive to the skin, eyes, and lungs. Ammonia is also flammable at concentrations of approximately 15 to 25 percent (by volume) in a mixture with air. Exposure to a concentration of approximately 300 parts per million (ppm) is considered to be immediately dangerous to life and health (IDLH). If the possibility of exposure to a concentration exceeding 300 ppm exists, a National Institute for Occupational Safety and Health (NIOSH) approved self-contained breathing apparatus (SCBA) is required. NIOSH sets the design, testing, and certification requirements for SCBA in the United States.

Responder Safety Tip

TRACEMP is a common acronym used to sum up a collection of potential types of harm an emergency responder may face.

- **T**hermal
- **R**adiological
- **A**sphyxiating
- **C**hemical
- **E**tiological/biological
- **M**echanical
- **P**sychogenic

Although SCBA will protect the responders from suffering an inhalation exposure, its use is only one piece of the PPE decision. Another question must be answered in this scenario: Will structural fire fighter's turnout gear provide sufficient skin protection? Knowing that ammonia presents a flammability hazard is important—and the turnout gear would address that potential—but that choice of equipment still does not address the hazard of skin irritation. The release is outside—would you make a different choice if the release were indoors, inside a poorly ventilated room? Responders should always consider the impact of the operating environment as part of the hazard evaluation.

The responders tasked with undertaking a medical reconnaissance mission in the ammonia scenario should balance these hazards and the risk of the mission with the potential gain. This scenario is based on an outdoor release, with unknown variables of wind speed, direction, and ambient air temperature, which may positively or negatively influence the decision to approach the cab of the truck. Unfortunately, it is impossible to decide on the right course of action based on a few sentences in this text—you must make that decision on the street, at the moment the emergency occurs. Unfortunately, not all structural fire fighters' gear is created equal, and not all types are intended to function in an environment that may contain a hazardous material. Structural fire fighters' gear that is 10 years old and well worn will certainly not provide the same level of protection as a new set of gear that meets the certification requirements for the latest edition of the National Fire Protection Association (NFPA) 1971, *Standard on Protective Ensembles for Structural Fire Fighting and Proximity Fire Fighting.*

To that end, it is incumbent on every emergency responder to understand the hazards that may be present on an emergency scene and to appreciate how those hazards may affect the PPE requirements and the mission the responders are tasked with carrying out.

Hazardous Materials: Specific Personal Protective Equipment

As discussed in the chapter *Estimating Potential Harm and Planning a Response*, different levels of PPE may be required at different hazardous materials incidents. This section reviews the protective qualities of various ensembles, from those offering the least protection to those providing the greatest protection.

At the lowest end of the spectrum are street clothing and normal work uniforms, which offer the least amount of protection in a hazardous materials emergency. Normal clothing (or flame resistant coveralls) may prevent a noncaustic powder from coming into direct contact with the skin, for example, but it offers no significant protection against many other hazardous materials. Such clothing is often used in industrial applications, such as oil refineries, rail yards, or city public works facilities as a general work uniform **FIGURE 8-2**. Police officers and emergency medical services providers typically wear this level of "protection." Most often, distance from the hazard is the best level of protection with this PPE.

The next higher level of protection is provided by structural firefighting protective equipment **FIGURE 8-3**. Such an outfit includes a helmet, a bunker coat, bunker pants, boots, gloves, a hood, SCBA, and a personal alert safety system (PASS) device. Standard firefighting turnout gear is not considered "chemical protection," because the fabric may break down when exposed to chemicals, and may not provide complete protection from the harmful gases, vapors, liquids, and dusts that could be encountered during hazardous materials incidents.

FIGURE 8-2 A Nomex jumpsuit.
Courtesy of the DuPont Company

FIGURE 8-3 Standard structural firefighting gear.
Courtesy of Rob Schnepp

Returning to the ammonia scenario, it may be safe and reasonable to carry out the patient assessment mission wearing this level of protection—again, based on a full risk assessment.

FIGURE 8-4 High temperature–protective equipment protects the wearer from high temperatures during a short-term exposure.
© Photodisc

Keep in mind that structural fire fighters' gear is primarily intended to protect the wearer from thermal hazards (predominantly encountered during firefighting), and mechanical hazards such as broken glass or other sharp objects. The same gear may be called upon for other reasons, such as for protecting the wearer against alpha and beta radiation sources, but that is not its primary function.

Responders wearing high temperature–protective equipment may best address unusually high thermal hazards, such as those posed by aircraft fires. This type of PPE shields the wearer during short-term exposures to high temperatures FIGURE 8-4. Sometimes referred to as a proximity suit, high temperature–protective equipment allows the properly trained fire fighter to work in extreme fire conditions. It provides protection against high temperatures only, however; it is not designed to protect the fire fighter from hazardous materials.

Chemical-Protective Clothing and Equipment

Chemical-protective clothing is unique in that it is designed to prevent chemicals from coming in contact with the body. Such equipment is not intended to provide high levels of protection from thermal hazards (heat and cold) or to protect the wearer from injuries that may result from torn fabric, chemical damage, or other mechanical damage (tears and abrasion) to the suit. Not all chemical-protective clothing is the same, and each type, brand, and style may offer varying degrees of protection and chemical resistance.

To help you safely estimate the chemical resistance of a particular garment, manufacturers supply compatibility charts

FIGURE 8-5 An example of a compatibility chart.

with all of their protective equipment FIGURE 8-5. These charts are designed to assist you in choosing the right chemical-protective clothing for the incident at hand. You must match the anticipated chemical hazard to these charts to determine the resistance characteristics of the garment.

Storage conditions, temperature, and resistance to cuts, tears, and abrasions are all factors that affect the chemical resistance of materials. Other factors include flexibility, shelf life, and sizing criteria. The bottom line is that chemical-resistant materials are specifically designed to inhibit or resist the passage of chemicals into and through the material by the processes of penetration, permeation, or degradation.

Penetration is the flow or movement of a hazardous chemical through closures such as zippers, seams, porous materials, pinholes, or other imperfections in the material. To reduce the threat of a penetration-related suit failure, responders should carefully evaluate their PPE prior to entering a contaminated atmosphere. Checking the garment fully—that is, performing a visual inspection of all its components—before donning the PPE is paramount. Just because the suit or gloves came out of a sealed package does not mean they are perfect! Also, a lack of attention to detail could result in zippers not being fully closed and tight. Poorly fit seams around ankles and wrists could allow chemicals to defeat the integrity of the garment. Use of the buddy system is beneficial in this setting because it creates the opportunity to have a trained set of eyes examine parts of the suit that the wearer cannot see. Prior to entering a contaminated atmosphere, or periodically while working, each member of the entry team should quickly scan the PPE of the other member(s) to see if anyone's suit has suffered any damage, discoloration, or other insult that may jeopardize the health and safety of the wearer.

Permeation is the process by which a hazardous chemical moves through a given material on the molecular level. It differs from penetration in that permeation occurs through the material itself rather than through openings in the material. Permeation may be impossible to identify visually, but it is important to note the initial status of the garment and to determine if any changes have occurred (or are occurring) during the course of an incident. Chemical compatibility charts are based on two properties of the material: its breakthrough time (how long it takes a chemical substance to be absorbed into the suit fabric and detected on the other side) and the permeation rate (how much of the chemical substance makes it through the material) FIGURE 8-6. The concept is similar to water saturating a sponge. Over time, a continuous drip of water will "fill up" the sponge and begin to seep through to the other side. When evaluating the effectiveness of a particular material against a given substance, responders should look for the longest breakthrough time available. For example, a good breakthrough time would be more than 480 minutes—a typical 8-hour workday.

Degradation is the physical destruction or decomposition of a clothing material owing to chemical exposure, general use, or ambient conditions (e.g., storage in sunlight). It may be evidenced by visible signs such as charring, shrinking, swelling, color changes, or dissolving. Materials can also be tested for weight changes, loss of fabric tensile strength, and other properties to measure degradation. Think about the rapid and destructive way in which gasoline dissolves a Styrofoam cup. When chemicals are so aggressive, or when the suit fabric is a poor match for the suspect substance, fabric degradation is possible. If the suit dissolves, the possibility of the wearer suffering an injury is high.

Types of Chemical-Protective Clothing

Chemical protective clothing can be constructed as a single or multi-piece garment. A single-piece garment may or may not completely enclose the wearer, and is often found as a coverall-type garment. A multi-piece garment typically has a jacket, pants, an attached or detachable hood, and perhaps attached fabric to cover the feet. Multi-piece garments are found as Level B and Level C protection. Level A protection is almost always built as an encapsulated one-piece suit with attached gloves and suit fabric that covers the feet. Chemical-resistant boots should be worn to offer protection from abrasion and mechanical hazards. Chemical protective equipment suited for law enforcement missions is becoming more popular and finding its way into traditional hazardous materials response. These protective ensembles typically offer protection against liquid and particulate forms of CBRN agents (chemical, biological, radiological, and nuclear) and are much cooler and more comfortable to wear for extended periods of time.

Chemical-protective clothing—both single- and multi-piece—are classified into two major categories: vapor-protective clothing and liquid splash–protective clothing. Both are described in this section. Many different types of materials are

Chemical Name	Concentration	Breakthrough Time	Permeation Rate
	(%)	Normalized (min)	(ug/cm2/min)
1,1,2,2-TETRACHLOROETHANE	95+	>480	0.0005
1,1,2-TRICHLOROETHANE	95+	>480	<0.01
1,3-DICHLOROACETONE (40°C)	95+	>480	<0.1
1,4-DIOXANE	95+	>480	<0.05
1,6-HEXAMETHYLENEDIAMINE	95+	>480	<0.01
2,2,2-TRICHLOROETHANOL	95+	>480	<0.01
2,2,2-TRIFLUOROETHANOL	95+	>480	<0.001
2,3-DICHLOROPROPENE	95+	>480	<0.08
2-CHLOROETHANOL	95+	>480	<0.008
2-METHYLGLUTARONITRILE	87	>480	<0.1
2-PICOLINE	95+	46	48
3,4-DICHLOROANILINE	95+	284	2.4
3-PICOLINE	95+	11	22
4,4'METHYLENE BIS(2-CHLOROANILINE)	95+	>480	<0.1
ACETALDEHYDE	95+	>480	<0.01
ACETIC ACID	95+	339	1.3
ACETIC ANHYDRIDE	95+	>480	<0.001
ACETONE	95+	>480	<0.001
ACETONITRILE	95+	>480	<0.01
ACETYL CHLORIDE	95+	181	2
ACROLEIN	95+	>480	<0.02
ACRYLAMIDE	50% in water	>480	<0.1
ACRYLIC ACID	95+	270	1.6
ACRYLONITRILE	95+	>480	<0.0003
ADIPONITRILE	95+	>480	<0.1
ALLYL ALCOHOL	95+	>480	<0.1
ALLYL CHLORIDE	95+	>480	<0.06
AMMONIA GAS	95+	46	0.62
AMMONIUM FLUORIDE	40	>480	<0.01
AMMONIUM HYDROXIDE	28-30	160	4.7
AMYL ACETATE	95+	>480	<0.003
ANILINE	95+	>480	<0.1
ARSINE	95+	>480	<0.01
BENZENE SULFONYL CHLORIDE	95+	>480	<0.1
BENZIDINE	25% in methanol	>480	<0.01
BENZONITRILE	95+	>480	<0.004
BENZONITRILE	95+	>480	<0.004
BENZOYL CHLORIDE	95+	>480	<0.05
BENZYL CHLORIDE	95+	>480	<0.01
BORON TRICHLORIDE	95+	>480	<0.02
BORON TRIFLUORIDE	95+	>480	<0.1

FIGURE 8-6 Breakthrough time is the time it takes a chemical substance to be absorbed into the suit fabric and detected on the other side.

Courtesy of The DuPont Company

manufactured for both categories; it is the AHJ's responsibility to determine which type of suit is appropriate for each situation. Be aware that no single chemical-protective garment (vapor or splash) on the market will protect you from everything.

Vapor-protective clothing, also referred to as *fully encapsulating protective clothing*, offers full body protection from highly toxic environments and requires the wearer to use an air-supplied respiratory device such as SCBA **FIGURE 8-7**. The wearer is completely zipped inside the protective "envelope," leaving no skin (or the lungs) accessible to the outside. If the ammonia scenario described at the beginning of the chapter were occurring in a different location—such as inside a poorly ventilated storage area within an ice-making facility—vapor-protective clothing might be required. Ammonia aggressively attacks skin, eyes, and mucous membranes such as the eyes and mouth and can cause severe and irreparable damage to the lungs. Hydrogen cyanide would be another example of a chemical substance that would require this level of protection. Hydrogen cyanide can be fatal if inhaled or absorbed through the skin, so the use of a fully encapsulating suit is required to adequately protect the wearer. NFPA 1991, *Standard on Vapor-Protective Ensembles for Hazardous Materials Emergencies*, sets the performance standards for vapor-protective garments.

Liquid splash–protective clothing is designed to protect the wearer from chemical splashes **FIGURE 8-8**. NFPA 1992, *Standard on Liquid Splash-Protective Ensembles and Clothing for Hazardous Materials Emergencies*, is the performance document that governs liquid splash–protective garments and ensembles. Equipment that meets this standard has been tested for penetration against a battery of five chemicals. The tests include no gases, because this level of protection is not considered to be vapor protection.

Responders may choose to wear liquid splash–protective clothing based on the anticipated hazard posed by a particular substance. As you learned in the chapter *Estimating Potential Harm and Planning a Response*, liquid splash–protective clothing does not provide total body protection from gases or vapors, and it should not be used for incidents involving liquids that emit vapors known to affect or be absorbed through the skin. This level of protection may consist of several pieces of clothing and equipment designed to protect the skin and eyes from chemical splashes. Some agencies, depending on the situation, choose to have their personnel wear liquid splash protection over or under structural firefighting clothing.

Responders trained to the operational level often wear liquid splash–protective clothing when they are assigned to enter the initial site, perform decontamination, or construct isolation barriers such as dikes, diversions, retention areas, or dams.

Hand Signals

Wearing PPE not only presents physical challenges, but also, in many cases, compromises communications. For this reason, even if radio communication is available, responders often resort to hand signals to communicate. Your safety briefing should always include a review of hand signals. Everyone should know the basic communication "fall back" position to ensure that in the event of an in-suit emergency (loss of

FIGURE 8-7 Vapor-protective clothing retains body heat, so it also increases the possibility of heat-related emergencies among responders.

FIGURE 8-8 Liquid splash–protective clothing is worn whenever there is the danger of chemical splashes.

breathing air, SCBA cylinder failure, or acute illness and injury) anyone having visual contact with the responders will quickly understand that a problem is occurring. **FIGURE 8-9** demonstrates common hand signals used by responders. Your AHJ may have different hand signals; it's the responsibility of all responders to know how to use them correctly.

Respiratory Protection

NFPA 1994, *Standard on Protective Ensembles for First Responders to CBRN Terrorism Incidents*, was developed to address the performance of protective ensembles and garments (including respiratory protection) specific to WMD. NFPA 1994 covers three classes of equipment: Class 2, Class 3, and Class 4. As discussed in the chapter *Estimating Potential Harm and Planning a Response*, as well as later in this chapter, the EPA and the OSHA HAZWOPER regulation classify ensemble levels as Level A, Level B, Level C, and Level D. The main difference between their system and the NFPA 1994 classification is that NFPA 1994 covers the performance of the garment *and* factors in the performance of the respiratory protection.

Simple asphyxiants such as nitrogen, argon, and helium, as well as oxygen-deficient atmospheres, are best handled by using a SCBA or another supplied air respirator (SAR). SCBA units that comply with the current version of NFPA 1981, *Standard on Open-Circuit Self-Contained Breathing*

Courtesy of Rob Schnepp

Courtesy of Rob Schnepp

Courtesy of Rob Schnepp

FIGURE 8-9 A. A hand on top of the head making a tapping motion indicates, "I'm okay."
B. Hands across the throat indicate "air problems."
C. Hands over the head in a waving motion or both hands tapping the head is a signal for "trouble. " This could indicate a suit problem or that something is amiss with the task or situation on the part of the responder or responders.

Apparatus (SCBA) for Emergency Services, are positive-pressure, chemical, biological, radiological, and nuclear (CBRN)-certified units that maintain a pressure inside the face piece, in relation to the pressure outside the face piece, such that the pressure is positive during both inhalation and exhalation. This is a very important feature when operating in airborne contamination.

The extra weight and reduced visibility associated with SCBA are factors to consider but are not reason enough to compromise a high level of respiratory protection. If an SCBA is indicated, it should be worn. As with any piece of PPE, there are as many positive benefits as there are negative points to consider in making this decision. Any responder called upon to wear an SCBA should be fully trained by the AHJ prior to operating in a contaminated environment. The OSHA HAZWOPER standard states that all employees engaged in emergency response who are exposed to hazardous substances *shall* wear a positive-pressure SCBA. Furthermore, the incident commander (IC) is *required* to ensure the use of SCBA. It is not just a good idea—it's the law. Additionally, all responders should follow manufacturers' recommendations for using, maintaining, testing, inspecting, cleaning, and filling the SCBA unit. Be sure to document all of these activities so there is a record of what has been done to the unit. Refer back to the chapter *Estimating Potential Harm and Planning a Response* for more information about the various types of respiratory protection.

Chemical-Protective Clothing Ratings

A variety of fabrics are used in both vapor-protective and liquid splash–protective garments and ensembles. Commonly used suit fabrics include butyl rubber, Tyvek, Saranex, polyvinyl chloride (PVC), and Viton. Protective clothing materials must offer acceptable resistance to the chemical substances involved, and the garments should be used within the parameters set by their manufacturer. The manufacturer's guidelines and recommendations should be consulted for material compatibility information.

The following EPA guidelines may be used by a responder to assist in determining the appropriate level of protection for a particular hazard. The procedures for the donning and doffing of equipment are described in the following sections.

Level A

A Level A ensemble consists of a fully encapsulating garment that completely envelops both the wearer and his or her respiratory protection **FIGURE 8-10**. Level A equipment should be used when the hazardous material identified requires the highest level of protection for skin, eyes, and lungs. Such an ensemble is effective against vapors, gases, mists, and dusts and is typically indicated when the operating environment exceeds IDLH values for skin absorption.

Level A protection, when worn in accordance with NFPA 1991, will protect the wearer against a transient episode of flash fire. To that end, thermal extremes should be approached with caution. Direct contact between the suit fabric and a cryogenic material, such as liquid nitrogen or liquid helium, may result in immediate suit failure. This type of ensemble more than addresses the asphyxiant threat—it's the temperature extreme that must be acknowledged. By contrast, a potentially flammable atmosphere should be considered an extremely dangerous situation. In such circumstances, Level A suits, even with the flash fire component of the suit in place, provide very limited protection. Moreover, it is difficult to see when wearing a Level A suit, which increases the possibility that the person may unknowingly bump into sharp objects or rub against materials that might puncture or abrade the suit's vapor protection. Therefore, some forethought about the operating environment should occur well before entering the contaminated atmosphere. As always, a risk-versus-reward thought process should prevail. The "best" level of protection is the one that is the most appropriate for the hazard and the mission.

FIGURE 8-10 A Level A ensemble envelops the wearer in a totally encapsulating suit.

Courtesy of Rob Schnepp

Level A protection is effective against alpha radiation, but because of the lack of fabric thickness (as compared to fire fighters' turnout gear) it may not offer adequate protection against beta radiation, and certainly is not a barrier to gamma radiation. Remember—thorough detection and monitoring actions will help you determine the nature of the operating environment.

Ensembles worn as Level A protection must meet the requirements outlined in NFPA 1991. A Level A ensemble also requires open-circuit, positive-pressure SCBA or an SAR for respiratory protection. The chapter *Estimating Potential Harm and Planning a Response* provides a list of the recommended and optional components of Level A protection.

To don a Level A ensemble, follow the steps in SKILL DRILL 8-1:

1. Conduct a pre-entry briefing, medical monitoring, and equipment inspection. (STEP 1)
2. While seated, pull on the suit to waist level and pull on the attached chemical boots. Fold the suit boot covers over the tops of the boots. (STEP 2)
3. Stand up and don the SCBA frame and SCBA face piece, but do not connect the regulator to the face piece. (STEP 3)
4. Place the helmet on the head. (STEP 4)
5. Don the inner gloves. (STEP 5)
6. Don the outer chemical gloves (if required by the manufacturer's specifications).
7. With assistance, complete donning the suit by placing both arms in the suit, pulling the expanded back piece over the SCBA, and placing the chemical suit over the head. (STEP 6)
8. Instruct the assistant to connect the regulator to the SCBA face piece and ensure that the airflow is working correctly. (STEP 7)
9. Instruct the assistant to close the chemical suit by closing the zipper and sealing the splash flap. (STEP 8)
10. Review hand signals and indicate that you are okay. (STEP 9)

SKILL DRILL 8-1 Donning a Level A Ensemble

NFPA 6.2.4.1

VIDEO

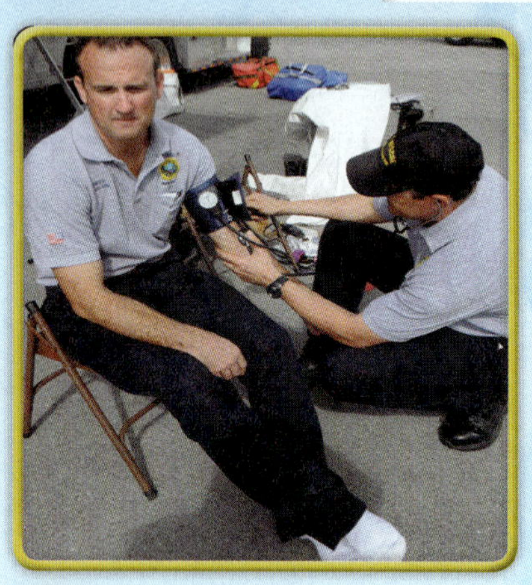

1 Conduct a pre-entry briefing, medical monitoring, and equipment inspection.

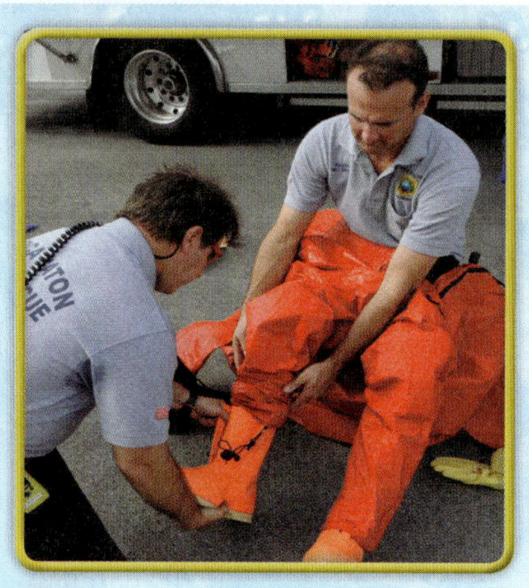

2 While seated, pull on the suit to waist level; pull on the chemical boots over the top of the chemical suit. Fold the suit boot covers over the tops of the boots.

3 Stand up and don the SCBA frame and SCBA face piece, but do not connect the regulator to the face piece.

4 Place the helmet on the head.

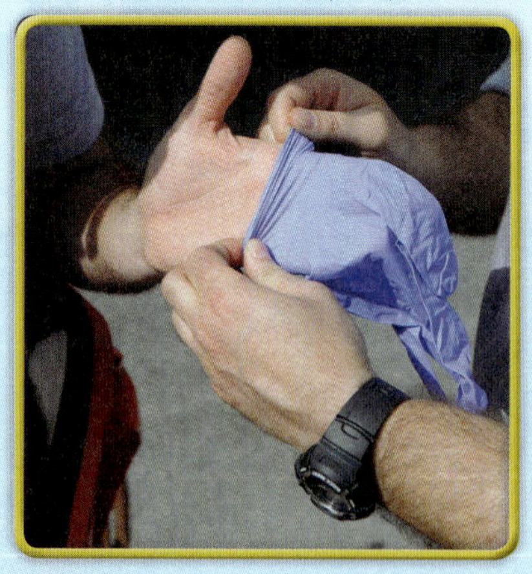

5 Don the inner gloves.

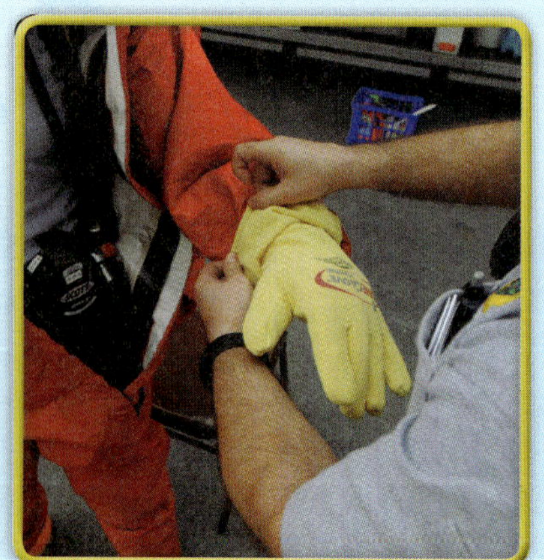

6 Don the outer chemical gloves (if required). With assistance, complete donning the suit by placing both arms in the suit, pulling the expanded back piece over the SCBA, and placing the chemical suit over the head.

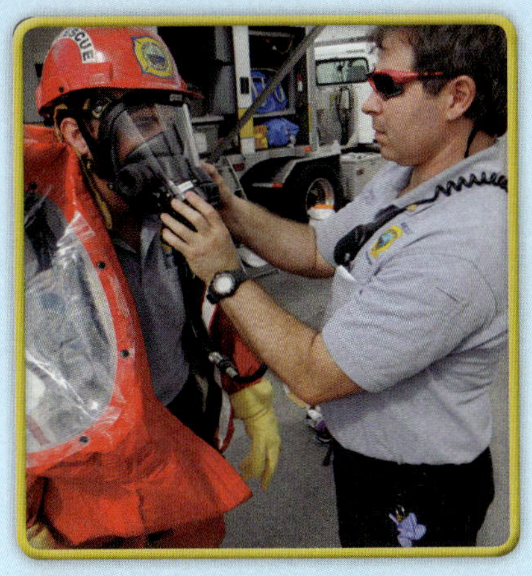

7 Instruct the assistant to connect the regulator to the SCBA face piece and ensure air flow.

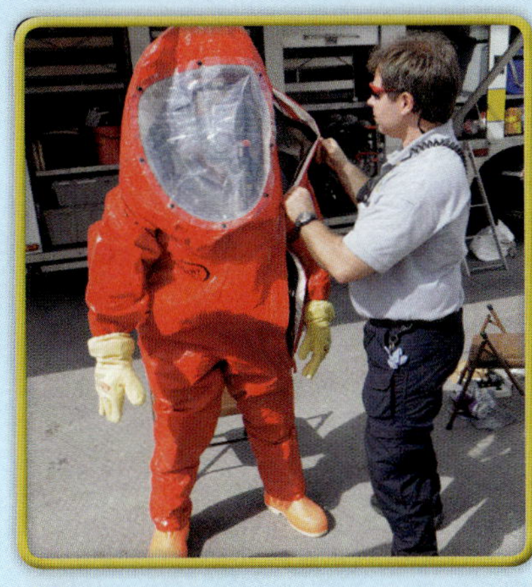

8 Instruct the assistant to close the chemical suit by closing the zipper and sealing the splash flap.

9 Review hand signals and indicate that you are okay.

To doff a Level A ensemble, follow the steps in SKILL DRILL 8-2:

1. After completing decontamination, proceed to the clean area for suit doffing.
2. Pull the hands out of the outer gloves and arms from the sleeves, and cross the arms in front inside the suit. (STEP 1)
3. Instruct the assistant to open the chemical splash flap and suit zipper. (STEP 2)
4. Instruct the assistant to begin at the head and roll the suit down and away until the suit is below waist level. (STEP 3)
5. Instruct the assistant to complete rolling the suit from the waist to the ankles and step out of the attached chemical boots and suit. (STEP 4)
6. Doff the SCBA frame. The face piece should be kept in place while the SCBA is doffed. (STEP 5)
7. Take a deep breath and doff the SCBA face piece; carefully remove the helmet, peel off the inner gloves, and walk away from the clean area.
8. Go to the rehabilitation area for medical monitoring, rehydration, and personal decontamination shower. (STEP 6)

SKILL DRILL 8-2 Doffing a Level A Ensemble

NFPA 6.2.4.1

VIDEO

1 After completing decontamination, proceed to the clean area for suit doffing. Pull the hands out of the outer gloves and arms from the sleeves, and cross the arms in front inside the suit.

2 Instruct the assistant to open the chemical splash flap and suit zipper.

3 Instruct the assistant to begin at the head and roll the suit down and away until the suit is below waist level.

(Continued)

SKILL DRILL 8-2 (CONTINUED)

Doffing a Level A Ensemble

NFPA 6.2.4.1

VIDEO

4 Instruct the assistant to complete rolling the suit from the waist to the ankles; step out of the attached chemical boot and suit.

5 Doff the SCBA frame. The face piece should be kept in place while the SCBA frame is doffed.

6 Take a deep breath and doff the SCBA face piece; carefully peel off the inner gloves and walk away from the clean area. Go to the rehabilitation area for medical monitoring, rehydration, and personal decontamination shower.

Level B

A Level B ensemble consists of multi-piece chemical-protective clothing, boots, gloves, and SCBA **FIGURE 8-11**. This type of protective ensemble should be used when the type and atmospheric concentration of identified substances require a high level of respiratory protection but less skin protection. The kinds of gloves and boots chosen will depend on the physical and chemical properties of the identified chemical. The SCBA components should be considered as well when wearing non-encapsulating PPE.

The Level B protective ensemble is the workhorse of hazardous materials response—it is a very common level of protection and is often chosen for its versatility. Personnel initially processing a clandestine drug laboratory, performing preliminary missions for reconnaissance, or engaging in detection and monitoring duties commonly wear such an ensemble. The typical Level B ensemble provides little or no flash fire protection, however. Thus it should be viewed in the same manner as Level A equipment when it comes to thermal protection and other considerations of use such as protection from mechanical hazards, radiation, or asphyxiants.

Garments and ensembles that are worn for Level B protection should comply with the performance requirements found in NFPA 1992. The chapter *Estimating Potential Harm and Planning a Response* provides a list of the recommended and optional components of a Level B protective ensemble.

You may also encounter single-piece garments that are worn as Level B protection. These suits, referred to in the field as encapsulating Level B garments, are not constructed to be "vapor tight" like Level A garments. Encapsulating Level B

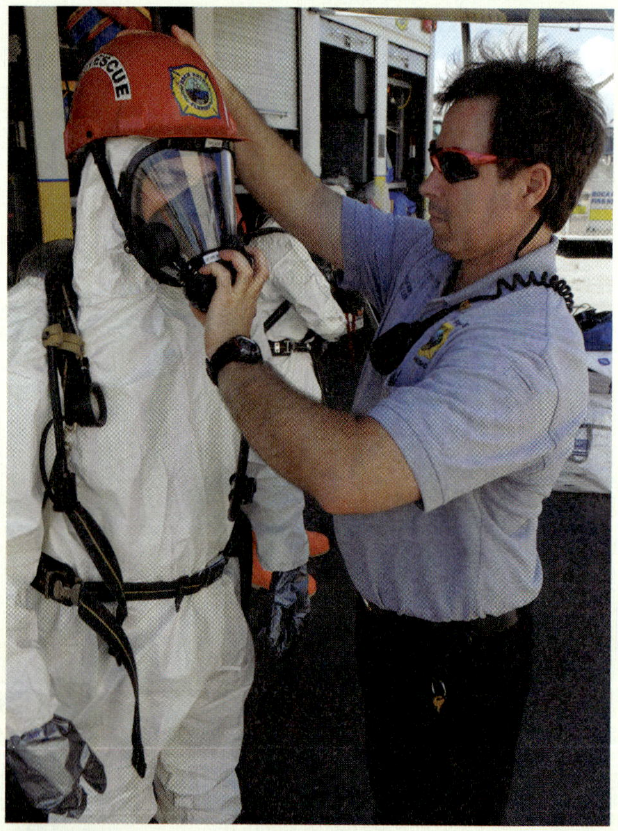

FIGURE 8-11 A Level B protective ensemble provides a high level of respiratory protection but less skin protection.

garments do not have vapor-tight zippers, seams, or one-way relief valves around the hood like Level A garments. Although the encapsulating Level B suit may look a lot like a Level A garment, it is not constructed similarly and will not offer the same level of protection.

To don and doff a Level B encapsulated chemical-protective clothing ensemble, follow the same steps found in Skill Drill 8-1 and Skill Drill 8-2. Remember, the difference between the Level A ensemble and Level B encapsulating ensemble is not the procedure—it is the construction and performance of the garment.

Responder Safety Tip

According to the OSHA HAZWOPER regulation, Level B is the minimum level of protection to be worn when operating in an unknown environment.

To don a Level B nonencapsulated chemical-protective clothing ensemble, follow the steps in SKILL DRILL 8-3:

1. Conduct a pre-entry briefing, medical monitoring, and equipment inspection. (STEP 1)
2. Sit down, and pull on the suit to waist level. Pull on the chemical boots over the top of the chemical suit. Pull the suit boot covers over the tops of the boots. (STEP 2)
3. Don the inner gloves. (STEP 3)
4. With assistance, complete donning the suit by placing both arms in the suit and pulling the suit over the shoulders.
5. Instruct the assistant to close the chemical suit by closing the zipper and sealing the splash flap. (STEP 4)
6. Don the SCBA frame and SCBA face piece, but do not connect the regulator to the face piece (STEP 5)
7. With assistance, pull the hood over the head and SCBA face piece.
8. Place the helmet on the head.
9. Pull the outer gloves over or under the sleeves, depending on the situation.
10. Instruct the assistant to connect the regulator to the SCBA face piece and ensure that the airflow is working correctly. (STEP 6)
11. Review hand signals and indicate that you are okay.

SKILL DRILL 8-3 Donning a Level B Nonencapsulated Chemical-Protective Clothing Ensemble

NFPA 6.2.4.1

VIDEO

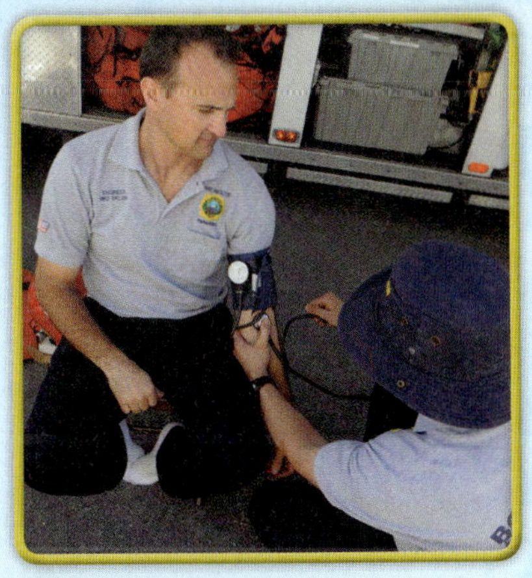

1 Conduct a pre-entry briefing, medical monitoring, and equipment inspection.

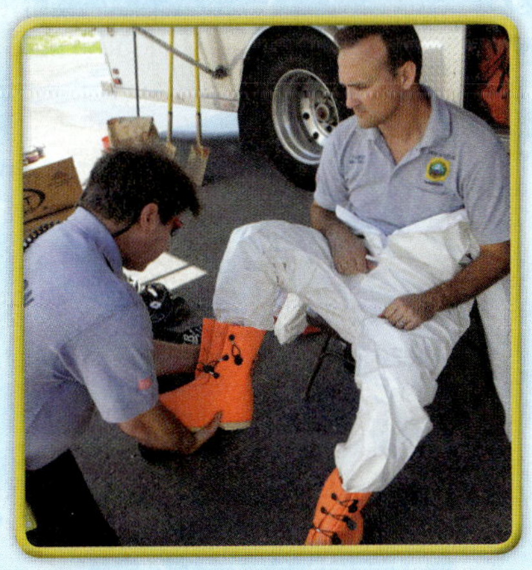

2 Sit down, pull on the suit to waist level; pull on the chemical boots over the top of the chemical suit. Pull the suit boot covers over the tops of the boots.

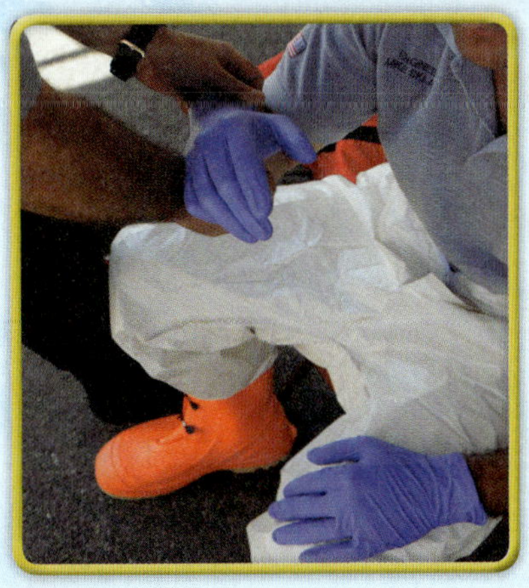

3 Don the inner gloves.

(Continued)

SKILL DRILL 8-3 (CONTINUED)

Donning a Level B Nonencapsulated Chemical-Protective Clothing Ensemble

NFPA 6.2.4.1

VIDEO

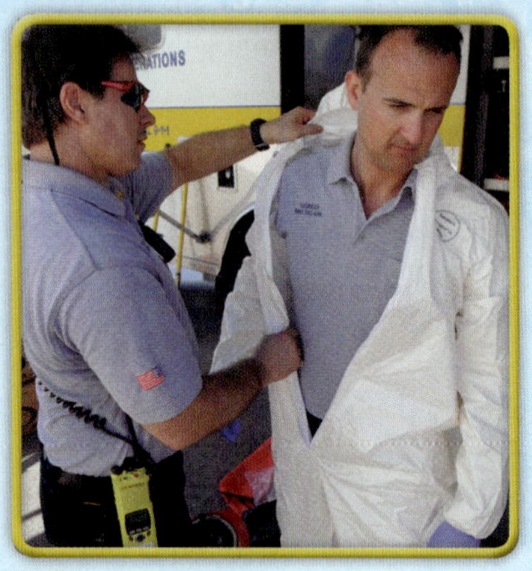

4 With assistance, complete donning the suit by placing both arms in suit and pulling suit over shoulders. Instruct the assistant to close the chemical suit by closing the zipper and sealing the splash flap.

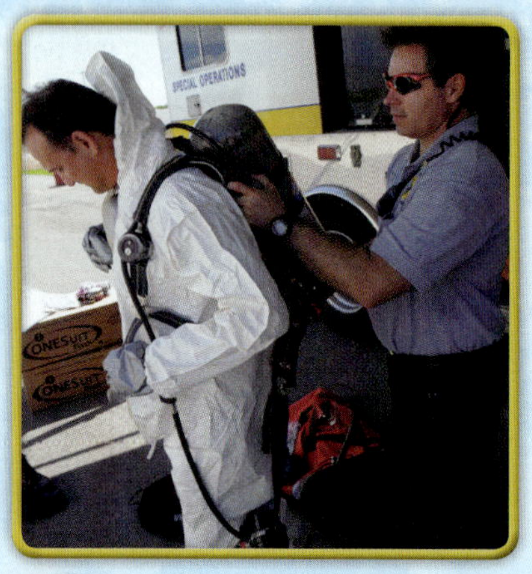

5 Don the SCBA frame and SCBA face piece, but do not connect the regulator to the face piece.

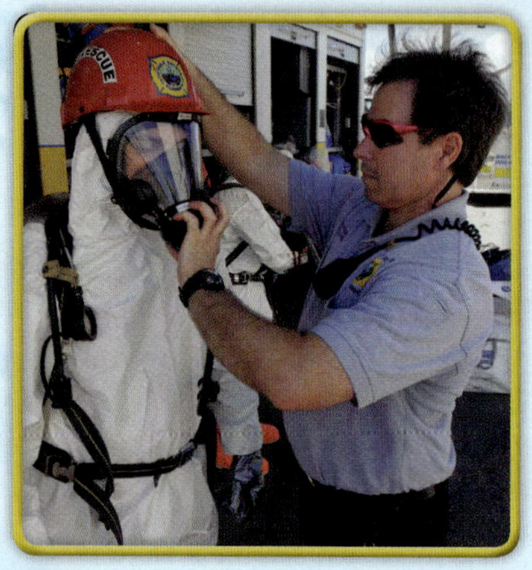

6 With assistance, pull the hood over the head and SCBA face piece. Place the helmet on the head. Put on the outer gloves. Instruct the assistant to connect the regulator to the SCBA face piece and ensure you have air flow. Review hand signals and indicate that you are okay.

To doff a Level B nonencapsulated chemical-protective clothing ensemble, follow the steps in SKILL DRILL 8-4:

1. After completing decontamination, proceed to the clean area for suit doffing.
2. Stand and doff the SCBA frame. Keep the face piece in place while the SCBA frame is placed on the ground. (STEP 1)
3. Instruct the assistant to open the chemical splash flap and suit zipper. (STEP 2)
4. Remove your hands from the outer gloves and arms from the sleeves, and cross your arms in front inside the suit.
5. Instruct the assistant to begin at the head and roll the suit down and away until the suit is below waist level. (STEP 3)
6. Sit down and instruct the assistant to complete rolling down the suit to the ankles. Step out of the outer boots and suit.(STEP 4)
7. Stand and doff the SCBA face piece and helmet. (STEP 5)
8. Carefully peel off the inner gloves and go to the rehabilitation area for medical monitoring, rehydration, and personal decontamination shower. (STEP 6)

Level C

A Level C ensemble is appropriate when the type of airborne contamination is known, its concentration is measured, and the criteria for using an air-purifying respirator (APR) are met. Typically, Level C ensembles are worn with an APR or powered air-purifying respirator (PAPR). The complete ensemble consists of standard work clothing, chemical-protective clothing, chemical-resistant gloves, and a form of respiratory protection other than an SCBA or SAR system. Level C equipment is appropriate when significant skin and eye exposure is unlikely FIGURE 8-12. In many cases, Level C ensembles are worn in low-hazard situations such as clean-up activities lasting hours or days; once an area is fully characterized and the hazards are found to be low enough to allow this level of protection; or after responders mitigate the problem to the extent that they can dress down to this lower level to complete the mission. Many law enforcement agencies have provided their officers with Level C ensembles to be carried in the trunk of patrol cars. Based on the mission of perimeter scene control, this may be a prudent level of protection.

The chapter *Estimating Potential Harm and Planning a Response* provides a list of the recommended and optional components of a Level C protective ensemble, and reviews the conditions of use for APRs and PAPRs. The garment selected must meet the performance requirements for NFPA 1992. Respiratory protection may be provided by a half-face (with eye protection) or a full-face mask.

SKILL DRILL 8-4

Doffing a Level B Nonencapsulated Chemical-Protective Clothing Ensemble

NFPA 6.2.4.1

VIDEO

1 After completing decontamination, proceed to the clean area for suit doffing. Stand and doff the SCBA frame. Keep the face piece in place while the SCBA frame is placed on the ground.

2 Instruct the assistant to open the chemical splash flap and suit zipper.

3 Remove your hands from the outer gloves and your arms from the sleeves of the suit. Cross your arms in front inside of the suit. Instruct the assistant to begin at the head and roll the suit down and away until the suit is below waist level.

4 Sit down and instruct the assistant to complete rolling down the suit to the ankles; step out of the attached chemical boots and suit.

5 Stand and doff the SCBA face piece and helmet.

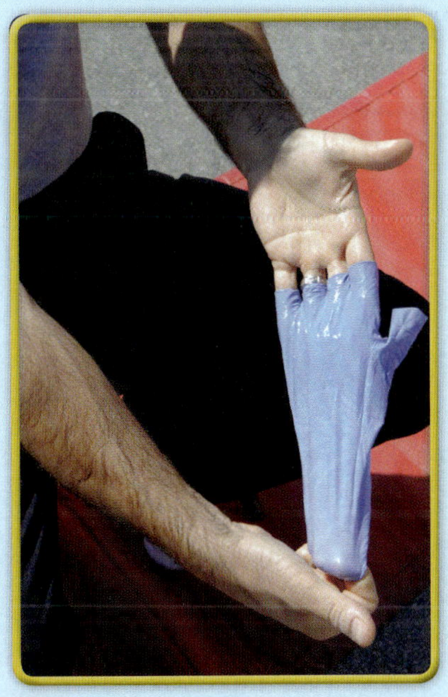

6 Carefully peel off the inner gloves and go to the rehabilitation area for medical monitoring, rehydration, and personal decontamination shower.

FIGURE 8-12 A Level C protective ensemble includes chemical-protective clothing and gloves as well as respiratory protection.

Courtesy of The DuPont Company

To don a Level C chemical-protective clothing ensemble, follow the steps in **SKILL DRILL 8-5**:

1. Conduct a pre-entry briefing, medical monitoring, and equipment inspection.
2. While seated, pull on the suit to waist level. Pull on the chemical boots over the top of the chemical suit. Pull the suit boot covers over the tops of the boots. (**STEP 1**)
3. Don the inner gloves. (**STEP 2**)
4. With assistance, complete donning the suit by placing both arms in the suit and pulling the suit over the shoulders.
5. Instruct the assistant to close the chemical suit by closing the zipper and sealing the splash flap. (**STEP 3**)
6. Don the APR/PAPR face piece.
7. With assistance, pull the hood over the head and the APR/PAPR face piece.
8. Place the helmet on the head.
9. Pull the outer gloves over or under the sleeves, depending on the situation.
10. Review hand signals and indicate that you are okay. (**STEP 4**)

SKILL DRILL 8-5

Donning a Level C Chemical-Protective Clothing Ensemble

NFPA 6.2.4.1

VIDEO

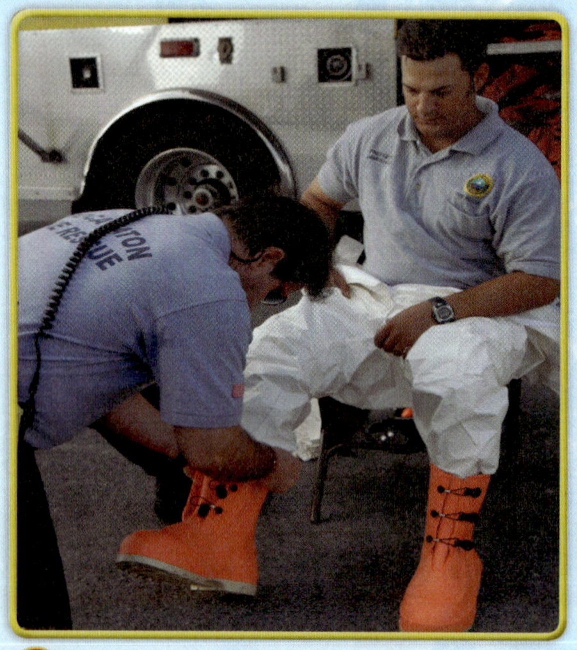

1. Conduct a pre-entry briefing, medical monitoring, and equipment inspection. While seated, pull on the suit to waist level; pull on the chemical boots over the top of the chemical suit. Pull the suit boot covers over the tops of the boots.

© Jones & Bartlett Learning. Photographed by Glen E. Ellman

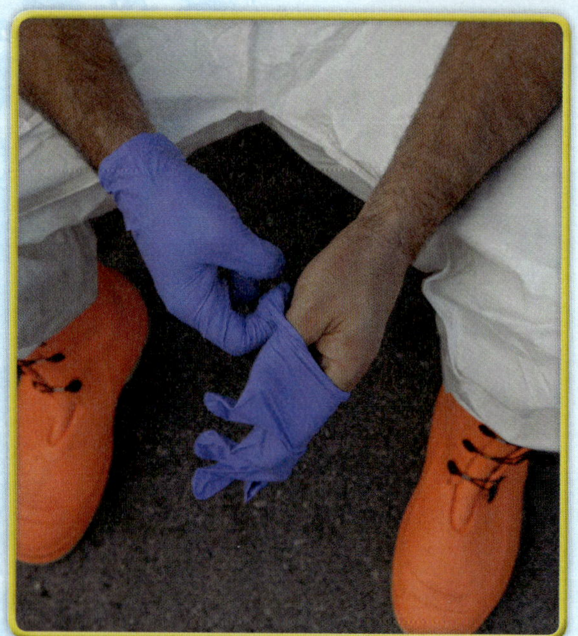

2. Don the inner gloves.

© Jones & Bartlett Learning. Photographed by Glen E. Ellman

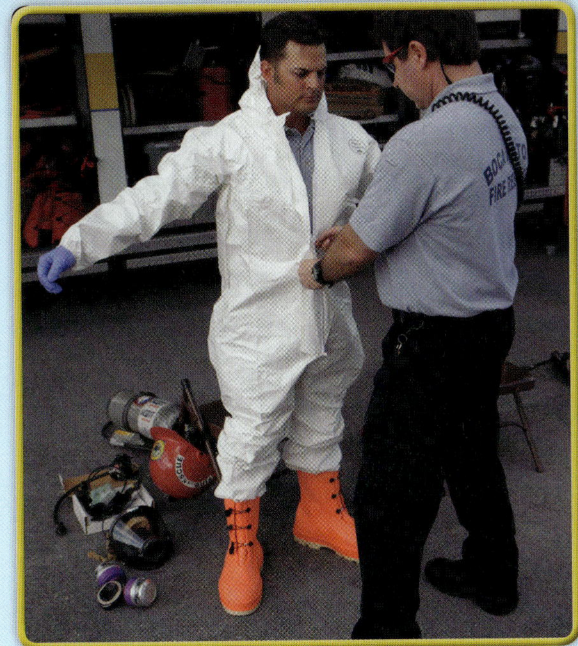

3 With assistance, complete donning the suit by placing both arms in suit and pulling suit over shoulders. Instruct the assistant to close the chemical suit by closing the zipper and sealing the splash flap.

4 Don the APR/PAPR face piece. With assistance, pull the hood over the head and the PAR/PAPR face piece. Place the helmet on the head. Pull on the outer gloves. Review hand signals and indicate that you are okay.

To doff a Level C chemical-protective clothing ensemble, follow the steps in SKILL DRILL 8-6:

1. After completing decontamination, proceed to the clean area for suit doffing.
2. As with Level B, instruct the assistant to open the chemical splash flap and suit zipper.
3. Remove the hands from the outer gloves and your arms from the sleeves.
4. Instruct the assistant to begin at the head and roll the suit down and away until the suit is below waist level.
5. Instruct the assistant to complete rolling down the suit and take the suit and boots away.
6. Instruct the assistant to help remove the inner gloves.
7. Remove the APR/PAPR. Remove the helmet. (STEP 1)
8. Go to the rehabilitation area for medical monitoring, rehydration, and personal decontamination shower. (STEP 2)

Level D

A Level D ensemble offers the lowest level of protection. It typically consists of coveralls, work shoes, hard hat, gloves, and standard work clothing. This type of equipment should be used only when the atmosphere contains no known hazard and when work functions preclude splashes, immersion, or the potential for unexpected inhalation of or contact with hazardous levels of chemicals. Level D protection should be used when the situation involves nuisance contamination (such as dust) only. It should not be worn on any site where respiratory or skin hazards exist. The chapter *Estimating Potential Harm and Planning a Response* provides a list of the recommended and optional components of Level D protection.

TABLE 8-1 describes the relationships among the NFPA hazardous materials protective clothing standards; the OSHA/EPA Level A, B, and C classifications; and the new NIOSH-certified respirator with CBRN protection standards. The table is intended to clarify the relationship between the NFPA guidelines and the OSHA standards and to summarize the expected performance of the ensembles.

Safety

There are many hazards associated with wearing PPE. These hazards are best addressed by understanding the NFPA performance requirements for chemical-protective ensembles and the safety considerations taken into account when wearing PPE.

Chemical-Protective Equipment Performance Requirements

Typical chemical-protective equipment performance requirements include tests for durability, barrier integrity after flex and abrasion challenges, cold-temperature flex, and flammability.

SKILL DRILL 8-6 Doffing a Level C Chemical-Protective Clothing Ensemble

NFPA 6.2.4.1

VIDEO

1 After completing decontamination, proceed to the clean area. As with Level B, the assistant opens the chemical splash flap and suit zipper. Remove the hands from the gloves and arms from the sleeves. Instruct the assistant to begin at the head and roll the suit down below waist level. Instruct the assistant to complete rolling down the suit and take the outer boots and suit away. The assistant helps remove inner gloves. Remove APR/PAPR. Remove the helmet.

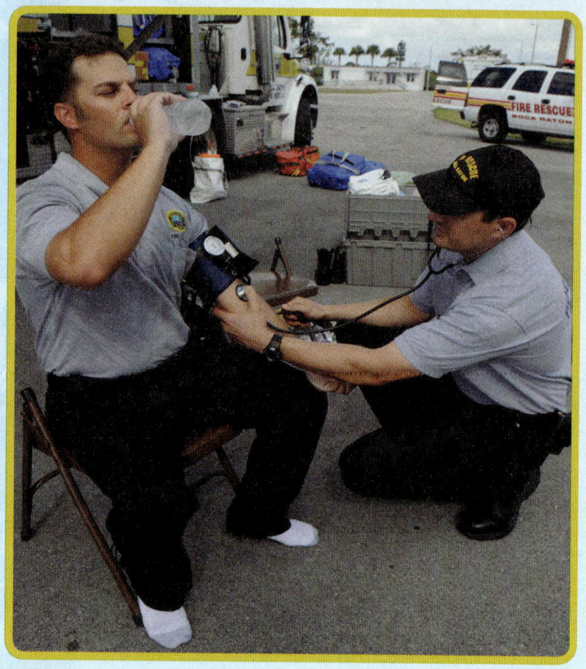

2 Go to rehabilitation area for medical monitoring, rehydration, and personal decontamination shower.

To don a Level D chemical-protective clothing ensemble, follow the steps in SKILL DRILL 8-7:

1. Conduct a pre-entry briefing and equipment inspection.
2. Don the Level D suit.
3. Don the boots.
4. Don safety glasses or chemical goggles.
5. Don a hard hat.
6. Don gloves, a face shield, and any other required equipment. (STEP 1)

Essentially, each part of the suit must pass a particular set of challenges prior to receiving certification based on the NFPA testing standards. Users of any garment that meets the performance requirements set forth by the NFPA standard can rest assured that the garment will withstand reasonable insults from most mechanical-type hazards encountered on the scene. Of course, achieving a third-party independent certification that uses the NFPA standard does not mean the suit is invincible and cannot fail; it simply means that the equipment will hold up under "normal" conditions. It is up to the user to be aware of the hazards and avoid situations that may cause the garment to fail.

Responder Safety Tip

As exemplified by the TRACEMP acronym, many hazards can be encountered during the course of a hazardous materials incident. Given this possibility, multiple layers or multiple types of protection may have to be used in some situations. You should also understand the working environment and match the right garment to the anticipated hazards.

TABLE 8-1 Levels of Protection

NFPA Standard	OSHA/EPA Level	NIOSH-Certified Respirator	NFPA Chemical Barrier Protection Method(s)	Chemical Vapor*	Chemical Liquid*	Particulate	Liquid-Borne Biological (Aerosol)
1991	A	CBRN SCBA (open recruit)	Protection against permeation and penetration*	X	X	X	X
1992	B	Non-CBRN SCBA (or CBRN SCBA)	Protection against permeation and penetration*	X			
	C	Non-CBRN APR or PAPR	Protection against penetration	X			
1994, Class 1	(Note: The NFPA 1994, Class 1 ensemble was removed in the 2006 edition of the standard because of its redundancy with NFPA 1991.)						
1994, Class 2	B	CBRN SCBA	Protection against permeation	X	X	X	X
1994, Class 3	C	CBRN APR or PAPR	Protection against permeation	X	X	X	X
1994, Class 4	B	CBRN SCBA	Protection against penetration	NA	NA	X	X
	C	CBRN APR or PAPR	Protection against penetration	NA	NA/NT	X	X

* Vapor protection for NFPA 1994 Class 2 and Class 3 is based on challenge concentrations established for NIOSH certification of CBRN open-circuit SCBA and APR/PAPR respiratory equipment. Class 2 and Class 3 do not require the use of totally encapsulating garments.

SKILL DRILL 8-7 Donning a Level D Chemical-Protective Clothing Ensemble

NFPA 6.2.4.1

VIDEO

1 Conduct pre-entry briefing, medical monitoring, and equipment inspection. Don the Level D suit. Don boots. Don safety glasses or chemical goggles. Don hard hat. Don gloves, a face shield, and any other required equipment.

Courtesy of Rob Schnepp

As described earlier, a variety of materials are used in both vapor- and splash-protective clothing. The most commonly used materials include butyl rubber, Tyvek, Saranex, polyvinyl chloride, and Vitron, either singly or in multiple layers consisting of several different materials. Special chemical-protective clothing is adequate for incidents involving some chemicals, yet useless for incidents involving other chemicals; no single fabric provides satisfactory protection from all chemicals.

All responders who may be called upon to wear any type of PPE should read and understand the manufacturer's specifications and procedures for the maintenance, testing, inspection, cleaning, and storage of PPE provided by the AHJ. The list of NFPA and NIOSH documents in TABLE 8-2 offers an overview of the testing and certification standards affecting the PPE currently on the market.

It is important for all responders to remember that some of the mission-specific competencies in this section are taken from competencies required of hazardous materials technicians. *That does not mean that operations level responders, with a mission-specific competency in PPE or any other mission-specific competency, are a replacement for a technician.*

Responder Safety

Working in PPE is a hazardous proposition on two different levels. First, simply by wearing PPE, the responder acknowledges that some degree of danger exists: If there were no hazard, there would be no need for the PPE! Second, wearing the PPE puts an inherent stress on the responder, separate and apart from the stress imposed by the operating environment. Much of this text is devoted to the "safety first" consideration. The next sections are devoted to raising your awareness of the issues that may arise from the very gear used to keep you safe.

The chapter *Implementing the Planned Response* outlines the various heat-related illnesses commonly experienced by responders. Those complications include heat exhaustion, heat cramps, and heat stroke, all of which are usually preceded by dehydration. Given that well-defined relationship, responders should be fully aware that their underlying level of hydration, prior to the response, may have an effect on their safety while

Responder Safety Tip

The authority having jurisdiction must properly outfit all responders expected to respond to a hazardous materials incident. The current OSHA HAZWOPER regulations [29 CFR 1910.120 (q)(3)(iii)] (and many local jurisdictional regulations) require the incident commander to ensure that the personal protective clothing worn at a hazardous materials emergency is appropriate for the hazards encountered.

TABLE 8-2 PPE Testing and Certification Standards

Agency	Standard Title	Description
NFPA 1994	*Standard on Protective Ensembles for First Responders to CBRN Terrorism Incidents*	For chemicals, biological agents, and radioactive particulate hazards. Certifications under NFPA 1994 are issued only for complete ensembles. Individual elements such as garments or boots are not considered certified unless they are used as part of a certified ensemble. Thus purchasers of PPE certified under NFPA 1994 should plan to purchase complete ensembles (or certified replacement components for existing ensembles).
NFPA 1992	*Standard on Liquid Splash–Protective Ensembles and Clothing for Hazardous Materials Emergencies*	For liquid or liquid splash threats.
NFPA 1991	*Standard on Vapor-Protective Ensembles for Hazardous Materials Emergencies*	Includes the now-mandatory requirements for CBRN protection for terrorism incident operations for all vapor-protective ensembles. It also includes the qualifications for the former NFPA 1994 Class 1 protective ensemble.
NFPA 1951	*Standard on Protective Ensembles for Technical Rescue Incidents*	For search and rescue or search and recovery operations where exposure to flame and heat is unlikely or nonexistent.
NFPA 1999	*Standard on Protective Clothing for Emergency Medical Operations*	For protection from blood and body fluid pathogens for persons providing treatment to victims after decontamination.
NFPA 1981	*Standard on Open-Circuit Self-Contained Breathing Apparatus (SCBA) for Emergency Services*	For all responders who may use SCBA; must be certified by NIOSH.
NIOSH	*Chemical, Biological, Radiological, and Nuclear (CBRN) Standard for Open-Circuit Self-Contained Breathing Apparatus*	To protect emergency responders against CBRN agents in terrorist attacks. Compliance with NFPA 1981.
NIOSH	*Standard for Chemical, Biological, Radiological, and Nuclear (CBRN) Full Facepiece Air-Purifying Respirator (APR)*	To protect emergency response workers against CBRN agents.
NIOSH	*Standard for Chemical, Biological, Radiological, and Nuclear (CBRN) Air-Purifying Escape Respirator and CBRN Self-Contained Escape Respirator*	To protect the general worker population against CBRN agents.

they are wearing PPE. The next section, which covers in-suit cooling measures, addresses that fact by revisiting dehydration and the various cooling technologies that may be used to reduce the effects of overheating inside a chemical-protective garment.

Responders should also be aware that their field of vision is compromised by the face piece of an SCBA or APR and even by the encapsulating suit. This factor may result in the responder slipping in a puddle of spilled chemicals or tripping on something. Moreover, the face piece often fogs up at some point, further limiting the responder's vision. This creates many problems, such as the inability to read labels, see other responders, see the screens on detection and monitoring devices, or quickly find an escape route in the event of an unforeseen problem in the hot zone. Wearing bulky PPE, such as an encapsulating suit, may inhibit the mobility of the wearer to the point that bending over becomes difficult or reaching for valves above head level is taxing. Furthermore, when gloves become contaminated with chemicals (especially solvents), they become slippery, making it difficult to effectively grip tools, handrails, or ladder rungs. All in all, the environment inside the PPE can be just as challenging as the conditions outside the suit.

To mitigate some of the potential safety considerations that arise when wearing PPE, responders can employ a variety of safety procedures and training. To begin, conducting a pre-entry medical evaluation is important to catch the medical indicators that may signal a responder should not wear PPE.

The chapter *Implementing the Planned Response* outlines the specific components of a pre-entry medical monitoring plan. Further guidance can also be found in NFPA 473, *Standard for Competencies for EMS Personnel Responding to Hazardous Materials/WMD Incidents*, in either the basic life support or advanced life support section. Keep in mind that the medical monitoring station may serve many purposes at the scene of a hazardous materials event. The primary role of the medical monitoring station is to evaluate the medical status of the entry team, the backup team, and those personnel assigned to decontamination duties. On the scene of larger incidents, a medical group or team may be required to obtain basic physiological information from each responder and plan to provide care in the event a responder becomes a patient.

The use of the buddy system is another way that responders can mitigate some of the hazards that may be encountered at the scene of a hazardous materials/WMD incident. As mentioned earlier in this text, the OSHA HAZWOPER regulation requires the use of the buddy system. Along with the buddy system comes the need to communicate—another potential safety issue on the scene. Prior to entry, all radio communications should be sorted out and tested. To back up that form of communication, all responders on the scene should have a method to communicate by universally accepted hand signals. These hand signals could be used to rapidly share messages about problems with an air supply, a suit problem, or any other problem that might occur in the hot zone. Communications are often problematic on emergency scenes, so take whatever steps you can to minimize problems before anyone enters a contaminated atmosphere.

In-Suit Cooling Technologies

Hazardous materials responders operating in protective clothing should be aware of the signs and symptoms of heat exhaustion, heat stress, heat stroke, dehydration, and illness caused by extreme cold. The chapter *Implementing the Planned Response* provides detailed information on heat and cold disorders. The most common malady striking anyone wearing PPE is heat related. If the body is unable to disperse heat because an ensemble of PPE covers it, serious short-and long-term medical issues could occur.

Most heat-related illnesses are typically preceded by dehydration. It is important for responders to stay hydrated so that they can function at their maximum capacity. As a frame of reference, athletes should consume approximately 500 mL (16 oz) of fluid (water) prior to an event and 200–300 mL of fluid at regular intervals during the event. Responders can be considered occupational athletes—so keep up on your fluids! OSHA Fact Sheet number 95-16, "Protecting Workers in Hot Environments," provides additional information about the dangers of heat-related illness in the workplace.

In an effort to combat heat stress while wearing PPE, many response agencies employ some form of cooling technology under the garment. These technologies include, but are not limited to, air-, ice-, and water-cooled vests, along with phase-change cooling technology. Many studies have been conducted on each form of cooling technology. Each of these approaches is designed to accomplish the same goal—to reduce the impact of heat stress on the human body. As mentioned earlier, the same suit that seals you up against the hazards also seals in the heat, defeating the body's natural cooling mechanisms.

Forced-air cooling systems operate by forcing prechilled air through a system of hoses worn close to the body. This is similar to the fluid-chilled system described later. As the cooler air passes by the skin, heat is drawn away—by convection—from the body and released into the atmosphere. Forced-air systems are designed to function as the first level of cooling the body would naturally employ. Typically, these systems are lightweight and provide long-term cooling benefits, but mobility is limited because the umbilical is attached to an external, fixed compressor.

Ice-cooled or gel-packed vests are commonly used due to their low cost, unlimited portability, and unlimited "recharging" by refreezing the packs. These garments are vest-like in their design and intended to be worn around the torso. The principle underlying this approach is that the ice-chilled vest absorbs the heat generated by the body. On the downside, this technology is bulkier and heavier than the aforementioned systems, and it may cause discomfort to the wearer due to the nature of the ice-cold vest near the skin. Additionally, the cold temperature near the skin may actually fool the body into thinking it is cold instead of hot, thereby encouraging retention of even more heat.

VOICES OF EXPERIENCE

In my years at the St. Joseph County Airport Public Safety Division, being cross-trained as a certified law enforcement police officer, a certified fire fighter and aircraft rescue fire fighter, EMT, hazardous materials technician, and hazardous materials instructor, all the basics were covered for responding to any type of police, fire, medical, airport, or hazardous materials incidents.

While on duty, I was called to assist the city fire department on an incident involving radioactive materials. Upon arriving at the emergency incident scene, the department's incident commander informed me that there was a spill of radioactive materials at a construction site. The spill involved the breaking of a portable density gauge metering instrument that was being used to measure the density of core samples of asphalt.Even though the fire department had radiological monitoring instruments on its rig, not one of the hazardous materials response team members nor any other fire fighters at this scene knew how to read or use the instruments.

To assist, I quickly checked the batteries of their radiological monitors and survey instruments. I performed the standard operational checks of their radiological survey meters, charged their personal dosimeters, and set the radiological monitoring probes in the correct position for monitoring and surveying. I assessed the hazardous materials team's Level A protective gear, as well as the backup team's personal protective equipment (PPE), and attached their personal dosimeters to their turnout PPE. I then finished donning my own PPE, put on a self-contained breathing apparatus (SCBA), attached a personal dosimeter to my turnout collar, and accompanied the team to the incident area so that I could read and monitor the radiological readings on the instruments for them.

It was determined that this hazardous materials incident involved a moisture density gauge that used two different radioactive sources to produce two different types of radiation. One of the radioactive sources, cesium 137, emits gamma ray photon radiation to determine density; the other radioactive source, americium 241 (combined with nonradioactive beryllium), emits neutron radiation to determine moisture content.

It was also determined that there were no injured victims at the scene. Proper exclusion zones were set up to consider both contamination and exposure control. A decontamination corridor was established and all decontamination protocols were followed.

Although early responders generally do not have extensive training or equipment to perform radioactive contamination surveys, they can significantly reduce the spread of contamination by applying relatively simple measures of contamination control. It is useful to have a checklist of information to be transmitted when requesting assistance or reporting a radiological incident. Knowing the principles of protection is vital: recognizing the hazards, time, distance, and shielding. First responders should be very familiar with their radiological monitoring instruments, including how to perform the basic operational check, how to accurately read the survey meters, and knowing the instruments' limitations.

Officer Judith Hoffmann (RET), CECM, EMT
Hoffmann Consulting
South Bend, Indiana

FIGURE 8-13 A fluid-chilled or water-cooled system.
Courtesy of DS Kool

Fluid-chilled systems operate by pumping ice-chilled liquids (water is often used, so these systems are referred to as "water-cooled") from a reservoir, through a series of tubes held within a vest-like garment, and back to the reservoir FIGURE 8-13. Mobility may be limited with some varieties of this system, because the pump may be located away from the garment. Some systems incorporate a battery-operated unit worn on the hip, but the additional weight may increase the body's workload and generate more heat, thereby defeating the purpose of the cooling vest.

Phase-change cooling technology operates in a similar fashion to the ice- or gel-packed vests FIGURE 8-14. The main difference between the two approaches is that the temperature of the material in the phase-change packs is chilled to approximately 60°F (15.5°C), and the fabric of the vest is designed to wick perspiration away from the body. The packs typically "recharge" more quickly than those of an ice- or gel-packed vest. Even though the temperature of the phase-change pack is higher than the temperature of an ice-or gel-packed vest, it is sufficient to absorb the heat generated by the body.

Responder Safety Tip

Approximately 90 percent of all body heat is generated by the organs and muscles located in your torso.

Responder Safety Tip

Remember to take rehabilitation breaks throughout the hazardous materials incident. Wearing any type of PPE requires a great deal of physical energy and mental concentration. Responders should also acknowledge the psychological stress that wearing PPE may present. Claustrophobia is a common problem when wearing chemical-protective equipment, especially encapsulated suits. This is one of the "P" (psychogenic) considerations in TRACEMP and can present a problem for responders.

FIGURE 8-14 Phase-change cooling technology.
Courtesy of Glacier Tek

Reporting and Documenting the Incident

As with any other type of incident, documenting the activities carried out during a hazardous materials/WMD incident is an important part of the response. Many responders may pass through the scene, and it could be quite difficult to sort everything out when it comes time to reconstruct the events for an accurate and legally defensible incident report. Good documentation after the incident is directly correlated with how well organized the response was.

Along with the formal written accounts of the event, some agencies require that personnel fill out exposure records that include information such as the name of the substances involved in the incident and the level of protection used. This information, coupled with a comprehensive medical surveillance program (discussed in the chapter *Estimating Potential Harm and Planning a Response*), provides a method to chronicle the exposure history of the responders over a period of time. Consult your AHJ for the exact details and procedures for reporting and documenting the incident.

Wrap-Up

Chief Concepts

- Using a risk-based approach is important when selecting personal protective equipment. All decisions should be well thought out and realistic, taking into account the positive and negative effects of the actions taken.
- Emergency responders should be familiar with the policies and procedures of the local jurisdiction so as to ensure a consistent approach is taken when selecting the proper PPE.
- Chemical-protective clothing is classified into two main categories: vapor-protective clothing and liquid splash–protective clothing.
- Unlike the OSHA HAZWOPER standard, NFPA 1994 covers the performance of the garment *and* factors in the performance of the respiratory protection.
- Typically, Level A protection is required when the operating environment exceeds IDLH values for skin absorption.
- According to the OSHA HAZWOPER regulation, Level B equipment is the minimum level of protection to be worn when operating in an unknown environment.
- Level C protection is appropriate when the type of airborne substance is known, its concentration is measured, and the criteria for using APRs are met.
- The most common malady striking anyone wearing PPE is heat related. In an effort to combat heat stress while wearing PPE, many response agencies employ some form of cooling technology under the garment.
- Manufacturers' guidelines for maintenance, testing, inspection, storage, and documentation procedures should be followed for all personal protective equipment provided by the AHJ.
- Along with the formal written accounts of the event, some agencies require personnel to fill out exposure records that include information such as the name of the substances involved in the incident and the level of protection used.

Hot Terms

Air-purifying respirator (APR) A device worn to filter particulates and contaminants from the air before it is inhaled. Selection of the filter cartridge for an APR is based on the expected contaminants.

Chemical-resistant materials Clothing (suit fabrics) specifically designed to inhibit or resist the passage of chemicals into and through the material by the process of penetration, permeation, or degradation.

Degradation The physical destruction or decomposition of a clothing material owing to chemical exposure, general use, or ambient conditions (such as storage in sunlight). Materials can also be tested for weight changes, loss of fabric tensile strength, and other properties to measure degradation.

Dehydration An excessive loss of body water. Signs and symptoms of dehydration may include increasing thirst, dry mouth, weakness or dizziness, and a darkening of the urine or a decrease in the frequency of urination.

Doffing The process of taking off an ensemble of PPE.

Donning The process of putting on an ensemble of PPE.

High temperature–protective equipment A type of personal protective equipment that shields the wearer during short-term exposures to high temperatures. Sometimes referred to as a proximity suit, this type of equipment allows the properly trained fire fighter to work in extreme fire conditions. It is not designed to protect against hazardous materials or weapons of mass destruction.

Level A ensemble Personal protective equipment that provides protection against vapors, gases, mists, and even dusts. The highest level of protection, it requires a totally encapsulating suit that includes a self-contained breathing apparatus.

Level B ensemble Personal protective equipment that is used when the type and atmospheric concentration of substances requires a high level of respiratory protection but less skin protection. The kinds of gloves and boots worn depend on the identified chemical.

Level C ensemble Personal protective equipment that is used when the type of airborne substance is known, the concentration is measured, the criteria for using an air-purifying respirator are met, and skin and eye exposure is unlikely. A Level C ensemble consists of standard work clothing with the addition of chemical-protective clothing, chemically resistant gloves, and a form of respirator protection.

Level D ensemble Personal protective equipment that is used when the atmosphere contains no known hazard and work functions preclude splashes, immersion, or the potential for unexpected inhalation of or contact with hazardous levels of chemicals. A Level D ensemble is primarily a work uniform that includes coveralls and affords minimal protection.

Liquid splash–protective clothing Clothing designed to protect the wearer from chemical splashes. It does not provide total body protection from gases or vapors and should not be used for incidents involving liquids that emit vapors known to affect or be absorbed through the skin. NFPA 1992 is the performance document pertaining to liquid-splash garments and ensembles.

National Institute for Occupational Safety and Health (NIOSH) The organization that sets the design, testing, and certification requirements for self-contained breathing apparatus in the United States.

Penetration The flow or movement of a hazardous chemical through closures such as zippers, seams, porous materials, pinholes, or other imperfections in a material. Liquids are most likely to penetrate a material, but solids (such as asbestos) can also penetrate protective clothing materials.

Permeation The process by which a hazardous chemical moves through a given material on the molecular level. Permeation differs from penetration in that permeation occurs through the material itself rather than through openings in the material.

Powered air-purifying respirator (PAPR) A type of air-purifying respirator that uses a battery-powered blower to pass outside air through a filter and then to the mask via a low-pressure hose.

Self-contained breathing apparatus (SCBA) A respirator with independent air supply used by fire fighters to enter toxic or otherwise dangerous atmospheres.

Supplied-air respirator (SAR) A respirator that obtains its air through a hose from a remote source such as a compressor or storage cylinder. A hose connects the user to the air source and provides air to the face piece. SARs are useful during extended operations such as decontamination, clean-up, and remedial work. Also referred to as positive-pressure airline respirators (with escape units).

Vapor-protective clothing Fully encapsulating chemical protective clothing that offers full-body protection from highly contaminated environments and requires air-supplied respiratory protection devices such as a self-contained breathing apparatus. NFPA 1991 sets the performance standards for these types of garments, which are commonly referred to as Level A ensembles.

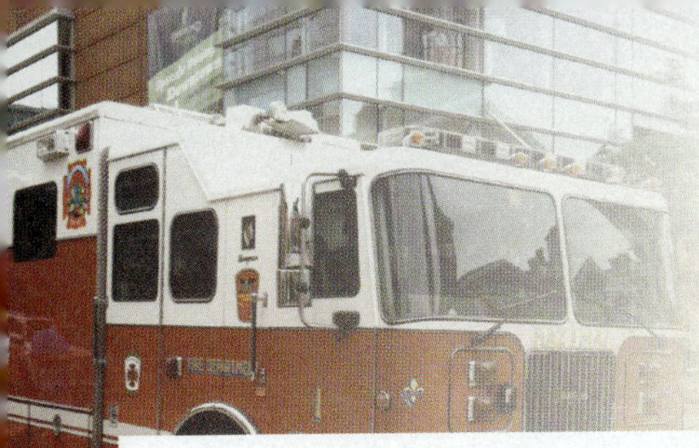

RESPONDER *in action*

The chief of your fire company asked you to give a brief presentation to the town council about the new Level A suits you are planning to purchase. You decide to use the example of an ammonia release at a local ice-making facility to underscore the reasons why your company needs this particular level of protection.

1. Which of the following NFPA standards would be the proper one to reference regarding your new Level A suits?
   - **A.** NFPA 1981, *Standard on Open-Circuit Self-Contained Breathing Apparatus for Emergency Services* (2013 edition)
   - **B.** NFPA 1951, *Standard on Protective Ensembles for Technical Rescue Incidents* (2013 edition)
   - **C.** NFPA 1999, *Standard on Protective Clothing for Emergency Medical Operations* (2013 edition)
   - **D.** NFPA 1991, *Standard on Vapor-Protective Ensembles for Hazardous Materials Emergencies* (2005 edition)
2. You also choose to mention that the OSHA HAZWOPER regulation requires the incident commander to ensure that appropriate personal protective equipment must be worn at a hazardous materials emergency. Which part of the OSHA HAZWOPER standard would be the proper piece to quote?
   - **A.** 29 CFR 1910.120(q)(3)(iii)
   - **B.** 29 CFR 1910.120(a)
   - **C.** 29 CFR 1910.134
   - **D.** 49 CFR 1910.22(f)
3. In addition to the garment, you also plan to purchase a forced-air cooling system. Which of the following gives the most accurate description of forced-air cooling technology?
   - **A.** Forced-air cooling systems operate by pumping ice-chilled liquids from a reservoir, through a series of tubes held within a vest-like garment, and back to the reservoir.
   - **B.** The principle of forced-air cooling systems is that an ice-chilled vest absorbs the heat generated by the body. This technology may cause discomfort to the wearer because the ice-cold vest is placed so close to the skin.
   - **C.** Forced-air cooling systems operate by forcing prechilled air through a system of hoses worn close to the body. As the cooler air passes by the skin, it is drawn away from the body and released into the atmosphere.
   - **D.** Forced-air cooling technology operates by chilling the hands and feet with ice-chilled liquids in an attempt to increase manual dexterity.
4. Which of the following is a true statement?
   - **A.** Forced-air cooling technology is bulky and heavy.
   - **B.** With forced-air cooling technology, the temperature of the material in the packs is approximately 60°F (15.5°C).
   - **C.** Forced-air cooling technology is lightweight and provides long-term benefits.
   - **D.** Forced-air cooling technology is recharged by refreezing the packs.

Smoke: © Greg Henry/ShutterStock, Inc.; Corner image: © Jones & Bartlett Learning. Photographed by William Larkin

Mission-Specific Competencies: Technical Decontamination

Greg Henry/ShutterStock, Inc.

© TFoxFoto/ShutterStock, Inc.

CHAPTER 9

Competencies for Operations Level Responders Assigned Mission-Specific Responsibilities

This chapter shall address competencies found in NFPA 472 for the following operations level responders assigned mission-specific responsibilities at hazardous materials/WMD incidents by the authority having jurisdiction beyond the core competencies at the operations level.

Knowledge Objectives

After studying this chapter, you will be able to:

- Plan a response by selecting a technical decontamination process that will minimize the hazard. (NFPA 6.4.1.2.2, p. 206–211)
- Identify the personal protective equipment required to support technical decontamination at hazardous materials/WMD incidents based on local procedures. (NFPA 6.4.3.1, p. 213)
- Identify and describe the limitations and advantages of technical decontamination operations and the technical decontamination methods. (NFPA 6.4.3.2(1)(2), p. 208–211)
- Identify resources for determining the correct technical decontamination procedure in a hazardous materials/WMD incident. (NFPA 6.4.3.2(3), p. 206–213)
- Identify the supplies and equipment needed for technical decontamination. (NFPA 6.4.3.2(4), p. 213)
- Identify tools, equipment, and procedures for processing evidence during technical decontamination. (NFPA 6.4.3.2(5), p. 206–211)
- Identify the role of the operations level responder assigned to technical decontamination. (NFPA 6.4.4.1(1), p. 206–215)
- Identify procedures, equipment and safety precautions for handling tools, equipment weapons, criminal suspects, and law enforcement canines brought to the decontamination corridor. (NFPA 6.4.3.2(6), p. 206–207)
- Identify procedures for evaluating the effectiveness of technical decontamination. (NFPA 6.4.5.1, p. 215)
- Identify the importance, steps, and requirements of maintaining records. (NFPA 6.4.6.1, p. 215)

Skills Objectives

After studying this chapter, you will be able to:

- Describe the procedures for implementing technical decontamination duties as assigned. (NFPA 6.4.4.1(2), p. 211–213)
- Demonstrate the ability to set up and implement technical decontamination operations in support of entry operations. (NFPA 6.4.4.2, p. 213)

You Are the Responder

Your truck company is dispatched to assist the regional hazardous materials team and law enforcement at the scene of a methamphetamine lab. The hazardous materials team is planning an entry to sample a number of unmarked glass bottles. Your officer receives orders to support the hazardous materials team by setting up a decontamination corridor. You and your crew are trained to all core competencies at the operations level, along with the mission-specific competencies of personal protective equipment (PPE) and technical decontamination.

1. Based on this scenario, should you be operating under the guidance of a hazardous materials technician, an allied professional, or standard operating procedures?
2. How would you handle the samples taken by the entry team as they move through the decontamination corridor?
3. A law enforcement canine is on scene and may have been contaminated during the search. What steps would you take to perform decontamination on the animal?

Introduction

According to NFPA 472, *Standard for Competence of Responders to Hazardous Materials/Weapons of Mass Destruction Incidents*, contamination is "the process of transferring a hazardous material, or the hazardous component of a weapon of mass destruction (WMD) from its source to people, animals, the environment, or equipment that can act as a carrier." Also by definition, decontamination is "the physical and/or chemical process of reducing and preventing the spread of contaminants from people, animals, the environment, or equipment involved at hazardous materials/weapons of mass destruction (WMD) incidents."

Those two definitions, when placed into context, should frame the concept of decontamination for any responder to a hazardous materials incident. Although there are many factors and influences surrounding the task of decontamination, the relationship of contaminant to decontamination is clear—if contamination of PPE is suspected or confirmed, it must be removed efficiently in a systematic fashion. In short, decontamination is a process driven activity, driven by the nature of the contaminant. The three major categories of decontamination are emergency decontamination, mass decontamination, and technical decontamination. Each category has its place in a response and a best practice associated with it. This chapter provides an overview of emergency decontamination and goes into detail on technical decontamination. The specifics of mass decontamination are covered in the chapter *Mission-Specific Competencies: Mass Decontamination*.

To begin, it's important to recognize the different types of decontamination processes and when to employ the proper type at the proper time. Emergency decontamination is used in potentially life-threatening situations to rapidly remove the contaminant from a person. It is quick and less "formal" than technical decontamination. The goal is to reduce the effect of an exposure and get a victim clean enough to receive medical care from first responders and, if needed, be admitted to a receiving hospital. Victims who receive emergency decontamination often must remove some or all of their clothing to reduce the harm posed by the contaminant.

Mass decontamination is a way of performing emergency decontamination on a large number of people; it can take place anywhere, with the same goal as emergency decontamination—remove the contaminants as quickly as possible to reduce the impact of a chemical exposure. The chapter *Mission-Specific Competencies: Mass Decontamination* covers this in more detail.

Technical decontamination is a different process, aimed at reducing or eliminating contamination from responder PPE prior to removing it. It is not intended for exposed persons, or as a way to reduce the health effects of an exposure. Technical decontamination is often a water-based process, using scrub brushes, some form of catch basin for water, tarps, and perhaps a cleaning solution. There are many ways to conduct technical decontamination. This chapter describes some of the more common methodologies, but it is up to you to know how technical decontamination, and any other type of decontamination, is accomplished in your authority having jurisdiction (AHJ).

Gross decontamination takes place within a controlled decontamination corridor; it consists of a prewash that occurs before technical decontamination takes place **FIGURE 9-1**.

In some cases, the decontamination process must be flexible and creative. For example, law enforcement or search canines may require some form of decontamination after going on a mission. In these instances, it is best to check with the canine handler to better understand the process. Federal Urban Search and Rescue (USAR) teams have procedures to perform canine decontamination. Make sure you understand the process, or check with your local jurisdiction for guidance, before undertaking this specialized process.

Additionally, it may be necessary to decontaminate a criminal subject, secured firearms, or other specialized pieces of equipment (detection and monitoring devices, hand tools, or other equipment used during the response). In these unique situations, it is wise to consult the owner of the piece of equipment, manufacturer guidelines, or an otherwise knowledgeable person, before performing decontamination on unfamiliar items. Unusual items may require unusual methods

FIGURE 9-1 An example of a gross decontamination setup. This apparatus might be placed at the entrance of the technical decontamination line.

Courtesy of Rob Schnepp

of decontamination—think about that possibility before you take any steps that may damage those items.

It may also be necessary to decontaminate a piece of evidence, or a sample taken at the scene, prior to analyzing it or turning it over to law enforcement officials. When evidence is processed through a decontamination corridor, chain of custody must be maintained. It is imperative to document the identity of any personnel who handled the evidence, the date and time that the decontamination occurred or the item was transferred from one person to another, and the reason for doing so FIGURE 9-2.

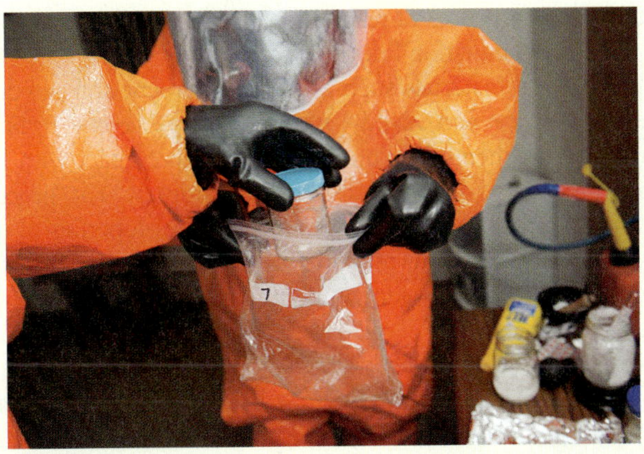

FIGURE 9-2 Evidence taken from the scene should be prepared for decontamination prior to leaving the hot zone. Documentation is the key to ensuring good chain of custody.

Courtesy of Rob Schnepp

The resulting record, which outlines the *chain of custody*, is discussed in more detail in the chapter *Mission-Specific Competencies: Evidence Preservation and Sampling*.

Responder Safety Tip

All attempts should be made to understand the physical and chemical properties of a contaminant prior to beginning decontamination. Some hazardous materials are water reactive and may require a special decontamination process. Remember that a good first step in technical decontamination is to avoid contact with the substance in the first place. A responder that avoids getting "dirty" in the hot zone makes the job of decontamination easier.

Technical Decontamination

NFPA 472, *Standard for Competence of Responders to Hazardous Materials/Weapons of Mass Destruction Incidents*, defines technical decontamination as "the planned and systematic process of reducing contamination to a level that is as low as reasonably achievable (ALARA)."

Technical decontamination may involve several stations or steps. During this decontamination process, multiple personnel (the decontamination team) typically use brushes to scrub and wash the person or object so as to remove as much of the contaminants as possible. Technical decontamination may involve water or a special cleaning solution, depending on the hazardous material, and takes place in the warm zone, within an established decontamination corridor. FIGURE 9-3 depicts a technical decontamination corridor. FIGURE 9-4 shows a close-up of one of the catch basins. You can see many components depicted: catch basins for water; hoses, sprayers and scrub brushes; a Stokes litter for possible victim rescue, and so on. The decontamination corridor is oriented with the hot zone to the right of the setup and the cold zone to the left. Notice the "tool drop" buckets in the right side of Figure 9-3. This is where a responder would drop contaminated tools or other equipment that might need to be decontaminated after the responders have made their way through the decontamination line.

FIGURE 9-3 Example of a technical decontamination corridor.

Courtesy of Rob Schnepp

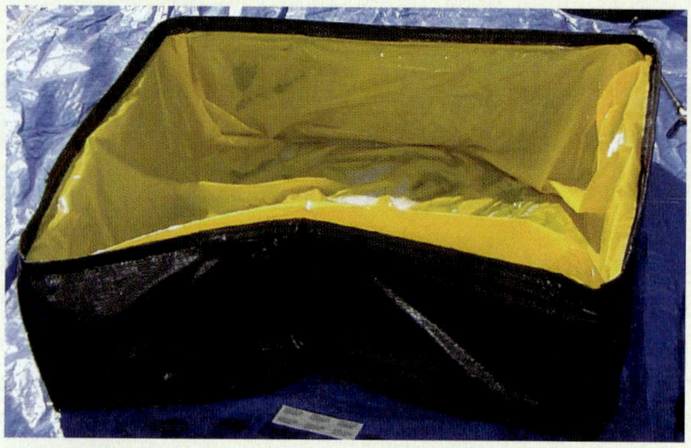

FIGURE 9-4 A close-up of a catch basin employed to catch the decontamination water. Each responder going through the decontamination line would step into the pool and be washed and rinsed.
Courtesy of Rob Schnepp

Not all decontamination processes require water or other wet decontamination solutions. In some cases, a dry decontamination method may be employed. Simply put, dry decontamination is accomplished by removing all PPE and placing it directly into bags for disposal. Often it may be necessary to lightly brush off visible contamination prior to bagging the items(s) as part of this process. Additionally there are next generation products that allow responders or victims to use a dry wipe that encapsulates the hazardous materials and is then disposed of. As the name implies, water or other wet solutions are not used during outer PPE removal, though they are used during the personal hygiene part of the technical decontamination. The process of dry decontamination may look similar to that followed in wet (i.e., water-based) decontamination FIGURE 9-5.

FIGURE 9-5 Technical decontamination is a systematic and thorough cleaning process.
© Jones & Bartlett Learning. Courtesy of MIEMSS

Methods of Technical Decontamination

In most cases, water is the preferred decontamination solution, but keep in mind that some hazardous materials have chemical properties that may require different methods of decontamination, or circumstances may dictate that responders use an alternative to water as a decontamination measure. Alternative technical decontamination procedures may include the following techniques:

- Physical techniques
  - Absorption
  - Adsorption
  - Vacuuming
  - Washing
- Chemical degradation
  - Dilution
  - Disinfection
  - Evaporation
  - Neutralization
  - Solidification
  - Sterilization
- Isolation and disposal

Physical Techniques

Physical methods of technical decontamination involve the actual removal of contaminant particles from the surfaces of responders and equipment. In most cases, decisions about which techniques to use will be based on the available equipment provided by the AHJ. The operations level responder with this mission-specific competency should be able to perform each of these technical decontamination methods while ensuring the appropriate level of protection.

Absorption

In absorption, a spongy material (natural soil, sawdust, or synthetic loose absorbents available from a variety of manufacturers) is mixed with a liquid hazardous material. The contaminated mixture is then collected and disposed of FIGURE 9-6. This technique is primarily used for decontaminating equipment and property; it has limited application for decontaminating personnel. Absorption may be used, for example, in a shuffle pit to clean the boots of responders before they enter the rest of the decontamination line FIGURE 9-7.

Although materials such as loose absorbent and sawdust are inexpensive and readily available, they become hazardous materials themselves once they come in contact with the contaminant and must be disposed of properly. The process of absorption does not change the chemical properties of the involved substance, but rather is used only to collect the substance for subsequent containment. Both the federal government and most states have laws and regulations that govern the disposal of used absorbent materials. Consult an expert knowledgeable about this type of disposal before discarding any materials suspected of being contaminated.

Adsorption

In adsorption, the contaminant adheres to the surface of an added material—such as sand or activated carbon—rather

FIGURE 9-6 Spongy materials are used to absorb liquid hazardous materials.

FIGURE 9-7 An example of a shuffle pit used to clean the responders' boots. In this case, the responder would step into a basin with dry absorbent and work both feet back and forth to remove as much of the contamination as possible. The absorbent would then be disposed of as hazardous waste.

than combining with it as in absorption FIGURE 9-8. In other words, the sorbent is placed on a contaminant, and the contaminant sticks to the outside surface of the sorbent; for example, sand could be used to adsorb motor oil.

Vacuuming

Vacuuming is the removal of dusts, particles, and some liquids by sucking them up into a container. A filtering system prevents the contaminated material from recirculating and reentering the atmosphere. A special high efficiency particulate air (HEPA) vacuum cleaner is used to remove hazardous dusts, powders, or fibers that are 0.3 micron or larger. HEPA filters allow air to pass through the filter but capture the particulate matter in the air. HEPA filters must be replaced regularly to maintain their effectiveness.

Washing

Washing is an effective, yet simple decontamination process that is ideal for removing oily substances and solvents from responder PPE, tools, and equipment. When washing is employed for technical decontamination, a simple soap-and-water mixture is usually the solution of choice. The PPE or other item is scrubbed with a brush or sponge, and then fully rinsed with water. The wash–rinse cycle can be repeated as many times as necessary until the item is free of contamination.

Chemical Degradation

Typically, chemical degradation occurs when a natural or artificial process causes the breakdown of a chemical substance.

FIGURE 9-8 Sand can be used as an adsorbent.

For example, the action of ultraviolet light on a dilute concentration of hydrogen peroxide may, over a period of time, cause the chemical breakdown of the chemical, thereby rendering it harmless. This form of technical decontamination, much like absorption, is more suited to environmental decontamination.

One example is the process of chemical degradation for the in situ (in place) remediation of underground soil contamination, perhaps from a leaking underground diesel storage tank. In this complex process, chemicals might be pumped into the underground contamination—with the intent of breaking the hazardous material down into less persistent chemical compounds—and then further broken down by the soil itself. When considering these kinds of measures for technical decontamination of responders, you must consult experts in the field who can fully appreciate the benefits and consequences of the proposed action.

Some chemical degradation methods, such as letting contaminated soil sit in the sun for long periods of time, are time consuming yet inexpensive ways of mitigating the dangers of chemicals. Heat may also be applied to some materials to promote their chemical degradation. Polyurethane-based plastics, for example, are susceptible to thermal degradation when exposed to certain temperatures over a given period of time.

Many forms of chemical degradation exist, some of which are discussed in the rest of this section. Be sure to consult experts in the field when considering an aggressive tactic such as chemical degradation to solve your problem, because this often-complex process is not within the scope of practice of awareness level personnel or operations level responders.

Dilution

Dilution most commonly uses plain water to fully rinse off a contaminated person or object in an attempt to weaken the concentration of the hazard. Dilution is both fast and economical as water is readily available from any domestic water supply, fire engine, or fire hydrant. The majority of gross decontamination, technical decontamination, and mass decontamination processes all use dilution—water—as the preferred rinsing agent. Before using water, however, consider whether the contaminant will react adversely or generate an unwanted by-product. Keep in mind that the larger the volume of water used, the more hazardous waste generated; these wastes may increase or spread the contamination and later must be disposed of safely.

Disinfection

Disinfection is the process used to destroy disease-carrying microorganisms, excluding spores (anthrax, for example). Commercial disinfectants are packaged with a detailed brochure that describes the limitations and capabilities of the product. Responders with medical research labs, hospitals, clinics, mortuaries, medical waste disposal facilities, blood banks, or universities in their response area should be familiar with the specific types of biological hazards present and the best disinfectants for each hazard. This is a specialized form of decontamination that requires the advice of a knowledgeable expert.

Evaporation

Evaporation is a natural form of chemical degradation. It is sometimes used as a safe, noninvasive way to allow a chemical substance to stabilize without human intervention. For example, a spill of a highly volatile liquid such as isopropyl alcohol, when it occurs on a hot sunny day, may evaporate on its own without any interventions required by responders. In some cases, when liquids with a high vapor pressure are spilled, responders may elect to take no direct action, but instead allow the substance to evaporate. This must be a well-thought-out decision (factoring in ignition sources, downwind consequences, time, and other circumstances), but in some cases doing nothing may be the best course of action.

Neutralization

Neutralization is typically used when the corrosivity of an acid or a base needs to be reduced. When neutralizing a strong acid (such as sulfuric, hydrochloric, or phosphoric acid), a weak base should be selected for the neutralization reaction. Conversely, when a strong base (such as sodium hydroxide or potassium hydroxide) requires neutralization, a weak acid should be selected. *This method of decontamination should never be selected as an option for skin decontamination*. The main reason for this warning is the heat generated when acids or bases are neutralized. Additionally, neutralizing a corrosive requires a good working knowledge of chemistry and should not be undertaken by any personnel without the proper training.

One example of a common neutralization reaction involves the combination of hydrochloric acid (HCl) and sodium hydroxide (NaOH):

$$\text{HCL} + \text{NaOH} \Rightarrow \Uparrow \Delta + \text{H}_2\text{O} + \text{NaCl}$$

The byproducts of a neutralization reaction are heat, indicated by the upward pointing arrow (⇧) next to the Δ (delta) symbol; water (H_2O); and a salt compound. In this case, the salt produced is sodium chloride (NaCl), commonly referred to as table salt. The chemical reaction shows that if this method were used on personnel it could cause injury.

Solidification

Solidification is a chemical process that causes a hazardous liquid to become a solid. This transformation makes the material easier to handle, but does not change the inherent chemical properties of the substance. Commercial products are available that cause certain liquids to solidify. Most commonly, they consist of cement-based products that are spread onto the spill, where they turn the liquid hazardous material into a solid, thereby quickly controlling the spill.

Sterilization

Biological agents are the most logical candidates for decontamination by sterilization. The process of sterilization—whether by heat, chemical means, or radiation—is intended to kill microorganisms, including spores such as anthrax. This process of decontamination is not intended for responders, but rather is primarily used for environmental decontamination and for decontamination of tools and equipment.

Isolation and Disposal

Isolation and disposal is a two-step removal process for items that cannot be completely decontaminated. First, the contaminated items (such as clothing, tools, and personal items) are removed from the personnel and/or the primary incident site and isolated in a designated area. The items can be segregated into logical groupings, if necessary; for example, you might put victims' clothing in one area, responder equipment in another area, and potential evidence in a third area. The items are also tagged, with the tag including the item name, date of collection, item description, location where it was found, and possible contamination. Next, the contaminated items are placed into a suitable container such as a bag, barrel, or bucket. They can then be legally transported to an approved treatment, storage, or disposal facility, where the items are stored, incinerated, buried in a hazardous waste landfill, or otherwise handled.

When this decontamination option is employed, you should consider the legal ramifications, costs, and responsibility for the decision. *Also keep in mind that any contaminated item removed from a person or the scene may be used later as evidence against the perpetrators of an incident with criminal intent.* Be mindful of destroying potential evidence!

The Technical Decontamination Process

The technical decontamination process should take place within a predesignated decontamination corridor located within the warm zone. That corridor should be set up and staffed prior to the entry team making access and going to work in the hot zone FIGURE 9-9. It would be a mistake to send an entry team in to work without some provisions having already been made for decontamination.

Decontamination corridors can be thought of as a transition between the hot zone and the cold zone. Decontamination operations take place in the warm zone, but in actuality, the warm zone becomes "warm" only upon the commencement of the decontamination process.

RESPONDER TIP

The warm zone is created when the personnel exiting the hot zone carry contamination into the decontamination corridor.

Prior to any contaminated responders passing through the decontamination corridor, the corridor can be considered cold. This explains why the decontamination corridor can be set up by responders wearing the type of PPE that is appropriate for cold zone operations.

A clearly marked, easily seen, and readily accessible entry point to the decontamination corridor should be established, as well as a clearly marked exit. If the decontamination operation occurs at night or in poor weather conditions, the area should be well marked and well lit. Remember that all

FIGURE 9-9 A decontamination corridor should be established prior to responders entering the hot zone.

VOICES OF EXPERIENCE

We normally associate technical decontamination with our bread-and-butter hazardous materials incident responses. At times, we become complacent or do not recognize the need for decontamination at other than hazardous materials operations. After the Oklahoma City bombing in April 1995, I was assigned as the task force leader for the FEMA Urban Search and Rescue Task Force NY TF-1, which responded within 24 hours of the incident. I will offer two instances where proper decontamination was either overlooked or improperly administered.

Upon our arrival, there were many people working in and around the site, and there were well-intentioned people in the area offering food and liquids to the operations personnel working in and around the area. Because scene security was not fully implemented yet, these well-intentioned people were able to bring the food in close proximity to where the rescuers were working. As rescuers were taking breaks, there was abundant food available in the form of pizzas, sandwiches, and other easily transportable foods. This food was either eaten immediately or brought back to rescuers working on the site. It didn't take long for rescuers to become overcome with intestinal ailments. This, in turn, reduced the operational effectiveness of the teams. The need for decontamination in the form of hand and face washing prior to eating was known; however, in the urgency of the moment decontamination was overlooked, resulting in the illness of rescuers. Scene security was later tightened, and food was controlled at central points away from the immediate area, with decontamination of personnel being performed at controlled entry and exit points at the work location.

Although decontamination points were set up, this, too, was not without issue. Someone on the scene had determined that decontamination would be performed at the controlled entry and exit points but that volunteers would be used to perform this function. As we all know, decontamination personnel must be trained and the decontamination methods must be appropriate for the circumstances. In this case, an unknown solution was placed into a sprayer, and the untrained volunteer was directed to spray all personnel leaving the area. Our task force hazardous materials specialist was astute enough to recognize the problems that would be encountered and set up a decontamination procedure and station for our members. It was shortly after this decision was made that a rescuer from another jurisdiction was sprayed by the volunteer at the other decontamination station and was severely burned on the face, particularly around the eyes. It turns out that the liquid in the sprayer was at full strength and inappropriate for the use intended. In the rush to set up decontamination, the wrong solution was used.

The lessons learned at this incident highlight the need to recognize when decontamination is required, even at "non-hazardous materials incidents," and then use appropriately trained personnel, protocols, and procedures. Either of the described instances could have had a more serious effect on the overall mission success. Thankfully, the adverse effects were not major in nature.

Craig H. Shelley (Retired)
City of Rutland Fire Department
Rutland, Vermont

responders leaving the hot zone must pass through the decontamination corridor.

Once the technical decontamination process begins, the level of PPE worn by the decontamination team is typically not less than one level below what the entry team is wearing. This is a flexible decision, however, and should be based on a risk-based thought process. If the expected contaminant is highly toxic and perhaps difficult to remove from PPE, the choice may be made to dress the decontamination team in the same level of protection as the entry team. There are no hard-and-fast rules here—the decisions made should be based on logic and the anticipated hazards of the released substance(s).

Inside the decontamination corridor, anyone who is contaminated may need to pass through several stations to complete the technical decontamination process. These stations may be set up using a variety of tools and equipment, including the following items (as shown in Figure 9-3):

- Collection basins to capture the water used during decontamination
- Portable sprayers to apply water and/or wash solutions
- Sponges for wiping off gloves or other PPE
- Buckets
- Long-handled scrub brushes
- Tarps

As with most other facets of emergency response, there is no single "right way" to do everything. Consult the standard operating procedures of your AHJ to better understand the tools, equipment, and procedures commonly used for performing technical decontamination. The technical decontamination process should be clearly laid out and easily understood by those being decontaminated. Keep in mind that any responders operating in PPE will be hot, fatigued, and in no frame of mind to have to figure out what the technical decontamination team wants them to do. It is the decontamination team's responsibility to guide the contaminated victims through the process, removing as much stress from the situation as possible.

Performing Technical Decontamination

The technical decontamination process actually starts before the entry team sets off for work in the hot zone. The pre-entry briefing should include information on the location of the decon corridor; the process to be used, and how the decontamination team will communicate with those moving through the process. Once work in the hot zone is complete, responders should notify the decontamination team (via radio or hand signals) that they are en route to the decontamination area. They should also take note of any obvious contamination on their or their partner's PPE—this should be relayed to the decontamination team so that special attention can be paid to those areas. If SCBA is worn, take note of the remaining amount of air pressure left in the cylinder of each responder. Once at the decontamination corridor, a decision must be made regarding which responder begins the process first. If one team member is grossly contaminated, and both responders have adequate air supply to complete decontamination, the more contaminated responder should enter first. If one responder is lower on air than the other, the one with the lowest air pressure remaining should enter the decontamination area first. Once at the entrance to the decontamination corridor, both responders should place any contaminated hand tools or other equipment in a tool drop area near the entrance of the decontamination corridor. (These items can be cleaned later, after the contaminated responders are taken care of.) This drop area can consist of a container, a recovery drum, a special tarp, or another collection device. Place sensitive electronic equipment such as radios and meters away from tools that could affect and damage this equipment. If another trip into the hot zone is required, subsequent teams may use the same tools. It would be at this point that samples taken or evidence collected should be handed over to the decontamination team to move through the decontamination process.

Once tools and equipment are placed in the tool drop area, the entry team proceeds into the decontamination corridor for gross decontamination (if required). The gross decontamination step is optional, depending on the amount and nature of the contaminant. A portable shower using a low-pressure, high-volume water flow may complete this step (the shower contains the water).

Technical decontamination typically involves one to three wash-and-rinse stations, again depending on the nature of the expected contamination. Only one contaminated responder is allowed in a wash-and-rinse station at a time. The decontamination team member who is scrubbing should pay special attention to the gloves, crevices in the PPE, and boot soles, because these are areas in which hazardous materials are likely to collect. The rinser should begin at the top of the head of the responders and thoroughly rinse downward into the collection pool.

Responder Safety Tip

To protect the respiratory system from potential injury during decontamination, protective respiratory equipment should remain on until the very end of the decontamination corridor.

After the chemical protective equipment is thoroughly scrubbed and rinsed, it can be safely removed from the responder. The SCBA face piece, full- or half-face air-purifying respirators, or PAPR face piece should remain in place (SCBA harness and cylinder or PAPR fan units can be removed and set off to the side) as long as possible. The members of the decontamination team who are responsible for assisting responders with doffing the PPE should fold or roll the PPE back so that the contaminated side of the garment contacts only itself. Outer chemical gloves (if worn) are carefully peeled away and off both hands of the responders. If the procedure is done properly, the contaminated side of the garment and gloves will not touch the person wearing it.

The responder then proceeds toward the cold zone end of the decontamination corridor, to an area where helmets, face piece, and any other ancillary equipment are removed and can be placed in a separate area or plastic bag (or other suitable container). Last, remove and discard inner gloves.

With decontamination complete, the responders can exit the decontamination area and enter the cold zone, where personal showers are taken and a fresh set of clothes are donned. Afterward, personnel should proceed to a medical station for evaluation.

RESPONDER TIP

Most chemical protective equipment in use today is considered to be disposable, with the decontamination process intended to render the garments safe enough to remove and discard, not rendered 100% clean. The concept of removing PPE should follow the same principle as applies when extricating a victim pinned inside an automobile: Remove the car from the person; do not take the person from the car. When it comes to decontamination and PPE removal, remove the PPE from the person; don't take the person out of the PPE.

To perform technical decontamination on a responder, follow the steps in SKILL DRILL 9-1:

1. Drop any tools and equipment into a tool drum or onto a designated tarp. (STEP 1)
2. Perform gross decontamination, if necessary. (STEP 2)
3. Perform technical decontamination. Wash and rinse the responder one to three times. The wash–rinse cycle is determined largely by the nature of the contaminant. Remember, the goal is to render the PPE safe to remove. (STEP 3)
4. Remove outer hazardous materials–protective clothing. (STEP 4)
5. Remove personal clothing.
6. Proceed to the rehabilitation area for medical monitoring, rehydration, and personal decontamination. (STEP 5)

SKILL DRILL 9-1

Performing Technical Decontamination on a Responder

NFPA 6.4.4.2

VIDEO

1 Drop any tools and equipment into a tool drum or onto a designated tarp.

2 Perform gross decontamination, if necessary.

3 Perform technical decontamination. Wash and rinse the responder one to three times. The wash–rinse cycle is determined largely by the nature of the contaminant. Remember, the goal is to render the PPE safe to remove.

4 Remove outer hazardous materials–protective clothing.

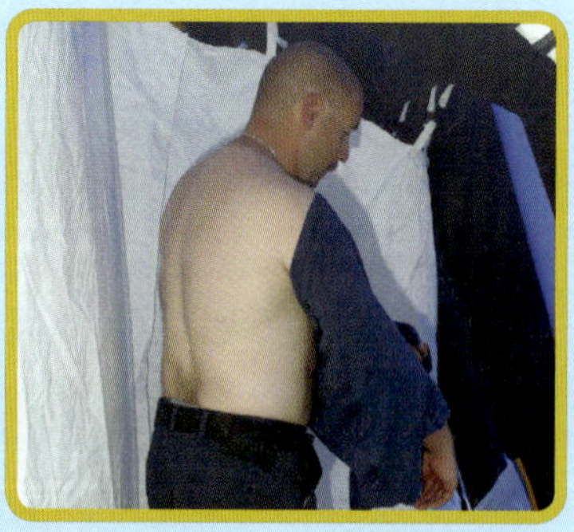

5 Remove personal clothing. Proceed to the rehabilitation area for medical monitoring, rehydration, and personal decontamination.

Technical decontamination can be performed in several ways, and your AHJ may have established a specific procedure for it. check your policies and procedures for instructions on the preferred way to carry out technical decontamination.

Evaluating the Effectiveness of Technical Decontamination

Evaluating the effectiveness of the decontamination is typically done at the end of the decontamination line, and should be based on the nature of the contaminant. The goal is to check for the effectiveness of decontamination using whatever method will offer the most accurate results. If the contaminant is corrosive, pH paper can be swiped along the PPE at various locations to ensure there is no corrosive residue left. A photo ionization detector (PID) might be used to determine if any residual organic compounds remain after the protective garment, gloves, or boots have been decontaminated. Radiation detectors might be passed over and around the responder to ensure no radiation contamination exists on the PPE. These are broad examples, and serve only to suggest some tools and actions that might be useful to ensure your decontamination efforts are successful. Again, creative thinking and well thought out actions are required when it comes to determining the effectiveness of your decontamination. In some jurisdictions, environmental safety and health representatives from municipal agencies may be available to assist with this process. Check your standard operating procedures for the proper ways to interact with those representatives if they are to be involved in your operation.

RESPONDER TIP

Responders should consult NFPA 473, *Standard for Competencies for EMS Personnel Responding to Hazardous Materials/ Weapons of Mass Destruction Incidents*, for guidance on medical monitoring.

Reports and Documentation

When it comes to decontamination, the person responsible for the decontamination corridor should complete any documentation and recordkeeping that is required by the emergency response plan or standard operating procedures. This information should be included in the overall documentation process for the incident. Items to record include the names of all persons arriving and processed through the decontamination corridor; information on the released substance; the potential for acute and chronic health effects of an accidental exposure; actions taken to limit exposures; a detailed description of the decontamination activities, including decontamination solutions and the overall effectiveness of those solutions; and any breaches or failures of the PPE noted during the decontamination process. As with any other type of incident, this report should be complete and accurate, and should stand as your legal account of the incident.

Wrap-Up

Chief Concepts

- Decontamination is the physical and/or chemical process of reducing and preventing the spread of contaminants from people, animals, the environment, or equipment involved at hazardous materials/weapons of mass destruction (WMD) incidents.
- Decontamination efforts should be matched to the known or anticipated physical and chemical properties of the released substance.
- The three major categories of decontamination are emergency decontamination, mass decontamination, and technical decontamination. Gross decontamination takes place within a controlled decontamination corridor and generally consists of a prewash before technical decontamination takes place.
- The decontamination corridor is a controlled area in the warm zone, for which access is limited to only those persons who have entered the hot zone or who are participating in decontamination.
- Responders may employ any number of technical decontamination methods for personnel, tools, and equipment, including absorption, adsorption, washing, dilution, and neutralization. It is up to the responder to match the best decontamination method with the physical and chemical characteristics of the contaminant.

- When it comes to PPE removal within the decontamination corridor, remove the PPE from the person; don't take the person out of the PPE.
- After personnel are thoroughly decontaminated and have showered and donned clean clothes, they should proceed to a medical station for evaluation.
- Use detection techniques to evaluate the effectiveness of decontamination.
- Document your account of the entire incident.

Hot Terms

Absorption The process of applying a material that will soak up and hold a hazardous material in a sponge-like manner, for collection and subsequent disposal.

Adsorption The process in which a contaminant adheres to the surface of an added material—such as silica or activated carbon—rather than combining with it (as in absorption).

Chemical degradation A natural or artificial process that causes the breakdown of a chemical substance.

Contamination The process of transferring a hazardous material from its source to people, animals, the environment, or equipment, all of which may act as carriers for the material.

Decontamination The physical and/or chemical process of reducing and preventing the spread of contaminants from people, animals, the environment, or equipment involved at hazardous materials/weapons of mass destruction incidents.

Decontamination corridor A controlled area within the warm zone where decontamination takes place.

Decontamination team The team responsible for reducing and preventing the spread of contaminants from persons and equipment used at a hazardous materials incident. The team members establish the decontamination corridor and conduct all phases of decontamination.

Dilution The process of adding some substance—usually water—in an attempt to weaken the concentration of another substance.

Disinfection The process used to destroy recognized disease-carrying (pathogenic) microorganisms.

Emergency decontamination The process of removing the bulk of contaminants off of a victim without regard for containment. It is used in potentially life-threatening situations, without the formal establishment of a decontamination corridor.

Evaporation A natural form of chemical degradation in which a liquid material becomes a gas, allowing for dissipation of a liquid spill. It is sometimes used as a safe, noninvasive way to allow a chemical substance to stabilize without human intervention.

Gross decontamination A technique for significantly reducing the amount of surface contaminant by application of a continuous shower of water prior to the removal of outer clothing. It differs from emergency decontamination in that gross decontamination is controlled through the decontamination corridor.

Isolation and disposal A two-step removal process for items that cannot be properly decontaminated. First, the contaminated article is removed and isolated in a designated area. Second, it is packaged in a suitable container and transported to an approved facility, where it is either incinerated or buried in a hazardous waste landfill.

Mass decontamination The physical process of reducing or removing surface contaminants from large numbers of victims in potentially life-threatening situations in the fastest time possible.

Neutralization The method used when the corrosivity of an acid or a base needs to be minimized. This process accomplishes decontamination by way of a chemical reaction that alters the material's pH.

Solidification The process of chemically treating a hazardous liquid so as to turn it into a solid material, thereby making the material easier to handle.

Sterilization A process using heat, chemical means, or radiation to kill microorganisms.

Technical decontamination A multistep process of carefully scrubbing and washing contaminants off of a person or object, collecting runoff water, and collecting and properly handling all items. It takes place after gross decontamination.

Vacuuming The process of cleaning up dusts, particles, and some liquids using a vacuum with high efficiency particulate air (HEPA) filtration to prevent recontamination of the environment.

Washing The process of dousing contaminated victims with a simple soap-and-water solution. The victims are then rinsed using water.

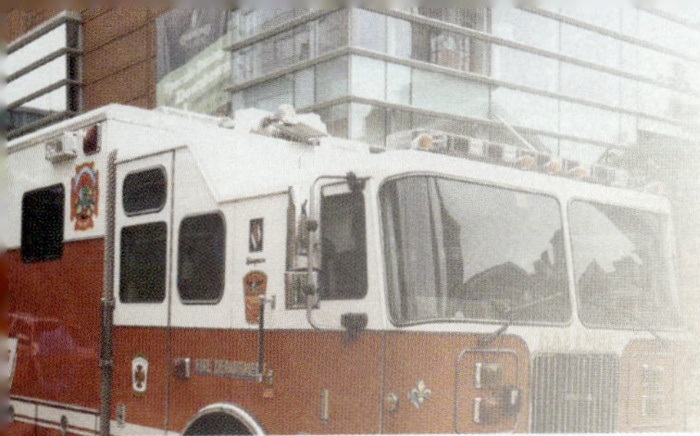

RESPONDER *in action*

It is Tuesday evening when your engine company is dispatched to a vehicle fire. Upon arrival, you find a fully involved van in a convenience store parking lot. The driver of the van is in custody away from the fire, sitting on the curb. Your lieutenant gives the order to pull a handline to extinguish the fire. You and your crew extinguish the fire and gain access to the van's interior through the rear doors. During overhaul, you find indicators that the van has been used as a mobile drug laboratory. Local law enforcement personnel and the regional hazardous materials team are also summoned to the scene.

1. Which of the following decontamination methods is likely to be the best choice for technical decontamination in this situation?
   - **A.** Absorption
   - **B.** Neutralization
   - **C.** Dilution
   - **D.** Solidification
2. The hazardous materials team arrives and enters the van for evidence collection on behalf of the local police. Which decontamination process should be used to decontaminate the evidence containers?
   - **A.** Mass
   - **B.** Technical
   - **C.** Gross
   - **D.** Emergency
3. If the entry team wore Level B ensembles to collect the evidence, which level of protection would be the most appropriate for the decontamination team?
   - **A.** Level A
   - **B.** Level B
   - **C.** Level C
   - **D.** Level D
4. Law enforcement personnel bring you the victim of the accident. He complains of burning eyes and skin. Which type of decontamination should be used for him?
   - **A.** Emergency
   - **B.** Technical
   - **C.** Mass
   - **D.** None

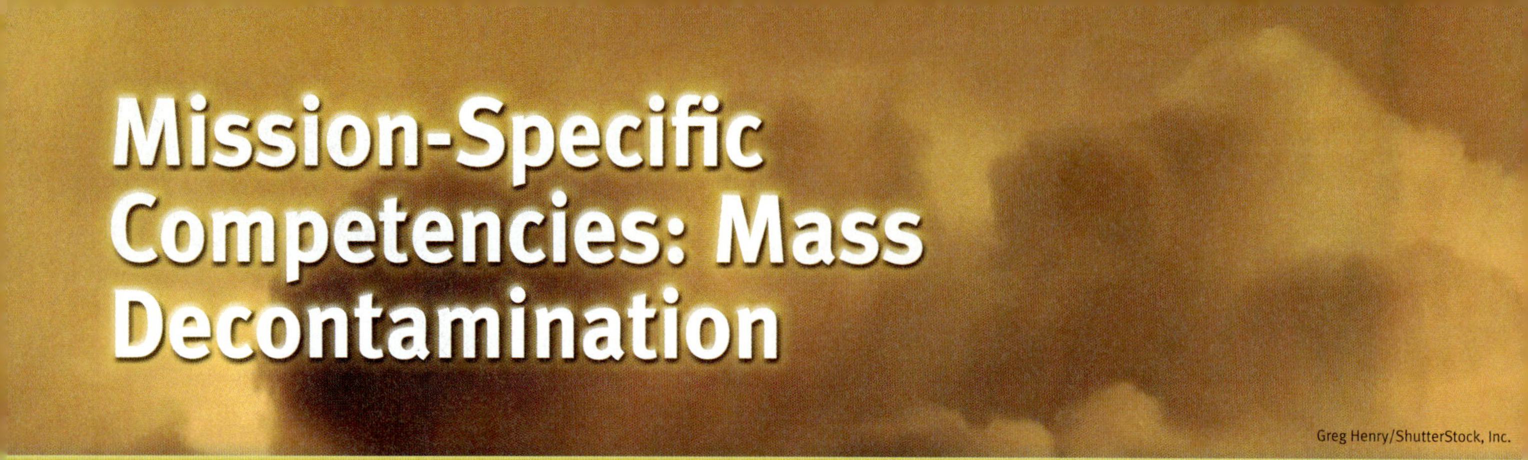

Mission-Specific Competencies: Mass Decontamination

Greg Henry/ShutterStock, Inc.

Courtesy of Emergency Film Group

CHAPTER 10

Competencies for Operations Level Responders Assigned Mission-Specific Responsibilities

This chapter shall address competencies found in NFPA 472 for the following operations level responders assigned mission-specific responsibilities at hazardous materials/WMD incidents by the authority having jurisdiction beyond the core competencies at the operations level.

Knowledge Objectives

After studying this chapter, you will be able to:

- Explain the advantages and limitations of mass decontamination operations. (NFPA 6.3.3.2(1), p. 220–222)
- Describe three ways to reduce or eliminate contamination on victims. (NFPA 6.3.3.2(2), p. 226–228)
- Describe the reference sources available for responders charged with performing mass decontamination. (NFPA 6.3.3.2(3), p. 228–231)
- Identify the equipment required to set up and implement mass decontamination operations. (NFPA 6.3.3.2(4), p. 222)
- Describe methods for crowd control. (NFPA 6.3.3.2(5), p. 231)
- Describe the steps required to perform mass decontamination on ambulatory and nonambulatory victims within the Incident Command System. (NFPA 6.3.4.1, 6.3.4.2, p. 222–223)
- Describe how to evaluate the effectiveness of a mass decontamination process. (NFPA 6.3.5.1, p. 224–225)
- Describe the importance of completing reports and documentation of mass decontamination operations. (NFPA 6.3.6.1, p. 231–232)

Skills Objectives

After studying this chapter, you will be able to:

- Plan a response within the capabilities of available personnel, personal protective equipment, and control equipment by selecting a mass decontamination process to minimize the hazard. (NFPA 6.3.4.1, p. 222–223)
- Set up and perform mass decontamination. (NFPA 6.3.4.2, p. 225–227)
- Evaluate the progress of the planned response by evaluating the effectiveness of the mass decontamination process. (NFPA 6.3.5.1, p. 224–225)

You Are the Responder

Your engine company has been called in on a mutual aid request to a neighboring jurisdiction. A chlorine release at a fixed facility has affected approximately 50 employees of an adjacent office complex. Upon your arrival, the incident commander (IC) directs your crew to assist with setting up a mass decontamination corridor using two fire engines. Your captain directs you to work with the fire fighters on scene to assist them with moving the victims through the mass decontamination corridor.

1. Which sources of information would be available to you for determining the correct mass decontamination procedure? Identify the methods you would use to access those resources in a hazardous materials/WMD incident.
2. Describe the supplies and equipment required to set up and implement mass decontamination operations in your jurisdiction.
3. Which methods, tools, or equipment would you use to evaluate the effectiveness of the decontamination operation?

Introduction

The success of mass decontamination—quickly removing contamination from a large population of people in a short amount of time—is conceptually similar to emergency decontamination in the following respects. First, it is important to identify the contaminant, if possible, and secure a source of water or other appropriate decontamination solution for use on humans. Second, the responders must select and use the proper level of personal protective equipment (PPE) to protect themselves. Third, it is important to have a predetermined process or procedure to perform decontamination. Lastly, it should all be coordinated using the Incident Command System (ICS) to keep track of resources and manage the situation.

Mass decontamination is fundamentally dissimilar to emergency decontamination because of the potential high victim count and the fact that there may not be enough trained personnel, at least early in the incident, to meet the need to quickly get a system in place. Add to this the fact that you will be dealing with people who are confused and scared, and it becomes clear that a real mass casualty decontamination situation will be challenging. Mass decontamination boils down to identifying the need for it in the first place—*quickly*; getting set up—*quickly*; running as many people through the process as you can—*quickly*; and incurring the least amount of risk to the responders. Keep in mind, however, that "mass casualty" is not a well-defined number. It does not always mean that several hundred contaminated people will need help. Even 10 to 15 victims will require you to shift your thinking from standard emergency decontamination to mass decontamination. The need for this procedure really depends on your particular agency and the tools, equipment, and personnel that are available.

The other thing to remember is that you will be dealing with all types of human beings who are involved in what might be a very dramatic situation. There could be elderly victims, non-English speakers, disabled persons, and children—all exhibiting various types of behavior. Stressful circumstances may cause people to behave erratically, which will undoubtedly complicate your situation. To that end, your ability to bring some level of order to the situation and communicate effectively with this population of people is critical to the success of the operation **FIGURE 10-1**.

FIGURE 10-1 Don't underestimate the human factor when thinking about the challenges of mass casualty decontamination.

Courtesy of Rob Schnepp

Mass Decontamination

Mass decontamination has a finite focus and intent. It is formally defined in NFPA 472, *Standard for Competence of Responders to Hazardous Materials/Weapons of Mass Destruction Incidents*, as "the physical process of reducing or removing surface contaminants from large numbers of victims in potentially life-threatening situations in the fastest time possible." This chapter discusses mass decontamination, the steps that are taken to perform mass decontamination, helpful reference sources responders might use to prepare for and carry out mass decontamination, and various methods of mass decontamination. Your agency may already have established procedures and equipment for mass decontamination. The examples in this chapter are intended only to provide conceptual guidance on the topic; they are not designed to be a complete reference or a set of fully developed procedures that can be implemented by your agency without review and refinement.

RESPONDER TIP

Life safety is the number one priority for both emergency decontamination and mass decontamination.

Viewed from a big-picture perspective, mass decontamination boils down to making a rapid assessment of the situation and the number of victims present, attempting to identify the contaminant, setting up some form of mass decontamination process, wearing the proper type and level of PPE, and getting to work FIGURE 10-2. This process can take place on any street, parking lot, or other area where fire apparatus or other response equipment can be deployed with a continuous, uncontaminated water supply. The standard thinking regarding decontamination—using minimal amounts of water and recovering the potentially contaminated runoff—becomes a secondary objective when mass decontamination is implemented. If lives are at stake, controlling runoff is not the responders' main concern.

The extent of mass decontamination required is largely driven by the contaminant and its effect on the people it interacts with. Not all contaminated persons will react the same way or require the same level of decontamination. Therefore, it is important to have some process in place to assign a priority to the population of victims that truly require your assistance. Mass decontamination is very labor and time intensive, so it's important to concentrate the workforce on those that are in need of your service. FIGURE 10-3 depicts a process for triaging your victim population. Those who are ambulatory and without obvious signs or symptoms of exposure could be directed to an area near the mass decontamination area (green box victims). They might go through the process eventually, but at least initially, it is not prudent to clog up your system with those who are not critical. Another choke point occurs when

FIGURE 10-2 Mass decontamination provides emergency decontamination for a large number of victims.

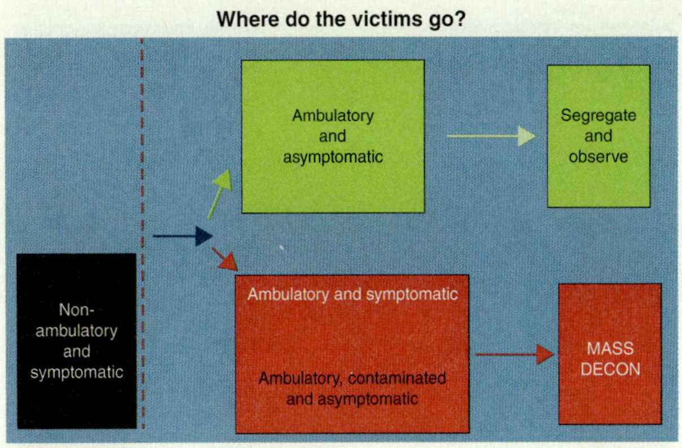

FIGURE 10-3 Victim triage.

contaminated ambulatory victims are mixed with nonambulatory victims. Those who cannot walk on their own or are critically ill (black box victims) might best be served by a system or part of your mass decontamination system dedicated to the slowest and most challenging group of victims. The target population for mass decontamination then becomes those who are able to walk on their own and are symptomatic or who are contaminated but not yet showing signs or symptoms of coming into contact with the substance (red and yellow box victims). Your AHJ may have a similar logic tree for determining the target population for mass decontamination. If not, think through one of your own and decide how you will approach the problem before the situation occurs.

Additionally, it is important to have a method for checking the effectiveness of decontamination at the back end of the process. If your victim population was exposed to a corrosive, perhaps swiping a piece of pH paper at various points on the body might confirm that the contaminant is completely washed off. Maybe a gas detection device could confirm that a solvent has been removed, or the use of M8 or M9 paper would confirm that a nerve agent is fully decontaminated FIGURE 10-4. If radioactive contamination is confirmed or suspected, radiation detection devices would be needed to evaluate the effectiveness of decontamination.

If a group of 50 civilians was exposed to carbon monoxide (CO), for example, would a water-based decontamination process be effective? This particular substance is gaseous at standard atmospheric temperature and pressure, and it will not leave any significant residue on the victims because it will not adhere for any length of time to anything it touches. Decontamination in this case would be geared toward letting the CO do what it naturally wants to do—be a gas. Water washing will not change the physical or chemical properties of CO, so the application of water in this case would yield minimum benefit.

RESPONDER TIP

Mass decontamination, even when performed perfectly, will be a difficult and chaotic situation.

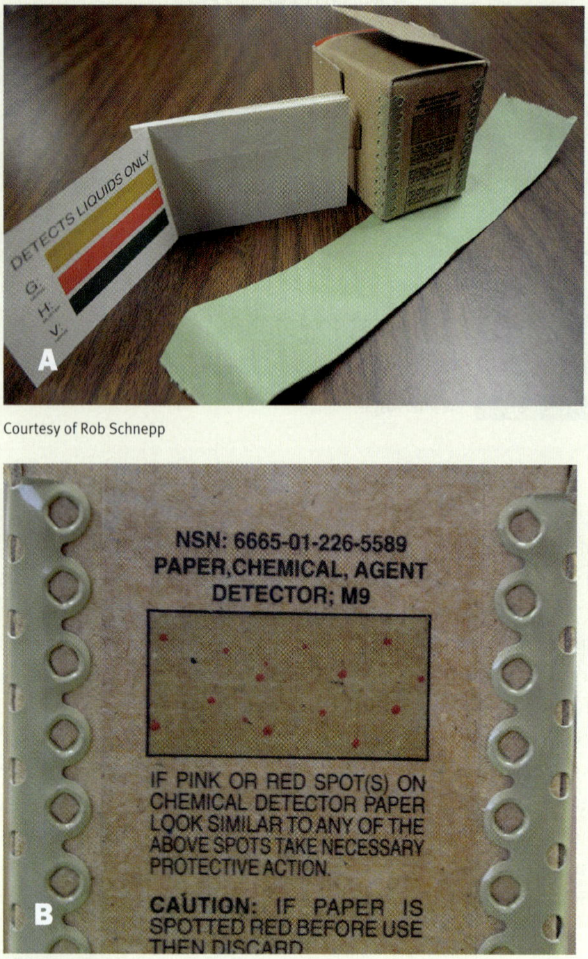

Courtesy of Rob Schnepp

FIGURE 10-4 **A.** M8 paper and **B.** M9 paper are rapid ways to check for the presence of a nerve agent.

Courtesy of Rob Schnepp

By contrast, if the chemical involved in the incident were VX—a thick, oily nerve agent—the decontamination process changes accordingly. In this case, the application of a water rinse–soap wash–water rinse would be the best method of removing the contaminant from affected victims. Clearly, it is important to understand the nature of the contaminant to devise a decontamination plan that will effectively and efficiently address the particular physical and chemical properties. Additionally, it is prudent to have a plan and method for ensuring the effectiveness of decontamination after the victims have been through the process. Again, this is driven by the nature of the contaminant—it might make sense to do a pH check in case the contaminant is corrosive or use M8 or M9 to check for the presence or absence of a nerve agent.

Mass decontamination can be performed in a number of ways, and your agency may have established a specific procedure for it **FIGURE 10-5**. In the fire service, this operation is sometimes accomplished by placing two fire apparatus side by side. Fog-type nozzles are attached to the pumpers opposite of each other, and a fog pattern is used to douse the victims as they walk between the two pieces of apparatus **FIGURE 10-6**. Pump pressures are set so that the fog patterns are effective but not overwhelming to the victims (usually between 30 and 50 psi). An aerial ladder device (either in conjunction with the pumpers or as a stand-alone unit) can also provide a complete overhead spray pattern by using preplumbed waterways or other configurations of hose lines. Check with your department regarding its specific policy and procedures for mass decontamination.

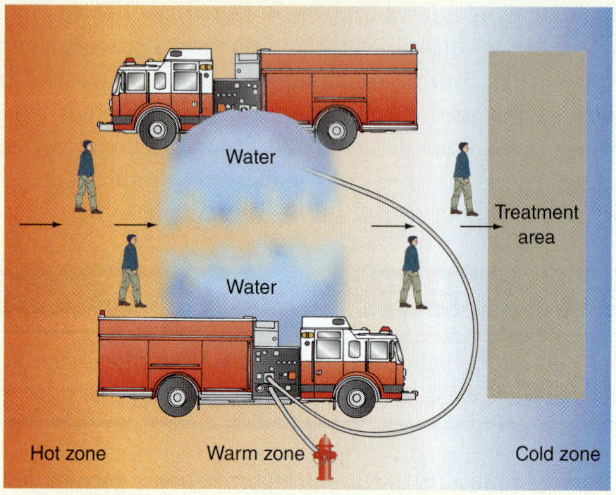

FIGURE 10-5 An example of a simple mass decontamination corridor using two fire engines.

Mass Decontamination Methods

Over the last several years, a number of mass decontamination methodologies have evolved. In today's marketplace, there is no shortage of prepackaged mass decontamination showers for ambulatory victims (able to walk) and nonambulatory victims (unable to walk without assistance). Several versions of preplumbed, rapid-deploy shelters are available that contain intricate showerheads and spray wands, along with segregated areas for gender-specific showering. Self-contained decontamination trailers with pop-out sides and overhead tents are also available, as are a wide array of portable showers. Some of these units come equipped with portable water heaters, space heaters, water collection bladders, and sections of rollered platforms for sliding nonambulatory victims (supine, on rigid backboards, or in Stokes baskets) through a series of wash–rinse stations.

Decontaminating nonambulatory victims is a much slower process than performing mass decontamination on ambulatory victims. Handling casualties who cannot walk, or who are unconscious or otherwise unresponsive, requires a significant number of emergency response personnel to complete the decontamination process. It is physically more taxing to carry unresponsive victims, and work times for emergency responders will be limited. Some manufacturers offer stretchers with wheels or other types of carts or sleds to carry those victims who cannot walk **FIGURE 10-7**, **FIGURE 10-8**, and **FIGURE 10-9**. Nevertheless, using these devices may not significantly speed up the mass decontamination process, but it does ease the workload on the responders.

FIGURE 10-6 Mass decontamination can be accomplished using fire apparatus.
Courtesy of Emergency Film Group

FIGURE 10-7 Rescue sled in use during a mass casualty simulation.
Courtesy of Rob Schnepp

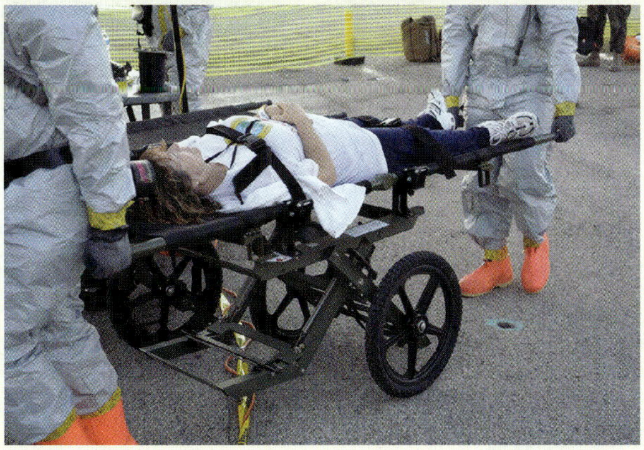

FIGURE 10-8 Wheeled litters or gurneys can relieve the workload of responders.
Courtesy of Rob Schnepp

Many jurisdictions set up two separate areas for mass decontamination: one for nonambulatory victims and one for ambulatory victims FIGURE 10-10 and FIGURE 10-11. It is up to your AHJ to determine the specific procedures for handling both types of victims.

From first responders' perspective, it is important to choose a system that fits the responding agency's needs based on staffing levels, anticipated numbers of casualties, topography, and proximity to other mass decontamination units in the region. There is no one perfect setup for all occasions. The AHJ must evaluate all operational facets of its own operations and choose a process that best suits the anticipated need.

FIGURE 10-9 A Stokes basket.

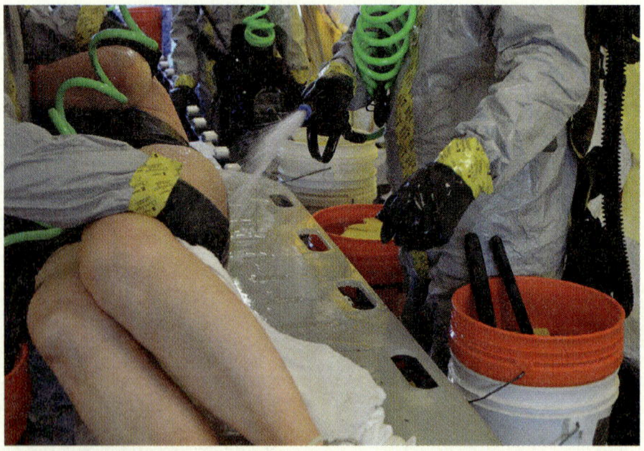

FIGURE 10-10 A mass decontamination configuration for a nonambulatory victim.

Evaluating the Effectiveness of Mass Decontamination

As previously mentioned, at the end of the mass decontamination process, an appropriate method should be implemented to evaluate the effectiveness of the operation **FIGURE 10-12**. The goal

FIGURE 10-11 A mass decontamination configuration for an ambulatory victim.

FIGURE 10-12 After victims are thoroughly decontaminated and showered, they should proceed to a medical station for evaluation. Emergency medical personnel (Advanced Life Support [ALS] level, if possible) should do a full patient assessment of each victim who was present at the scene.

is to determine the effectiveness of decontamination using whatever method will offer the most accurate results. In some cases, monitoring devices may be used to evaluate the completeness of decontamination, or radiological detection devices can be employed if the situation calls for it. In some jurisdictions, environmental safety and health representatives from municipal agencies may be available to assist with postdecontamination monitoring. Check your standard operating procedures for the proper ways to coordinate your agency's efforts with those representatives if they are to be involved in your operation.

You should be familiar with the procedures and policies mandated by your AHJ when it comes to evaluating the effectiveness of decontamination. Poorly performed decontamination will leave responders with a false belief that the victims are safe to treat medically or transport. Eventually, inadequate decontamination efforts may adversely affect transport ambulances, hospitals, law enforcement officers, or anyone who comes into contact with a supposedly "clean" victim.

Because water is a good general-purpose solvent, washing off as much of the contaminant as possible with a massive water spray is the best and quickest way to decontaminate a large group of people.

To set up and use a mass decontamination system on ambulatory victims, follow the steps in SKILL DRILL 10-1:

1. Ensure you have the appropriate PPE to protect against the chemical threat.
2. Stay clear of the product, and d contact with it.
3. Direct victims out of the hazar able location for decontaminatio female lanes or areas are preferable). (STEP 1)
4. Set up the appropriate type of mass decontamination system based on the nature of the contaminant, type of apparatus, equipment, and/or decontamination system available.(STEP 2)
5. Instruct victims to remove their contaminated clothing and walk through the decontamination corridor (personal belongings should be bagged and marked with victims' names).
6. Flush the contaminated victims with water. (A water temperature of 70°F [21°C] is ideal but may not be possible. Try to avoid using water that is uncomfortably hot or cold.) Employ the appropriate method and/or technology to check for the effectiveness of decontamination upon exit. (STEP 3)
7. Direct the contaminated victims to the triage area for medical evaluation, which may include on-scene treatment and/or transport to an appropriate receiving hospital. (Many agencies provide modesty/comfort packages after decontamination that includes gowns, booties, towels, and other pertinent items). (STEP 4)

SKILL DRILL 10-1

Performing Mass Decontamination on Ambulatory Victims

NFPA 6.3.4.2

VIDEO

1 Ensure that you have the appropriate PPE to protect from the chemical threat. Stay clear of product and do not make physical contact with it. Make an effort to contain runoff by directing victims out of the hazard zone and into a suitable location.

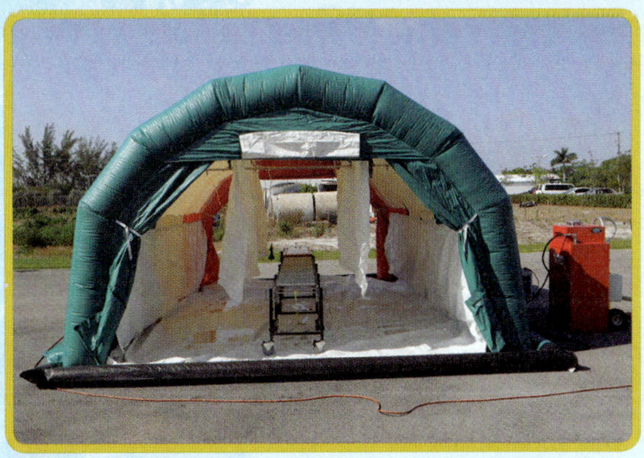

2 Set up the appropriate type of mass decontamination system based on the type of apparatus, equipment, and/or system available.

3 Instruct victims to remove all contaminated clothing and walk through decontamination process. Flush the contaminated victims with water.

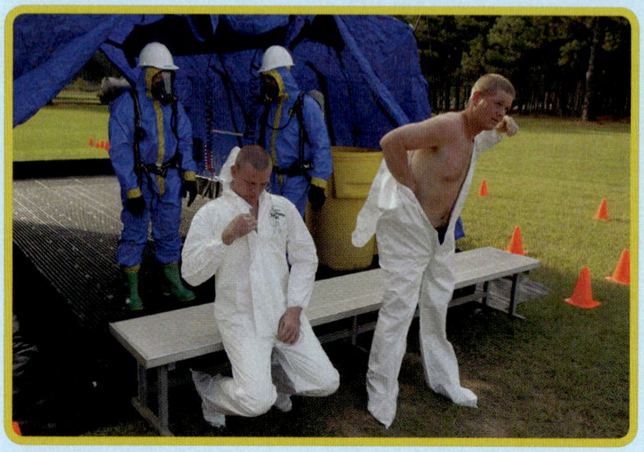

4 Direct the contaminated victims to the triage area.

RESPONDER TIP

Training is a key component in ensuring an effective mass decontamination operation. Without regular and frequent training, it is impossible to remain proficient in this infrequently used, but potentially high-impact technique. Contaminated victims will not wait around for responders to establish a formal decontamination area. Instead, they will hurry to the hospital by self-transport while still contaminated, which will ultimately affect the ability of the hospital-based providers to render proper medical care.

Regardless of the mass decontamination methodology employed by the responding agency, the theory behind the work focuses on one of three ways to reduce or eliminate contamination: dilution, isolation, and washing.

Dilution

Dilution is the process of adding some substance—usually water—to a contaminant to decrease its concentration. Dilution is both fast and economical. Water is readily available from any fire engine or fire hydrant. However, before using water, consider whether the contaminant will react adversely with water, whether the contaminant will be soluble, or whether the application of water will spread the contamination to a larger area.

Water does have some limitations when it comes to removing viscous, oily liquids. VX, as mentioned earlier, would be difficult to remove from a population of victims using only water. Use of soap and water would be more effective in this situation, albeit perhaps more time-consuming. The dilution method of decontamination works best with acids and bases (to dilute the substance and reduce the negative effects of its strong pH)

To perform mass decontamination on nonambulatory victims, follow the steps in **SKILL DRILL 10-2**:*

1. Set up the appropriate type of mass decontamination system based on the nature of the contaminant, type of apparatus, equipment, and/or decontamination system available. (**STEP 1**)
2. Ensure you have the appropriate PPE to protect against the chemical threat.
3. Remove the appropriate amount of the victim's clothing. Do not leave any clothing underneath the victim; these items may wick the contamination to the victim's back and hold it there, potentially worsening the exposure. (Medical trauma scissors are a helpful and rapid way to accomplish this step.) (**STEP 2**)
4. Flush the contaminated victims with water. (A water temperature of 70°F [22°C] is ideal but may not be possible. Try to avoid using water that is uncomfortably hot or cold.) Make sure to rinse well under and around the straps that may be holding the victim to a backboard or other extrication device. *Take care to avoid compromising the victim's airway with water during the process.* (**STEP 3**)
5. Move the victims through the decontamination corridor and into the triage area for medical evaluation, which may include on-scene treatment and/or transport to an appropriate receiving hospital. Gowns or other suitable privacy garments should be available. In most cases, significant medical treatment should be provided after decontamination, in a designated medical treatment area. (**STEP 4**)

*This skill drill begins after the victims have been extricated from the contaminated environment and transported in some manner to the mass decontamination corridor.

SKILL DRILL 10-2 Performing Mass Decontamination on Nonambulatory Victims

NFPA 6.3.4.2

VIDEO

1. Set up the appropriate type of mass decontamination system based on the type of equipment available.

© Jones & Bartlett Learning. Photographed by Glen E. Ellman

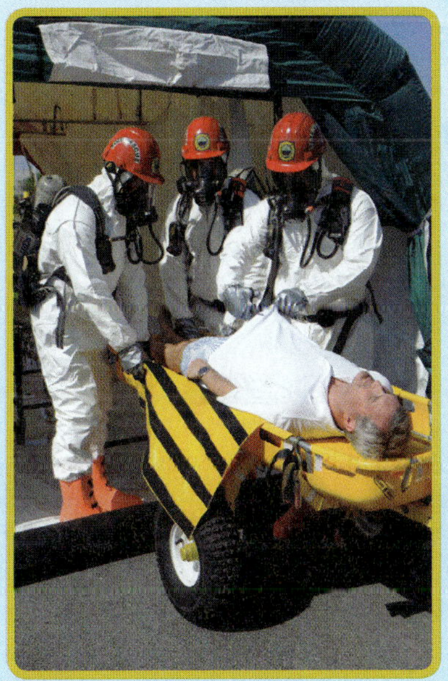

2. Ensure that you have the appropriate PPE to protect from the chemical threat. Remove the victim's clothing. Do not leave any clothing underneath the victim; these items may wick the contamination to the victim's back and hold it there, potentially worsening the exposure.

© Jones & Bartlett Learning. Photographed by Glen E. Ellman

3. Flush the contaminated victim with water.

© Jones & Bartlett Learning. Photographed by Glen E. Ellman

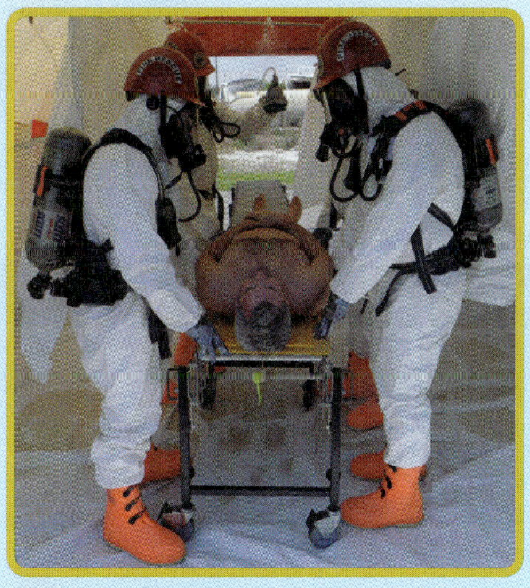

4. Move the victim to a designated triage area for medical evaluation.

© Jones & Bartlett Learning. Photographed by Glen E. Ellman

and other water-soluble substances. Insoluble substances (for example, benzene, toluene, and most nerve agents) *would not* be the best candidates for a straight water wash.

Isolation and Disposal

As discussed in the chapter *Mission-Specific Competencies: Personal Protective Equipment*, isolation and disposal is a two-step process for removing items that cannot be properly decontaminated from the incident scene. The isolation step involves removing the contaminated items (e.g., clothing, tools, or personal items) and *isolating* them in a designated area. The items can then be segregated into logical groupings if necessary; for example, victims' clothing might be placed in one area, responder equipment in another, and potential evidence in yet another area. These items are then tagged (with information such as the item name, date of collection, item description, location where it was found, possible contamination) and placed in a suitable container such as a bag, barrel, or bucket. Subsequently, those isolated items may be retained as evidence, tested further to positively identify the released substance, or transported to an approved disposal facility.

Washing

Washing is an effective yet simple decontamination process, ideal for removing most harmful substances. When washing is employed for mass decontamination, a simple soap-and-water solution is made. Victims are doused with this solution and fully rinsed with water.

Keep in mind that the goal of every mass decontamination process is to quickly reduce the effects of the exposure. The intent is not to concentrate on any one victim at the expense of the others. Washing may not be a complete solution to the decontamination problem you are facing. As with dilution, some viscous chemicals cannot be completely removed from the skin by washing alone.

RESPONDER TIP

Clothing removal is an excellent first step in reducing the amount of contamination on a victim. Make sure to instruct ambulatory victims to remove their clothing prior to entering any mass decontamination process. Preserving the modesty of a person, while attempting to perform effective decontamination, can be a delicate and somewhat uncomfortable part of the decontamination process. Responders have a duty to be prudent when ordering contaminated victims to remove the appropriate amount of clothing (sometimes all of it) and comply with the mass decontamination process. It may be wise to separate groups by gender to ease some of the victims' anxiety.

Responder Safety Tip

Never neutralize chemicals on human skin. The primary reason for this edict is the heat and/or toxic gases that may be generated when acids and bases are neutralized. The risk of causing more damage is not worth the potential benefit from using this approach. In such cases, water washing is the quickest, safest, and most reliable method of decontamination.

The Role of Reference Sources in Mass Decontamination

From the clues presented during dispatch, response, and approach to the scene, the operations level responder should be able to determine whether a hazardous material is present and, if released, whether the released material has affected a significant number of victims. Bystanders or witnesses, placards, the normal occupancy of buildings at the scene (such as chemical storage buildings), the types of containers involved, and the presence of fires or explosions are a few typical indicators of the presence and/or release of a hazardous material FIGURE 10-13.

FIGURE 10-13 Look carefully for indicators of a hazardous material.
© Chase Jarvis/age fotostock

Understanding the physical properties and the associated mechanisms of injury, along with the health implications associated with the released substance, will assist in formulating a sound plan for mass decontamination. Once the nature of the contaminant is known, appropriate reference sources should quickly be consulted to learn more about the hazardous material. Many of these reference sources were discussed in the chapter *Recognizing and Identifying the Hazards*. As emphasized previously, mass decontamination is a time-critical operation. The exposed victims will not be willing to wait around while you research the chemical for an extended period of time.

When a quick reference is required, the *Emergency Response Guidebook* (*ERG*) may prove useful for responders operating at a hazardous materials incident FIGURE 10-14.

Fire fighters, police, and other emergency services personnel—all of whom may be the first responders to arrive at the scene of a transportation incident involving a hazardous material—are the primary audience for the *ERG*. This reference identifies and outlines predetermined evacuation distances and basic action plans for the chemicals that are highlighted in the yellow or blue sections of the book, based on spill size estimates. See the chapter *Recognizing and Identifying the Hazards* for more information about the *ERG*.

Other sources of information about the hazardous material may include placards—diamond-shaped indicators (10 3/4 inches on each side) that must be placed on all four sides of highway transport vehicles, railroad tank cars, and other forms of transportation carrying hazardous materials. Labels (4-inch diamond-shaped indicators) are smaller versions of placards; they are placed on the four sides of individual boxes and smaller packages being transported. Placards and labels are intended to give responders a general idea of the hazard inside a particular container or cargo tank. Keep in mind that a placard identifies only the broad hazard class (e.g., flammable, poison, corrosive) to which the material inside the transport vehicle belongs; there may also be secondary hazards that are not noted on the placard FIGURE 10-15. A label on a box inside a delivery truck, for example, relates only to the potential hazard inside that particular package FIGURE 10-16.

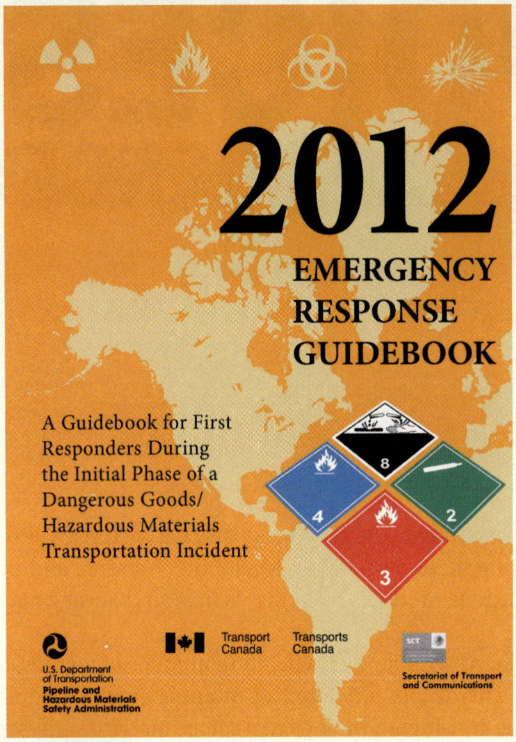

FIGURE 10-14 The *Emergency Response Guidebook.*

Courtesy of the U.S. Department of Transportation

FIGURE 10-15 A placard identifies the broad hazard class for materials carried by a transport vehicle.

© Mark Winfrey/ShutterStock, Inc

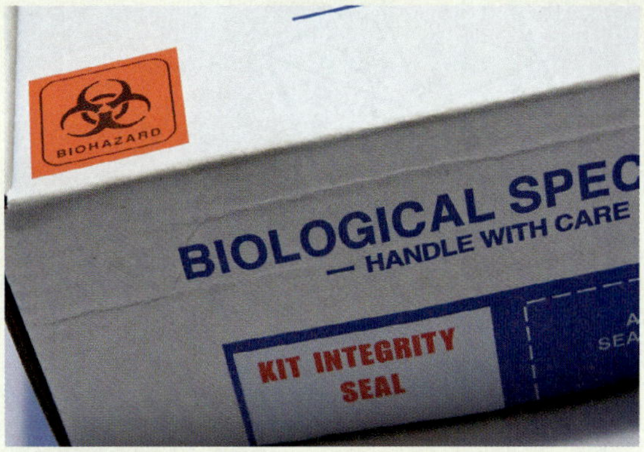

FIGURE 10-16 A label relates to the potential hazard inside a particular package or box.

© JHP Public Safety/Alamy Images

FIGURE 10-17 The NFPA 704 hazard identification system is designed for fixed-facility use.

© Jones & Bartlett Learning. Photographed by Glen E. Ellman

Online databases and other medically based resources may provide useful information when it comes to understanding the health effects of a particular substance. Additional sources of information include medical reference books, poison control centers, and the Agency for Toxic Substances and Disease Registry (ATSDR), which is part of the U.S. Centers for Disease Control and Prevention (CDC). Telephone numbers for voice and data communications with these agencies should be made available to the operations level responder.

NFPA 704, *Standard System for the Identification of the Hazards of Materials for Emergency Response*, delineates a marking system that is designed for fixed-facility use. The markings covered by this standard are found on the outsides of buildings, on doorways to chemical storage areas, and on fixed storage tanks. The NFPA 704 hazard identification system uses a diamond-shaped symbol of any size, which is itself broken into a set of four smaller diamonds, each representing a particular property or characteristic **FIGURE 10-17**. The chapter *Recognizing and Identifying the Hazards* explains the characteristics of each of the four diamonds. If this type of signage is present, responders may find the NFPA diamonds useful when attempting to understand the broad hazards posed by chemicals stored in a building or part of a building. Comparatively speaking, the information provided by the NFPA marking system is less comprehensive than what can be found in the *ERG*. The NFPA system is useful only for understanding the broad hazards of a particular facility or material. More definitive reference sources, including a safety data sheet (SDS) where applicable, should be consulted when more detailed information is required.

VOICES OF EXPERIENCE

Today there are many concerns raised with regard to the decontamination of mass-casualty victims. We must prepare ourselves for the task of effectively decontaminating these persons by developing, training, and practicing methods with which we can safely remove the hazardous products from the victims so that they may be triaged, treated, and transported to medical facilities.

The first-arriving emergency response units may be faced with an overwhelming number of victims, and the officer in charge will be faced with many decisions. The decontamination of multiple civilian casualties requires the implementation of effective methods with due consideration of citizen safety, modesty, and sensibility. The idea that first-arriving units will simply open up on the crowd with a deck gun is wrong and must be dispelled.

The mass decontamination of civilians is a critical public safety capability. Whether these people are the victims of a terrorist act or of an industrial accident, the effective and efficient decontamination of exposed persons has been demonstrated to result in fewer and less severe casualties. Mass decontamination, however, is only one part of a rational decontamination plan. To protect exposed persons and provide for the survivability of the emergency healthcare system, we cannot rely exclusively on field mass decontamination of victims at a single place. The concept of "decontamination in depth" provides for multiple opportunities to decontaminate victims—implemented in the field; in hospital emergency rooms, clinics, and trauma centers; and at other casualty collection points. "Decontamination in depth," with field mass-casualty decontamination as a critical centerpiece, is the only system imaginable for dealing effectively with a mass contamination incident.

Glen Rudner
Transportation Technology Center, Inc.- SERTC
Pueblo, Colorado

In some cases, it may be possible to access information from CHEMTREC in the United States, CANUTEC in Canada, or SETIQ in Mexico, depending on your location. Review the chapter *Recognizing and Identifying the Hazards* for the specifics on each of these resources.

Crowd Control

Any event requiring mass casualty decontamination will be chaotic. As a consequence, crowd control is one of the most significant challenges first responders will face. In almost all cases, there will be more frightened and possibly agitated victims than there will be calm, cool, and collected responders. Because responders will be outnumbered, they must conduct themselves in a way that commands respect and establishes them as symbols of authority. Imagine arriving at a crowded shopping mall 4 minutes after a significant explosion produced a high number of casualties and generated widespread panic. Simply put, you must corral the victims in this scenario and "convince" them to follow your instructions—a considerable challenge.

The badge you wear may go only so far in persuading victims to follow your instructions. You must also present yourself in such a way that the panicked victims will *choose* to follow your directions. When possible, put yourself in a position above the group. Standing on the tail step of a fire engine, or a planter box, or the hood of a squad car will place you in a position to be seen and heard. Speak plainly in a loud voice and make decisive hand gestures. If you aren't wearing a badge or a uniform, try holding a portable radio high in the air when you speak. The fact that you have this type of communication device sets you apart from the general civilian population and implies a certain amount of authority. Remember—if you seem panicked, the crowd will pick up on it.

Try to use naturally occurring barriers to your advantage. It's far easier to direct a moving group when they must travel around barriers, especially when fire engines are being used for mass casualty decontamination. Create a traffic flow that will harness the population of victims and create a pattern of movement that is advantageous to your operation. Perhaps the goal is to end the decontamination line near a treatment or transport area that is bordered by fences or other physical barriers. Once law enforcement becomes available, use uniformed officers to direct the flow of victims and/or keep victims away from particular areas. The general population is more accustomed to following the direction of police officers when it comes to directing traffic. Also, once a crowd begins to move in a particular direction, it is difficult to get them to change direction. To gain the attention of the group and encourage its members to move in a different direction or to stop moving, it may be beneficial to use a megaphone or external speaker on a fire engine.

Ideally, all crowd control efforts will be aimed at getting the crowd to behave in a manner that is beneficial to the mass decontamination operation. Once you lose control, any notion of returning order to the chaos will be met with stiff resistance. Remember—the victims are scared and may be irrational.

Responder Safety Tip

Resist the urge to touch anyone during the decontamination operation. You may end up becoming contaminated yourself.

RESPONDER TIP

Ambulatory victims who are following instructions correctly will serve as a good example to the other victims. Calmness and orderly behavior are just as contagious as panic. Remember, however, that you may never gain complete control of a crowd regardless of what you do.

Reports and Documentation

After the incident has been terminated, an incident report should be written **FIGURE 10-18**. As with any other type of incident, this report should be complete and accurate and should stand as your legal account of the incident. It may be necessary to submit worker's compensation claims for injured or exposed responders, and other medical reports may be required for any injured or exposed civilians. If the event involves a weapon of mass destruction (WMD), your reports may be used in the prosecution of the suspects. Any ICS forms should be completed and turned in to the IC.

Additionally, a postincident analysis or some other form of after-action review should be arranged by the AHJ for responders. These event reviews are vital to all personnel involved and can be used as a constructive learning tool.

When it comes to decontamination, the person responsible for the mass decontamination corridor should complete any documentation and recordkeeping required by the emergency response plan or standard operating procedures. This information should be folded into the overall documentation process for the incident and should include the following items:

- The names of all persons decontaminated (Collection of this data is difficult and may not happen given the circumstances.)
- Any information known about the released substance

FIGURE 10-18 Record the information from the incident in a complete and accurate manner.

- The level of protection worn by the responders
- Actions taken to limit exposures to personnel while performing decontamination
- A detailed description of the decontamination activities, including decontamination solutions and the overall effectiveness of those solutions
- Any evidence collected or samples taken
- Any other observations made about the scene in general

It is equally as important to file the documentation in an appropriate place so it can be accessed easily at a later date. If mass decontamination was performed, rest assured that your documentation will need to be very detailed, thorough, and easy to find.

Evidence Preservation

Although life safety is the first priority in mass decontamination, it is also important for first responders to avoid destroying any potential evidence found on contaminated victims and medical patients during the decontamination process. Keep in mind that the first responders—you—may be the people who put the pieces together and realize the incident is criminal in nature. Typically, the first responders are closest to the incident and have immediate access to victims. That access may provide clues in the form of victims' comments or the signs and symptoms of exposure that they demonstrate.

When possible, make every attempt to track the valuables and clothing taken from the victims. You may consider using small bags (tagged with the appropriate information) to secure valuables and clothing for later testing, as part of any subsequent legal action, and ultimately for reuniting valuable items with their owner.

Any incident plan should provide a procedure for securing evidence during mass decontamination operations at hazardous materials/WMD incidents. In addition, all responders should ensure that any information regarding suspects or observations made during mass decontamination are documented and communicated to the law enforcement agency having investigative jurisdiction.

Wrap-Up

Chief Concepts

- The purpose of mass decontamination is to perform emergency decontamination quickly on a large number of victims.
- Mass decontamination can take place on any street, parking lot, or area where equipment or apparatus can be deployed with a continuous, uncontaminated water supply; environmental concerns are a secondary concern in this setting.
- It is important to know the nature of the contaminant to devise an effective and efficient decontamination plan.
- There are three ways to reduce or eliminate contamination: dilution, isolation and disposal, and washing.
- Many reference sources are available to help identify hazardous materials and delineate their effects and risks.
- Crowd control will be necessary during mass decontamination.
- It is critical to evaluate the effectiveness of a mass decontamination process to avoid further complications, such as ongoing spread of the contaminant.
- Documentation and reporting is an important final step of decontamination operations because responders may be asked to provide the paperwork at a later date or to provide specific details about the event in civil or criminal proceedings.
- Evidence preservation should be considered during the mass decontamination effort.

Hot Terms

Ambulatory victims Victims who can walk on their own and can usually perform a self-rescue with direction and guidance from rescuers.

Dilution The process of adding a substance—usually water—in an attempt to weaken the concentration of another substance.

Isolation and disposal A two-step removal process for items that cannot be properly decontaminated. The contaminated article is removed and isolated in a designated area, and then packaged in a suitable container. The second step involves transport of the item to an approved facility, where it is either incinerated or buried in a hazardous waste landfill.

Mass decontamination The physical process of reducing or removing surface contaminants from large numbers of victims in potentially life-threatening situations in the fastest time possible.

Nonambulatory victims Victims who cannot walk under their own power. These types of victims require more time and personnel when it comes to rescue.

Washing The process of dousing contaminated victims with a simple soap-and-water solution; the victims are then rinsed using water.

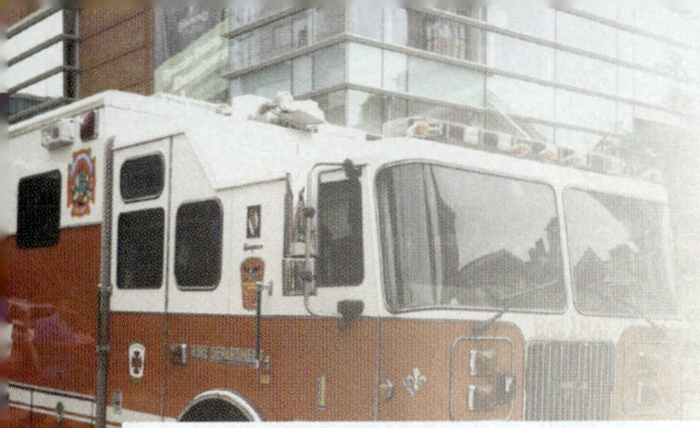

RESPONDER *in action*

Your ambulance arrives at a large regional shopping mall where a suspected terrorist attack has occurred. It's suspected that the chemical used may be the nerve agent VX. You and your paramedic partner normally work in a city located several miles away from the site, and you responded to the scene as part of a mutual aid request. Upon your arrival, you see fire engines everywhere, as well as approximately 30 people in various stages of going through a mass decontamination corridor. Over the radio, you receive an order to report to a location adjacent to the mass decontamination corridor. When you reach this site, the medical group supervisor quickly directs you to set up a medical treatment area, designates you as the treatment group leader, and directs you and your partner to medically evaluate the victims after decontamination.

1. While you are performing a medical evaluation of one of the victims, he tells you that just before he started feeling sick, he noticed a group of men open a briefcase near one of the stores and quickly walk away. What should you do?
   - **A.** Do nothing. The victim is probably hysterical and the information is unreliable.
   - **B.** Stop the evaluation immediately and tell the victim to find a law enforcement officer to whom he should relay this information.
   - **C.** Jot down the victim's name and contact information and make a note of the details of his story, while continuing the evaluation.
   - **D.** Tell the victim to find the incident commander and tell him what was observed.
2. After evaluating several victims, you notice a trend of pinpoint pupils and complaints of mild nausea. From a medical standpoint, these findings raise your index of suspicion for the presence of a nerve agent. Which on-scene reference source would be the quickest and most appropriate to consult, given the circumstances?
   - **A.** The NFPA 704 diamond on the side of the building.
   - **B.** A printed version of the SDS for the nerve agent VX.
   - **C.** There are no readily available resources to consult on scene.
   - **D.** The *Emergency Response Guidebook* if one is available.
3. If you have any questions about the effectiveness of the mass decontamination activities or how the decontamination team is verifying the effectiveness of decontamination, which of the following people would you seek out and question?
   - **A.** The decontamination group leader
   - **B.** The incident commander
   - **C.** The medical group supervisor
   - **D.** The medical branch director

Mission-Specific Competencies: Evidence Preservation and Sampling

Greg Henry/ShutterStock, Inc.

Courtesy of Rob Schnepp

CHAPTER 11

Competencies for Operations Level Responders Assigned Mission-Specific Responsibilities

This chapter shall address competencies found in NFPA 472 for the following operations level responders assigned mission-specific responsibilities at hazardous materials/WMD incidents by the authority having jurisdiction beyond the core competencies at the operations level.

Knowledge Objectives

After studying this chapter, you will be able to:

- Describe the potential criminal violation and identify when a hazardous materials/WMD incident could be a violation of criminal law. (NFPA 6.5.2.1, p. 237–238)
- Identify the various scenarios where hazardous materials/WMD agents could be encountered and identify the appropriate law enforcement agencies that could be involved in an investigation. (NFPA 6.5.2.1(1)(2), p. 237–238)
- Describe the chain of custody and its importance. (NFPA 6.5.3.1(1), p. 242)
- Understand the difference between evidence preservation, public safety sampling, and evidence collection. (NFPA 6.5.4.1, p. 236)
- Describe the hazards, safety procedures, and tactical guidelines for incidents involving each of the following: illicit laboratories, environmental crimes, suspicious packages, and hazardous materials/WMD agents. (NFPA 6.5.3.1(5), p. 237)
- List the factors to consider in selecting sampling equipment, detection devices, packaging, and transport containers. (NFPA 6.5.3.1, p. 240–248)
- Describe sampling options associated with liquid and solid evidence collection. (NFPA 6.5.3.1(1), p. 246–247)
- Describe field screening protocols for collected samples and evidence. (NFPA 6.5.3.1(1), p. 248)
- Describe the types of field screening tools that can be used for screening corrosivity, flammability, oxidation, radioactivity, and VOC. (NFPA 6.5.3.1(7), p. 246)
- Explain the potential negative effects of destructive field screening methods. (NFPA 6.5.3.1(8), p. 246)
- Explain the procedures for maintaining the evidentiary integrity of objects removed from the scene. (NFPA 6.5.3.1(9), p. 246–249)
- Describe the key concepts to be taken into consideration when analyzing, planning, and implementing an evidence preservation and sampling response consistent with the response plan of the AHJ or standard operating procedures. (NFPA 6.5.3.2, 6.5.4.1, 6.5.4.2, p. 237–249)

Skills Objectives

After studying this chapter, you will be able to:

- Demonstrate how to secure the scene and characterize and preserve evidence at the scene. (NFPA 6.5.4.1(1), p. 243–244)
- Document personnel and scene activities related to the incident. (NFPA 6.5.4.1(2), p. 244)
- Determine whether responders are within their authority to collect evidence. (NFPA 6.5.4.1(3), p. 244)
- Describe the procedure to notify the agency with investigative authority. (NFPA 6.5.4.1(4), p. 246)
- Identify potential samples and evidence to be collected. (NFPA 6.5.4.1(6), p. 246)
- Demonstrate procedures to protect samples and evidence from secondary contamination. (NFPA 6.5.4.1(7), p. 246–247)
- Demonstrate correct techniques to collect samples using the equipment provided. (NFPA 6.5.4.1(8), p. 246–247)
- Demonstrate documentation procedures. (NFPA 6.5.4.1(9), p. 247–248)
- Demonstrate sampling protocols. (NFPA 6.5.4.1(10), p. 248)
- Demonstrate field screening protocols for samples and evidence collected. (NFPA 6.5.4.1(11), p. 248)*
- Demonstrate evidence/sample labeling and packaging procedures. (NFPA 6.5.4.1(12), p. 248–249)
- Demonstrate evidence/sample decontamination procedures. (NFPA 6.5.4.1(13), p. 248–249)**
- Demonstrate evidence/sample packaging procedures for evidence/sample transportation. (NFPA 6.5.4.1(14), p. 248–249)
- Explain the chain of custody for evidence/sample preservation. (NFPA 6.5.4.1(15), p. 242)

**See also Mission-Specific Competency: Detection and Monitoring*

*** See also Mission-Specific Competency: Response to Illicit Laboratories and Mission-Specific Competency: Technical Decontamination*

You Are the Responder

Your crew has been dispatched to a multiple vehicle accident on the interstate. Upon arrival, you observe that several passenger vehicles and a tractor trailer have made impact with one another. Smoke and flames can be seen coming from several vehicles. Witnesses advise you that the driver of the tractor trailer fled on foot immediately after the crash occurred. The rear doors to the tractor trailer are open, and you observe numerous unmarked 50-gallon drums in the trailer; an unknown fluid is now leaking from some of the drums. There are no placards on the trailer, and no visible identification of the trucking company name. There is not an SDS sheet or a bill of lading to be found.

1. Which clues do you have to identify the potential hazards? Is this a law enforcement scene, a fire department scene, or both?
2. Which criminal violations of law, if any, could be occurring here?
3. If a crime has occurred, which potential steps could you take to help preserve any evidence within the scene?

Introduction

Responders must maintain an awareness of the potential causation of hazardous materials/WMD events and work to identify and protect any evidence that may be discovered during the response efforts. Regardless of the type of attack, weapon dissemination method, or other factors that might surround an incident with criminal intent, it is imperative that evidence be preserved and collected properly so that the person or persons responsible for the event can be identified, captured, and prosecuted. For this to happen successfully, all responders on scene must be cognizant of how their presence might affect evidence and/or evidence collection. Every responder should be diligent in remembering that his or her actions, observations, and preservations play a vital role in this process **FIGURE 11-1**.

FIGURE 11-1 Even a minute piece of evidence could lead investigators to the cause of the incident.

Evidence refers to all of the information/items gathered and used by an investigator in determining the cause of an incident. When evidence can be used in the legal process to establish a fact or prove a point, it is often referred to as forensic evidence. To be admissible in court, evidence must be gathered and processed under strict procedures. To that end, nonsworn responders are not the ideal candidates to take "evidence samples" in the hopes that law enforcement can use the samples later. It is important to define the mission of sampling prior to making entry. Non–law enforcement personnel typically take samples for identification and less often take samples for eventual prosecution. Based on that fact, coordination with law enforcement is a critical component of sample plan development.

The priority of first responders is to protect lives, and often the requirements of evidence collection can create conflict with this mandate. Public safety responders must often collect samples of materials for the purposes of safety, control zones, patient care, and as part of hazard risk analysis. These samples should not be taken under the auspices of being evidence unless the collectors are sworn law enforcement with the authority to do so; rather, samples should be collected as what they are: public safety samples.

These public safety samples may be used during testing and analysis by public safety personnel for the purposes of classifying or presumptively identifying unknown materials. The disposition of these samples as potential evidence must be the decision of sworn law enforcement based on numerous factors, including availability of additional materials.

RESPONDER TIP

Preplanning efforts have allowed local, state, and federal law enforcement agencies to develop collaborative response plans and a system of shared intelligence and resources. Although plans vary from state to state, a regional response plan for a criminal hazardous materials/weapon of mass destruction (WMD) incident in your area has likely been developed.

Analyzing the Incident

Responders should maintain situational awareness and observe the potential for criminal and terrorist activity involving hazardous materials by monitoring previously documented cases on local and federal intelligence bulletins. Having such awareness will better equip the responder to identify indicators when they are encountered. General indicators that a crime is involved may include anonymous threats leading up to the incident, nearby notes or graffiti claiming responsibility for the incident, or suspicious activity on the scene indicating that the criminals are still present.

Numerous criminal cases have involved explosives, chemicals, biological agents, and radioactive materials being sent illegally in letters and packages. Indicators that a letter or package may be suspicious may include excessive postage (to ensure that the package is not returned to the sender), threatening messages on the exterior of the package, and visible leaks or stains **FIGURE 11-2**.

Although a criminal often uses an illicit laboratory to produce methamphetamine and other drugs, other types of crime may take place at such locations. A laboratory or manufacturing process can be set up just about anywhere to construct explosive devices, manufacture chemical agents, or culture biological agents. Indicators of an illicit laboratory may include efforts taken to conceal the presence of the lab, such as fences, excessive window coverings, or enhanced ventilation and air filtration systems. Alarms and counter-surveillance systems may also be found at such locations **FIGURE 11-3**.

An intentional release or attack using a WMD agent may not always be obvious (see chapter on *Terrorism* for more information on WMD agents). For example, a structure fire could be the unintended secondary effect of an explosive device that was detonated for the purpose of disseminating a chemical agent. The unsuspecting responder should pay careful attention to the symptoms experienced by victims on scene, because they may provide tremendous insight into the types of hazardous materials that may be present. Review the chapter on terrorism for examples of the signs and symptoms associated with various hazardous materials/WMD agents. Other indicators of this type of crime may include the presence of suspicious devices or containers or fragmented pieces of such items. Although fire and explosive incidents are often seen as potential criminal acts, it is also important to consider this possibility at incidents involving leaks and spills. Indicators that legitimate toxic industrial chemicals were released intentionally to cause harm may include opened valves, punctures or cuts in containment vessels, or cut chains, locks, or other safety devices that would normally be in place to contain the hazard.

Environmental crimes include the intentional release or disposal of hazardous materials, their byproducts, and waste into the environment **FIGURE 11-4**. These releases may occur into the air, into the ground, or into natural or human-made water systems. Indicators that such crimes have occurred may include containers (either labeled or unlabeled) that have been

FIGURE 11-2 It is important to check suspicious packages for visible leaks or stains.

FIGURE 11-3 Some other obvious indicators of an illicit laboratory could be the presence of chemical storage cylinders and glass bottles, excessive electrical wiring, laboratory glassware, electric heat sources, and Bunsen burners.

Courtesy of Rob Schnepp

FIGURE 11-4 Crop-dusting equipment can be used as a means to commit environmental crimes.

Courtesy of Tim McCabe/USDA

discarded at the site, staining or odors near street drainage systems, and dead or dying plants, insects, or animals in the nearby area.

Evidence preservation and evidence sampling are also important at many scenes where a crime has not been committed because lawsuits may subsequently be filed by victims to claim or recover damages that resulted from an incident. The evidence recovered is essential to the processing of these claims.

Regardless of whether the hazardous material evidence originated from an illicit laboratory, a suspicious letter or package, an environmental crime, or an intentional release of a hazardous material or WMD agent, responders should assess the need for personal protective equipment (PPE) based on intelligence and warning signs present at the scene. Monitoring and detection equipment should also be used upon arrival at the scene. Selection and use of monitoring and detection equipment may vary based on the hazards identified, but at a minimum will normally include an oxygen-level meter, a combustible-gas indicator, radiological monitoring devices, pH paper, and a photo-ionization detector. These devices are covered in more detail in the chapter *Mission-Specific Competencies: Air Monitoring and Sampling*.

Finally, responders should realize that standard decontamination procedures apply to these types of crime scenes as well. Every piece of equipment and item of evidence that is seized must pass through the decontamination process.

Investigative Jurisdictions

Hazardous materials incidents are often very complex and dynamic situations. It is common for a unified command, consisting of multiple disciplines and jurisdictional agencies, to be established at such events. In fact, law enforcement agencies are often present during even routine hazardous materials incidents to assist with traffic and crowd control operations.

Every hazardous materials–related crime will be different, with the variables including the substance, the manner in which it was disseminated, the effects the substance has on the victims or the environment, and, of course, the intent of the suspects involved. Generally all criminal investigations start at the local level with the authority having jurisdiction (AHJ) for that geographical area. Investigators from those agencies will help determine who ultimately has investigative authority to assume the case. Investigative authority means that the agency has the legal jurisdiction to enforce a local, state, or federal law or regulation and that it is the most appropriate law enforcement organization to ensure the successful investigation and prosecution of such a case. Quite often, in these types of cases, multiagency task forces are formed to maximize the talent, experience, and resources of several agencies. At this type of incident, responders should not be surprised to find themselves working alongside a local police investigator who is working under the direction of a federal agent.

In the United States, any suspicious letter or package that is sent through the postal system is investigated by the Postal Inspection Service. If drugs are involved, depending on the quantity, the package will be investigated either by local or state agencies or by the Drug Enforcement Administration (DEA). An intentional release or attack involving hazardous materials or a WMD would be investigated by the Federal Bureau of Investigation (FBI). Depending on the quantity of hazardous material or the scope of the impact, environmental crimes may be investigated either by local and state agencies or by the Environmental Protection Agency (EPA). If responders have identified a pattern of indicators suggesting that the incident is criminal in nature, it is vitally important that the information be shared with the law enforcement agencies on scene or those having jurisdiction in that geographical area.

The planning stage of the response to a potentially criminal hazardous materials/WMD incident should involve collaboration with the law enforcement AHJ. The investigating law enforcement officers will likely consult with the prosecuting attorney to determine which evidence must be collected. Issues regarding proper search and seizure, search warrants, and witness and suspect interviews should be discussed in the planning stage to ensure that evidence is gathered properly and follows the rules of evidence collection. The first step in the development of the evidence preservation and/or sampling response plan should include identification of a method to adequately secure the crime scene. Such measures may include minimizing the number of responders within the scene to only those still involved in life-saving or hazard mitigation operations, denying access to the scene to anyone not involved in the collection or sampling process, and physically protecting the evidence by placing barriers over or around those items so they cannot be disturbed.

Types of Evidence

For an alleged suspect to be prosecuted successfully, the investigating law enforcement agency and the prosecuting attorney must present a case to a judicial body, such as a judge and jury. Their goal is to re-create the crime scene as they observed it, by presenting evidence in several forms.

Physical evidence consists of items that can be observed, photographed, measured, collected, examined in a laboratory, and presented in court to prove or demonstrate a point **FIGURE 11-5**. Examples of physical evidence include burn patterns on a wall or an empty gasoline can left at the scene of an incident.

FIGURE 11-5 Physical evidence can be observed.

Courtesy of Unified Investigations & Sciences, Inc.

Trace (transfer) evidence consists of a minute quantity of physical evidence that is conveyed from one place to another. For example, a suspect's clothing may contain the residue of the same ignitable liquid found at the scene of a fire, or the cleat of a suspect's shoe could hold traces of the same soil that is found at the crime scene FIGURE 11-6.

Demonstrative evidence is anything that can be used to validate a theory or to show how something could have occurred. For example, to demonstrate how a fire could spread, an investigator may use a computer model of a burned building. To match the impressions found at the scene with a specific tool, an investigator may use a cast to demonstrate the match FIGURE 11-7.

Evidence used in court can be considered either direct or circumstantial. Direct evidence includes facts that can be observed or reported firsthand. Examples include statements made by suspects, victims, or witnesses. Another example is a videotape from a security camera showing a person committing a crime FIGURE 11-8. Direct evidence may also include physical evidence—that is, items that can be collected, photographed, measured, or examined and then presented in court.

Circumstantial evidence is information that can be used to prove a theory based on facts that were observed firsthand. For example, an investigation might show that the gasoline

FIGURE 11-7 A cast of a tool mark.

Courtesy of Sirchie Finger Print Labs, Inc.

FIGURE 11-6 A side-by-side comparison of the color and texture of soil can eliminate a large percentage of samples as not being matches.

© Jones & Bartlett Learning. Photographed by Kimberly Potvin

FIGURE 11-8 A videotape of a person committing a crime is considered to be direct evidence.

© Corbis

from a container found at the scene was used to start the fire. Two different witnesses might testify that the suspect purchased a container of gasoline before the fire and walked away from the fire scene without the container a few minutes before the fire department arrived. Such circumstantial evidence clearly places the suspect at the fire site with an ignition source at the time when the fire started. Investigators must often work with this type of evidence at fire scenes.

Preservation of Evidence

Responders have a responsibility to perform evidence preservation. Evidence preservation is the process of protecting evidence until it can be documented, sampled, and collected appropriately. The preservation of evidence could indicate the cause or point of origin of an incident. Responders who discover an item that could potentially be evidence should leave it in place, make sure that no one interferes with it or the surrounding area, and notify a law enforcement officer or investigator immediately. For example, evidence is frequently found during the salvage and overhaul phases of a fire scene. For this reason, salvage and overhaul should always be performed carefully and, in some cases, can be delayed until an investigator has examined the scene. Do not move debris any more than is absolutely necessary and never discard debris until the investigator gives his or her approval to do so.

Responders at the scene are not in the position to decide whether the evidence they find will be admissible in court and, therefore, is worthy of preservation; that is also the investigator's decision. As a first responder, your responsibility is to make sure that potential evidence is not destroyed or lost. Too much evidence is better than too little, so no piece of potential evidence should ever be considered insignificant.

If you find that evidence could be damaged or destroyed during or after incident operations, cover it with a salvage cover or some other type of protection, such as garbage can, a clean piece of plastic, or some other type of protection that will not contaminate the evidence. Use barrier tape and/or the physical presence of a sworn public safety responder or law enforcement officer to keep others from accidentally walking through or otherwise contaminating or disturbing evidence. Before moving an object to protect it from damage, make sure that witnesses are present, that a sketch of the evidence's location is made, and that a photograph is taken. Sketches do not need to be complicated. They should be simple in nature, and include exact measurements indicating where the evidence is located taken from at least two fixed points. This information will allow investigators to re-create the crime scene more accurately at a later date. Photographs should be taken from different angles to show all sides of an evidentiary item and should include both close-up and wide-angle views of the evidence and its original location. Ideally, all these actions should be taken by trained law enforcement crime scene investigators.

Evidence should not be altered from its original state in any way. To avoid contamination from any other products, investigators use special tools and containers to collect and store evidence.

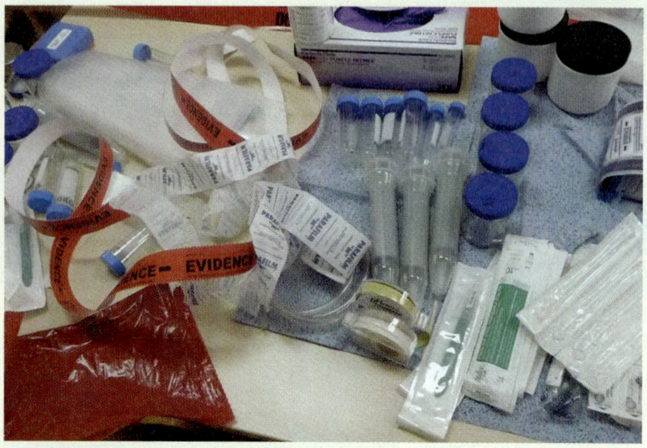

FIGURE 11-9 Examples of sealing tape used for sealing and marking sample containers.

Courtesy of Rob Schnepp

Investigators and collectors of samples must use sterile collection tools and discard used tools after each sample is taken, using new ones for each sampling or collection procedure. This step ensures that material from one piece of evidence will not be unintentionally transferred to another piece. Investigators should also change their gloves each time they take a sample of evidence, and place evidence only in clean containers, sealed with a special type of sealing tape (parafilm is commonly used) and labeled **FIGURE 11-9**. Special collection container consideration must be taken for biological and chemical evidence, based on consultation with appropriate scientists or other subject matter experts. Before entering the scene, investigators often decontaminate their boots to keep from transporting any contamination into the scene.

Common standard operating procedures (SOPs) for collecting and processing evidence generally include the steps given in **SKILL DRILL 11-1**, a common procedure used in fire investigations. Evidence collection for CBRN events may be more complicated. Designated law enforcement officers from the AHJ will be trained on the specifics of evidence collection for these situations.

1. Take photographs of each sample as it is found and collected. If possible, photograph the item exactly as it was found, before it is moved or disturbed. (**STEP 1**)
2. Sketch, mark, and label the location of the sample. Sketch the scene as near to scale as possible. (**STEP 2**)
3. Place the sample in appropriate containers to ensure its safety and prevent contamination. Unused paint cans with lids that automatically seal when closed are the best containers for transporting samples and potential evidence. Clean, unused (certified clean) glass jars sealed with sturdy sealing tape (e.g., parafilm) are appropriate for transporting smaller quantities of materials. Plastic containers and plastic bags should not be used to hold petroleum-based materials because these chemicals may lead to deterioration of the plastic. Paper bags can be used for storage of dry clothing or metal articles, matches, or

papers. Soak up small quantities of liquids with either a cellulose sponge or cotton batting. Protect partially burned paper and ash by placing them between layers of glass (assuming that small sheets or panes of glass are available at the scene). (STEP 3)

4 Label the sample. Anything being transported to the laboratory should include a label with the date, time, location, discoverer's name, and witnesses' names. (STEP 4)

5 Record the time when the evidence/sample was discovered, the location where it was found, and the name of the person who found it. Keep a record of each person who handled the sample. (STEP 5)

6 Keep a constant watch on the sample/evidence until it can be stored in a secure location.

7 Preserve the chain of custody in handling samples/evidence. A broken chain of custody may result in a court ruling that the evidence is inadmissible.

SKILL DRILL 11-1 Collecting Samples and Preserving Evidence

NFPA 6.5.4.1

VIDEO

1 Take photographs of the evidence.

2 Sketch, mark, and label the locations of samples taken and evidence observed.

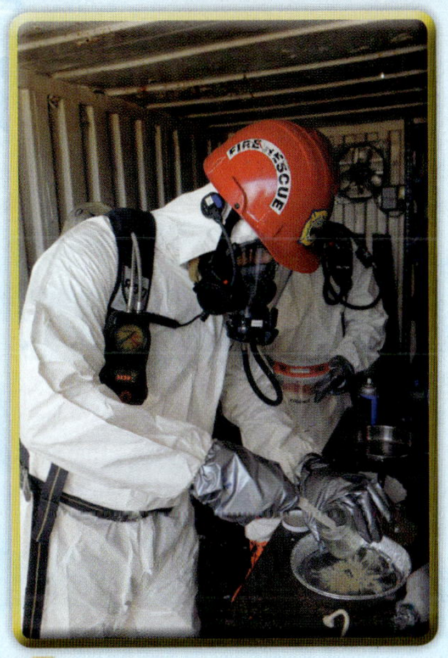

3 Place the sample in the appropriate container.

4 Tag the container.

5 Document your findings.

Chain of Custody

To be admissible in a court of law, physical evidence must be handled according to certain prescribed standards. Because the cause of the incident may not be known when evidence is first collected, all evidence should be handled according to the same procedure. For example, evidence found at an arson fire should be handled in exactly the same way as evidence found at an illicit laboratory.

Chain of custody (also known as *chain of evidence* or *chain of possession*) is a legal term that describes the process of maintaining continuous possession and control of the evidence from the time it is discovered until the time it is presented in court. Every step in the capture, movement, storage, and examination of the evidence must be properly documented. For example, if a gasoline can is found in the debris of a suspected arson fire, documentation must record the name of the person who found the can and where and when the can was found. Photographs should be taken to show the location where the gasoline can was found and its condition. In court, the investigator must be able to show that the gas can presented is the same can that was found at the fire site.

The person who takes initial possession of the evidence must keep it under his or her personal control until that item is turned over to another official. Each successive transfer of possession must be properly recorded. If evidence is stored, documentation must indicate where and when it was placed in storage, whether the storage location was secure, and when the evidence was removed from that location. Often, evidence is maintained in a secured evidence locker to ensure that only authorized personnel have access to it.

Everyone who had possession of the evidence must be able to attest that it has not been contaminated, damaged, or changed. If evidence is examined in a laboratory, the laboratory tests must be documented. The documentation for chain of custody must establish that the evidence was never out of the control of the responsible agency and that no one could have tampered with it. To monitor the path taken by evidence, law enforcement investigators use a chain-of-custody log form. This form allows the investigator to capture the name and identity of each person who handled the evidence and the date and time at which the evidence was transferred from one person to another. Anyone whose name is listed in the chain-of-custody log is subject to testifying in court.

Responders may be the first link in the chain of custody, depending on the policy of the law enforcement agency having jurisdiction. Their responsibility in protecting the integrity of the chain is relatively simple: Report everything to a senior official and disturb nothing needlessly. The individual who finds the evidence should remain with it until the material is turned over to the investigator **FIGURE 11-10**. Remember, as a responder, you could be called as a witness to state that you were the person who discovered a particular piece of evidence at a specific location.

Only one person should be responsible for collecting and taking custody of all evidence at a scene, no matter who discovers it. If someone other than the assigned evidence collector must seize the evidence, that person must photograph, mark, and preserve the evidence properly and turn it over to the evidence collector custodian as soon as possible. The evidence collector must also document all evidence that is collected, including the date, time, and location of discovery; the name of the finder; the known or suspected nature of the evidence; and the evidence number (taken from the evidence tag or label). A log of all photographs taken should be created at the scene as well.

FIGURE 11-10 Evidence should remain where you find it until you can turn it over to an officer or investigator.

Identifying Witnesses

Interviews with witnesses should be conducted by the incident investigator or by a law enforcement officer. If the investigator is not on the scene or does not have the opportunity to interview the witness, the responder should obtain the witness's name, address, and telephone number and give that information to the investigator. A witness who leaves the scene without providing this information could be difficult or impossible to locate later **FIGURE 11-11**.

Responders should pay attention to their surroundings and the situation, make mental notes about their observations, and tell the investigator about any odd or unusual happenings. Information, suspicions, or theories about the incident should be shared only with the investigator and only in private.

Responders should not make any statements of accusation, personal opinion, or probable cause to anyone other than the investigator. Comments that are overheard by the property owner, the occupant, a news reporter, or a bystander can impede the efforts of the investigator to obtain complete and accurate information. A witness who is trying to be helpful might report an overheard comment as a personal observation. In this way, inaccurate information can generate a rumor, which then becomes a theory, which ultimately turns into a reported "fact" as it passes from person to person.

Never make jesting remarks, jokes, or careless, unauthorized, or premature remarks at the scene. For example, statements to news reporters about a suspicious package found at a federal building should be made only by an official spokesperson after the investigator and the incident commander have agreed on their accuracy and validity **FIGURE 11-12**. In this situation, a sufficient reply to any questions concerning the suspicious package is simply, "The incident is under investigation."

FIGURE 11-11 Witnesses can provide invaluable information.

© Louis Brems, The Herald-Dispatch/AP Photos

FIGURE 11-12 Statements to news reporters about an incident's cause should be made only by an official spokesperson after the investigator has confirmed their accuracy.

© Mark C. Ide

Taking Action

Evidence sampling is the process of collecting portions of a hazardous material or WMD for the purposes of field screening, laboratory testing, and, ultimately, criminal prosecution. Most law enforcement agencies in the United States follow the 12-step process recommended by the FBI regarding the collection or sampling of evidence:

1. Preparation.
2. Approach the scene.
3. Secure and protect the scene.
4. Initiate a preliminary survey.
5. Evaluate physical evidence possibilities.
6. Prepare a narrative description.
7. Depict the scene photographically.
8. Prepare a diagram or sketch of the scene.
9. Conduct a detailed search.
10. Record and collect physical evidence.
11. Conduct the final survey.
12. Release the scene.

A sampling team typically consists of three people: a "sampler," who conducts field screening and handles the evidence prior to packaging; an "assistant," who facilitates proper packaging of evidence and ensures that cross-contamination of samples does not occur; and a "documenter," who makes notes during the process and may photograph items at the scene. Once again, the collection process for chemical, biological, radiological, and nuclear (CBRN) incidents will be determined by the AHJ. Sampling teams may include personnel from a variety of public safety agencies depending on the availability and capabilities of the units involved. Although all responders have it within their authority to protect or preserve evidence, taking samples or collecting evidence should be done either by a law enforcement officer or by other nonsworn personnel working under the direction of a law enforcement officer. Each step of this process should be discussed prior to entry and included in the response plan.

Securing, Characterizing, and Preserving the Scene

Once an incident is identified as being criminal in nature and/or potential evidence is observed, the scene must be secured immediately. Securing a crime scene may involve placing hazard tape around the scene's boundaries or assigning personnel to limit access of other nonessential responders and particularly the public.

Early characterization of the scene will also be important. Attempt to identify, at least in broad terms, the type of crime suspected, any major hazards associated with the scene, and the location of significant pieces of evidence. Once suspected evidence is identified, responders should take steps to preserve it. Contaminated clothing is also considered evidence and should be tracked and identified by responders.

Simultaneously with their efforts to secure the crime scene, responders should notify the law enforcement agency with investigative authority, if members of that agency are not already on scene. Communication with the investigative authority is also essential whenever an explosive device is suspected to be involved, to ensure that the appropriate hazardous devices (bomb squad) personnel are notified.

To secure, characterize, and preserve the scene, follow the steps in SKILL DRILL 11-2:

1. Characterize the scene by assessing the number of victims and property damage, if any, and the type (such as a liquid or a solid) and quantity of evidentiary materials on site. (STEP 1)
2. Secure the scene by placing caution or hazard tape so as to limit access to the scene. (STEP 2)
3. Preserve any suspected evidence by protecting it from being disturbed. (STEP 3)

SKILL DRILL 11-2 Securing, Characterizing, and Preserving the Scene

NFPA 6.5.4.1

VIDEO

1 Observe the scene for certain characteristics that could lead to the discovery of evidence.

2 Place caution tape around the scene to limit access.

3 Preserve suspected evidence by protecting it from being disturbed.

Documenting Personnel and Scene Activity

Document the identity and purpose of all personnel who are on scene when you arrive, as well as every person who enters the crime scene once it is so characterized. "Tag in/tag out" records kept by the incident commander are often a good way to capture that information. The evidence preservation plan should allow time for all first responders involved in the incident to make notes of their observations upon arrival or any statements overheard, and to document any evidence that may have been disturbed during life-saving or hazard mitigation operations. Once responders get a break from the initial response, they should sit down and make some written notes on what they did and saw, even if no one asks them for that information right away. They may be asked for it at a later point, once the investigators determine what their involvement was.

Responders' observations during the initial phases of the incident may play a key role in the planning of the evidence sampling and collection missions that follow. Responders are within their authority to collect evidence if such collection is done at the direction of the law enforcement AHJ over the investigation.

VOICES OF EXPERIENCE

Although we all can learn from our mistakes, the margins for error in hazardous materials operations are significantly smaller, and the consequences can be significantly larger than in many other areas of our lives. I learned this as my meandering career path led me into fire investigations and hazardous materials response—two fire service disciplines that complement each other in practical application. Although the experience I have gained from my personal mistakes could fill a book, I want to focus on a short story related to evidence collection and preservation.

A few years ago, I responded to the aftermath of a restaurant fire in a well-known resort community. The fire was out, but the ramifications of the event were just beginning. Fire fighters had been injured, and an entire block of upscale businesses was forced to close their doors. Subsequently, the financial losses climbed into the millions of dollars, and everyone wanted to know who or what was responsible.

The investigation involved identifying the cause of the fire, critiquing the fire department's operations, understanding the applicable codes, initiating hazardous materials abatement, reviewing the city's jurisdictional responsibilities, and considering any criminal culpability, among other things. During the course of the investigation, a literal truckload of evidence was collected.

Each piece of evidence was unique and each item required its own specialized method of identification, documentation, collection, and preservation. Depending on their properties, smaller samples were collected in jars, cans, or boxes. Some evidence items were hand-carried and some were removed by a crane. Subsequently, everything was packaged, logged, and transported to a warehouse in another city.

A small army of investigators prepared their reports, and the litigation process slowly rolled into motion. Ultimately, the case revolved around an improperly constructed and maintained hood system in one of the restaurants. During the painful deposition process, one of the attorneys focused on our failure to sample grease deposits inside the ductwork—grease deposits that were crucial to our fire ignition hypothesis. We had visually inspected them, but no one thought to collect a sample for laboratory analysis. I felt quite stupid for not thinking of it and hoped that I had not damaged our case.

Fortunately, we had collected the suspect ductwork and the "unconfirmed" substance could still be recovered and tested—we had a second chance. Through additional testing, we were able to demonstrate that the substance inside the ductwork was a combustible material. The collected evidence had encompassed the first fuel, the ignition source, and the faulty construction that brought the two of them together. Our fire cause hypothesis had been proven by a thorough investigation and evidence collection.

I have found that I can learn as much from minor mistakes as from the big ones. I believe that it is always profitable for us to translate our mistakes into knowledge, training, and experience for life's long resume.

Kevin Hammons
Metro State University of Denver
Denver, Colorado

To document the activity of personnel at the scene of a hazardous materials incident, follow the steps in SKILL DRILL 11-3 (NFPA 6.5.4.1):

1. Keep a written log or record of the name of each responder who enters the scene.
2. Keep a written log or record of the actions of each responder who enters the scene.

Notifying Investigative Authority and Hazardous Devices Personnel

Upon arrival, if a responder suspects that an explosive device or material is present, immediate notification should be made to the appropriate hazardous devices team having jurisdiction at that scene FIGURE 11-13. Hazardous devices personnel will coordinate their actions with those of other responders to conduct an assessment of the scene and plan a response. Once it is determined or suspected that a crime has occurred, the law enforcement agency having jurisdiction at that scene should be notified immediately. Early law enforcement presence at a crime scene will ensure that evidence identification and collection are conducted properly.

To implement response actions, follow the steps in SKILL DRILL 11-4 (NFPA 6.5.4.1):

1. Identify the presence of suspected explosive materials or devices.
2. Establish a safe perimeter and consider the need for evacuations.
3. Notify the appropriate hazardous devices unit through dispatch or direct contact.
4. Once on scene, meet with hazardous devices personnel to provide information about the type and location of the material or device.

Identifying Samples and Evidence to Be Collected

Once potential evidence is observed, if feasible, efforts should be made to indicate where the evidence is located. Colored cones or tape can be placed at the site of each piece of evidence so that it may be found more easily by the sampling or collection teams. Remember, however, that any markings or identification must be nondestructive and cannot alter the state of the evidence—evidence must remain in the state it was discovered.

To identify samples and evidence to be collected, follow the steps in SKILL DRILL 11-5 (NFPA 6.5.4.1):

1. Note the location and physical characteristics of any suspected hazardous materials evidence.
2. If possible, mark the locations with a unique numbering or color-coded identifier.

Collecting Samples and Preventing Secondary Contamination

At incidents involving large quantities of materials, it may not always be necessary or practical for the evidence sampling and collection team to collect all of the material as evidence.

FIGURE 11-13 Hazardous devices personnel are specially trained responders and should be summoned to the scene when explosives are suspected.

Courtesy of Rob Schnepp

Instead, team members will likely estimate the amount of material, take measurements, photograph or video the material, and then collect several samples to characterize the overall scene.

Sampling techniques may include drawing up liquid samples through a pipette, syringe, or pump; collecting solid material samples by scooping, swabbing, or scraping; or even collecting vapor samples through a vacuum tube. If the substance is unknown, air monitoring and sampling may be conducted for the purpose of field screening. Field screening procedures are initial tests conducted to classify substances and identify immediate hazards to the responders, such as determining if the substance is a corrosive, an oxidizer, a volatile organic compound (VOC), flammable, or a radioactive substance. These methods do not yield conclusive results about the identity of the material, but rather help narrow the list of candidates and provide the evidence sampling team with a piece of information to be considered along with the other dynamics of the situation. Effort should be made to ensure that all field screening methods are nondestructive when possible. If field screening techniques destroy the remaining material, there will be nothing left to submit for laboratory testing or to present as evidence during a criminal proceeding.

RESPONDER TIP

Any photographs, videos, or notes that you take—even when doing so in an unofficial capacity—are subject to the rules of evidence. As a consequence, you may be required to turn them over to law enforcement so that they may be produced during the court proceedings.

During evidence sampling, all members of the evidence collection team must also ensure that secondary contamination does not occur. Any material or substance that is inadvertently added to the collected sample will contaminate it. It is vitally important that samples collected are free of any outside contaminant and that cross-contamination between samples is prevented.

To collect samples using the equipment provided and to prevent secondary contamination of those samples, follow the steps in SKILL DRILL 11-6:

1. When working as a sampler under the direction of law enforcement, conduct field screening and collect a sample. Be careful not to touch any other object or substance prior to placing the sample in the appropriate packaging. (STEP 1)
2. When working as an assistant, hold the sample package or container open so the sampler can place the item safely within it without cross-contaminating the evidence with other samples or objects. (STEP 2)

SKILL DRILL 11-6

Collecting Samples Using Equipment and Preventing Secondary Contamination

NFPA 6.5.4.1

VIDEO

1 The sampler obtains two samples: one sample to use for field screening and another to preserve as evidence.

2 The assistant holds open the evidence package or container so the sampler can place the evidence inside without cross contaminating the evidence.

Documentation of Sampling

The law enforcement officers involved in the investigation will ultimately have to re-create the scene in a court of law. To properly enter evidence into a judicial process, they must be able to show that the sample was collected, documented, and maintained properly during the entire time it was in custody. The documentation procedures for the collection of samples will typically include labeling of those packages to include a unique number, the name of the person who collected the item initially, and the date and time the sample was obtained and packaged.

The documentation of the physical state of the agent (liquid, solid, or vapor), the estimated quantity present, and the size and condition of the container will also help the sampling and collection team determine the most appropriate sampling equipment and tools, as well as the size and quantity of sample containers to bring into the hot zone. For example, telling the sampling and collection team that the agent is in a liquid form may not be enough information. The tools required to take a liquid sample from a puddle on the floor are completely different than those required to take a liquid sample from the bottom of a 50-gallon drum.

To demonstrate the documentation of sampling, follow the steps in SKILL DRILL 11-7:

1. When working as the documenter, photograph and/or videotape the sampling and collection process, if appropriate. (STEP 1)
2. Make notes about the name of the person collecting or sampling the material.
3. Make notes about the physical location and time that the sample was collected.
4. Make notes about the physical state of the agent (liquid, solid, or vapor), the estimated quantity present, and the size and condition of the container. (STEP 2)

SKILL DRILL 11-7 Documentation of Sampling
NFPA 6.5.4.1

VIDEO

1 The documenter photographs and/or videotapes the sampling and collection process.

2 The documenter makes notes about the name of the person collecting the sample, the physical location of the agent, the state of the agent, the quantity present, the time of the sample collection, and the size and condition of the container.

Sampling and Field Screening Protocols

Most laboratories require that field screening procedures include, at a minimum, nondestructive testing to identify the presence of explosive devices, radiological materials, flammable materials, toxic materials, strong oxidizers, or corrosives as a preventive safety measure before the samples are allowed to enter the laboratory. Therefore, the sampling plan must include steps to ensure that such hazards are rendered as safe as practical, that the samples are packaged in appropriate containers suitable for safe transportation, and that those containers are labeled according to the laboratory's requirements.

Labeling, Packaging, and Decontamination

Once the sample is collected, it must be placed in an appropriate container that is suitable for labeling, capable of maintaining the integrity of the sample, and appropriate to contain the specific hazard. For example, biological agents should be packaged only in certified, sterile containers; this restriction is meant to ensure cross-contamination with other microorganisms does not occur. Likewise, corrosive agents should be packaged only in containers that will not react violently or degrade once exposed to the material.

The sampling plan should, therefore, include a procedure that allows the containers to be decontaminated appropriately and transported safely and securely to the laboratory. Technical decontamination must be conducted on each piece of packaged material as it leaves the hot zone and prior to its submission to a laboratory or evidence storage facility. Technical decontamination procedures will vary based on the hazardous material to which the container has been exposed. Procedures to document the chain of custody, including the identity of each person in possession of the sample and the times of transfer, will be part of the plan.

To demonstrate appropriate labeling, packaging, and decontamination procedures, follow the steps in SKILL DRILL 11-8:

1. Seal the initial container with the appropriate sealing tape to ensure that the sample cannot escape from the container and that no foreign substances can enter it. Place your initials on the seal to prevent tampering with the sample. (STEP 1)
2. Place the initial container in a secondary container, and seal this container in the same manner as the first container. Label the secondary container with a unique exhibit number, the name of the person who collected the item, and the location and time the sample was seized. (STEP 2)
3. Place the secondary container in a tertiary container so that the exhibit can pass through the decontamination procedure without being affected.

Another important part of the sampling plan is identifying where the collected sample will go for more conclusive analysis, such as a laboratory, and adhering to the packaging and submission requirements of that laboratory. In the case of criminal incidents, this process will be guided by law enforcement.

SKILL DRILL 11-8 Labeling, Packaging, and Decontamination

NFPA 6.5.4.1

VIDEO

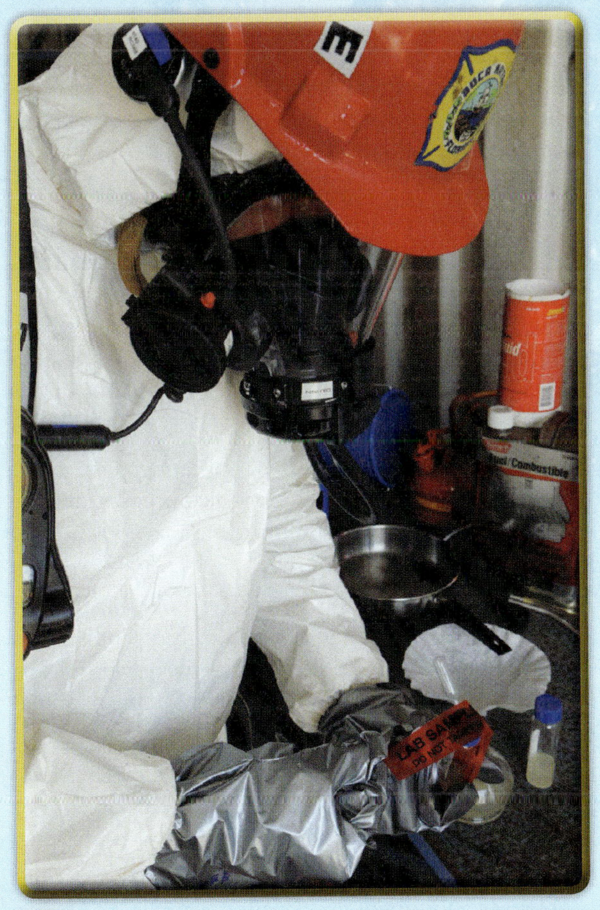

1 Seal the initial container with tape, and place your initials on the tape or seal. This step will prevent tampering.

2 Place the initial container in a secondary container and label it with a unique exhibit number, the name of the person who collected the item, and the location, time, and date of the evidence collection.

Wrap-Up

Chief Concepts

- Evidence refers to all of the information that is gathered and used by an investigator in determining the cause of a hazardous materials/WMD incident.
- Responders have to consider whether the calls they are responding to are a result of criminal activities.
- Responders must notify the appropriate local law enforcement agency as soon as practical when a hazardous materials/WMD incident occurs and work with those agencies to develop an evidence preservation and sampling plan.
- It is imperative that evidence be preserved, sampled, and collected properly so that the person responsible for the event can be identified, captured, and prosecuted or in case victims of the incident file lawsuits to claim or recover damages.
- Physical evidence consists of items that can be observed, photographed, measured, collected, examined in a laboratory, and presented in court to prove or demonstrate a point.
- Trace (or transfer) evidence consists of a minute quantity of physical evidence that is conveyed from one place to another.
- Demonstrative evidence is anything that can be used to validate a theory or to show how something could have occurred.
- Evidence used in court can be considered either direct or circumstantial.
- Evidence preservation is the process of protecting potential evidence until it can be documented, sampled, and collected appropriately.
- Evidence sampling is the process of collecting portions of a hazardous material/WMD for the purposes of field screening, laboratory testing, and, ultimately, criminal prosecution.
- Public safety sampling is the process of collecting portions of a hazardous material/WMD for the purpose of determining the threat to the public and responders.
- Document the identity and purpose of all personnel who are on scene when you arrive.
- If a responder suspects that an explosive device or material is present, the appropriate hazardous devices team should be notified immediately.
- Once the sample or actual evidence is collected, it must be placed in an appropriate container suitable for labeling, decontaminated, and transported safely and securely to the laboratory.

Hot Terms

Chain of custody A record documenting the identities of any personnel who handled the evidence, the date and time that contact occurred or the evidence was transferred from one person to another, and the reason for doing so.

Circumstantial evidence Information that can be used to prove a theory based on facts that were observed firsthand.

Contaminated The process of transferring a hazardous material from its source to people, animals, the environment, or equipment, all of which may act as carriers for the material.

Demonstrative evidence Materials used to demonstrate a theory or explain an event.

Direct evidence Statements made by suspects, victims, and witnesses.

Drug Enforcement Administration (DEA) Established in 1973, the federal agency charged with enforcing controlled substances laws and regulations in the United States.

Environmental Protection Agency (EPA) Established in 1970, the U.S. federal agency charged with ensuring safe manufacturing, use, transportation, and disposal of hazardous substances.

Evidence Information that is gathered and used by an investigator in determining the cause of an incident. When evidence is used in or suitable to courts to answer questions of interest, it is referred to as forensic evidence.

Evidence preservation The process of protecting potential evidence until it can be documented, sampled, and collected appropriately.

Evidence sampling The process of collecting portions of a hazardous material/WMD for the purposes of field screening, laboratory testing, and, ultimately, criminal prosecution.

Federal Bureau of Investigation (FBI) Established in 1908, the federal agency charged with defending and protecting the United States against terrorist and foreign intelligence threats and with enforcing criminal laws including acts of terrorism.

Hazardous devices personnel Personnel trained to detect, identify, evaluate, render safe, recover, and dispose of unexploded explosive devices.

Illicit laboratory Any unlicensed or illegal structure, vehicle, facility, or physical location that may be used to manufacture, process, culture, or synthesize an illegal drug, hazardous material/WMD device, or agent.

Investigative authority The agency that has the legal jurisdiction to enforce a local, state, or federal law or regulation. It is the most appropriate law enforcement organization to ensure the successful investigation and prosecution of a case.

Methamphetamine A psychostimulant drug manufactured illegally in illicit laboratories.

Physical evidence Items that can be observed, photographed, measured, collected, examined in a laboratory, and presented in court to prove or demonstrate a point.

Postal Inspection Service Established in 1737, the federal agency charged with securing and managing the U.S. mail system.

Trace (transfer) evidence A minute quantity of physical evidence that is conveyed from one place to another.

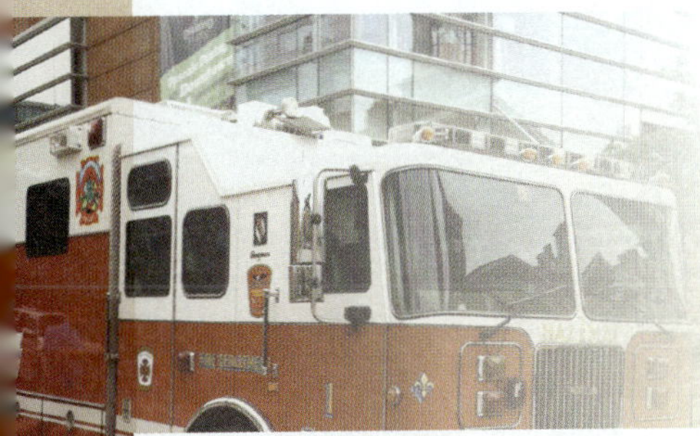

RESPONDER *in action*

Your crew has been dispatched to a county government building in your jurisdiction. The caller advised that numerous employees are complaining of a set of common symptoms: nausea, headaches, vomiting, and dizziness. Your crew arrives and establishes a position at an upwind, uphill location a safe distance from the hazard. You see many people outside the main entrance and collecting at a nearby parking lot. Some victims are lying motionless on the ground. The scene is chaotic. An unaffected employee from the building approaches your lieutenant and advises that the building managers have been receiving threats from a known environmental activist group over the past few weeks.

1. Based on the information that you have, whom should you notify immediately?
   - **A.** The building security manager
   - **B.** The local law enforcement agency having jurisdiction in that area
   - **C.** The Environmental Protection Agency
   - **D.** The Federal Bureau of Investigation
2. Assume that victim rescue, evacuation, and emergency decontamination operations are now underway. The appropriate investigative authority has advised you that they are responding. What should your next priority be until representatives of that agency arrive?
   - **A.** Begin interviewing potential witnesses
   - **B.** Secure the crime scene
   - **C.** Collect samples of evidence to show investigators when they arrive
   - **D.** Do nothing until the authority provides further advice
3. Your lieutenant has assigned you to keep a log of personnel names at the scene to provide to the investigators upon their arrival. Which individuals should you include in that log?
   - **A.** The fire fighters from your agency
   - **B.** Any person who was at the scene upon your arrival
   - **C.** Any person who entered the scene after your arrival
   - **D.** Both B and C
4. You are now assigned to assist with the decontamination process for the victims of the incident. During that process, the victims' clothing and possessions are removed and placed in individual bags and labeled. The law enforcement investigators have asked for a chain-of-custody record for those bags. Who should be listed on that record?
   - **A.** The person labeling the bag
   - **B.** The victim
   - **C.** The last person in possession of the bag
   - **D.** All of the above

Mission-Specific Competencies: Product Control

Greg Henry/ShutterStock, Inc.

© Jones & Bartlett Learning. Photographed by Glen E. Ellman

CHAPTER 12

Competencies for Operations Level Responders Assigned Mission-Specific Responsibilities

This chapter shall address competencies found in NFPA 472 for the following operations level responders assigned mission-specific responsibilities at hazardous materials/WMD incidents by the authority having jurisdiction beyond the core competencies at the operations level.

Knowledge Objectives

After studying this chapter, you will be able to:

- Describe and identify the control options available to operations level responders. (NFPA 6.6.3.1, p. 254–255)
- Describe and identify the control options available for flammable liquid and flammable gas incidents. (NFPA 6.6.3.1, p. 254–255)
- Describe the purpose, equipment, and precautions associated with control options. (NFPA 6.6.3.1, 6.6.3.2, p. 255–267)
- Describe the applicability and characteristics of aqueous film-forming foam, alcohol-resistant concentrates, fluoroprotein foams, protein foams, and high-expansion foams. (NFPA 6.6.4.1(1)(2), p. 265)
- Identify the location and describe the use of emergency remote shut-off devices on MC306/DOT406, MC307/DOT407, and MC331 cargo tanks containing flammable liquids or gases. (NFPA 6.6.4.1(4)(5), p. 262–263)
- Describe the recovery phase, and the transition from emergency to clean-up. (NFPA 6.6.4.2, p. 267–268)

Skills Objectives

After studying this chapter, you will be able to:

- Perform the following control activities:
  - Absorption (NFPA 6.6.4.1(3)(a), p. 255–257)
  - Adsorption (NFPA 6.6.4.1(3)(b), p. 255–257)
  - Damming (NFPA 6.6.4.1(3)(c), p. 258–259)
  - Diking (NFPA 6.6.4.1(3)(d), p. 259–260)
  - Dilution (NFPA 6.6.4.1(3)(e), p. 259–261)
  - Diversion (NFPA 6.6.4.1(3)(f), p. 260–261)
  - Retention (NFPA 6.6.4.1(3)(g), p. 260–262)
  - Remote valve shut-off (NFPA 6.6.4.1(3)(h), p. 262–263)
  - Vapor dispersion (NFPA 6.6.4.1(3)(i), p. 263–264)
  - Vapor suppression (NFPA 6.6.4.1(3)(j), p. 263–265)
  - Perform the following methods of applying foam: (NFPA 6.6.4.1)
    - Rain-down method (p. 266)
    - Roll-in method (p. 266–267)
    - Bounce-off method (p. 267)

You Are the Responder

Your engine company has been dispatched to assist a rescue company on the scene of a rolled over diesel truck pulling a flatbed trailer. The trailer was carrying several lengths of large steel pipe, which are now lying on the road. The fuel tanks on the diesel truck are leaking. Upon arrival, your officer receives an assignment to protect a curbside drain located downslope from the leaking tanker. He assigns the task to you and two other fire fighters. There is a slight breeze blowing away from you, back toward the leaking tanker. The product is confirmed to be diesel fuel. The concrete roadway ahead of the spill is completely dry. You have access to three plastic shovels and several bags of loose absorbent.

1. Would full structural fire fighter's turnout gear and SCBA offer adequate protection in this situation?
2. Based on your level of training (i.e., operations core competencies and all of the mission-specific competencies), would you be qualified to perform this task?
3. Describe the steps you would need to take, including obtaining information about diesel fuel, to complete the assignment.

Introduction

It is not uncommon for a hazardous materials incident to require some form of product control. Scenarios like the one described in the chapter-opening vignette are relatively common, as are incidents involving leaking drums, spills into waterways, and other types of releases involving flammable liquids or gases. Many of these incidents require emergency responders to intervene by shutting off valves or applying loose absorbents or various kinds of foam to mitigate the situation.

In most cases, the best course of action is to confine the problem to the smallest area possible. Confinement is the process of attempting to keep the hazardous material within the immediate area of the release. This goal can usually be accomplished by damming or diking a material, or by suppressing vapor with an appropriate type of foam. Containment, by contrast, refers to actions that stop the hazardous material from leaking or escaping its container. Examples of containment include patching or plugging a breached container, or righting an overturned container to stop a slow leak. Sometimes it is necessary to first stop the leak (contain), and then confine whatever has been released. In other cases, it may be impossible to stop the leak and all that can be done is to confine the hazardous material as much as possible.

There is no clear-cut sequence or single best way to handle every problem involving product control issues. You must size up the situation as you would any other problem and employ the best methods available to handle the incident safely. When considering a control option, certain factors must be evaluated, such as the maximum quantity of material that can be released and the likely duration of the incident without intervention.

It is vital in these situations, as with any other type of release, for responders to understand the nature of the release, wear the proper level of personal protective equipment (PPE), and have all the tools and equipment—including air monitoring and detection equipment—available to accomplish the task at hand. In many cases, product control measures bring responders in close proximity to the released product. To work safely in a contaminated (or potentially contaminated) atmosphere, responders must stay informed about their working environment. Readings from monitoring and detection devices should always be interpreted in the context of the specific event. The chapter *Mission-Specific Competencies: Air Monitoring and Sampling* provides more information on detection and monitoring. Additionally, all emergency responders should be familiar with the policies and procedures of the local jurisdiction to ensure that they employ a consistent approach to the selected control option.

RESPONDER TIP

Some of the mission-specific competencies described in this section are taken from competencies required of hazardous materials technicians. That does not mean, however, that operations level responders, with a mission-specific competency in product control or any other mission-specific competency, are replacements for hazardous materials technicians. Operations level responders should operate under the guidance of a hazardous materials technician, an allied professional, or standard operating procedures.

Control Options

The most challenging aspect of mitigating a hazardous materials emergency is arriving at a solution that can be employed quickly and safely, while minimizing the potential negative effects on people, property, and the environment. If that sounds like a tall order, it is. Handling a hazardous materials incident is a bit like a chess game: One move sets up and (either positively or negatively) influences the next move or series of moves. As mentioned earlier, it is vital to use a risk-based thought process when choosing to employ any control

FIGURE 12-1 Bales of hay procured from a local feed store. In this instance, the bales were used to absorb some of the downstream flow of a gasoline spill (floating on water) from an MC306/DOT406 cargo trailer.

Courtesy of Rob Schnepp

FIGURE 12-2 Highly volatile flammable liquids like methyl alcohol may be left to evaporate naturally without taking offensive action to clean them up.

Courtesy of Rob Schnepp

option. Remember—a solid response objective should be well thought out and realistic, taking into account both the positive and negative effects of the actions taken. Think creatively and know what resources are available in your jurisdiction—sometimes you have to think outside the box **FIGURE 12-1**.

Sometimes, no action is the safest course of action. Unfortunate as it may seem, occasionally the situation is so extreme, or the responders cannot be properly protected, or the product will evaporate quickly, cool down, or solidify, such that it may be prudent to create a safe perimeter and let the problem stabilize on its own. Choosing not to intervene may be challenging to a culture of aggressive fire fighters, but there are times when taking a hands-off approach may be the correct course of action. As an example, a fire in a gasoline tanker may have to burn itself out if not enough foam is available to fight the fire effectively.

RESPONDER TIP

An accurate size-up must be accomplished prior to taking any action! You must understand the nature of the problem and the ramifications of the chosen control option.

Similarly, when incompatible chemicals are mixed, it may be wise to let the reaction run its course before intervening. Imagine a spill of a highly volatile flammable liquid like isopropyl alcohol on a hot summer day. It would likely evaporate before you could make an entry and attempt to clean it up. Think about the chemical and physical properties of the spilled material and how those properties may be used to your advantage regarding confinement or containment **FIGURE 12-2**. This is only one example of a situation when no action is a safer way to handle the problem.

As emergency responders approach a hazardous materials incident, they should be aware of natural control points—areas in the terrain or places in a structure where materials might be contained or confined. Natural barriers that might be used include doors to a room, doors to a building, designated areas for secondary containment, and curbed areas of roadways. A number of control measures may be available if responders are thinking creatively and are aware of their surroundings.

Absorption and Adsorption

Absorption is the process whereby a spongy material (e.g., soil or loose absorbents such as vermiculite, clay, or peat moss) or specially designed spill pads are used to soak up a liquid hazardous material **FIGURE 12-3**. The contaminated mixture of absorbent material and chemicals are then collected, and all of the materials are disposed of together. Most states have laws and regulations that dictate how to dispose of used absorbent materials.

Absorption minimizes the spread of liquid spills, but is effective only on flat surfaces. Although absorbent materials such as soil and sawdust are inexpensive and readily available, they become hazardous materials themselves once they come in contact with the spilled liquid; therefore, they must be disposed of properly. The process of absorption does not change the chemical properties of the involved substance. It is only a method to collect the substance for subsequent containment.

FIGURE 12-3 Spill pads are often used to soak up a liquid hazardous material.

FIGURE 12-5 A spill boom deployed by boats to contain a fuel spill in a harbor.

Courtesy of Rob Schnepp

FIGURE 12-4 Gasoline floating on water in a creek.

Courtesy of Rob Schnepp

FIGURE 12-6 A spill boom can be used to confine a liquid.

The technique of absorption may prove challenging for personnel because it requires them to be in close proximity to the spilled material. It also involves the addition of material to a spilled product, which adds volume to the spill. The absorbent material may also react with certain hazardous substances, so it is important to determine whether an absorbent substance is compatible with the hazardous material. Hydrofluoric acid, for example, is not compatible with the silica-based absorbents commonly found in some types of spill booms.

Some absorbent materials repel water while still absorbing a spilled liquid that does not mix with water—an effective property under the right circumstances (such as when the goal is to mitigate an oil spill on a body of water) FIGURE 12-4 . In these situations, spill booms are deployed as floating barriers used to obstruct passage. These booms are typically constructed of an outer mesh covering and filled with a material such as polypropylene or silica. Many different sizes and types of spill booms are on the market today. Some will float on water, whereas others are designated for use only on dry land FIGURE 12-5 and FIGURE 12-6. It is up to your authority having jurisdiction (AHJ) to decide which types of booms are appropriate for your jurisdiction.

Another example where the technique of absorption may prove challenging for personnel is an incident involving

Responder Safety Tip

It is important to determine whether an absorbent substance is compatible with the hazardous material before it is used.

nitric acid at a concentration higher than 72%. In this situation, the nitric acid becomes such an aggressive oxidizer that using an organic-based absorbent may result in the acid igniting the absorbent material. To reiterate the point made earlier, not all absorbent materials are created equal, and using the wrong one might end up unduly complicating your situation.

The opposite of absorption is adsorption. In adsorption, the contaminant *adheres* to the surface of an added material—such as silica or activated carbon—rather than combining with it (as in absorption). In some cases, the process of adsorption can generate heat—a key point you should consider when use of adsorbent materials is proposed. The concept of adsorption is analogous to the mechanism underlying Velcro: Adsorbents are "sticky" and grab onto whatever substance they are designed to be used for. A common adsorbent material is activated carbon, which is found in the filter cartridges used for air-purifying respirators.

To use absorption/adsorption to manage a hazardous materials incident, follow the steps in SKILL DRILL 12-1:

1. Collect the basic materials.
   A. Absorbents
   B. Adsorbents
   C. Absorbent pads
   D. Absorbent booms
2. Decide which absorbent/adsorbent is best suited for use with the spilled product and the physical characteristics of the spill (i.e., a spill in a lake, pond, or creek versus on wet or dry ground).
3. Assess the location of the spill and stay clear of any spilled product. (STEP 1)
4. Use detection and monitoring devices to determine whether airborne contamination is present. Consult reference sources for the physical and chemical properties of the spilled material.
5. Apply the appropriate absorbent/adsorbent material to control the spilled material. (STEP 2)
6. Maintain control of the absorbent/adsorbent materials and take appropriate steps for their disposal. (STEP 3)

SKILL DRILL 12-1

Using Absorption/Adsorption to Manage a Hazardous Materials Incident

NFPA 6.6.4.1

VIDEO

1 Decide which material is best suited for use with the spilled product. Assess the location of the spill and stay clear of any spilled product.

2 Apply the appropriate material to control the spilled product.

3 Maintain control of the absorbent/adsorbent materials and take appropriate steps for their disposal.

Damming

Damming is a containment technique that is used when liquid is flowing in a natural channel or depression and its progress can be stopped by blocking the channel. Three kinds of dams are typically used in such circumstances: a complete dam, an overflow dam, or an underflow dam.

A *complete dam* is placed across a small stream or ditch to completely stop the flow of materials through the channel. This type of dam is used only in areas where the stream or ditch is basically dry and the amount of material that needs to be controlled is relatively small.

For hazardous materials spills in streams or ditches with a continuous flow of water, an overflow or underflow dam must be constructed. An *overflow dam* is used to contain materials heavier than water (specific gravity > 1). It is constructed by building a dam base up to a level that holds back the flow of water **FIGURE 12-7**. Polyvinyl chloride (PVC) pipe or hard suction hose is installed at a slight angle to allow the water to flow "over" the released liquid, thereby trapping the heavier material at a low level at the base of the dam.

An *underflow dam* is used to contain materials lighter than water (specific gravity < 1) and is basically constructed in the opposite manner of the overflow dam. The piping on the underflow dam is installed near the bottom of the dam so the water flows "under" the dam, thereby allowing the materials floating on the water to accumulate at the top of the dam area **FIGURE 12-8**. It is critical to have sufficient pipes and hoses to allow enough water to flow past the dam without overflowing over the top of the dam.

To construct an overflow dam, follow the steps in **SKILL DRILL 12-2**:

1. Based on the spilled material having a specific gravity greater than 1, determine the need and the best location for an overflow dam.
2. Build a dam with sandbags or other available materials. (**STEP 1**)
3. Install two to three lengths of 3- to 4-inch plastic pipe horizontally on top of the dam. Add more sandbags or other material on top of the dam, if available.
4. Complete the dam installation and ensure that the piping will allow the proper flow of water without allowing the heavier-than-water material to pass through the pipes. (**STEP 2**)

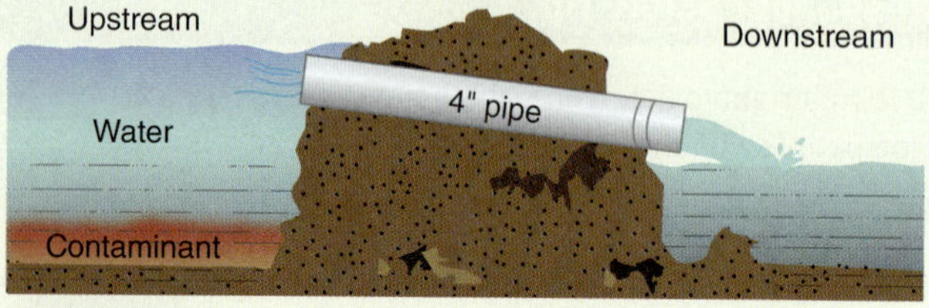

FIGURE 12-7 An overflow dam is used to contain materials that are heavier than water.

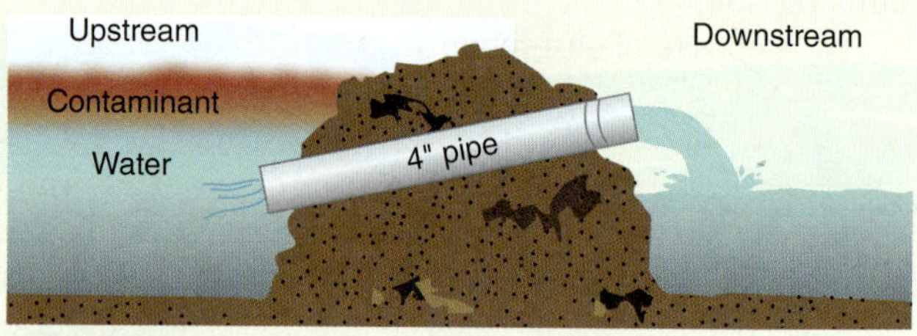

FIGURE 12-8 An underflow dam is used to contain materials that are lighter than water.

SKILL DRILL 12-2 Constructing an Overflow Dam

NFPA 6.6.4.1

VIDEO

1 Determine the need for and location of an overflow dam. Build a dam with sandbags or other available materials.

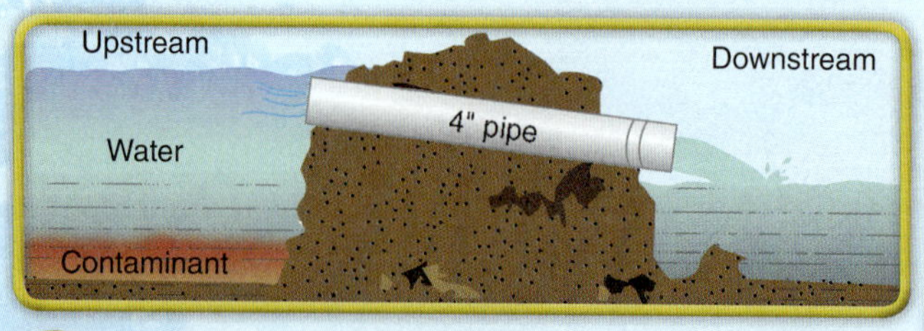

2 Install appropriate number of 3- to 4-inch plastic pipes horizontally on top of the dam and add more sandbags on top of the dam. Complete the dam installation and ensure that the piping will allow the proper flow of water without allowing the heavier-than-water material to pass through the pipes.

To construct an underflow dam, follow the steps in SKILL DRILL 12-3:

1. Based on the spilled material having a specific gravity less than 1, determine the need and best location for an underflow dam.
2. Build a dam with sandbags or other available materials. (STEP 1)
3. Install two to three lengths of 3- to 4-inch plastic pipe at a 20- to 30-degree angle on top of the dam. Add more sandbags or other material on top of the dam if available.
4. Complete the dam installation and ensure that the size will allow the proper flow of water underneath the lighter-than-water liquid. (STEP 2)

SKILL DRILL 12-3 Constructing an Underflow Dam

NFPA 6.6.4.1

VIDEO

1 Determine the need for and location of an underflow dam. Build a dam with sandbags or other available materials.

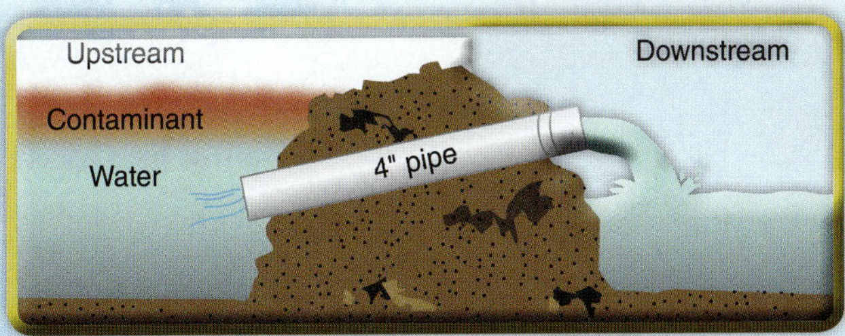

2 Install appropriate number of 3- to 4-inch plastic pipes at 20- to 30-degree angle on top of the dam and add more sandbags on top of the dam. Complete the dam installation and ensure that the piping will allow the proper flow.

Diking

Diking is the placement of a selected material such as sand, dirt, loose absorbent, or concrete so as to form a barrier that will keep a hazardous material (in liquid form) from entering an unwanted area or to hold the material in a specific location. Before contemplating construction of a dike, you must confirm that the material used to control the hazard will not react adversely with the spilled material.

To construct a dike, follow the steps in SKILL DRILL 12-4:

1. Based on the spilled material and the topography conditions, determine the best location for the dike.
2. Dig a depression (if possible) 6 to 8 inches deep.
3. Ensure that plastic will not react adversely with the spilled chemical. Use plastic to line the bottom of the depression, and allow sufficient plastic to cover the dike wall. (STEP 1)
4. Build a short wall with sandbags or other available materials. (STEP 2)
5. Install plastic to the top of the dike and secure it with additional sandbags.
6. Complete the dike installation and ensure that its size will contain the spilled product. It may be necessary to construct a series of dikes if there is a concern about the amount of product being released. A wise incident commander will consider backing up the original structure to ensure that the product will be controlled effectively. (STEP 3)

Dilution

Dilution is the addition of water or another substance to weaken the strength or concentration of a hazardous material (typically a corrosive). Dilution can be used only when the identity and properties of the hazardous material are known with certainty. One concern with dilution is that water applied to dilute a hazardous material may simply increase the total volume; if the volume increases too much, it may overwhelm the containment measures implemented

SKILL DRILL 12-4 Constructing a Dike

NFPA 6.6.4.1

VIDEO

1. Determine the best location for the dike. If necessary, dig a depression in the ground 6 to 8 inches deep. Ensure that plastic will not react adversely with the spilled chemical. Use plastic to line the bottom of the depression and allow for sufficient plastic to cover the dike wall.

2. Build a short wall with sandbags or other available materials.

3. Complete the dike installation and ensure that its size will contain the spilled product.

by responders. For example, if 1 gallon of water were added to 3 gallons of spilled hydrochloric acid (pH = 3), the result of this action would be the creation of 4 gallons of hydrochloric acid (pH = 3); it takes a *tremendous* amount of water to effectively dilute such a low-pH acid. Dilution should be used with extreme caution and only on the advice of those knowledgeable about the nature of the chemicals involved in the incident.

To use dilution to manage a hazardous materials incident, follow the steps in SKILL DRILL 12-5:

1. Based on the spilled material's characteristics and the size of the spill, determine the viability of a dilution operation.
2. Obtain guidance from a hazardous materials technician, technical specialist such as a chemist, or other qualified allied professional.
3. Ensure that the water used will not overflow and affect other product control activities.
4. Add small amounts of water from a distance to dilute the product. If applicable, check the pH periodically during this operation.
5. Contact the hazardous materials technician (or other qualified professional) if any additional issues arise. (STEP 1)

Diversion

Diversion techniques in general are intended to redirect the flow of a liquid away from an endangered area to an area where it will have less impact. In many cases, however, responders do not need to build elaborate dikes to divert the flow of a spilled liquid. Instead, existing barriers—such as curbs or the curvature of the roadway—may be effective ways to divert liquids away from storm drains or other unwanted destinations. Additionally, dirt berms, spill booms, and plastic tarps filled with sand, dirt, or clay may provide a rapidly employed diversion mechanism. These diversion methods are not as "permanent" as a dike, and they can be constructed fairly quickly. Remember—the intent of diversion is to protect sensitive environmental areas or any other areas that would be unsuitable for the spilled liquid.

To employ a diversion technique to manage a spilled liquid, follow the steps in SKILL DRILL 12-6:

1. Based on the spilled material and the topography conditions, determine the best location for the diversion.
2. Use sandbags or other materials to divert the product flow to a more desirable location. This location may not be perfect, but may be more advantageous to the overall control efforts.
3. Stay clear of the product flow.
4. Monitor the diversion channel to ensure the integrity of the system. (STEP 1)

SKILL DRILL 12-5 Using Dilution to Manage a Hazardous Materials Incident

NFPA 6.6.4.1

VIDEO

1. Determine the viability of a dilution operation. Obtain guidance from a hazardous materials technician, specialist, or professional. Ensure that the water used will not overflow and affect other product-control activities. Add small amounts of water from a distance to dilute the product. Contact the hazardous materials technician, specialist, or other qualified professional if any additional issues arise.

SKILL DRILL 12-6 Constructing a Diversion

NFPA 6.6.4.1

VIDEO

1. Determine the best location for the diversion. Use sandbags or other materials to divert the product flow to an area with fewer hazards. Stay clear of the product flow. Monitor the diversion channel to ensure the integrity of the system.

Retention

Retention is the process of creating a defined area to hold hazardous materials. For instance, it may involve digging a depression in the ground and allowing material to pool in the depression. The material is then held there until a clean-up contractor can recover it. In some cases, some sort of diversion technique may be required to guide the spilled liquid into the retention basin. As an example, a city's public works department or an independent contractor might use a backhoe to create a retention area for runoff at a safe downhill distance from the release.

To use retention, follow the steps in SKILL DRILL 12-7:

1. Based on the spilled material and the topography conditions, determine the best location for the retention system.
2. Dig a deep depression (by hand or with heavy equipment such as a tractor or backhoe) to serve as a retention basin.
3. Use plastic to line the bottom of the depression (if time permits and the spilled liquid is compatible with plastic).
4. Use sandbags or other materials to hold the plastic in place.
5. Stay clear of the product flow.
6. Monitor the retention system to ensure its integrity. (STEP 1)

SKILL DRILL 12-7 **Using Retention to Manage a Hazardous Materials Incident**

NFPA 6.6.4.1

VIDEO

1 Determine the best location for the retention system. Dig a depression in the ground to serve as a retention area. Use plastic to line the bottom of the depression. Use sandbags or other materials to hold the plastic in place. Stay clear of the product flow. Monitor the retention system to ensure its integrity.

Remote Valve Shut-off

A protective action that should always be considered—especially with transportation emergencies or incidents at fixed facilities—is the identification and isolation of the remote valve shut-off. Many chemical processes, and many piped systems that carry chemicals, provide a way to remotely shut down a system or isolate a leaking fitting or valve. This action may prove to be a much safer way to mitigate the problem FIGURE 12-9.

Fixed ammonia systems provide a good example of the effectiveness of using remote valve shut-offs. In the event of a large-scale ammonia release, most current systems have a main dump valve that will immediately redirect anhydrous ammonia into a water deluge system or diffuser. These systems are designed to knock down the vapors and collect the resulting liquid in a designated holding system.

Many types of cargo tanks have emergency remote shut-off valves. MC306/DOT406 cargo tanks, for example, carry flammable and combustible liquids and Class B poisons. These single-shell aluminum tanks may hold as much as 9200 gallons of product at atmospheric pressure. The tanks have an oval/elliptical cross-section and overturn protection that serves to protect the top-mounted fittings in the event of a rollover. In the event of fire, a fusible link is designed to melt at a predetermined temperature (typically around 250°F [121°C]); its collapse causes the closure of the internal

FIGURE 12-9 When available, consider using remote shut-off valves to mitigate the problem.

product discharge valve. A baffle system within the cargo tank compartment reduces the surge of the liquid contents. Remote shut-off valves are located at the front of the cargo tank on the driver's side or at the rear of the cargo tank on the passenger side **FIGURE 12-10**.

MC-307/DOT-407 cargo tanks carry chemicals that are transported at low pressure, such as flammable and combustible liquids as well as mild corrosives and poisons. These vehicles may carry as much as 7000 gallons of product and are outfitted with many safety features similar to those found on the MC306/DOT406 cargo tanks **FIGURE 12-11**.

MC-331 cargo tanks carry compressed liquefied gases such as anhydrous ammonia, propane, butane, and liquefied petroleum gas (LPG). The tanks, which are not insulated, have rounded ends, baffles, and a circular cross-section. MC-331 cargo tanks have a carrying capacity between 2500 and 11,500 gallons. These types of cargo tanks have remote shut-off valves located at both ends of the tank, internal shut-off valves, a rotary gauge depicting product pressure, and top-mounted vents **FIGURE 12-12**.

FIGURE 12-11 The remote shut-off valve is typically found near the front of the cab, adjacent to the driver's door, or at the rear of an MC307/DOT407 cargo tank.

Courtesy of Glen Rudner

Responder Safety Tip

Participating in product control operations may require a great deal of physical energy and mental concentration. Remember to take rehabilitation breaks throughout the incident to minimize the effects of heat stress and fatigue. Refer to NFPA 473, *Standard for Competencies for EMS Personnel Responding to Hazardous Materials/WMD Incidents*, for more information on heat stress.

FIGURE 12-12 The MC331 cargo tank has remote shut-off valves at both ends of the tank, internal shut-off valves, a rotary gauge depicting product pressure, and two top-mounted vents.

Courtesy of Rob Schnepp

FIGURE 12-10 The remote shut-off valve is typically found near the front of the cab, adjacent to the driver's door, or at the rear of an MC306/DOT406 cargo tank.

Courtesy of Glen Rudner

Vapor Dispersion and Suppression

When a hazardous material produces a vapor that collects in an area or increases in concentration, vapor dispersion or suppression measures may be required.

Vapor dispersion is the process of lowering the concentration of vapors by spreading the vapors out. Vapors can be dispersed with hose streams set on fog patterns, large displacement fans (being mindful of the fan itself becoming a source of ignition), or other types of mechanical ventilation found in fixed hazardous materials handling systems. Before dispersing vapors, consider the consequences of this action. If the vapors are highly flammable, an attempt to disperse them may ignite the vapors. The dispersed vapors can also contaminate areas outside the hot zone. Vapor dispersion with fans or a fog stream should be attempted only after the hazardous material is safely identified and in weather conditions that will promote the diversion of vapors away from any populated areas. Conditions must be constantly monitored during the entire operation.

Vapor suppression is the process of controlling fumes or vapors that are given off by certain materials, particularly flammable liquids, in an attempt to prevent their ignition. It is accomplished by covering the hazardous material with foam or some other material, or by reducing the temperature of the material. For example, gasoline gives off vapors that present a danger because they are flammable. To control these vapors, a blanket of firefighting or vapor-suppressing foam is layered on the surface of the liquid.

Not all vapor-suppressing foams are appropriate for all types of applications. Much like the concept of choosing the right PPE to protect you from a hazardous substance, responders must choose a type of foam that is designed to work in certain situations. For example, Class A foam is typically used to fight fires involving ordinary combustible materials such as wood and paper. In most situations involving hazardous materials or weapons of mass destruction (WMD) agents, Class A foam is *not* the most appropriate type of foam to use. Class B foam, by contrast, is used to fight fires and suppress vapors involving flammable and combustible liquids such as gasoline and diesel fuel. This type of foam seals the surface of a spilled liquid and prevents the vapors from escaping, thereby reducing the danger of vapor ignition. If the vapors are burning, a gently applied foam blanket will also spread out over the surface of the liquid and work to smother the fire.

Responders should be aware of the nature of the materials that are released and the way in which particular foam concentrates behave when they are applied. Take the time to become familiar with the types of foam concentrates that may be used by your agency.

Reducing the temperature of some hazardous materials will also suppress vapor formation. Unfortunately, there is no easy way to accomplish this goal except in cases of small spills.

Employ this control measure only after consulting with a hazardous materials technician or another subject matter expert in the field.

To use vapor dispersion to manage a hazardous materials release, follow the steps in SKILL DRILL 12-8:

1. Based on the characteristics of the released material, the size of the release, and the topography, determine the viability of a dispersion operation.
2. Use the appropriate monitoring instrument, such as a combustible gas indicator, photo-ionization detector, or multi-gas meter, to determine the boundaries of a safe work area.
3. Ensure that ignition sources in the area have been removed or controlled. (STEP 1)
4. Apply a water spray (from firefighting hose lines) from a distance to disperse vapors.
5. Monitor the environment until the vapors have adequately dispersed. (STEP 2)

SKILL DRILL 12-8 Using Vapor Dispersion to Manage a Hazardous Materials Incident

NFPA 6.6.4.1

VIDEO

1 Determine the viability of a dispersion operation. Use the appropriate monitoring instrument to determine the boundaries of a safe work area. Ensure that ignition sources in the area have been removed or controlled.

2 Apply water from a distance to disperse vapors. Monitor the environment until the vapors have been adequately dispersed.

To perform vapor suppression actions using foam, follow the steps in SKILL DRILL 12-9:

1. Based on the characteristics of the released material, the size of the leak, and the topography, determine the viability of a vapor suppression operation.
2. Use the appropriate instrument to determine the boundaries of a safe work area.
3. Ensure that ignition sources in the area have been removed or safely controlled.
4. Apply the appropriate type of foam from a safe distance to suppress vapors.
5. Monitor the environment until the vapors have been adequately suppressed. (STEP 1)

SKILL DRILL 12-9 Using Vapor Suppression to Manage a Hazardous Materials Incident

NFPA 6.6.4.1

VIDEO

1 Determine the viability of a vapor suppression operation. Use the appropriate instrument to determine a safe work area. Ensure that ignition sources in the area have been removed or controlled. Apply foam from a safe distance to suppress vapors. Monitor the environment until the vapors have been adequately suppressed.

Always remember that as foam is applied, more volume will be added to the spill. All of the following foams suppress the ignition of flammable vapors and may be used to extinguish fires in flammable or combustible liquids:

- Aqueous film-forming foam (AFFF) can be used at a 1%, 3%, or 6% concentration. This type of foam is designed to form a blanket over spilled flammable liquids to suppress vapors or on actively burning pools of flammable liquids. Foam blankets are designed to prevent a fire from reigniting once extinguished. Most AFFF foam concentrates are biodegradable. AFFF is usually applied by way of in-line foam eductors, foam sprinkler systems, and portable or fixed proportioning systems.
- Alcohol-resistant concentrates have properties similar to AFFF, except that they are formulated so that alcohols such as methyl alcohol and isopropyl alcohol and other polar solvents will not dissolve the foam. (Regular protein foams cannot be used on these types of products.)
- Fluoroprotein foam contains protein products mixed with synthetic fluorinated surfactants. These foams can be used on fires or spills involving gasoline, oil, or similar products. Fluoroprotein foams rapidly spread over the fuel, ensuring fire knockdown and vapor suppression. Fluoroprotein foams, like many of the synthetic foam concentrates and other types of special-purpose foams, are resistant to polar solvents such as alcohols, ketones, and ethers.
- Protein foam concentrates are made from hydrolyzed proteins (animal by-products), along with stabilizers and preservatives. These types of foams are very stable and possess good expansion properties. They are quite durable and resistant to reignition when used on Class B fires or spills involving nonpolar substances such as gasoline, toluene, oil, and kerosene.
- High-expansion foam is used when large volumes of foam are required for spills or fires in warehouses, tank farms, and hazardous waste facilities. It is not uncommon to have a yield of more than 1000 gallons of finished foam from every gallon of foam concentrate used. The expansion is accomplished by pumping large volumes of air through a small screen coated with a foam solution. Because of the large amount of air in the foam, high-expansion foam is referred to as "dry" foam. By excluding oxygen from the fire environment, this agent smothers the fire, leaving very little water residue.

There are several ways to apply foam. Foam concentrates (other than high-expansion foam) should be gently applied or bounced off another adjacent object so that they flow down across the liquid and do not directly upset the burning surface. Foam can also be applied in a rain-down method by directing the stream into the air over the material and letting the foam gently fall onto the surface of the liquid, as rain would.

To apply the rain-down method, follow the steps in SKILL DRILL 12-10:

1. Open the nozzle and test to ensure that foam is being produced.
2. Move within a safe range of the product and open the nozzle.
3. Direct the stream of foam into the air so that the foam gently falls onto the pool of the product.
4. Allow the foam to flow across the top of the pool of product until it is completely covered. (STEP 1)

The rain-down method is less effective when the fire is creating an intense thermal column. When the heat from the thermal column is high, the water content of the foam will turn to steam before it comes in contact with the liquid surface.

SKILL DRILL 12-10 Performing the Rain-Down Method of Applying Foam

NFPA 6.6.4.1

VIDEO

1 Open the nozzle to ensure that foam is being produced. Move within a safe range of the product and open the nozzle. Direct the stream of foam into the air so that the foam gently falls onto the pool of the product. Allow the foam to flow across the top of the pool of the product until it is completely covered.

Foam can also be applied via the roll-in method by bouncing the stream directly into the front of the spill area and allowing it to gently push forward into the pool rather than directly "splashing" it in.

The bounce-off method is used in situations where the fire fighter can use an object to deflect the foam stream and let it flow down onto the burning surface. For example, this method could be used to apply foam to an open-top storage tank or a rolled-over transport vehicle. The foam should be swept back and forth against the object while the foam flows down and spreads back across its surface. As with all foam application methods, it is important to let the foam blanket flow gently on the surface of the flammable liquid to form a blanket.

To apply foam using the roll-in method, follow the steps in SKILL DRILL 12-11:

1. Open the nozzle and test to ensure that foam is being produced.
2. Move within a safe range of the product and open the nozzle.
3. Direct the stream of foam onto the ground just in front of the pool of product.
4. Allow the foam to roll across the top of the pool of product until it is completely covered. (STEP 1)

SKILL DRILL 12-11 Performing the Roll-In Method of Applying Foam

NFPA 6.6.4.1

VIDEO

1. Open the nozzle and test to ensure that foam is being produced. Move within a safe range of the product and open the nozzle. Direct the stream of foam onto the ground just in front of the pool of product. Allow the foam to roll across the top of the pool of product until it is completely covered.

To perform the bounce-off method of applying foam, follow the steps in SKILL DRILL 12-12:

1. Open the nozzle and test to ensure that foam is being produced. (STEP 1)
2. Move within a safe range of the product and open the nozzle.
3. Direct the stream of foam onto a solid structure such as a wall or metal tank so that the foam is directed off the object and onto the pool of product.
4. Allow the foam to flow across the top of the pool of product until it is completely covered.
5. Be aware that the foam may need to be banked off of several areas of the solid object so as to extinguish the burning product. (STEP 2)

Recovery

The recovery phase of a hazardous materials incident occurs when the imminent danger to people, property, and the environment has passed or is controlled, and clean-up begins. During the recovery phase, local, state, and federal agencies may become involved in cleaning up the site, determining the responsible party, and implementing cost-recovery methods. The recovery phase in large-scale incidents can go on for days, weeks, or even months and may require large amounts of resources and equipment FIGURE 12-13.

FIGURE 12-13 The recovery phase involves clean-up, determination of the responsible party, and implementation of cost recovery.

Courtesy of Captain David Jackson, Saginaw Township Fire Department

SKILL DRILL 12-12 Performing the Bounce-Off Method of Applying Foam

NFPA 6.6.4.1

VIDEO

1 Open the nozzle and test to ensure that foam is being produced.

2 Move within a safe range of the product and open the nozzle. Direct the stream of foam onto a solid structure such as a wall or metal tank so that the foam is directed off the object and onto the pool of product. Allow the foam to flow across the top of the pool of product until it is completely covered. Be aware that the foam may need to be banked off of several areas of solid object so as to extinguish the burning product.

In some situations there is a clear transition between the emergency phase and the clean-up phase. The decision to declare a change from the emergency phase to the recovery phase is typically made by the incident commander. For example, a distinct hand-off between on-scene public safety responders and private-sector commercial clean-up companies would fit this description. It is not uncommon for public-sector responders to hand off the clean-up operations but remain on scene to ensure that the operation is carried out properly. As long as public safety personnel and responders remain on the scene, they should maintain their vigilance for safety to ensure that no new hazards are created or injuries incurred during the recovery phase.

RESPONDER TIP

The ultimate goal of the recovery phase is to return the property or site of the incident to its preincident condition and to return the facility or mode of transportation to the responsible party.

RESPONDER TIP

After performing any product control procedure, follow your local procedures for technical decontamination.

At other times, the initial responders perform both the emergency response and the clean-up. For example, responders often handle in their entirety small gasoline spills resulting from auto accidents or spills at gas stations.

The transition from emergency to clean-up should be carried out in a manner consistent with your agency's standard operating procedures and the guidelines determined by your AHJ. The recovery phase also includes completion of the records necessary for documenting the incident. Additionally, responders should be aware of cost recovery policies and procedures in their AHJ. There are certain instances when the spiller or responsible party for the release may be financially responsible for the costs incurred to mitigate the problem.

VOICES
OF EXPERIENCE

There are times where all the aspects of what you have learned just come together. This happened to me on one very cold February night.

I had been an officer for a couple of years at the time of this incident, and was feeling comfortable in my knowledge base. We were called to a report of a multiple-vehicle crash, including one semi, at a particular mile marker on Interstate 25. The semi was reported to be leaking diesel fuel. We assembled our plan en route, including scene safety, security, patients, patching the hole in the tank, and throwing down "kitty litter" and pads to clean up what had spilled.

We didn't realize that this particular mile marker included a bridge over a lake. The semi was on its side and the MC-307/DOT-407 was leaking diesel fuel from the top of the tanker trailer. The fuel hadn't begun dripping off the side of the bridge, but we knew we had less than 10 minutes before it did.

While two fire fighters checked the scene and assessed patients, one fire fighter grabbed what we had of kitty litter and pads to create a small diversion dam and a small sponge area to give us a little time. We knew it wouldn't hold, and we needed more material to make it last. Help was quickly called for the department's complete stock of "kitty litter," public works personnel with lots of sand (which I was told would be at least 45 minutes in arriving), and the regional hazardous materials team.

Using the kitty litter we had, we built a long diversion barrier to allow the diesel fuel to slowly flow down the length of the bridge toward its end, away from its edge. We also used the natural traffic ruts in the interstate to help us with this task. Where the flow reached the end of the bridge, we had several fire fighters working with picks and axes to break up the ground. These fire fighters created a dam to retain the flow in place with the dirt they were able to break up.

Once the public works department arrived, we used the equipment and truckloads of dirt they brought to reinforce both our diversion barrier and the dam. This completely contained the spill until the clean-up and offloading crews arrived, along with representatives of the regional hazardous materials team. Upon their arrival, we turned the scene over to them and went back in service (another crew remained on scene to help with offloading the tanker).

On the way back, it occurred to us that the information we learned in hazardous materials classes really does work, and that multiple techniques and scenarios can easily apply to just one incident (recognition, diking, absorption, diversion, containment, damming, recovery phase, and retention). By taking the appropriate steps, remaining flexible, and using all the knowledge we had gained, we had prevented this incident from becoming a serious environmental nightmare.

Philip Oakes
Laramie County Fire District #4
Cheyenne, Wyoming

Wrap-Up

Chief Concepts

- When considering a control option, a variety of factors must be evaluated, such as the maximum quantity of material that can be released and the likely duration of the incident if no intervention is made.
- Sometimes the situation is so extreme, or the proposed action is so risky, that it may be prudent to create a safe perimeter and let the problem stabilize on its own.
- Control techniques or options generally aim to contain, redirect, or lower the concentration of the hazardous material involved and/or to prevent the ignition of flammable liquids or gases.
- Responders can employ a variety of product-control options, such as absorption, diversion, damming, diking, or isolating a leak with remote shut-off valves.
- Special foams can help both in vapor suppression actions and in extinguishment of fires associated with flammable liquid releases.
- Responders should be aware of the locations of emergency shut-off valves at fixed facilities and on the various types of cargo tanks typically found in their response area.
- The recovery phase may last for days or months and aims to return the exposure area to its original condition and return the facility or mode of transportation to the responsible party.

Hot Terms

Absorption The process of applying a material that will soak up and hold a hazardous material in a sponge-like manner for collection and subsequent disposal.

Adsorption The process in which a contaminant adheres to the surface of an added material—such as silica or activated carbon—rather than combining with it (as in absorption).

Alcohol-resistant concentrate A foam concentrate with properties similar to those of aqueous film-forming foam, except that the concentrate is formulated so that alcohols and other polar solvents will not break down the foam.

Aqueous film-forming foam (AFFF) A water-based extinguishing agent used on Class B fires that forms a foam layer over the liquid and stops the production of flammable vapors.

Confinement The process of keeping a hazardous material within the immediate area of the release.

Containment Actions relating to stopping a leak from a hazardous materials container, such as patching, plugging, or righting the container.

Damming The product-control process used when liquid is flowing in a natural channel or depression, and its progress can be stopped by constructing a barrier to block the flow.

Diking The placement of materials to form a barrier that will keep a hazardous material in liquid form from entering an area or that will hold the material in an area.

Dilution The process of adding a substance—usually water—in an attempt to weaken the concentration of another substance.

Diversion The process of redirecting spilled or leaking material to an area where it will have less impact.

Fluoroprotein foam A blended organic and synthetic foam that is made from animal by-products and synthetic surfactants.

High-expansion foam A foam created by pumping large volumes of air through a small screen coated with a foam solution. Some high-expansion foams have expansion ratios ranging from 200:1 to approximately 1000:1.

Protein foam A foam that is made from organic materials. It is effective on non-alcohol-type fuels such as gasoline and diesel fuel.

Recovery phase The stage of a hazardous materials incident after imminent danger has passed, when clean-up and the return to normalcy have begun.

Remote valve shut-off A type of valve that may be found at fixed facilities utilizing chemical processes or piped systems that carry chemicals. These remote valves provide a way to remotely shut down a system or isolate a leaking fitting or valve. Remote shut-off valves are also found on many types of cargo containers.

Retention The process of purposefully collecting hazardous materials in a defined area.

Vapor dispersion The process of lowering the concentration of vapors by spreading them out, typically with a water fog from a hose line.

Vapor suppression The process of controlling vapors given off by hazardous materials, thereby preventing vapor ignition, by covering the product with foam or other material or by reducing the temperature of the material.

RESPONDER *in action*

Your engine company has arrived on the scene of an MC306/DOT406 flammable liquid cargo tanker parked on the side of the interstate. Prior to your arrival, the dispatch center advised you that the tanker is carrying gasoline. Upon arrival, you notice the beginnings of a stream of released liquid flowing from the underside of the tank, away from your location. The driver approaches you and confirms the product to be gasoline, and states that he has no idea why the leak is occurring. The air temperature is approximately 90°F (32°C) and the winds are calm. You are considering the options available to control the released liquid.

1. Where would be the most likely places to find the remote shut-off valves on this type of cargo tanker?
   - **A.** Underneath the truck, near the front and rear wheels
   - **B.** Adjacent to the driver's door or at the rear of the cargo tank
   - **C.** Inside the cab next to the safety data sheet (SDS) or under the passenger seat
   - **D.** On top of the cargo tank; one on the front of the tank and the other at the rear
2. Assuming that the leak has been stopped and approximately 500 gallons of liquid is now retained in a catch basin, which type of foam might be applied effectively to this type of product (gasoline) to suppress the vapors?
   - **A.** High-expansion foam
   - **B.** Foam eduction solution
   - **C.** Surfactant foam concentrate
   - **D.** AFFF
3. You notice the AFFF foam blanket appears to be breaking down too quickly. Which of the following statements best describes a likely cause?
   - **A.** The pavement temperature is too hot; the foam is breaking down due to heat.
   - **B.** The AFFF is incompatible with the spilled liquid; you should consider switching to alcohol-resistant-type foam.
   - **C.** The pool of spilled liquid is too deep; foam works only on shallow pools of liquid.
   - **D.** You should change nothing; keep applying the same foam.
4. Imagine that the leak worsens, to the point that the flowing liquid has now reached an adjacent 15-foot-wide drainage canal with water flowing in it. Your team decides to build a dam to capture the flowing gasoline. Which of the following types of dams would be appropriate given the physical properties of gasoline?
   - **A.** A diversion dam
   - **B.** An underflow dam
   - **C.** An overflow dam
   - **D.** A retention dam

Mission-Specific Competencies: Victim Rescue and Recovery

Greg Henry/ShutterStock, Inc.

Courtesy of Rob Schiepp

© Photos.com

CHAPTER 13

Competencies for Operations Level Responders Assigned Mission-Specific Responsibilities

This chapter shall address competencies found in NFPA 472 for the following operations level responders assigned mission-specific responsibilities at hazardous materials/WMD incidents by the authority having jurisdiction beyond the core competencies at the operations level.

Knowledge Objectives

After studying this chapter, you will be able to:

- Describe tactical considerations such as attempting to make a rescue without the proper PPE or without backup personnel, or deciding whether a rescue attempt has a good chance of success. (NFPA 6.8.3.1, 6.8.3.2, p. 274–275)
- Describe entry team and backup team responsibilities (p. 275–279).
- Describe the difference between ambulatory and nonambulatory victims, and considerations for each. (NFPA 6.8.3.1(2), p. 275–276)
- Describe the difference between rescue mode and recovery mode. (NFPA 6.8.3.1(2), p. 277–278)
- Describe considerations in providing medical care and/or decontamination to victims during rescue mode or recovery mode. (NFPA 6.8.3.1, p. 278–279)
- Describe the equipment needed for search, rescue, and recovery operations. (NFPA 6.8.4.1(2), p. 281–282)
- Describe the assists, lifts, and carries that are commonly used during rescue operations. (NFPA 6.8.4.1, p. 283–297)
- Describe the benefits of sheltering-in-place. (NFPA 6.8.4.1, p. 283)
- Describe the process of triage. (NFPA 6.8.4.1(4), p. 276–277)

Skills Objectives

After studying this chapter, you will be able to:

- Demonstrate the one-person walking assist. (NFPA 6.8.4.1, p. 283–284)
- Demonstrate the two-person walking assist. (NFPA 6.8.4.1, p. 283–285)
- Demonstrate the two-person extremity carry. (NFPA 6.8.4.1, p. 284–286)
- Demonstrate the two-person seat carry. (NFPA 6.8.4.1, p. 285–286)
- Demonstrate the two-person chair carry. (NFPA 6.8.4.1, p. 287)
- Demonstrate the cradle-in-arms carry. (NFPA 6.8.4.1, p. 288)
- Demonstrate the clothes drag. (NFPA 6.8.4.1, p. 289)
- Demonstrate the blanket drag. (NFPA 6.8.4.1, p. 290)
- Demonstrate the standing drag. (NFPA 6.8.4.1, p. 291)
- Demonstrate the webbing sling drag. (NFPA 6.8.4.1, p. 292)
- Demonstrate the fire fighter drag. (NFPA 6.8.4.1, p. 293)
- Demonstrate the one-person emergency drag from a vehicle. (NFPA 6.8.4.1, p. 294)
- Demonstrate the long backboard rescue. (NFPA 6.8.4.1, p. 295–297)

You Are the Responder

You are a fire fighter/paramedic assigned to a rescue ambulance. You and your EMT partner arrive first on the scene of a medical response for a 35-year-old male with breathing difficulties. The incident is located in a rural area, on the property of a well-known local farmer. Upon your arrival, the farmer tells you he returned from an errand to find one of his workers lying next to a weed sprayer. The farmer thought the man might have been sleeping, but once he got closer he could see several bottles of Sevin insecticide nearby. The man is lying in the dirt, in a puddle of liquid, and appears to have vomited. The farmer tells you he knows the smell of Sevin, and he is certain that the puddle of liquid must be from the nearby insecticide containers. From your position, the man looks like he is twitching. You have no chemical-protective equipment or turnout gear on the ambulance.

1. What is your first impression in terms of the severity of the victim's signs and symptoms?
2. Would full structural fire fighter's turnout gear and self-contained breathing apparatus (SCBA) offer adequate protection in this situation?
3. Assuming you had the proper training, would you make the rescue and provide medical treatment before any help arrived? Why or why not?

Introduction

The first priority of a responder is to protect life—in other words, the first goal is to separate the people from the problem. This statement does not mean that emergency responders should risk their lives for little gain, however. When it comes to saving lives, sometimes the risk outweighs the benefit, thereby rendering the act of victim rescue impractical. Unfortunately, on some occasions emergency responders may not have enough training, personnel, resources, or protective equipment to positively effect a rescue.

The determination of whether a rescue is feasible is not an exact science, and there are no clear-cut guidelines to follow. The choice to act—like all other decisions—must be based on sound information, good training, adequate personal protective gear, availability of enough trained personnel to accomplish the task, and a reasonable expectation of a positive outcome. It is unwise to arrive on a scene and become part of the problem yourself. Think before you act: It could mean the difference between life and death.

Despite this warning about some of the general pitfalls of performing a victim rescue, you must accept the fact that there may come a time in your career when you have to put yourself at risk to save a life. If that day arrives, you should approach the situation with a calm mind, a critical eye to notice and interpret critical clues, and a detached, nonemotional demeanor that enables you to make the best risk-based decision possible.

Refer back to the chapter-opening scenario. The insecticide Sevin is a member of the carbamate family of chemicals, and it will affect humans much like a nerve agent would. Armed with this knowledge, and having a reasonable belief that the victim was exposed to this insecticide, would you attempt a rescue without any protective gear other than your normal work uniform? Would you perform the rescue if you had full structural fire fighter's gear and SCBA? Would you perform the rescue if you had a full Level A ensemble but no partner, so that you had to perform the rescue alone? What if you had no idea what the product was? Would you assume no hazard was present and immediately enter the area to rescue the victim?

As you can see, there are many variables to consider. In each case, you would need more information about the particular situation. In some cases the parameters of the situation may be set for you: You may lack the proper training to carry out the rescue or be unable to identify the product. Perhaps you lack the proper personal protective equipment (PPE) or have no PPE, or perhaps not enough personnel are available to make a safe entry into a contaminated atmosphere. Occasionally, the victim may be in such a dire situation that a successful rescue seems unlikely. Again, the parameters of the situation will have a tremendous influence on your decision to attempt a victim rescue. That's an absolutely critical point to remember—making a victim rescue is a choice. Choose wisely.

Tactical Considerations

When faced with a potential victim rescue, emergency responders should first ensure that enough responders are on the scene to make the attempt **FIGURE 13-1**. The notion that a single responder, or even two responders, should attempt to make a rescue without the proper PPE or anyone backing them up should be banished from your thinking. In most life-threatening hazardous materials situations, at least five trained responders—not including the supervisor—are required to make a rescue attempt. This number is based on filling the following positions: two responders to make the initial entry (the **entry team**), two responders to back them up (a **backup team**), and one responder to staff an emergency decontamination process. All things considered, a five-person team is quite an abbreviated crew for making a rescue, and five people will not be nearly enough when operating at a mass-casualty event. This does not include the position of incident commander or the responders needed for medical care and transport of victims. Again, this crew of five should be supervised

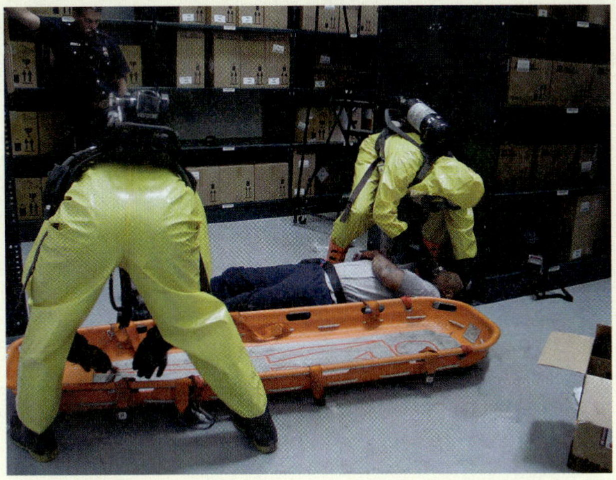

FIGURE 13-1 Responders practicing a simulated rescue of an exposed victim.
Courtesy of Rob Schnepp

in whatever way is customary in your own authority having jurisdiction (AHJ).

Early on in the rescue effort, a decision must be made about the viability of the victim(s). For example, a worker at a food-processing plant who becomes caught in a significant ammonia release—in an enclosed and poorly ventilated space—and who has been trapped for more than a few minutes without any PPE may have a small chance of survival. Countless variables may influence the outcome of an incident, and we cannot account for them all here, but this illustration serves to reinforce a key point: First decide if the rescue attempt has a good chance of success. Ask yourself this question: Can your team gain access to the viable victim, carry or package him or her, and get out of the hazard zone without anyone else being exposed to the hazard? If the answer is yes, make sure the entry team has the proper PPE, a readily available and properly protected backup team, some form of decontamination in place, radio communications, and at least one person supervising the entire activity.

It is imperative that emergency responders maintain good situational awareness when responding to incidents where injuries or deaths related to an exposure have occurred. Take the time to size up the scene and understand the hazards present. Make sure you fully understand the potential threats to the health and safety of the responders and the civilian population.

RESPONDER TIP

Responders working in the hot zone for any reason should work in at least two-member teams (an Occupational Safety and Health Administration [OSHA] requirement), stay in close proximity to each other, and have the ability to communicate with each other and another responder outside of the hot zone.

Entry Team Responsibilities

The entry team is just that—a *team*, consisting of at least two appropriately trained responders, wearing the proper level and type of PPE, equipped with radio communications and appropriate tools (which may include detection and monitoring equipment, small hand tools, and flashlights), operating under the direction of a supervisor. The entry team can also consist of more than just two members; there could be multiple teams of two or even an odd number of responders working together (along with the appropriate number of backup team members).

The entry team may be tasked to go into a contaminated atmosphere to do reconnaissance, map the scene, perform search and rescue, or begin triaging victims or directing victims out of the contaminated environment. That part of the mission—the task—is subject to the requirements of the situation. The manner in which that entry is made—that is, working in pairs and having a backup team in place—should not be compromised. Entry teams should also be in the habit of giving a report on the conditions they encounter, such as the number of victims and their medical status, to the supervisor as soon as possible. This information enables the supervisor to assemble more resources if necessary and begin to plan for treatment and/or transport of the victim population.

RESPONDER TIP

Just as with any of the other mission-specific competencies, the PPE requirements for victim rescue should be based on the mission.

Responder Safety Tip

If the entry team is making an entry for the sole reason of attempting a rescue, they should take whatever equipment may be required, such as a Stokes basket, rigid backboard, or rescue stretcher (specialized rescue equipment).

Non-line-of-sight situations with either ambulatory victims (able to walk on their own) or nonambulatory victims (unable to walk under their own power) require responders to perform the sometimes time-consuming and dangerous act of *searching* before the rescue. Sometimes the search may be more dangerous than the process of removing the victims. In these situations, responders may be operating in reduced visibility environments, under hostile conditions, with little knowledge of the structure in which they are searching. When victims are located and are able to walk under their own power, responders have more options to remove them from harm's way.

In some cases, when the means of egress is clear and out of danger, it may be wise to appoint one of the victims as the ad hoc group leader and ask this person to guide/direct the victims outside to safety. When this is not possible, it might be better to have the trained responders lead the victims to safety. In these dynamic situations, you must use your best judgment to make a sound decision. In any case, when the victims can help you by helping themselves to some degree (walking under their own power is a great start!), the process becomes somewhat easier.

Ambulatory victims who are within the line of sight of the entry team, conscious, able to follow verbal commands, and able to walk are clearly the easiest to rescue. They most often need some direction and encouragement to leave the area, but can do so under their own power.

RESPONDER TIP

When an extensive search precedes a rescue, the entire process will be more time consuming and perhaps more complicated.

Victims may be nonambulatory for a number of reasons, including trauma, fear, or incapacitation by a chemical substance. Those victims may prove problematic to rescue. First, this type of rescue is labor and equipment intensive. Having to carry adult victims is physically taxing **FIGURE 13-2**. When victim rescue also requires decontamination, the process becomes even more complicated, especially with nonambulatory victims. The ensuing decontamination efforts go much more slowly than normal because of the need to carry the victims, remove their clothing, and move them through the entire decontamination process **FIGURE 13-3**. See the chapter *Mission-Specific Competencies: Mass Decontamination* for a discussion of the process of moving nonambulatory victims through mass decontamination.

FIGURE 13-3 Some level of decontamination is necessary for all victims when a chemical exposure is suspected or confirmed.

Courtesy of 1st Lt. Toni Tones/U.S. Air Force

FIGURE 13-2 Carrying or dragging victims while wearing PPE is hard work and physically challenging.

© Jones & Bartlett Learning. Photographed by Glen E. Ellman

Second, it must first be determined whether the nonambulatory victim can be saved from a medical standpoint. As callous as it may seem, it is pointless to "rescue" the dead. Therefore, nonambulatory victims must be sorted out based on their medical priority using the triage method approved in your AHJ. Especially in multiple-casualty situations, it is imperative to first rescue live victims who have the best chance of survival.

RESPONDER TIP

Triage is essential at all mass-casualty incidents. Triage is the process of sorting two or more victims based on the severity of their conditions to establish priorities for care based on available resources.

In a smaller-scale mass-casualty incident, the first responders on the scene who have the highest level of medical training usually begin the triage process. Victims are ranked in order of the severity of their conditions. The victim with the most severe injuries is given the highest-priority attention. Triage at a large-scale mass-casualty incident should occur in several steps, as in the well-recognized START (Simple Triage and Rapid Treatment) triage system **FIGURE 13-4**. This system prompts the responder to assess the patient's breathing rate, pulse rate, and mental status so as to assign a treatment priority to the victim.

Most triage methods incorporate the following aspects:

- Use a color-coding system to classify victims found during search efforts so as to indicate priorities for treatment and transportation from the scene. Typically, red-tagged victims are the first priority, yellow-tagged

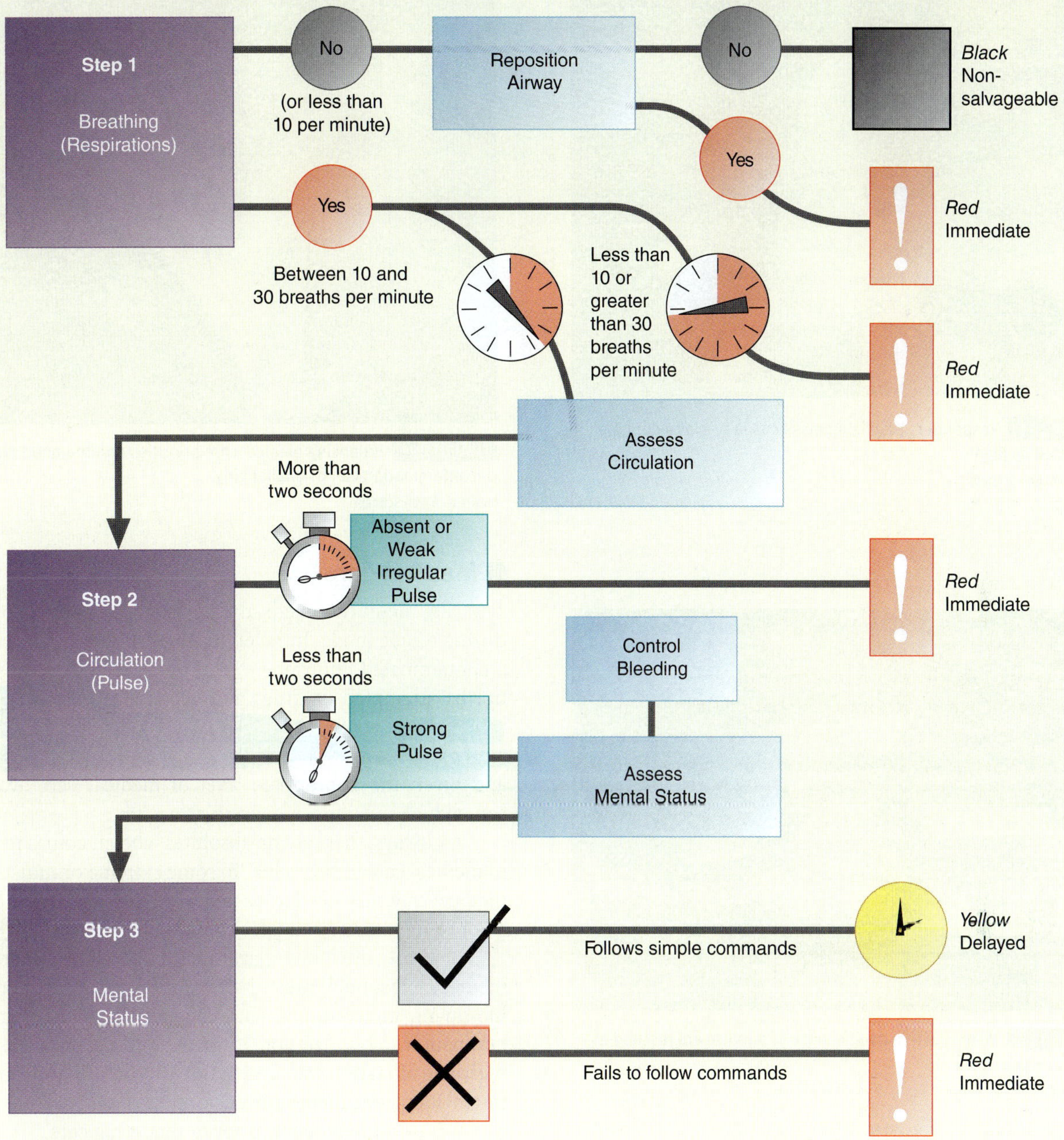

FIGURE 13-4 The START triage method.

victims are the second priority, and victims tagged with green or black are the lowest priority FIGURE 13-5.

- Rapidly remove red-tagged victims for field treatment and transportation as ambulances become available.
- Use separate treatment areas to care for red-tagged victims if transport is not immediately available. Yellow-tagged victims can also be monitored and managed in the treatment area while waiting for transportation.
- When there are more victims waiting for transport than there are ambulances, the transportation group supervisor decides which victim is the next to be loaded.
- Specialized transportation resources (e.g., air ambulance, paramedic ambulances) require separate decisions when these resources are available but limited.

When victims are present and determined to have a good chance for survival, the incident is considered to be in rescue mode FIGURE 13-6. At some point, however, the operation may transition from rescue mode to recovery mode.

FIGURE 13-5 Responders practicing mass casualty triage during a simulated incident.

Courtesy of Rob Schnepp

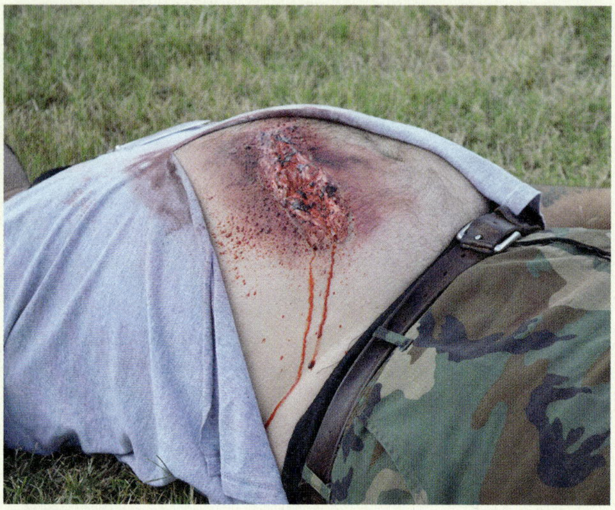

FIGURE 13-7 Balancing medical care priorities and the need for decontamination can be challenging.

Courtesy of Rob Schnepp

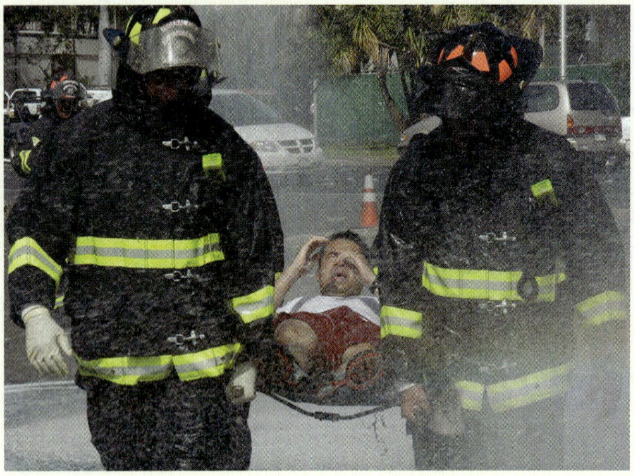

FIGURE 13-6 Victims with a good chance of survival are rescued as quickly as possible.

Courtesy of Journalist 3rd Class Ryan C. McGinley/U.S. Navy

Recovery mode entails a more systematic search and removal of bodies after the initial rescue mode has been completed. During recovery mode there is no chance of rescuing a victim alive. The responders' efforts move at a slower pace than when the incident was still considered an emergency, during which the focus was on live victims. In some cases, deceased persons will need to be decontaminated prior to their bodies' removal from the scene. Evidence may need to be collected from the clothing and/or the victims' bodies in the event the incident is deemed to be criminal in nature. How the recovery efforts are carried out will depend largely on the circumstances of the incident; the incident commander will be making many difficult decisions about removing the deceased in an efficient and sensitive manner.

Medical Care

Typically, definitive medical care is not rendered to victims during rescue mode. Intravenous lines should not be started in a contaminated atmosphere; endotracheal intubations should not be attempted or drugs administered. The best thing that can be done for patients is to get them out of the hazardous atmosphere and to ensure they are properly decontaminated and given the appropriate level of medical care available—somewhere outside the hot or warm zone.

As always, there are no absolutes when it comes to delivery of medical care to save a life. In some extreme or unusual cases, medical care may need to be delivered in the warm or hot zone prior to or concurrent with decontamination **FIGURE 13-7**.

In those situations, hazardous materials responders and medical personnel must balance the need for performing life-saving interventions with the need for ensuring decontamination. This decision is made on a case-by-case basis, with consideration being given to the nature and severity of the incident, the medical needs of the patient, and the need to perform decontamination prior to rendering care.

For example, think about rescuing a person from a structure fire. Would it be customary to locate the individual, place him or her on oxygen, establish an intravenous line, and begin treating the patient for smoke inhalation before the patient is removed from the fire building? Absolutely not—victims rescued from structure fires must be quickly located and then removed from the environment as soon as possible, for the safety of both the fire fighters and the victims. Definitive treatment then occurs outside the fire building, in the cold zone. In most situations, a victim of a chemical exposure who is being rescued from a hazardous environment should be treated in exactly the same way.

Backup Team

For each person on the entry team, one responder is held in reserve as a backup. If six responders will make up the initial

FIGURE 13-8 One backup entry team member should be provided for each entry team member.

entry team, there should be six responders on the backup team. *In hazardous materials response, the entry team/backup team ratio is always 1:1.*

The backup team members are dressed in the same level of PPE as the entry team members. Backup teams are staged adjacent to the same access point the entry team used, and they should be ready to deploy in a matter of seconds FIGURE 13-8. Other than zipping up a suit or putting on a mask and turning on the air, the backup team should be ready to go at a moment's notice.

The backup team should monitor the radio traffic of the entry team and make every attempt to keep track of their progress and location in the hot zone. If called into service to assist the entry team, perhaps in the process of performing a rescue, the backup team will certainly have their hands full. To that end, the members of this team should always be formulating a plan of action as the entry team progresses through their mission. The backup team should be responsible for having a ready and appropriate cache of equipment on hand in the event they are called upon.

Emergency Decontamination

As discussed in previous chapters, emergency decontamination is performed in potentially life-threatening situations to rapidly remove the bulk of the contamination from an individual FIGURE 13-9. When emergency decontamination is required,

FIGURE 13-9 Emergency decontamination involves the immediate removal of contaminated clothing.

the process usually involves removing contaminated clothing (when the victim is nonambulatory or otherwise unable to do so) or directing the ambulatory victim to remove this clothing, and then dousing the victim with adequate quantities of water. If possible, efforts should be made to contain the water runoff in some fashion, or to prevent the water from flowing into drains, streams, or ponds. Do not delay decontamination to implement these containment measures when it comes to exposed persons, however: Human life always comes first. The consequences of the runoff can be addressed later.

All attempts should be made to understand the physical and chemical properties of a substance before beginning any form of decontamination.

Responder Safety Tip

In hazardous materials response, the entry team/backup team ratio is always 1:1.

RESPONDER TIP

Hazardous materials teams practiced the concept of rapid intervention before it was adopted in structural firefighting.

Search, Rescue, and Recovery

Search, rescue, and recovery efforts carried out at a hazardous materials/weapons of mass destruction (WMD) incident can be time-consuming, dangerous for the responder, and very

VOICES OF EXPERIENCE

When it comes to hazardous materials emergency response, Elvis Presley was right: "Only fools rush in." There are so many considerations to bear in mind before attempting to affect the rescue of a potentially contaminated victim from an uncontrolled environment. Just as an emergency responder needs to weigh the risks versus the benefits in deciding whether to enter a fully involved structure fire, confined space, or whitewater to attempt a rescue, so, too, should the responder decide whether he or she has the proper personal protective equipment, training, time, and ability to enter a hazardous environment to remove a victim who has been contaminated.

Although structural fire fighters' protective clothing and self-contained breathing apparatus may provide limited protection and allow a rapid "in-and-out" effort in some circumstances, the responder needs to be sure that the benefit will outweigh the risk. Where structural fire fighters' protective clothing and self-contained breathing apparatus are not adequate for the hazardous material involved, proper chemical-protective clothing must be used, even if it means delaying the rescue.

I had responded on a call that came in as a gaseous chlorine release in a residential area. On arrival, I was able to use binoculars to see two large gas cylinders—one of which was obviously leaking—that were marked with "chlorine gas" and "inhalation danger" labels. Trees and vegetation in the immediate area were quickly turning brown. We quickly learned that this incident was the result of an experimental pesticide project being done inappropriately in a residential location. My first thought was that we needed to secure the leak, because we didn't know if there were victims involved. I donned the appropriate PPE, went in, and successfully turned off the valve to control the leak. The hazard abated, but my rush to act gnawed at me. We had not yet set up decontamination equipment, and we didn't yet have a full backup team in position. The OSHA permissible exposure limit (PEL) was 1 part per million (ppm) and the immediately dangerous to life and health (IDLH) level was 10 ppm, yet I had no way of knowing the amount of gas in the air.

In some cases it may be possible to direct victims to safety without the responder actually having to enter the hot zone and come in contact with contaminants or be exposed themselves. In other cases, the incident commander may have to make a judgment call and permit responders to perform an expedient entry and exit. The incident commander needs to size up the situation; weigh the risks and benefits; assess the various factors involved, such as the materials encountered, the number of victims, the duration of the rescue, and the likelihood of success; and plan for the aftermath of the decision made before risking the safety of the responders. In this case, we did not have decontamination equipment or a backup entry team in place.

Far too many rescuers become victims themselves, so assess the situation carefully before gambling on the safety of emergency personnel. The first priority in hazardous materials emergency response is the safety of the emergency responders! I got away with it that day—but only because I was lucky, not because I was smart.

Leo DeBobes
Suffolk County Community College
Selden, New York

labor intensive. In many cases, the responders who conduct these missions during a working incident may rapidly become fatigued simply by the act of aggressively searching a building. In addition to the stress of the incident, working in any level of PPE raises the wearer's blood pressure and heart rate, increases body temperature, and leads to an elevated breathing rate that may exhaust the air supply quickly. To that end, it is important for responders undertaking search, rescue, or recovery to adhere to the old adage of "Work smarter, not harder." Moving efficiently through a building or other designated search area, in an organized search pattern, may assist with covering the most ground possible in the shortest amount of time. Also, when it comes to carrying living or deceased victims, it is important to remember that two rescuers, carrying an average-size adult over any distance, will quickly run through an air supply and become fatigued. For this reason, responders might need to employ special methods of carrying victims or use some mechanical means (such as rescue sleds or Stokes baskets) to aid in that movement.

As in structural firefighting, searching an area may be done in stages. The first stage, or *primary search*, is a rapidly conducted search, focusing on those victims who are easily seen, who are lightly trapped or disoriented, or who are otherwise found in easily accessible locations. For example, suppose a search team (or multiple search teams) is tasked with conducting a primary search of a shopping mall after a suspected terrorist attack with a nerve agent. They might start at one end of the building and make their way quickly to an end point, looking for people who are ambulatory and found in easily accessible locations. The team should wear the appropriate type and level of PPE; be equipped with triage tags, radio communications, and basic personal hand tools; and use monitoring and detection devices to determine the characteristics of the atmosphere. Subsequent search/rescue/recovery teams might then adjust their level of PPE based on the primary search team's findings. This team would not render significant medical aid or spend time to free any victims who might be heavily trapped or obviously dead. Instead, they would escort ambulatory victims toward exits or other collection points established at the incident, perhaps as a prelude to decontamination. The team members' path of travel would likely include only the main halls and/or corridors of the mall, and they would not be expected to carry or otherwise transport nonambulatory victims outside the building (a very tiring activity).

The primary search team should also make mental notes of the scene to pass along to subsequent teams, which would conduct a more thorough *secondary search*. The secondary search is more detailed. For example, it might include excursions into the individual businesses found within the shopping mall, or deeper into other parts of the structure such as warehouse areas, employee break rooms, and other rooms connected to the individual businesses. This team may also conduct a more detailed evaluation of heavily trapped victims, but would not get caught up in a rescue that would require significant time, tools, or personnel.

The latter effort should be left to rescue teams, which can be directed to specific locations and should be prepared to spend some time extricating those persons who are in need. As always, there is no immediate need for recovery of a nonsalvageable or deceased victim; instead, time should be spent on those persons who have a good chance of survival.

These examples provide an overview of the concept of search and rescue; they are not meant to serve as a specific plan. Each incident will have unique factors that will require sound decision making from all responders working at the scene.

Search and Rescue/Recovery Equipment

To perform search, rescue, and recovery tasks properly, responders must have the appropriate equipment. In some cases, the equipment carried may be similar to what fire fighters routinely carry:

- Appropriate PPE
- Portable radio
- Hand light or flashlight
- Forcible-entry (exit) tools
- Thermal imaging devices (if available)
- Long rope(s) in some cases
- A piece of tubular webbing or short rope (16 to 24 feet)

In other cases, these efforts may require heavy tools such as sledgehammers, long pry bars, concrete breaching tools, or air-powered lifting devices. The details of the specific situation will dictate the personnel and equipment needs, including the need for rescue sleds, stretchers, evacuation chairs, spine boards, or wheeled carts to transport victims, or other useful adjuncts for transporting victims outside the hot zone FIGURE 13-10 through FIGURE 13-13.

Full PPE is essential for search, rescue, and recovery operations. At least one member of each search team should be equipped with a portable radio. Ideally, each individual should have a radio. If a responder gets into trouble, the radio is the best means to obtain assistance.

When using SCBA, responders must pay attention to their air supplies during these operations. All responders must have an adequate amount of air to conduct a reasonable search, travel to the anticipated location of the victim(s), make a safe

Responder Safety Tip

During search, rescue, and recovery operations, responders should follow these guidelines:

- Work from a single plan.
- Maintain radio contact with the incident commander (IC), the hazardous materials group supervisor, or whoever is designated as the contact through the chain of command. The important point to remember is that you should always have radio contact with someone outside the hazard zone.
- Monitor environmental conditions during the search. This includes paying attention to what is going on around you, including potential collapse situations, the possibility of secondary devices, or other responders working in close proximity. Use information from detection and monitoring devices to understand the current and changing conditions of the workplace.
- Adhere to the personal accountability system in your AHJ.
- Stay with a partner.

FIGURE 13-10 Responders using a rescue sled to extricate a victim.
Courtesy of Rob Schnepp

FIGURE 13-13 A collection of backboards ready for multiple victims.
Courtesy of Rob Schnepp

FIGURE 13-11 A stretcher.
Courtesy of Ferno Washington, Inc.

FIGURE 13-12 A wheeled victim litter.
Courtesy of Rob Schnepp

exit, and perform technical decontamination, if required. Responders must start to exit *before* they hear the low-pressure alarm on the SCBA. It is important to plan for an adequate supply of air to ensure a safe exit, because this consideration limits the amount of time that can be spent searching and the distance that responders can penetrate into a building.

RESPONDER TIP

The IC must always balance the risks involved in an emergency operation with the potential benefits:

- Actions that present a high level of risk to the safety of responders are justified only if there is potential to save lives.
- It is not acceptable to risk the safety of responders when there is no chance of saving lives.

Search and Rescue/Recovery Methods

As an emergency responder, you must learn and practice the various types of assists, carries, drags, and other methods used to rescue or recover victims from dangerous situations. After mastering these techniques, you will be able to rescue/recover victims from a wide variety of life-threatening situations. *Not all techniques shown in this chapter are appropriate in all circumstances. It is up to you to determine which lift/carry might be applicable in any given rescue or recovery situation.*

As previously mentioned, rescue/recovery is the second component of response, following search. When you locate a victim during a search, you must direct, assist, or carry that victim to a safe area. This action may be as simple as verbally directing an occupant toward an exit, or it can be as demanding as extricating a trapped, unconscious victim or deceased victim and physically carrying that person out of the building.

Most people who realize that they are in a dangerous situation will attempt to escape on their own. Children and elderly persons, physically or developmentally handicapped persons, and ill or injured persons, however, may be unable to escape and will need to be rescued. Toxic gases or other chemical exposures may incapacitate even healthy individuals before they can reach an exit. Victims also may become trapped when a rapidly spreading fire, an explosion, or a structural collapse cuts off potential escape routes.

When an emergency situation threatens the lives of both victims and rescuers, the first priority is to remove the victim from the dangerous area as quickly as possible, remembering that it is usually better to move the victim to a safe area first, and only then provide any necessary medical treatment. The assists, lifts, and carries described in this chapter should not be used if you suspect that the victim has a spinal injury, unless there is no other way to remove him or her from the life-threatening situation.

Always use the safest and most practical means of egress when removing a victim from a dangerous area. A building's normal exit system, such as interior corridors and stairways, should be used if those areas are open and safe. If the regular exits cannot be used, an outside fire escape, a ladder, or some other method of egress must be found. Ladder rescues/recoveries, which are not covered in this chapter, can be both difficult and dangerous, and should be conducted by trained personnel such as fire fighters. Sometimes, simply keeping victims away from the hazard, but leaving them inside a structure (one form of sheltering-in-place), may be a viable option.

Sheltering-in-Place

In some situations, the best option is to shelter the victims in place instead of trying to remove them from a building. This option should be considered when the occupants are conscious and are found in a part of the building that is adequately protected from the hazard, whether by distance or by fire-resistive construction and/or fire suppression systems. If smoke and fire conditions block the exits, victims might be safer staying in the sheltered location than attempting to evacuate the structure by going through a hazardous environment.

For example, such a situation might occur in a high-rise apartment building when a fire, explosion, chemical release, or some other problem is confined to one specific part of the building. In this scenario, the stairways and corridors might be filled with smoke or a released gas, but the occupants who are remote from the problem would be safe on their balconies or in their apartments. These persons would be exposed to greater levels of risk if they attempted to exit than if they remained in their apartments until the fire is extinguished or the release is over or has been stopped. The decision to follow a sheltering-in-place strategy in this case must be made by the IC.

When sheltering-in-place is not a feasible option, rescue operations may be required. Keep in mind that carrying victims, even when they are not particularly heavy, is more complicated when you are wearing chemical-protective equipment.

Chemical gloves reduce dexterity and make both clothing and skin difficult to grasp. Gloves tend to slide around, and the smooth surface of the garments do not allow for any friction between the victim's clothing and your own. Additionally, the increased physical effort will cause you to generate more body heat, which also makes wearing the PPE less bearable.

> The following types of rescue techniques are demonstrated in standard fire fighter bunker gear. The illustrations are intended only to demonstrate the method, not imply that this level of PPE is recommended for a certain type of rescue.

Exit Assist

The simplest rescue is the exit assist, in which the victim is responsive and able to walk without assistance or with very little assistance. The responder may simply need to guide such a person to safety or provide a minimal level of physical support. Even if the victim can walk without assistance, the responder should take the person's arm or use the one-person walking assist to make sure the victim does not fall or become separated from the responder.

Two types of assists can be used to help responsive victims exit a situation:

- One-person walking assist
- Two-person walking assist

One-Person Walking Assist

The one-person walking assist can be used if the person is capable of walking.

To perform a one-person walking assist, follow the steps in SKILL DRILL 13-1:

1. Help the victim stand next to you, facing the same direction. (STEP 1)
2. Have the victim place his or her arm behind your back or around your neck. Hold the victim's wrist as it drapes over your shoulder.
3. Put your free arm around the victim's waist and help the victim to walk. (STEP 2)

Two-Person Walking Assist

The two-person walking assist is useful if the victim cannot stand and bear weight without assistance. The two responders completely support the victim's weight. It may be difficult to walk through doorways or narrow passages using this type of assist, however.

To perform a two-person walking assist, follow the steps in SKILL DRILL 13-2:

1. Two responders stand facing the victim, one on each side of the victim. (STEP 1)
2. Both responders assist the victim to a standing position. (STEP 2)

SKILL DRILL 13-1 Performing a One-Person Walking Assist

NFPA 6.8.4.1

VIDEO

1 Help the victim stand.

2 Have the victim place his or her arm around your neck, and hold on to the victim's wrist, which should be draped over your shoulder. Put your free arm around the victim's waist and help the victim to walk.

3. Once the victim is fully upright, place the victim's right arm around the neck of the responder on the right side. Place the victim's left arm around the neck of the responder on the left side. The victim's arms should drape over the responders' shoulders. Each responder holds one of the victim's wrists in one hand. (STEP 3)
4. Both responders put their free arms around the victim's waist, grasping each other's wrists for support and locking arms together behind the victim. (STEP 4)
5. Both responders slowly assist the victim to walk. Responders must coordinate their movements and move slowly. (STEP 5)

Simple Victim Carries

Four simple carry techniques can be used to move a victim who is conscious and responsive, but incapable of standing or walking:

- Two-person extremity carry
- Two-person seat carry
- Two-person chair carry
- Cradle-in-arms carry

Two-Person Extremity Carry

The two-person extremity carry, also known as the sit pick, requires no equipment and can be performed in tight or narrow spaces, such as corridors of mobile homes, small hallways, or narrow spaces between buildings. The focus of this carry is on the victim's extremities.

To perform a two-person extremity carry, follow the steps in SKILL DRILL 13-3:

1. Two responders help the victim to sit up. (STEP 1)
2. One responder kneels behind the victim, reaches under the victim's arms, and grasps the victim's wrists. (STEP 2)
3. The second responder backs in between the victim's legs, reaches around, and grasps the victim behind the knees. (STEP 3)
4. At the command of the first responder, both responders stand up and carry the victim away, walking straight ahead. The responders must coordinate their movements. (STEP 4)

SKILL DRILL 13-2 Performing a Two-Person Walking Assist

NFPA 6.8.4.1

VIDEO

1. Two responders stand facing the victim, one on each side of the victim.

© Jones & Bartlett Learning. Photographed by Glen E. Ellman

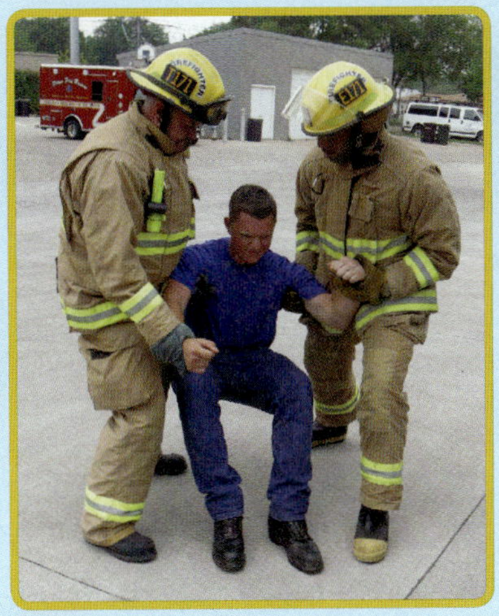

2. The responders assist the victim to a standing position.

© Jones & Bartlett Learning. Photographed by Glen E. Ellman

3. Once the victim is fully upright, drape the victim's arms around the necks and over the shoulders of the responders, each of whom holds one of the victim's wrists.

© Jones & Bartlett Learning. Photographed by Glen E. Ellman

4. Both responders put their free arm around the victim's waist, grasping each other's wrists for support and locking their arms together behind the victim.

© Jones & Bartlett Learning. Photographed by Glen E. Ellman

5. Assist walking at the victim's speed.

© Jones & Bartlett Learning. Photographed by Glen E. Ellman

Two-Person Seat Carry

The two-person seat carry is used with victims who are disabled or paralyzed. This type of carry requires the assistance of two responders, and moving through doors and down stairs may be difficult.

To perform a two-person seat carry, follow the steps in SKILL DRILL 13-4:

1. Two responders kneel near the victim's hips, one on each side of the victim. (STEP 1)
2. Both responders raise the victim to a sitting position and link arms behind the victim's back. (STEP 2)
3. The responders place their free arms under the victim's knees and link them together. (STEP 3)
4. If possible, the victim puts his or her arms around the responders' necks for additional support. (STEP 4)

SKILL DRILL 13-3 Performing a Two-Person Extremity Carry

NFPA 6.8.4.1

VIDEO

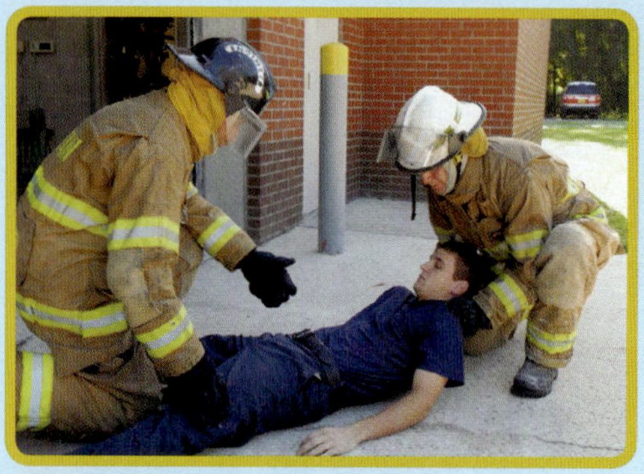

1. Two responders help the victim to sit up.

2. The first responder kneels behind the victim, reaches under the victim's arms, and grasps the victim's wrists.

3. The second responder backs in between the victim's legs, reaches around, and grasps the victim behind the knees.

4. The first responder gives the command to stand and carry the victim away, walking straight ahead. Both responders must coordinate their movements.

SKILL DRILL 13-4 Performing a Two-Person Seat Carry

NFPA 6.8.4.1

VIDEO

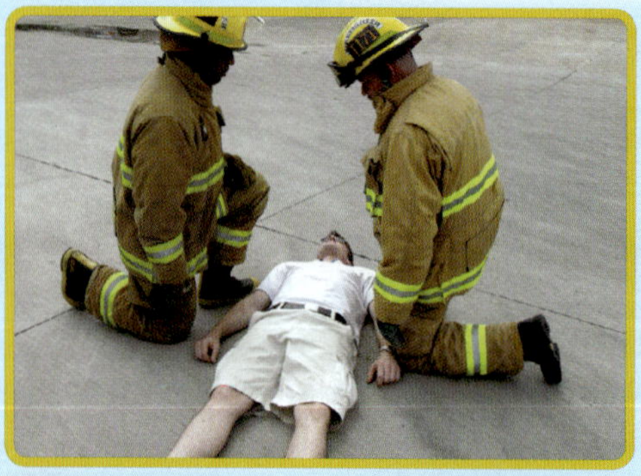

1. Kneel beside the victim near the victim's hips.

2. Raise the victim to a sitting position and link arms behind the victim's back.

3 Place your free arms under the victim's knees and link arms.

4 If possible, the victim puts his or her arms around your necks for additional support.

Two-Person Chair Carry

The two-person chair carry is particularly suitable when a victim must be carried through doorways, along narrow corridors, or up or down stairs. In this technique, two rescuers use a chair to transport the victim. A folding chair cannot be used for this purpose, and the chair must be strong enough to support the weight of the victim while he or she is being carried. The victim should feel much more secure with this carry than with the two-person seat carry, and he or she should be encouraged to hold on to the chair.

To perform a two-person chair carry, follow the steps in SKILL DRILL 13-5:

1. Ensure the victim is seated in a suitable chair. Instruct the victim to grasp his or her hands together.
2. One responder stands behind the seated victim, reaches down, and grasps the back of the chair. (STEP 1)
3. The responder tilts the chair slightly backward on its rear legs so that the second responder can step between the legs of the chair and grasp the tips of the chair's front legs. The victim's legs should be between the legs of the chair. (STEP 2)
4. When both responders are correctly positioned, the responder behind the chair gives the command to lift and walk away.
5. Because the chair carry may force the victim's head forward, watch the victim for airway problems. (STEP 3)

SKILL DRILL 13-5 Performing a Two-Person Chair Carry

NFPA 6.8.4.1

VIDEO

1 One responder stands behind the seated victim, reaches down, and grasps the back of the chair.

2 The responder tilts the chair slightly backward on its rear legs so that the second responder can step back between the legs of the chair and grasp the tips of the chair's front legs. The victim's legs should be between the legs of the chair.

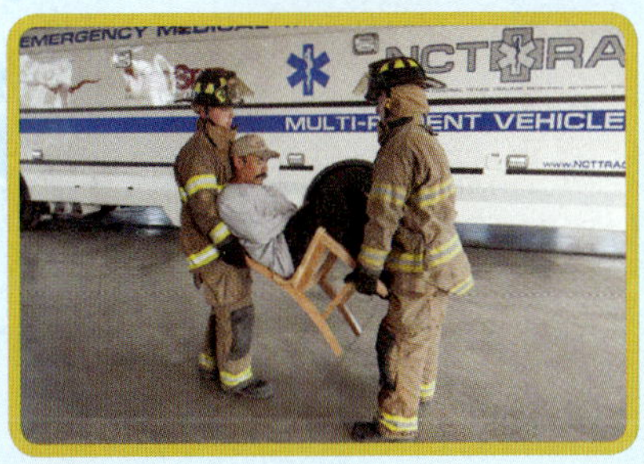

3 When both responders are correctly positioned, the responder behind the chair gives the command to lift and walk away. Because the chair carry may force the victim's head forward, watch the victim for airway problems.

Responder Safety Tip

Keep your back muscles as straight as possible and use the large muscles in your legs to do the lifting!

Cradle-in-Arms Carry

The cradle-in-arms carry can be used by one responder to carry a child or a small adult. With this technique, the responder should be careful of the victim's head when moving through doorways or down stairs.

To perform the cradle-in-arms carry, follow the steps in **SKILL DRILL 13-6**:

1. Kneel beside the child, and place one arm around the child's back and the other arm under the thighs. (**STEP 1**)
2. Lift slightly and roll the child into the hollow formed by your arms and chest. (**STEP 2**)
3. Be sure to use your leg muscles to stand. (**STEP 3**)

SKILL DRILL 13-6 Performing a Cradle-in-Arms Carry

NFPA 6.8.4.1

VIDEO

1 Kneel beside the child, and place one arm around the child's back and the other arm under the thighs.

2 Lift slightly and roll the child into the hollow formed by your arms and chest.

3 Be sure to use your leg muscles to stand.

Emergency Drags

An efficient method to rapidly move a victim from a dangerous location is a drag. Six emergency drags can be used to remove unresponsive or deceased victims from a dangerous situation:

- Clothes drag
- Blanket drag
- Standing drag
- Webbing sling drag
- Fire fighter drag
- Emergency drag from a vehicle

When using an emergency drag, the responder should make every effort to pull the victim in line with the long axis of the body to provide as much spinal protection as possible. The victim should be moved head first to protect the head.

Clothes Drag

The clothes drag is used to move a victim who is on the floor or the ground and is too heavy for one responder to lift and carry alone. In this technique, the responder drags the person by pulling on the clothing in the neck and shoulder area. The responder should grasp the clothes just behind the collar, use the arms to support the victim's head, and drag the victim away from danger.

To perform the clothes drag, follow the steps in SKILL DRILL 13-7:

1. Crouch behind the victim's head, grab the shirt or jacket around the collar and shoulder area, and support the victim's head with your arms. (STEP 1)
2. Lift with your legs until you are fully upright. Walk backward, drag the victim to safety. (STEP 2)

SKILL DRILL 13-7 Performing a Clothes Drag
NFPA 6.8.4.1

VIDEO

1 Crouch behind the victim's head, and grab the shirt or jacket around the collar and shoulder area.

2 Lift with your legs until you are fully upright. Walk backward, drag the victim to safety.

Blanket Drag

The blanket drag can be used to move a victim who is not dressed or who is dressed in clothing that is too flimsy for the clothes drag (for example, a nightgown). This procedure requires the use of a large sheet, blanket, curtain, or rug. Place the item on the floor and roll the victim onto it, and then pull the victim to safety by dragging the sheet or blanket.

To perform a blanket drag, follow the steps in SKILL DRILL 13-8:

1. Lay the victim supine (face up) on the ground. Stretch out the material for dragging next to the victim. (STEP 1)
2. Roll the victim onto his or her right or left side. Neatly bunch one third of the material against the victim so the victim will lie approximately in the middle of the material. (STEP 2)
3. Lay the victim back down (supine) on the material. Pull the bunched material out from underneath the victim and wrap it around the victim. (STEP 3)
4. Grab the material at the head and drag the victim backward to safety. (STEP 4)

SKILL DRILL 13-8 Performing a Blanket Drag

NFPA 6.2.4.1

VIDEO

1 Stretch out the material you are using next to the victim.

2 Roll the victim onto one side. Neatly bunch one third of the material against the victim's body.

3 Lay the victim back down (supine). Pull the bunched material out from underneath the victim and wrap it around the victim.

4 Grab the material at the head and drag the victim backward to safety.

Standing Drag

The standing drag is a physically taxing but effective way to move patients who are unconscious or unable to walk on their own. It works best when you are taller and/or heavier than the victim because you are required to walk backwards and pull the entire weight of the person being rescued. If the victim is wearing clothing that is fragile or if the victim is not wearing clothing, this drag can be even more difficult.

The standing drag is performed by following the steps in SKILL DRILL 13-9:

1. Kneel at the head of the supine victim. (STEP 1)
2. Raise the victim's head and torso by 90 degrees, so that the victim is leaning against you. (STEP 2)
3. Reach under the victim's arms, wrap your arms around the victim's chest, and lock your arms. (STEP 3)
4. Stand straight up using your legs.
5. Drag the victim out. (STEP 4)

SKILL DRILL 13-9 Performing a Standing Drag

NFPA 6.8.4.1

VIDEO

1 Kneel at the head of the supine victim.

2 Raise the victim's head and torso by 90 degrees, so that the victim is leaning against you.

3 Reach under the victim's arms, wrap your arms around the victim's chest, and lock your arms.

4 Stand straight up using your legs. Drag the victim out.

Webbing Sling Drag

The webbing sling drag ensures the responder has a secure grip around the upper part of a victim's body, allowing for a quick and efficient exit from the dangerous area. In this drag, a sling is placed around the victim's chest and under the armpits, and then used to drag the victim. The webbing sling helps support the victim's head and neck. A webbing sling can be rolled and kept in the pocket of any garment. A carabiner can be attached to the sling to secure the straps under the victim's arms and provide additional protection for the head and neck.

To perform the webbing sling drag, follow the steps in SKILL DRILL 13-10:

1. Using a prepared webbing sling, place the victim in the center of the loop so the webbing is behind the victim's back in the area just below the armpits. (STEP 1)
2. Take the large loop over the victim and place it above the victim's head. Reach through, grab the webbing behind the victim's back, and pull through all the excess webbing. This creates a loop at the top of the victim's head and two loops around the victim's arms. (STEP 2)
3. Adjust your hand placement to protect the victim's head while dragging. (STEP 3)

SKILL DRILL 13-10 Performing a Webbing Sling Drag

NFPA 6.8.4.1

VIDEO

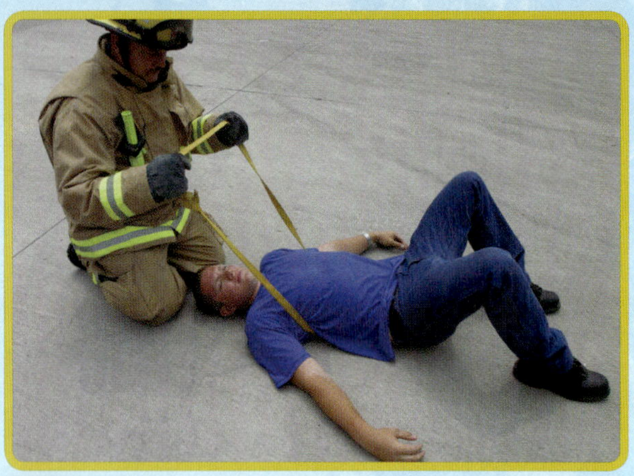

1 Place the victim in the center of the loop so the webbing is behind the victim's back.

2 Take the large loop over the victim and place it above the victim's head. Reach through, grab the webbing behind the victim's back, and pull through all the excess webbing. This creates a loop at the top of the victim's head and two loops around the victim's arms.

3 Adjust your hand placement to protect the victim's head while dragging.

Fire Fighter Drag

The fire fighter drag can be used if the victim is heavier than the rescuer because it does not require lifting or carrying the victim.

To perform the fire fighter drag, follow the steps in SKILL DRILL 13-11:

1. Tie the victim's wrists together with anything that is handy, such as tubular webbing, a cravat (a folded triangular bandage), gauze, or a belt. (STEP 1)
2. Get down on your hands and knees and straddle the victim. (STEP 2)
3. Pass the victim's tied hands around your neck, straighten your arms, and drag the victim across the floor by crawling on your hands and knees. (STEP 3)

SKILL DRILL 13-11 Performing a Fire Fighter Drag

NFPA 6.8.4.1

VIDEO

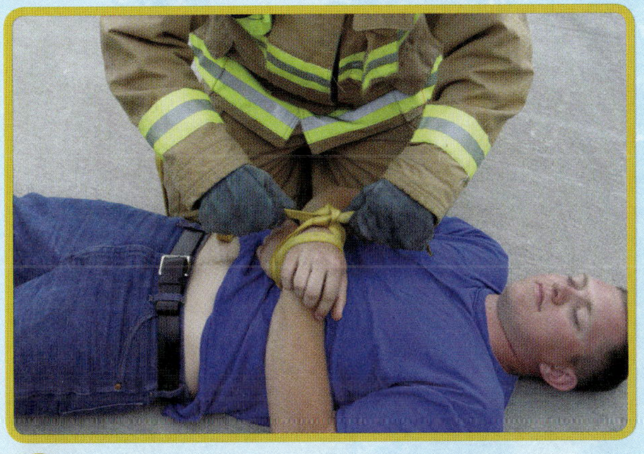

1 Tie the victim's wrists together with anything that is handy.

2 Get down on your hands and knees and straddle the victim.

3 Pass the victim's tied hands around your neck, straighten your arms, and drag the victim across the floor by crawling on your hands and knees.

Emergency Drag from a Vehicle

An emergency drag from a vehicle is performed when the victim must be quickly removed from a vehicle to save his or her life. The one-responder method or the long backboard rescue might be used, for example, if the vehicle is on fire or the victim requires cardiopulmonary resuscitation.

One Responder

It is almost impossible for one person to remove a victim from a vehicle without some movement of the neck and spine; however, it is important to prevent excess movement of the victim's neck (unless the victim is deceased).

To perform an emergency drag from a vehicle with only one responder, follow the steps in SKILL DRILL 13-12:

1. Grasp the victim under the arms and cradle his or her head between your arms. (STEP 1)
2. Gently pull the victim out of the vehicle. (STEP 2)
3. Lower the victim down into a horizontal position in a safe place. (STEP 3)

SKILL DRILL 13-12 Performing a One-Person Emergency Drag from a Vehicle

NFPA 6.8.4.1

VIDEO

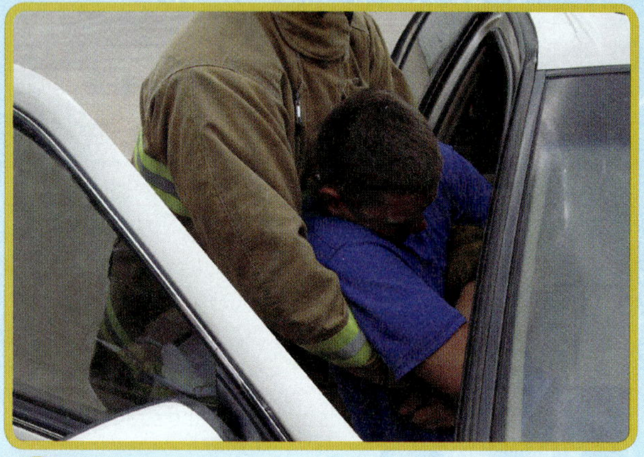

1 Grasp the victim under the arms and cradle his or her head between your arms.

2 Gently pull the victim out of the vehicle.

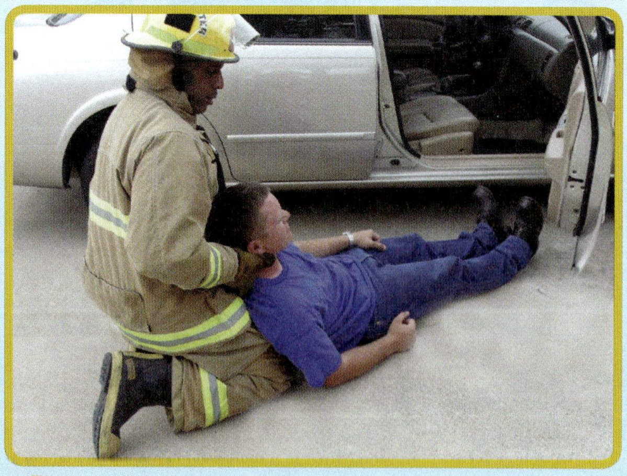

3 Lower the victim down into a horizontal position in a safe place.

Long Backboard Rescue

If four or more responders are present, one responder can support the victim's head and neck, while the second and third responders can move the victim by lifting under the victim's arms. The victim can then be moved in line with the long axis of the body, with the head and neck stabilized in a neutral position. Whenever possible, a long backboard should be used to remove a victim from the vehicle.

Follow the steps in SKILL DRILL 13-13 to perform a long backboard rescue:

1. The first responder supports the victim's head and cervical spine from behind. Support may be applied from the side, if necessary, by reaching through the driver's side doorway. (STEP 1)
2. The second responder serves as team leader and, as such, gives commands until the patient is supine on the backboard. Because the second responder lifts and turns the victim's torso, he or she must be physically capable of moving the victim. The second responder works from the driver's-side doorway. If the first responder is also working from that doorway, the second responder should stand closer to the door hinges toward the front of the vehicle. The second responder applies a cervical collar. (STEP 2)
3. The second responder provides continuous support of the victim's torso until the victim is supine on the backboard. Once the second responder takes control of the torso, usually in the form of a body hug, he or she should not let go of the victim for any reason. Some type of cross-chest shoulder hug usually works well, but you will have to decide which method works best for you with any given victim. You cannot simply reach into the car and grab the victim, because this will twist the victim's torso. Instead, you must rotate the victim as a unit.
4. The third responder works from the front passenger's seat and is responsible for rotating the victim's legs and feet as the torso is turned, ensuring that they remain free of the pedals and any other obstruction. With care, the third responder should first move the victim's nearer leg laterally without rotating the victim's pelvis and lower spine. The pelvis and lower spine rotate only as the third responder moves the second leg during the next step. Moving the nearer leg early makes it much easier to move the second leg in concert with the rest of the body. After the third responder moves the legs together, both legs should be moved as a unit. (STEP 3)
5. The victim is rotated 90 degrees so that his or her back is facing out the driver's door and his or her feet are on the front passenger's seat. This coordinated movement is done in three or four "eighth turns." The second responder directs each quick turn by saying, "Ready, turn" or "Ready, move." Hand position changes should be made between moves.
6. In most cases, the first responder will be working from the back seat. At some point—either because the door post is in the way or because he or she cannot reach farther from the back seat—the first responder will be unable to follow the torso rotation. At that time, the third responder should assume temporary support of the victim's head and neck until the first responder can regain control of the head from outside the vehicle. If a fourth responder is present, he or she stands next to the second responder. The fourth responder takes control of the victim's head and neck from outside the vehicle without involving the third responder. As soon as the change has been made, the rotation can continue. (STEP 4)
7. Once the victim has been fully rotated, the backboard is placed against the victim's buttocks on the seat. Do not try to wedge the backboard under the victim. If only three responders are present, place the backboard within arm's reach of the driver's door before the move so that the board can be pulled into place when needed. In such cases, the far end of the board can be left on the ground. When a fourth responder is available, the first responder exits the rear seat of the car, places the backboard against the victim's buttocks and maintains pressure in toward the vehicle from the far end of the board. (Note: When the door opening allows, some responders prefer to insert the backboard onto the car seat before rotating the victim.)
8. As soon as the victim has been rotated and the backboard is in place, the second and third responders lower the victim onto the board while supporting the head and torso so that neutral alignment is maintained. The first responder holds the backboard until the victim is secured. (STEP 5)
9. The third responder moves across the front seat to be in position at the victim's hips. If the third responder stays at the victim's knees or feet, he or she will be ineffective in helping to move the body's weight, because the knees and feet follow the hips.
10. The fourth responder maintains support of the victim's head and now takes over giving the commands. The second responder maintains direction of the extrication. This responder stands with his or her back to the door, facing the rear of the vehicle. The backboard should be immediately in front of the third responder. The second responder grasps the victim's shoulders or armpits. Then, on command, the second and third responders slide the victim 8 to 12 inches along the backboard, repeating this slide until the victim's hips are firmly on the backboard. (STEP 6)
11. The third responder gets out of the vehicle and moves to the opposite side of the backboard, across from the second responder. The third responder now takes

control at the shoulders, and the second responder moves back to take control of the hips. On command, these two responders move the victim along the board in 8- to 12-inch slides until the victim is placed fully on the board. (STEP 7)

12 The first (or fourth) responder continues to maintain support of the victim's head. The second and third responders grasp their side of the board, and then carry it and the victim away from the vehicle and toward the prepared cot nearby. (STEP 8)

SKILL DRILL 13-13 Performing a Long Backboard Rescue

NFPA 6.8.4.1

VIDEO

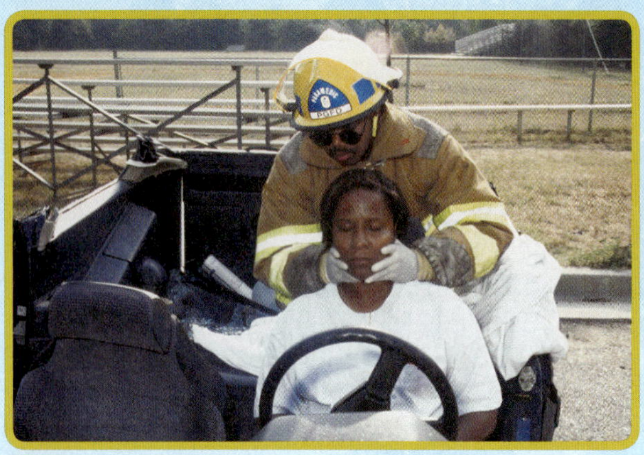

1 The first responder provides in-line manual support of the victim's head and cervical spine.

2 The second responder gives commands and applies a cervical collar.

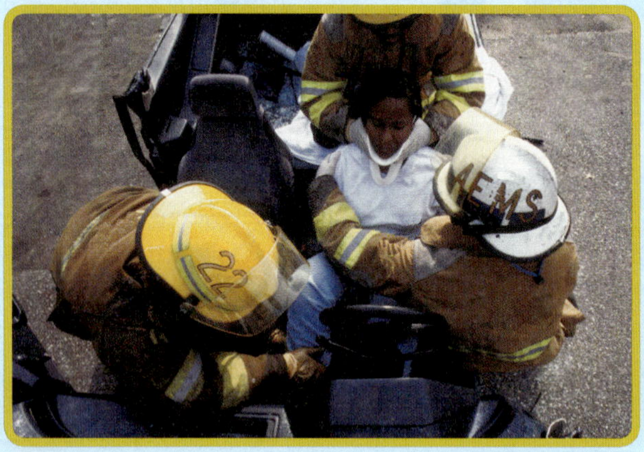

3 The third responder frees the victim's legs from the pedals and moves the legs together without moving the victim's pelvis or spine.

4 The second and third responders rotate the victim as a unit in several short, coordinated moves. The first responder (relieved by the fourth responder as needed) supports the victim's head and neck during rotation (and later steps).

5 The first (or fourth) responder places the backboard on the seat against the victim's buttocks. The second and third responders lower the victim onto the long backboard.

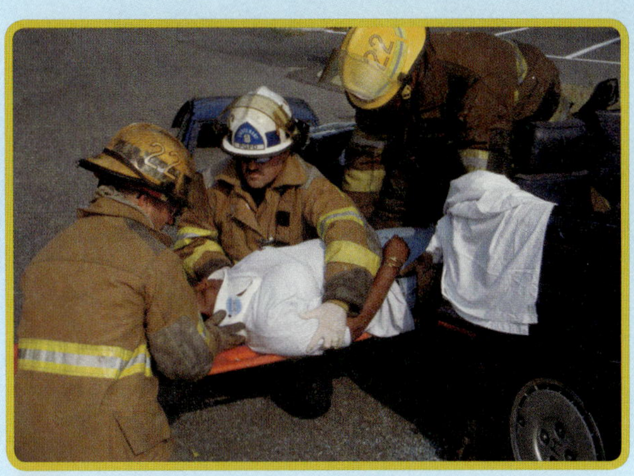

6 The third responder moves to an effective position for sliding the victim. The second and third responders slide the victim along the backboard in coordinated, 8- to 12-inch moves until the victim's hips rest on the backboard.

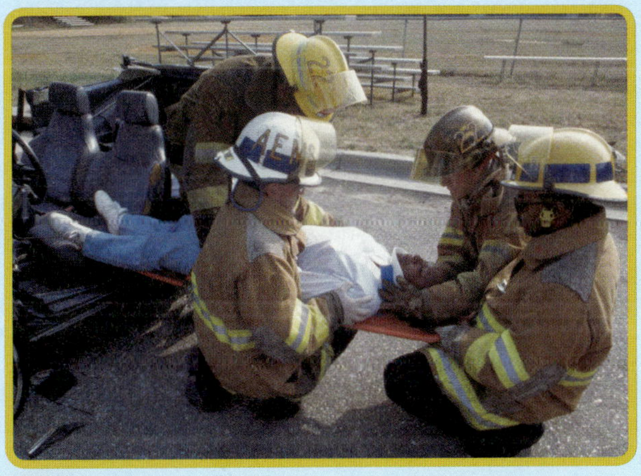

7 The third responder exits the vehicle and moves to the backboard opposite the second responder. Working together, they continue to slide the victim until the victim is fully on the backboard.

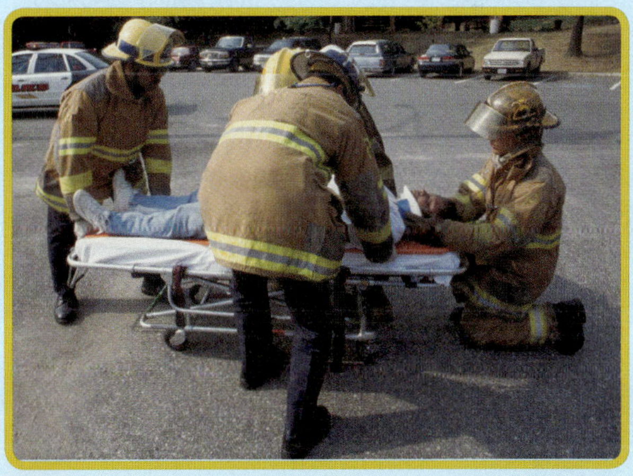

8 The first (or fourth) responder continues to stabilize the victim's head and neck while the second, third, and fourth responders carry the victim away from the vehicle.

These steps must be considered a general procedure to be adapted as needed based on the type of vehicle, type of victim, and rescue crew available. Two-door cars differ from four-door models; larger cars differ from smaller, compact models; pickup trucks differ from full-size sedans and four-wheel drive vehicles. Likewise, you will handle a large, heavy adult differently than you will handle a small adult or child. Every situation is different—a different car, a different patient, and a different crew. Your resourcefulness and ability to adapt are necessary elements to successfully perform this rescue technique.

Wrap-Up

Chief Concepts

- The determination of whether a rescue is feasible is not an exact science; there are no clear-cut guidelines to follow.
- Rescue attempts should be based on sound information, good training, adequate personal protective gear, the availability of enough trained personnel to accomplish the task, and a reasonable expectation of a positive outcome.
- The information discovered during size-up will have a tremendous impact on whether you choose to make a victim rescue.
- Ambulatory victims are easier to rescue than nonambulatory victims.
- When considering a rescue attempt in a hazardous materials incident, it is recommended to use a team consisting of at least five trained responders—not including the supervisor.
- In most cases, definitive medical care is delivered outside the hot zone.
- When making a rescue attempt, always have a backup team in position and ready to respond at an instant's notice.
- To perform a rescue properly, responders must have the appropriate equipment.
- You should learn and practice the various types of assists, carries, drags, and other techniques used to rescue people from dangerous situations.
- Everyone on the scene should know when and if an operation has transitioned from rescue mode to recovery mode.
- Always have a decontamination plan in place prior to executing the rescue attempt.

Hot Terms

Ambulatory victim A victim who can walk on his or her own and can usually perform a self-rescue with direction and guidance from rescuers.

Backup team Individuals who function as a standby rescue crew of relief for those entering the hot zone (entry team). Also referred to as backup personnel.

Entry team A team of fully qualified and equipped responders who are assigned to enter into the designated hot zone.

Nonambulatory victim A victim who cannot walk under his or her power. These types of victims require more time and personnel when it comes to rescue.

Recovery mode A shift in mindset and tactics whereby the incident response reflects the fact that there is no chance of rescuing live victims.

Rescue mode Those activities directed at locating endangered persons at an emergency incident, removing those persons from danger, treating injured victims, and providing for transport to an appropriate healthcare facility.

Triage The process of sorting victims based on the severity of their injuries and medical needs to establish treatment and transportation priorities.

RESPONDER *in action*

You are the shift battalion chief conducting an informal training session at the jurisdiction's designated hazardous materials station. A new captain has been assigned to the hazardous materials company, and you want to touch base with him and talk about the concept of rescue in the hazardous materials environment.

1. You explain to the new officer that rescue at a hazardous materials incident is conceptually similar to rescue at a structure fire. Which of the following points most closely illustrates your point?
   - **A.** Just as in a structure fire, your main goal is to remove the victim from the environment as soon as possible, then render medical care outside the hazard zone.
   - **B.** Hazardous materials incidents are like structure fires in that you must sound the building alarm when the rescue mode is completed.
   - **C.** It is common to encounter extreme temperatures at hazardous materials incidents, so you must plan to treat any victim for thermal burns.
   - **D.** Most chemical exposure victims are nonsalvageable; rescue is generally not needed.
2. The officer is unclear about the definition of an ambulatory victim. Which of the following statements would you choose to explain the concept?
   - **A.** Ambulatory victims should be instructed to report to local hospitals on their own to free up emergency crews.
   - **B.** Ambulatory victims are those persons found at least 100 yards away from an incident scene.
   - **C.** Ambulatory victims are usually unconscious and unresponsive.
   - **D.** Ambulatory victims can walk on their own and can usually perform a self-rescue with direction and guidance from the rescuers.
3. The officer asks about the difference between gross decontamination and emergency decontamination. Which of the following statements best defines emergency decontamination?
   - **A.** Emergency decontamination is used in potentially life-threatening situations to rapidly remove the bulk of the contamination from an individual.
   - **B.** Emergency decontamination is performed on responders while they are inside a formal decontamination corridor.
   - **C.** Emergency decontamination should be performed only in the cold zone, near transport ambulances.
   - **D.** First responders never perform emergency decontamination; you must be a member of an organized hazardous materials team to perform this skill.
4. The officer asks you for some points to consider if he is ever faced with the prospect of rescuing a victim at a hazardous materials incident. You tell him that the choice to act should be based on at least which of the following points?
   - **A.** Adequate training to perform the mission
   - **B.** The right PPE for the mission
   - **C.** The availability of enough trained responders to safely carry out the task
   - **D.** All of the above

Mission-Specific Competencies: Response to Illicit Laboratories

Greg Henry/ShutterStock, Inc.

© Rain Smith, Washington County Times/AP Photos

© Photos.com

CHAPTER 14

Competencies for Operations Level Responders Assigned Mission-Specific Responsibilities

This chapter shall address competencies found in NFPA 472 for the following operations level responders assigned mission-specific responsibilities at hazardous materials/WMD incidents by the authority having jurisdiction beyond the core competencies at the operations level.

Knowledge Objectives

After studying this chapter, you will be able to:

- Understand the role that all first responders have when encountering an illicit laboratory (p. 306–309)
- Describe how to recognize an illicit laboratory. (NFPA 6.9.3.2.1, p. 302–305)
- Identify the manufacturing process and common chemical hazards associated with methamphetamine production. (NFPA 6.9.3.2.1, 6.9.4.1.2, p. 303–305)
- Identify the different law enforcement agencies that could be involved in an investigation of an illicit laboratory. (NFPA 6.9.2.1(5), 6.9.3.3, p. 308)
- Understand the tactical considerations of securing an illicit laboratory while using joint hazardous materials and hazardous device technicians to assess potential hazards. (NFPA 6.9.3.2.2, 6.9.3.4.1, 6.9.3.4.2, 6.9.4.1.3, p. 305–306)
- Describe the key concepts to be taken into consideration when analyzing, planning, and implementing a response to an incident involving an illicit laboratory. (NFPA 6.9.2.1, 6.9.3.1, 6.9.3.5, 6.9.4.1, 6.9.4.1.4, 6.9.4.1.5, p. 306–310)

Skills Objectives

After studying this chapter, you will be able to:

- Identify or avoid unique safety hazards. (NFPA 6.9.4.1.1(1)(3), p. 306–308)
- Conduct a joint hazardous materials operation. (NFPA 6.9.4.1.1(4), p. 305–306)
- Decontaminate tactical law enforcement personnel such as canines and SWAT teams. (NFPA 6.9.4.1.1(2), p. 309–310)

You Are the Responder

Your engine company has been dispatched to a residential structure fire. Upon your arrival you encounter neighboring residents, who report hearing a series of loud "popping" sounds before they noticed smoke coming from the back of the building. They also report suspicious activity at the house, strange odors frequently in the air, and people coming and going at all hours of the night. The house has boarded up windows with heavy smoke pushing from the eaves.

1. Is your index of suspicion elevated by the information provided by the neighbors?
2. Would you attack the fire?
3. Do you suspect the presence of criminal activity?

Introduction

The term illicit laboratory refers to any unlicensed or illegal structure, vehicle, facility, or physical location that may be used to manufacture, process, culture, or synthesize an illegal drug, hazardous material/weapon of mass destruction (WMD) device, or improvised explosive device. In the past, such laboratories were commonly associated with the production of methamphetamine, an illegal stimulant, but that is changing and expanding, as should your idea of what an illicit laboratory looks like **FIGURE 14-1**. Methamphetamine production has evolved over the years to become less dependent on the traditional "cooking" process. Along with this evolution, the potential for the existence of other types of laboratories has grown. Synthetic illegal drugs such as LSD (lysergic acid diethylamide), Ecstasy (methylenedioxymethamphetamine [MDMA]), and GHB (gamma-hydroxybutyric acid) are also commonly produced in illicit laboratories, adding to the list of illegal substances produced by clandestine operations. Some type of illicit laboratory is likely operating in your area at this very moment.

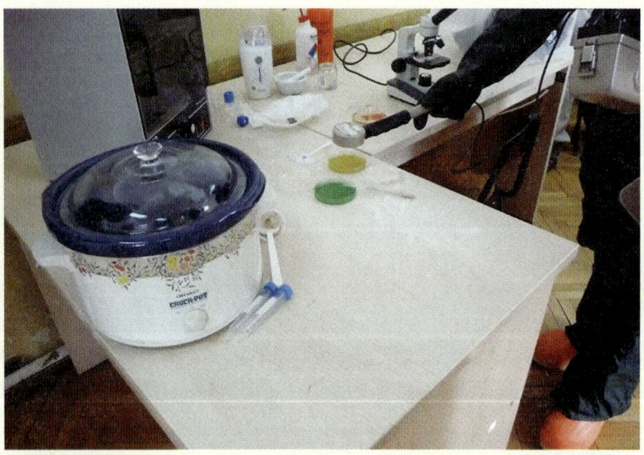

FIGURE 14-1 Pay attention to the presence of unusual items or dual-use items that appear out of place at the emergency scene. Good situational awareness is critical to hazard recognition.
Courtesy of Rob Schnepp

Illicit laboratories also include sites being used to manufacture chemical or biological agents or even explosives. The current threat of foreign and domestic terrorism should prompt every responder to consider the possibility that an illicit laboratory may contain hazards far more harmful than illegal drugs. For example, with the appropriate knowledge and equipment, bacterial agents such as *Bacillus anthracis* (anthrax) can be cultured in an illicit laboratory. Recipes and instructions for manufacturing a variety of crude chemical warfare agents, improvised explosive devices, and drugs such as methamphetamine are readily available on the Internet.

> **Responder Safety Tip**
>
> Dangerous chemicals are not the only hazard at illicit laboratories. Methamphetamine laboratory operators are often their own best customers. Methamphetamine is a powerful stimulant, and its users often suffer from extreme paranoia, sleeplessness, and anxiety. Use extreme caution when dealing with these individuals, who may behave quite erratically.

Many of the precursor chemicals or ingredients that are needed in the illicit manufacturing process are regulated by federal and state laws, making it difficult for laboratory operators to obtain them. For this reason, the precursors are often stolen or obtained illegally, thereby tipping off law enforcement investigators to the possible existence of an illicit laboratory. Follow-up investigations or tips from concerned citizens regarding suspicious activity at a home often lead investigators to the laboratories themselves. Equally as frequently, however, local fire fighters and EMS personnel are discovering these laboratories. For example, when neighbors report odd chemical odors, illegal chemical container disposal, illnesses caused by chemical vapors, or sudden explosions or fires, these responders are the first ones called to the incident—only to discover the incident was caused by an illicit laboratory.

Identifying Illicit Laboratories

Illicit laboratories vary in size and can involve a variety of chemical processing methods. Some processing methods require electricity or other power supplies; others do not.

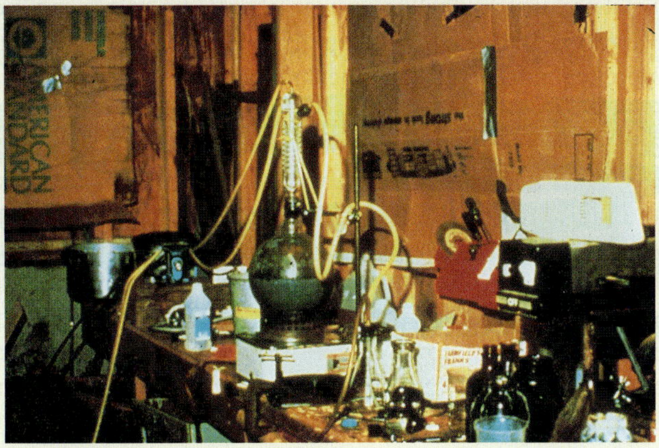

FIGURE 14-2 Illicit laboratories often have windows that are partially or completely obscured.
Courtesy of Oregon State Fire Marshal's Office, Salem, OR

Some methods require ventilation to remove vapors that could be harmful to those doing the manufacturing. Some processes generate a lot of waste—others generate little to no waste or byproducts. To that end, it is difficult to identify a "typical" laboratory or laboratory location. Laboratories can be small enough to fit into the trunk of a car or large enough to fill an entire residential home or commercial building. Locations with certain characteristics are favored spots as sites of illicit laboratories. These locations include basements or large attics, residences obscured by trees or toward the back of a deep lot, buildings with windows obscured or boarded up, or storage units FIGURE 14-2. Other locations often used as sites of illicit laboratories include motor homes, delivery-type vans, or trucks FIGURE 14-3.

People working in illegal laboratory settings may exhibit a certain degree of suspicious behavior, keeping odd hours or appearing nervous and exhibiting a high level of anxiety. They may be very protective of the laboratory area—guard dogs are common and may be encountered in or around the building, or there may be surveillance cameras tucked away in discreet or unusual places. Clearly those engaging in clandestine activities are not interested in allowing free access to their lab.

Drug Laboratories

Although many precursor chemicals are regulated and difficult to obtain in bulk, other items used to manufacture the drugs will be common, everyday items— jars, bottles, glass cookware, and coolers are frequently used to produce the illicit drugs. Additionally, laboratory glassware and tubing may be found at drug laboratories FIGURE 14-4. Specific chemicals and materials that may be present include large quantities of pill bottles, hydrochloric or sulfuric acid, paint thinner, drain cleaners, iodine crystals, table salt, aluminum foil, lithium camera batteries, blenders, food processors, strainers, coffee filters, glass cookware, and heat sources such as propane stoves, household stovetops, or electric boiler plates FIGURE 14-5. The strong smell of an unusual chemical, such as ether, ammonia, or acetone, is also very common.

FIGURE 14-3 Motor homes are sometimes used as mobile cooking laboratories.
Courtesy of Oregon State Fire Marshal's Office, Salem, OR

The term "laboratory" may actually be misleading, because many of these types of environments lack most—if not all—of the health and safety protocols that are typically associated with legitimate laboratory work. The inexperienced chemists who run these laboratories take many shortcuts and often disregard typical safe work practices. The Drug Enforcement Administration (DEA) defines a clandestine drug laboratory as "an illicit operation consisting of a sufficient combination of apparatus and chemicals that either has been or could be used in the manufacture or synthesis of controlled substances." The "cook," or person responsible for operating the laboratory, will quite often handle volatile, flammable, or corrosive substances in close proximity to open flames and heat sources. Personal protective equipment (PPE), such as respiratory protective equipment, eye protection, gloves, or chemical-protective garments, is rarely used. If protective equipment is used, it is often inappropriate for the job, not maintained, or used only occasionally.

Methamphetamine

Methamphetamine, also referred to as crank or ice, is a psycho-stimulant drug manufactured illegally in illicit laboratories. Methamphetamine can be produced using a variety of methods—again, the ability to make small batches with common household chemicals is becoming more popular, reducing the likelihood that you will encounter laboratories employing some of the more traditional "cooking" methods. One of the traditional processes to make methamphetamine is known as a "Red P" or red phosphorus lab. This type of production involves the use of red phosphorus, iodine, lye, and sulfuric acid, and requires electricity and a heating source. Another common method, known as an anhydrous ammonia lab, is popular because it is faster than other methods and requires little knowledge of chemistry. This process involves lithium strips extracted from batteries, anhydrous ammonia, starter fluid, and drain cleaners. Another method is referred to as a "P2P" or phenyl-2-propanone; this process involves chemicals such as phenyl-2-propanone, aluminum, methylamine, and mercuric acid and often yields lower-quality methamphetamine.

A common characteristic of each of these laboratories is the use of ephedrine and pseudoephedrine (cold medicine) tablets. For the reduction method of manufacturing, the cook may obtain large amounts of over-the-counter or prescription ephedrine/pseudoephedrine tablets. The tablets are then crushed and subjected to a chemical and heating process that causes the chemicals within the tablets to separate from the inert, or unwanted, ingredients FIGURE 14-6.

FIGURE 14-4 Materials used to manufacture drugs include items such as laboratory glassware and tubing.
Courtesy of Oregon State Fire Marshal's Office, Salem, OR

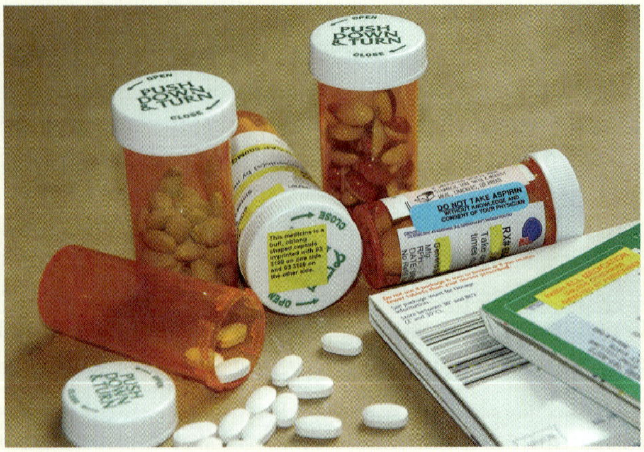

FIGURE 14-5 Large quantities of pills or bottles may be present at the site of an illicit drug laboratory.
© Anne Kitzman/ShutterStock, Inc.

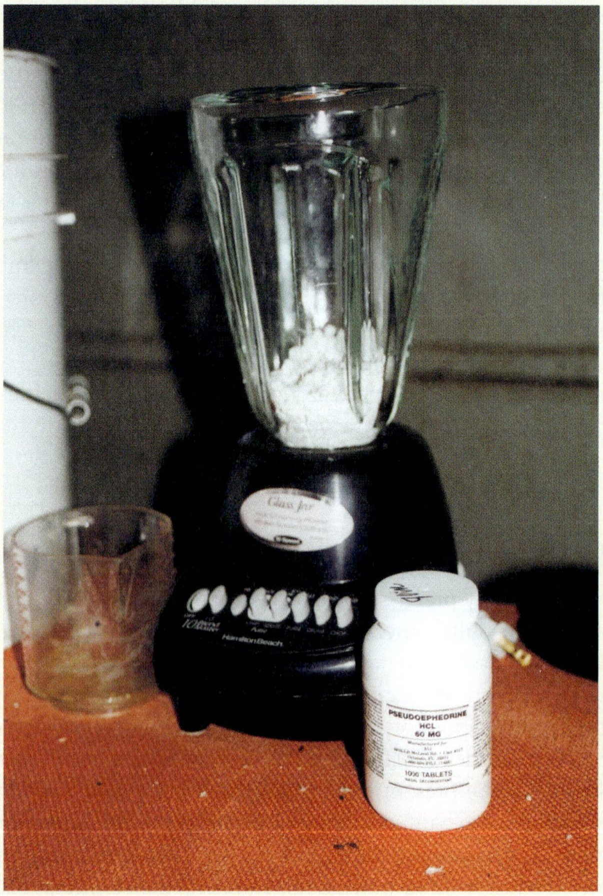

FIGURE 14-6 Ephedrine and pseudoephedrine, available at most pharmacies, are ground in household blenders as the first step in methamphetamine production.
Courtesy of Dennis Krebs

Again, some materials commonly used to produce methamphetamine often have legitimate uses TABLE 14-1.

RESPONDER TIP

Although methamphetamine laboratories are the most commonly encountered illicit laboratories, never make the assumption that a site is a drug laboratory until detection, sampling, and further investigation operations are conducted.

Table 14-1 Methamphetamine Chemicals and Their Legitimate Uses

Chemical	Legitimate Use
Anhydrous ammonia	Fertilizer
Methanol	Gasoline additive
Ether	Engine starter
Toluene	Brake cleaner
Sulfuric acid	Brake cleaner
Muriatic acid	Pool supply
Iodine	Medical use
Kerosene	Camp stove fuel
Acetone	Paint remover
Lithium	Batteries
Sodium hydroxide	Lye, drain cleaner
Red phosphorus	Matches, flare igniters

Responder Safety Tip

The lithium metal found inside batteries is generally coated with an oily film. Once this coating is removed, the lithium becomes highly volatile when exposed to water or moist air. Sodium metal may be substituted in this process in illicit laboratories. Upon contact with water, both of these metals can produce flammable gases that can ignite even without an independent ignition source. Fires involving lithium and sodium metals may produce irritating, toxic, or corrosive gases. For this reason, responders must use extreme caution in determining the appropriate fire suppression methods to employ at suspected methamphetamine laboratories involving lithium and sodium metals.

Weapons of Mass Destruction Laboratories

In some cases there may be overt indicators of possible criminal or terrorist activity involving illicit laboratories; for example, terrorist paraphernalia such as ideological propaganda and documents indicating affiliation with known terrorist groups may be found at the site. Other equipment that may be present in illicit laboratory areas includes surveillance materials (such as videotapes, photographs, maps, blueprints, or time logs of the target hazard locations), nonweapon supplies (such as identification badges, uniforms, and decals that would be used to allow the terrorist to access target hazards), and weapon supplies (such as timers, switches, fuses, containers, wires, projectiles, and gunpowder or fuel) FIGURE 14-7.

In addition, guns, knives, explosives, or booby traps may be present at WMD production sites, as well as at all other types of illicit laboratories. Much like the operators of drug laboratories, the operators of WMD laboratories do not want to be discovered and may not hesitate to use violence in an attempt to evade capture. If the laboratory operators are present, regardless of whether they are compliant, exit the scene and notify the appropriate law enforcement authorities. If it is suspected that explosive materials or devices are present, also exit the area immediately and notify the appropriate hazardous device personnel (bomb squad) or team. Those personnel may be trained to the mission-specific competencies found in NFPA 472, 6.10, *Mission-Specific Competencies: Disablement/Disruption of Improvised Explosives Devices (IEDs), Improvised WMD Dispersal Devices, and Operations at Improvised Explosives Laboratories.*

Upon discovery of any scene that appears to be an illicit laboratory, it is important to make some level of early identification of the general hazards present. Take care to not disturb evidence or interrupt any process currently in operation until you fully understand what you are doing. *In some cases it may be the best course of action to simply recognize that you are in an illicit lab of some type—get out safely, isolate and deny entry, call your local law enforcement agency, and get some expert advice on how to proceed.* Barring any threat to life, there is no reason to rush headlong into a situation you do not fully understand. You may have stumbled into what could end up being a high-profile federal crime scene.

FIGURE 14-7 Detonating cords may look like rope to an uninformed observer.
Courtesy of Dennis Krebs

In some cases it may be possible to identify the chemicals present by reading labels. Handmade labels or notes could provide clues, but never assume that the label is guaranteed to reflect the contents of the container. Keep in mind that the criminals doing the manufacturing are not playing by the same rules that govern a legitimate manufacturing process.

Chemical Laboratories

With the appropriate knowledge and the right laboratory equipment, committed criminals can manufacture a variety of crude chemical warfare agents in a garage or basement laboratory. Simple forms of blister agents such as sulfur mustard, blood agents such as cyanide, and choking agents such as chlorine can be manufactured for subsequent use in a criminal or terrorist act. The manufacture of many of these agents requires acquisition of ingredients such as solvents, acids, oxidizers, and other volatile chemicals. These precursor chemicals and finished agents, which are designed to kill or seriously incapacitate human beings, can be extremely dangerous. The hazards associated with such agents include respiratory and dermal exposures, which can result in difficulty in breathing, nausea, dizziness, nerve damage, and even death.

Chemical laboratories may also be characterized by the presence of either legitimate or improvised laboratory equipment such as heating sources, glass jars and beakers, and a vapor collection hood.

Biological Laboratories

Biological warfare agents will generally be categorized in one of four ways: bacterial agents (such as anthrax), fungal agents (aflatoxin), viral agents (such as Ebola virus or smallpox), and toxins (such as ricin or botulinum). The first three categories describe forms of living microorganisms; the last comprises by-products of living organisms.

Bacteria, viruses, and fungi all require living conditions somewhat similar to those that support humans. For example, these pathogens need a food source, warmth, and moisture to develop properly and survive. Viruses also require a host to survive for prolonged periods. To meet these needs, an illicit biological laboratory may contain laboratory glassware, Petri dishes, growth medium and protein supplements, industrially manufactured incubators, crude homemade heating sources, and slow-speed mixers or agitators. If viruses are being cultivated, hosts such as chicken eggs may also be used.

Biological toxins are typically extracted from a plant or an animal. Ricin, for example, is extracted from the castor bean plant; it is extremely toxic to humans, and poisoning with this agent has no known cure FIGURE 14-8.

Once a toxic chemical is extracted from the plant or animal, it may be dried and ground down into smaller particles for dissemination. Pharmaceutical grinders may, therefore, be present in laboratories in which toxins are being produced.

Biological laboratories are just as dangerous as chemical laboratories. Depending on the type, a biological agent may be able to enter the human body through the respiratory tract, digestive tract, or even the skin. Biological agents, however, typically have a greater time until the onset of symptoms is noticed, possibly as long as 72 hours in some cases. Symptoms of a biological agent exposure may include nausea, fever, skin blisters or rashes, fatigue, and even death.

FIGURE 14-8 Castor beans ready to be used in the production of ricin.
Courtesy of Rob Schnepp

Tasks and Operations

Regardless of whether you encounter an illicit drug or WMD laboratory, you should realize that the persons responsible for its operation have numerous reasons for wanting it to not be discovered. You should also realize that, if the laboratory is discovered, its operators will go to great lengths to ensure that incriminating evidence is destroyed and may make any attempt possible to delay their capture. Although WMD laboratories are apparently rare in the United States, they do exist and should not be dismissed as something that couldn't happen in your AHJ. Illicit drug laboratories, however, are far more prevalent and have offered a great deal of intelligence as to what laboratory operators are capable of when it comes to protecting their manufacturing sites. It is not uncommon to encounter improvised explosive devices and/or other booby traps, along with dogs intended to provide security and/or early warning of intruders, trying to make entry into the laboratory or cause harm to responders once they gain access to the inside of the site FIGURE 14-9. Laboratory operators may also try to destroy the laboratory by mixing incompatible chemicals or setting fires if they believe they have been discovered and want to burn the evidence of their crimes or divert the responders' attention while they attempt to flee.

Because of the nature of these types of criminal laboratories, it is always considered a prudent practice to include personnel properly trained in the recognition of explosive devices as part of your hazard assessment plan. If you encounter any device or substance that you suspect may be explosive, back away and notify an explosive device technician.

VOICES OF EXPERIENCE

In the late 1980s and early 1990s, methamphetamine laboratories plagued the Northwest, especially the county to which our hazardous materials team routinely responded. It was estimated that each year the county had as many as 1000 laboratories of varying sizes operating at any one time.

This particular day we were backing up the county sheriff's clandestine laboratory team as they conducted a tactical entry to serve a warrant on a suspected cook. The county sheriff's office would be responsible for the tactical entry and securing the scene and the suspect(s). Our hazardous materials team would stand by to offensively control any materials that would have the potential to be released during the entry, assist the law enforcement team in locating and identifying the hazardous materials in the occupancy, and provide decontamination support for the team.

As we looked through a bedroom of the house, we noticed a green military ammunition can under the bed with the lid partially open. Using a video camera with night-vision capability, we were able to see what appeared to be wires attached to a camera flash cell going down to what appeared to be a detonator in the bottom of the can. We evacuated the house immediately, and the device was removed by the bomb squad. It turned out to be a detonator inserted into a block of plastic explosive. The intent of the camera was to provide the electrical charge necessary to set off the detonator, causing the bomb to explode. It is unclear if this would have worked, or if the device was to be a booby trap or a weapon.

After the bomb squad cleared the rest of the house for other devices, we continued down into the basement, where the main portion of the laboratory was located. To my amazement, we found a tremendous amount of hazardous substances being stored, including multiple propane cylinders—some with altered valves storing anhydrous ammonia, some still containing propane; gallons of sulfuric acid in plastic bottles; several cans of lye (sodium hydroxide); cases of starting fluid (ether); and about 10 one-gallon cans of toluene. In large glass containers on top of a rickety aluminum shelf, we found one container with sodium metal and one container with lithium strips being stored under mineral spirits. The only ventilation was a 19-inch box fan that was pointed toward a 4-inch piece of dryer venting. The dryer venting ran outside through a piece of cardboard that had been taped over the broken basement window. Outside the basement window was a garden hose with a small cut in it that created a sprinkler effect intended to capture or help disperse the vapors from the basement.

Standing in that basement, I could only imagine the fate of fire fighters who might have found themselves inside that structure during a fire: the explosive materials, acids and bases, flammable liquids and gases, anhydrous ammonia, and the possibility that with a single sweep of a hose stream, the containers with the sodium or lithium metal could be broken, allowing the metals to come in contact with water, creating an explosion that could level the entire building.

Extreme caution should be exercised during incidents involving illicit laboratories. Without specific knowledge of exactly what is stored inside an illicit laboratory, there is no way to quantify the risk.

Bret Stohr
McChord Air Force Base
Tacoma, Washington

FIGURE 14-9 Mines have been used to provide perimeter defense around the outside of laboratories.

Courtesy of Dennis Krebs

To identify and/or avoid potential unique safety hazards, follow the steps in SKILL DRILL 14-1 (NFPA 6.9.4.1.1):

VIDEO

1. Visually assess the structure or property that is suspected to contain a laboratory operation for outward warning signs, such as boarded up windows, the presence of security or surveillance systems (including triggering devices and booby traps) unusual or extreme for the occupancy, precursor chemical containers, laboratory equipment, or hostile dogs or occupants.
2. Establish a safe containment perimeter based on the hazards identified.
3. Notify the appropriate law enforcement personnel, technicians, and allied professionals based on the hazards identified.
4. Make an assessment of any victims who may be present and any symptoms they are presenting.

Notifying Authorities

Once you have determined or suspect that you have encountered an illicit drug or WMD laboratory, you should begin by establishing a perimeter and notifying the local law enforcement agency having jurisdiction for that geographic area.

If it is a drug laboratory, the local law enforcement agency may have its own drug investigation unit or participate in a regional multiagency unit that is properly trained and equipped to handle such investigations. Depending on the quantity of drugs involved or the capacity of the illicit laboratory, the incident may be investigated by many levels of governmental law enforcement.

If you encounter a known or suspected WMD laboratory, your local law enforcement agency will likely contact state, provincial, and federal resources, such as the Federal Bureau of Investigation, because of the complexity of the crimes being committed.

Both illicit drug and WMD laboratories can have significant effects on the environment, and it is not uncommon for additional criminal charges to be brought against the offenders for the illegal storage and disposal of hazardous materials. Local and state agencies or the Environmental Protection Agency (EPA) may investigate this phase of the investigation, depending on the hazardous materials and quantities.

Coordinating Joint Agency Operations

Coordination must occur between fire service and law enforcement agencies throughout the incident. Unified command is the appropriate way to include all stakeholders in the response. Sharing information and working cooperatively will undoubtedly assist with making the mitigation efforts run smoothly. Each discipline should respect and understand the operational needs of the others. At times, the pendulum of responsibility will shift from discipline to discipline. There may be a greater emphasis on the fire department end of things at some point, switching to EMS if there are injuries or illness caused by the event, moving back to a law enforcement emphasis when the life safety elements have stabilized. Hazardous materials teams may need to make entry to take samples for identification, then re-enter to work with law enforcement for evidence collection. Your authority having jurisdiction (AHJ) will dictate procedure in these cases, but operationally, it is best if all agencies work together and avoid territorial confrontations.

Law enforcement investigators will likely interview all responders on scene and will be interested in which actions they took prior to their arrival, which evidence may have been disturbed, and why. It will be beneficial for all responders to take a few minutes and make personal notes on their own observations and involvement in the incident as soon as practical.

In addition, responders may need to establish a decontamination operation to support a forensic crime scene processing team if the team cannot provide its own decontamination. All suspects, evidence, and equipment utilized to collect that evidence may need to be decontaminated.

To conduct joint hazardous materials/hazardous device team operations, follow the steps in SKILL DRILL 14-2 (NFPA 6.9.4.1.1):

VIDEO

1. Discuss with appropriate personnel those materials or devices that are potentially explosive and/or hazardous.
2. Develop a joint response plan, if necessary, to render the device or materials safe for collection as evidence. This should include establishing safe distances; perimeter support; detection and monitoring strategies, if applicable; a communication plan; emergency medical and evacuation plans; and the like.
3. Develop a decontamination plan to support all personnel and equipment.

Determining Response Options

Once it is determined or suspected that you are dealing with an illicit drug or WMD laboratory and the proper notifications have been made to the appropriate enforcement agency, you must begin planning your response operations. This statement is not meant to suggest that life-saving or fire suppression operations should be delayed; however, as soon as the scene has begun to stabilize or it is practical, careful consideration must be given as to which response actions will be taken next. Consider implementation of the following progressive response operations.

Securing and Preserving the Scene

When you know or strongly suspect that you are dealing with some type of criminal violation, securing the integrity of the scene and preserving any evidence that is present should be a priority. Consider how your operations and mere presence may affect the evidence, and attempt to minimize the destruction or degradation of such evidence. Even if investigators aren't on scene yet, the incident is still a crime scene. As a responder, you are responsible for preserving the evidence until it can be properly documented and collected.

When possible, minimize the number of responders who enter the laboratory interior until a thorough hazard analysis can be conducted. Remember that the threat of explosives, live human threats, and WMD agents may be outside the scope of your normal duties. It is best to assemble a group of multidisciplinary personnel, such as a hazardous materials technician, a special weapons and tactics (SWAT) operator, and a hazardous device (bomb squad) technician, who can conduct a joint hazardous materials/explosives analysis. The function of this team, which consists of highly trained and properly equipped personnel, is to identify and mitigate the hazards associated with improvised explosive devices as well as to conduct unknown agent detection, air monitoring, and sampling. During this identification and initial mitigation process, every effort should be made to follow applicable evidence preservation guidelines. The members of the multidisciplinary team will ensure that the scene is rendered as safe as practical for responders to conduct further investigation.

Any type of illicit laboratory may produce a significant amount of chemical waste that can be toxic to the environment. Part of securing and preserving the scene, therefore, focuses on preventing the scene from growing larger. Illicit laboratory waste is often flushed down toilets, dumped in sewers or drainage ditches, or even dumped on the side of the road. Efforts should be made to stop the flow of toxic waste immediately and to minimize runoff caused by responders during life-saving and rescue operations.

Documenting Scene Activities

Ultimately, an illicit laboratory will be treated as a crime scene by investigators. They will need to know the identity of every responder, victim, and witness who was on scene. The incident commander (IC), or the IC's designee, should document that information as soon as possible. If practical, a rough sketch and photographs of the scene, along with any notes provided by responders, should be provided to the arriving investigators.

Personal Protective Equipment

It is imperative that responders wear PPE that is appropriate for the hazards being encountered at the illicit laboratory FIGURE 14-10. Many of the substances used in these operations may be highly toxic; in addition, some materials may cause great harm if they come in contact with skin. Structural fire fighter's gear may not provide sufficient protection against the hazards encountered during responses to illicit laboratories. PPE selection should be based on detection and sampling results and indicators observed on scene.

Detection Devices

Law enforcement agencies may not have the equipment needed to conduct thorough and effective monitoring of atmospheric conditions, such as the detection of chemical agents and the determination of upper and lower explosive limits (UELs and LELs), or to conduct decontamination and site remediation efforts. As a consequence, hazardous materials responders' support may be required throughout the course of the investigation.

FIGURE 14-10 Responders must wear the appropriate personal protective equipment while operating at illicit laboratories.
Courtesy of Rob Schnepp

Selection and use of monitoring and detection equipment at this type of scene may vary based on the precise hazards identified, but at a minimum will normally include an oxygen monitoring device, a combustible gas indicator, radiation detection devices, pH paper, and a photoionization detector (PID). The chapter *Mission-Specific Competencies: Air Monitoring and Sampling* discusses these devices in further detail.

Decontamination

Decontamination areas and equipment should be established prior to any responder entering the illicit laboratory scene. Discussions among all responders, prior to entering a hot zone, should address specific issues related to decontamination including prisoners or canines. If suspects are encountered in a contaminated area, they will need to be decontaminated while in custody. Additionally, law enforcement functions such as SWAT, hazardous device team, canine teams, and forensic evidence collection teams, as well as the evidence they collect, will all have to be decontaminated. Each of these circumstances presents unique challenges for both hazardous materials decontamination teams and law enforcement agencies. For example, procedures should be carefully followed to ensure that firearms are rendered safe prior to being decontaminated and that evidence is properly packaged, labeled, and documented as it is handled by the decontamination team.

To decontaminate tactical law enforcement personnel such as canines and SWAT teams, follow the steps in SKILL DRILL 14-3 (NFPA 6.9.4.1.1).

1. Provide clear instructions to the law enforcement officers entering the decontamination line. Realize that although they may be properly trained to wear their PPE, they may not be trained in decontamination procedures.
2. Instruct law enforcement officers to make any weapons safe by pointing them in a safe direction, unloading them completely, and locking back the firing mechanism and engaging the safety selector switch.
3. Consult with canine officers to determine the best way to handle the animal if decontamination is required.

VIDEO

Instruct law enforcement officers handling a canine to maintain control of the animal during the entire process. Have them apply a muzzle to the animal if necessary.

4. Instruct law enforcement officers handling prisoners to maintain control of the prisoners at all times. It is recommended that two law enforcement officers control each prisoner during this process, so that the officers can be decontaminated as well.
5. Begin decontamination procedures as necessary.

RESPONDER TIP

Although situations vary from state to state, more progressive agencies are now developing joint response plans that ensure specialized units such as hazardous materials technicians, SWAT, bomb squads, and forensic evidence teams can operate cohesively. This concept is often supported by shared training opportunities, full-scale exercises, and the purchase of similar or compatible equipment by the various agencies.

Remediation Efforts

Processing of illicit laboratories may not be completed in a single operational period; indeed, management of such a site could become a long-term event that may very well end up gathering widespread media attention. Imagine having to conduct multiple entries at a federal crime scene under the constant scrutiny of the national media.

The documentation and collection of evidence can, in some cases, take a considerable amount of time. In addition, chemical and biological agents may have contaminated the structure as well as soil and water on site and in the surrounding area. Detection and sampling efforts can be expanded to determine the scope of contamination and factored into the remediation plan. Remediation may include the technical decontamination of a structure or possibly the removal of a structure or soil, depending on the level of contamination.

Wrap-Up

Chief Concepts

- The need for life-saving and fire suppression operations should be weighed against the hazards posed to responders at an illicit laboratory scene.
- Illicit laboratories vary in size and can involve a variety of chemical processing methods.
- Upon discovery of any scene that appears to be an illicit laboratory, early identification of the hazards present will be vitally important.
- A criminal can manufacture a variety of crude chemical warfare agents in a garage or basement laboratory.
- An illicit laboratory is a crime scene. Make efforts to protect the area, properly preserve the evidence, and minimize your impact on any potential evidence.
- Coordination must occur between fire service and law enforcement agencies throughout the incident.
- Document important information as soon as possible.
- PPE selection should be based on detection and sampling results and indicators observed on scene.
- Selection and use of monitoring and detection equipment at this type of scene may vary based on the hazards identified.
- Decontamination procedures during illicit laboratory incidents often present unique challenges involving the need to decontaminate the environment, firearms, evidence, suspects, and canines.

Hot Terms

Clandestine drug laboratory An illicit operation consisting of a sufficient combination of apparatus and chemicals that either has been or could be used in the manufacture or synthesis of controlled substances.

Hazardous device personnel (bomb squad) Personnel trained to detect, identify, evaluate, render safe, recover, and dispose of unexploded explosive devices. Also known as a hazardous device technician.

Illicit laboratory Any unlicensed or illegal structure, vehicle, facility, or physical location that may be used to manufacture, process, culture, or synthesize an illegal drug, hazardous material/WMD device, or agent.

Methamphetamine A psychostimulant drug manufactured illegally in illicit laboratories.

RESPONDER *in action*

Your engine company has been dispatched to a residence following a report of multiple victims with chemical burns. Upon your arrival, you observe a mother and two children being given aid by neighbors in the front yard. All three victims are coughing and vomiting. One of the children has what appears to be a liquid-splash chemical burn covering most of his upper body. The upset mother manages to tell you that her husband was making "crank" in the kitchen when "things just started blowing up." She also tells you that her husband and one other child are still inside the residence.

1. Based on the information that you have, whom should you notify immediately?
   - **A.** Additional fire service units
   - **B.** The local law enforcement agency having jurisdiction in that area
   - **C.** Environmental Protection Agency
   - **D.** Federal Bureau of Investigation
2. Which of the following terms best defines an illicit drug produced by the "Red P" method?
   - **A.** Ecstasy
   - **B.** LSD
   - **C.** GHB
   - **D.** Methamphetamine
3. Law enforcement teams have arrived on scene and want to confirm the existence of an illicit laboratory. Unified law/fire command agrees that a joint hazard assessment should be performed while conducting a thorough search of the residence. Who should be involved in this team of personnel?
   - **A.** Tactical law enforcement officers
   - **B.** Hazardous materials technicians
   - **C.** Hazardous device technicians
   - **D.** All of the above
4. The father is located by law enforcement officers and is subsequently arrested. Who is responsible for decontaminating the father now that he is in custody?
   - **A.** The law enforcement agency making the arrest
   - **B.** The fire service agency having jurisdiction on scene
   - **C.** The staff at the jail to which the suspect will be transported
   - **D.** Both fire and law enforcement may have a shared responsibility to ensure proper decontamination occurs.

Mission-Specific Competencies: Air Monitoring and Sampling

Greg Henry/ShutterStock, Inc.

© Huntstock/Thinkstock

CHAPTER 15

Competencies for Operations Level Responders Assigned Mission-Specific Responsibilities

This chapter shall address competencies found in NFPA 472 for the following operations level responders assigned mission-specific responsibilities at hazardous materials/WMD incidents by the authority having jurisdiction beyond the core competencies at the operations level.

Knowledge Objectives

After studying this chapter, you will be able to:

- Plan and implement air monitoring and sampling activities. (p. 325–328)
- Select equipment suitable for detecting or monitoring solids, liquids, or gaseous hazardous materials/WMD. (**NFPA 6.7.3.1**, p. 314–315)
- Describe the operation, capabilities and limitations, local monitoring procedures, field testing, and maintenance procedures associated with each detection/monitoring device. (**NFPA 6.7.3.2**, p. 321–325)
- Describe the local procedures for responder decontamination as well as the decontamination procedures for detection/monitoring devices upon completing the air monitoring mission. (**NFPA 6.7.3.3, 6.7.4.2**, p. 318–319)

Skills Objectives

After studying this chapter, you will be able to:

- Demonstrate the field test and operation of each detection/monitoring device, and interpret the readings based on local procedures. (**NFPA 6.7.4.1**, p. 316)
- Implement the 10 basic rules for detection and monitoring. (p. 316)
- Perform a start-up procedure with a bump test. (p. 317)

You Are the Responder

Your engine company has arrived on the scene of an overturned propane tanker. The accident occurred on a city street, and several bystanders in the immediate area relay to the incident commander (IC) that they can smell gas. While the hazardous materials team prepares to address the tanker, your crew is ordered to evacuate several businesses near the accident site and determine whether the concentration of propane gas has reached a dangerous level in or around any of the buildings. Your quick research of propane reveals that it is a heavier than air gas, with a flammable range of 2.1 to 9.0 percent. You take a combustible gas indicator and begin the task.

1. Would structural fire fighter's turnout gear and SCBA offer adequate skin and respiratory protection from propane?
2. Which steps would you take if the alarm on the gas indicator instrument went off and told you the concentration of propane is above the lower explosive limit?
3. Would any additional detection and/or monitoring devices be especially useful in this situation?

Introduction

Detection and monitoring devices have greatly improved over the years, increasing the ability of first responders to assess potentially hazardous atmospheres more effectively. The first and most popular instrument used in the fire service was the MSA 2a combustible gas indicator (CGI) **FIGURE 15-1**.

This detector, introduced in 1935, was originally designed for the mining industry to detect methane gas in coal mines. The fire service discovered the benefit of detecting the presence of flammable gasses and eventually embraced the same technology. Soon thereafter, industry and the fire service began using CGIs in scenarios involving natural gas leaks and other situations where flammable gases might be present.

Today, detection and monitoring technology has advanced well past the ability to simply detect the presence of a flammable gas. Currently, emergency responders have access to a vast array of sophisticated instrumentation. For example, some of today's instruments can analyze airborne/vapor, liquid, and solid samples of a hazardous material and determine the exact identity of the substance. Some devices even allow the user to download and/or customize a unique and vast library of chemical profiles to use as a reference when analyzing unknown materials.

FIGURE 15-1 The MSA 2a combustible gas indicator was the first detection instrument used in the fire service.
Courtesy of MSA - The Safety Company

Some detection devices can pick up the presence of volatile organic compounds (VOCs), even when their concentrations are in the range of parts per billion (ppb); other detectors are designed to identify the presence of a nerve agent or some other weaponized material. Many hazardous materials response teams and fire departments carry multi-gas detection equipment suitable for use in confined space entry, rescue situations, and general-purpose detection/monitoring. Additionally, there are instruments designed to detect single hazards, such as carbon monoxide (CO), hydrogen sulfide (H_2S), and oxygen.

Regardless of the nature of the device, it is important to remember that a particular instrument, or combination of instruments, serves a unique purpose. *There is no single detection and/or monitoring device on the market today that can do it all.* For this reason, it is vitally important to understand the various types of technologies on the market and to assemble a cache of instruments that will provide the best coverage for all the different types of hazards you may encounter **TABLE 15-1**. These devices are described in detail later in this chapter.

RESPONDER TIP

Volatile organic compounds are those organic compounds (i.e., compounds containing carbon) that are capable of vaporizing into the atmosphere under normal environmental conditions. Examples include benzene, acetone, and ethyl alcohol.

To use any instrument effectively, responders must understand the operating principles of each machine, its limitations, the benefits of using the machine, and the best approach for incorporating the machine into existing response procedures.

Table 15-1 Types of Detection and Monitoring Devices

Product	Purpose/Use
Photo-ionization detector (PID)	An instrument that provides measurements of airborne VOC levels.
Combustible gas indicator (CGI)	A device designed to detect flammable gases and vapors. Also referred to as a flammable gas detector.
Gas chromatography (GC)	A sophisticated instrument used to identify the components of a sample substance that may consist of a mixture of several different chemicals.
Flame ionization detector (FID)	A detector that is similar in operational concept to the PID. A key difference is that FIDs can detect methane, whereas PIDs cannot.
Oxygen (O_2) monitoring device	A single-gas device intended to measure high (greater than 23.5 percent) and low (less than 19.5 percent) levels of oxygen in the air.
Carbon monoxide (CO) detector	A single-gas device that uses a specific toxic gas sensor to detect and measure levels of CO in an airborne environment.
Hydrogen sulfide (H_2S) detector	A single-gas device that uses a specific toxic gas sensor to detect and measure levels of H_2S in an airborne environment.
Multi-gas meter	A versatile detection device typically equipped with a combination of toxic gas sensors along with the ability to detect flammable gases and vapors. A typical four-gas configuration commonly found in hazardous materials response might include a flammable gas sensor (to detect gases at their lower explosive limit [LEL], also referred to as the lower flammable limit [LFL]), an oxygen sensor, a hydrogen sulfide sensor, and a carbon monoxide sensor. There are many different ways to configure a multi-gas meter.
Colorimetric tubes	A reagent-filled tube designed to draw in a sample of air by way of a manual or automatic handheld pump. The reagent will undergo a particular color change when exposed to the contaminant it is intended to detect. A wide array of tubes are available on the market, including those targeting ammonia, chlorine, and acetone.
pH paper	Also called litmus paper. This paper is used to measure the pH of corrosive liquids and gases.
Chemical test strips	Also known as chemical classifier strips. These test strips give the responder the ability to test for several different classifications of liquids at one time. The test strip is dipped into the unknown liquid, allowing the chemical to come into contact with several small "windows," each of which contains specific reagents designed to identify the presence of different chemical classes. Examples of the different classes detected might include halogens (chlorine, bromine, iodine), fluoride compounds, acids and bases, and solvents and oxidizers.
Specialized detection devices	Unique instruments designed to specifically identify a sample substance. These devices are capable of naming the substance in question, sometimes even by its trade name. For example, the device might identify the presence of "Gold Bond Baby Powder," instead of just saying "cornstarch." Fourier transform infrared spectroscopy (FTIR) is an example of such a technology. Raman spectroscopy is another specialized detection instrument that may be used to identify chemical substances in the field. (Both FTIR and Raman technology are discussed later in this chapter.)
Radiation detection devices	These instruments employ a variety of technologies to detect the presence of radiation. They include personal dosimeters (which measure a potential dose of radiation) as well as general survey meters designed to detect alpha, beta, and/or gamma radiation. Some specialized radiation detectors are even capable of identifying and naming the radioactive isotope present.
Personal dosimeter	A small unit that is clipped to a front shirt pocket and is capable of measuring a specific contaminant. A radiation dosimeter, for example, will record the potential dose of radiation the wearer might have received. Other personal dosimeters are designed to measure the amount of a particular chemical contaminant (to which the wearer may have been exposed) over a specified time period.

Do not become enamored with the *features* of a particular machine, such as its flashing lights, backlit display, or data logging capabilities. Although important, the features are not indicative of how the device will fit into your mission or address the potential risks your team may encounter. Instead, consider the *benefits* of each machine to determine which one is appropriate for use in a particular situation. There is a distinct difference between these two aspects of any piece of equipment: The *features* are the unique operating characteristics of the machine that primarily affect the user, whereas the *benefits* more closely reflect how the device fits into your operational plan and regional response activities. Think of it this way: Features are felt in the hand; benefits are realized in the plan. A great device with wonderful features is useless unless it fits into the detection/monitoring doctrine of the team and is based on expected operational needs.

RESPONDER TIP

There is no single detection/monitoring device on the market that can do everything. Make sure you understand how an instrument will fit into your standard operating procedures and response plans. An excellent device that is implemented incorrectly or that is purchased for the wrong reasons is useless!

Using detection and monitoring equipment requires some technical expertise, a lot of common sense, and a commitment to continual training. All too often, responders believe they can simply turn on a machine, wait for it to beep, and expect it to solve the problem. This is absolutely not the case. A reading from any detection/monitoring device, if taken out of context or misinterpreted, may cause an entire response to head off in the wrong direction, leading to an unsafe decision or a series of inefficient tactics. Using a detector or monitor entails more than just reading the screen or waiting for an alarm to sound. The responder must correctly interpret the information the instrument is providing and make sound decisions based on that information. Achieving these goals requires continual training and a commitment to becoming—and remaining—proficient with the instruments. Think of it this way: the devices provide data. The decision is a human function—made by you!

Additionally, most instruments need to be checked and maintained on a regular basis. The manufacturer will make recommendations on the best practices and procedures for testing and maintenance, and all responders who are required to use the devices should feel confident that they are being properly maintained. These routine checks need to be documented in an instrument log; your standard operating procedures will often determine how this documentation is done. Keep in mind that each instrument will require a unique maintenance plan. Additionally, many machines on the market require an annual trip back to the manufacturer for in-depth servicing. This process can be both costly and time consuming, so make sure that these costs are understood up front before your agency purchases any instrument.

Detection and Monitoring

Detection and monitoring activities, as with every other aspect of hazardous materials/WMD response, are based in part on the nature of the materials involved and the factors surrounding the release. Additionally, using a risk-based approach to understanding the problem at hand will help you maintain good situational awareness of the incident. Situational awareness (SA) is about focus and observation. It's about understanding the visual cues available, orienting yourself and perhaps other responders to those inputs relative to the current situation, and making sound decisions based on those inputs. When pilots talk about SA, they describe an ability to "get ahead of the airplane," which is good, and if all things are going well, to "stay ahead of the airplane," which is even better. Good SA is achieved by seeing and synthesizing—that is, by staying a step ahead of the game. Nothing is worse than playing catch-up when you're trying to manage an emergency.

On the scene of a hazardous materials incident, SA can be improved by combining the information obtained from resources such as detection/monitoring devices, references, and witnesses with what is observed and learned about the incident as a whole. This blend of information is then used to make well-informed strategic and tactical decisions that affect both the short- and long-term outcomes.

Good SA doesn't just happen. It is a by-product of good training, a focused demeanor, and clear thinking. When it comes to detection and monitoring, SA includes understanding the operational principles of the devices used in your authority having jurisdiction (AHJ), recognizing the terms and definitions related to detection and monitoring, and developing a sound plan to carry out the detection/monitoring actions. This plan should include some standard actions to ensure you have covered all the bases before you or your colleagues set off on a detection/monitoring mission. Following is a list of 10 suggested actions that should be considered at any hazardous materials/WMD incident. These general steps offer conceptual guidance for approaching such a scenario, while recognizing that each incident will require you to choose the most appropriate tools or combination of tools tailored to the particular details.

VIDEO

To complete the 10 basic actions for detection and monitoring, follow the steps in SKILL DRILL 15-1 (NFPA 6.7.4.1):

1. Attempt to identify the source and nature of the released substance prior to entry.
2. Research and understand the nature of any known identified atmospheric contamination. *Do not assume that only one hazard exists.*
3. Select the proper personal protective equipment (PPE) for the task.
4. Select the appropriate instrument(s) for the task.
5. Properly prepare the instrument(s) for use.
6. Prioritize your monitoring areas.
7. Develop an overall monitoring plan.
8. Confirm all readings when obtained and record when appropriate.
9. Establish action levels (readings from the devices) that could dictate tactics or other actions.
10. At the conclusion of the incident, survey the areas again to confirm that the hazard has been mitigated.

Detection and Monitoring Terminology

In broad terms, calibration is the process of setting or perhaps correcting a measuring device by adjusting it to match a known "source." For example, radiation detection equipment must be properly calibrated to provide accurate readings, and flammable gas detectors must be calibrated so the sensors "see" the intended substances at predetermined concentrations FIGURE 15-2. Initially, an instrument is calibrated at the factory prior to being shipped to the buyer. From then on, the owner of the machine must calibrate the instrument (on a regular basis) according to the manufacturer's specifications. Again, an accurately calibrated machine ensures that the device is detecting what it is intended to detect, *at a given level.*

Calibration of a gas detection device that uses an electrochemical sensor or multiple sensors FIGURE 15-3 is done by challenging the sensor(s) with a known concentration of calibration gas. To calibrate the H_2S sensor, for example, a small sample cylinder of calibration gas (which may be a mixed gas intended to challenge the entire configuration of sensors) is attached to the meter with the proper fittings. A given concentration of H_2S in the sample gas is then taken into the device to challenge the H_2S sensor. If the sensor is functioning properly, it will cause the device to respond and display the level of H_2S appropriately, demonstrating the sensor can accurately detect H_2S at a given level. Generally speaking, this same procedure is carried out for all types of toxic gas sensors. To properly document these activities, an instrument calibration history needs to be maintained throughout the lifespan of the machine.

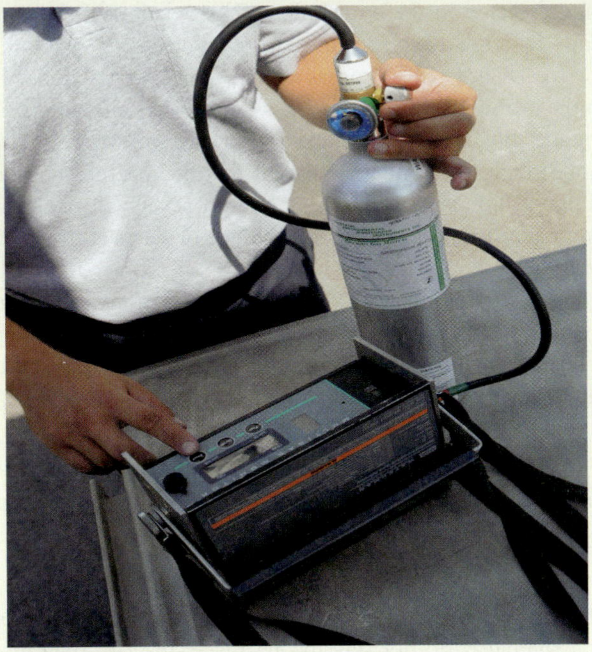

FIGURE 15-2 Calibration is a mandatory process to ensure that monitoring/detection equipment will respond appropriately during an incident. © Jones & Bartlett Learning. Photographed by Glen E. Ellman

FIGURE 15-3 A typical four-gas meter, configured as described earlier in the chapter, is safe to operate only when it is properly calibrated. Courtesy of Rob Schnepp

A bump test is a quick test carried out in the field to ensure the meter is operating correctly prior to entering a contaminated atmosphere. To perform a bump test, the device is exposed to a gas that is designed to elicit a reading from the device, possibly causing an alarm to go off. A bump test may be carried out by simply cracking open the valve of a bump gas cylinder (which can be purchased from the manufacturer of the device) near the inlet port of the detector, thereby allowing the machine to take in a small amount of gas and show some response, perhaps even to the point the machine goes into alarm mode. The gas source is then removed and the machine is allowed to recover back to normal levels. If all of this activity happens as described, the machine has been successfully bump tested and is ready for use. A bump test is typically carried out each time the machine is used. Some hazardous materials response teams bump test all of their instruments on a daily basis.

RESPONDER TIP

Calibration ensures the instrument will detect a given concentration of gas at a certain level—that is, it confirms the device will respond appropriately. *Bump tests* are quick field tests intended to make sure the device will "see" the gases it's supposed to see.

VIDEO

To perform a typical start-up procedure for a multi-gas meter, including a bump test, follow the steps in **SKILL DRILL 15-2**:

1. Turn the device on.
2. Check the status of the battery.
3. Allow the device to warm up (15 minutes is a reasonable amount of time in most cases, but you should always refer to the manufacturer's guidelines).
4. Ensure the device is "zeroed," or not picking up any readings while in a clean environment.
5. Perform the appropriate bump test.
6. Allow the device to return to zero.
7. The device is ready to use.

The period of time from when an air sample is drawn into the machine until the machine processes the sample and gives a reading is referred to as reaction time (or response time). Each detector differs from the others based on the way the sample is introduced into the instrument and the manner in which it internally processes that sample. Some gas detection devices have pumps that actively draw a sample of air into the machine; others are diffusion devices in which a target gas must permeate the sensor through a membrane. Depending on the type of device and the nature of the sensor, reaction times can be as short as 1 or 2 seconds or as long as 30 to 60 seconds—in some cases, even longer (HCN can be up to 200 seconds). To ensure that the exposure of the device to the potential contaminant is sufficient, responders must avoid surveying an atmosphere too quickly. If a sensor has a 2-minute reaction time to thoroughly evaluate an airborne environment for a particular gas, a responder must move slowly and methodically through an area. Moving too quickly will provide false readings.

Some devices may allow for the use of a long, small-diameter pick-up tube to extend the reach of the device **FIGURE 15-4**. When such tubes are used, reaction times may be increased.

VIDEO

Recovery time is another important facet of air monitoring activities that responders must understand. The recovery time of a particular device is a function of how much time it takes a detector or monitor to clear so a new reading can be taken. It is affected by many factors, including the physical properties of the substance being sampled. Some recovery times are very

FIGURE 15-4 Small-diameter pick-up tubes extend the reach of certain detection devices. Courtesy of Rob Schnepp

fast and go largely unnoticed by the user. For other devices, recovery may take anywhere from several seconds to a couple of minutes. Some complex sampling devices can take as long as 15 minutes to clear and be ready for a new sample.

You may hear the term "zero" used when discussing the concept of recovery time. A device is "zeroed" when it begins its operational period in a clean atmosphere by displaying normal values (or no values) or when the device recovers to that same baseline state after an exposure to a gas or vapor. A device cannot be accurately zeroed unless or until it is operated in a clean atmosphere.

Flammable gas detectors are calibrated to a specific gas. Of course, not every incident will involve the same gas that was used to initially calibrate the flammable gas detector. (Pentane is a common calibration gas.) To that end, each monitor will have a relative response curve that accounts for the different types of gases that might be encountered, other than the one used for its calibration. The manufacturer tests the monitor against various gases and vapors and provides a relative response factor that can be used to determine the correct percentage of the gas being monitored for. The relative response factor is a mathematical computation that must be carried out to correlate the differences between the gas that has been used to calibrate the machine and the gas that is actually being detected in the atmosphere. This relationship, in turn, allows for a relative response chart to be created.

To see how this process works, consider the following simple example: A flammable gas detector is calibrated using pentane, but it is now being used at a scene where toluene has been released. The issue here is that the machine has been calibrated for one substance (pentane), so it will not give a completely accurate reading for toluene or alarm at the proper levels of the lower explosive limit (LEL)/lower flammable limit (LFL) of toluene. Therefore, a correction factor must be employed to obtain an accurate reading at the real-world scene. This mathematical computation involves using specific values established by the manufacturer. All responders being trained to this mission-specific competency should be fully trained on the specific monitoring/detection devices used by their agency.

Detection and Monitoring Concepts

When they are released into the atmosphere, vapors and gaseous chemicals do not stay in one place. Instead, they move through the area, with air currents, ventilation systems, and other influences creating a constantly changing environment. Although you may take a particular reading in one part of a room or one part of a building, all other parts of that same room or building will not necessarily have the same atmosphere. The presence and concentration of lighter-than-air substances, in particular, are very difficult to pinpoint. In large or complicated incidents, a monitoring diagram may be needed to help keep track of areas that have been monitored and the readings that have been obtained. Additionally, it is prudent to think through what technologies are appropriate for the situation and what you are trying to accomplish with the detection and monitoring plan. Again, recognizing that one device cannot detect everything, detection and monitoring actions may need the input of a technician level responder or other allied specialist. You may end up employing several devices and/or a variety of technologies to adequately cover all your detection and monitoring bases FIGURE 15-5.

RESPONDER TIP

Each incident is different, so it is important to develop a monitoring plan that makes sense for the given situation.

Before monitoring inside a building, for example, it is prudent to start monitoring the outside atmosphere prior to making entry. Start by making a complete circle of the building, taking time to check all natural openings—windows, doors, air intakes, drains, or any other place that "pokes through" the building. This approach may help to identify a general area that warrants further monitoring inside. The same holds true for monitoring a particular room, or around a drum, or inside a cargo delivery van: Start from an outer perimeter and work inward, being mindful to monitor natural openings or other places where gaseous substances may escape; however, this is

FIGURE 15-5 When developing a detection/monitoring plan for an unknown atmosphere, it is advisable to use several different types of devices. You may need to bring a number of devices to cover all anticipated hazards. Courtesy of Rob Schnepp

only a general tip. Keep in mind that the gas or vapor may actually be migrating to the opening from somewhere deep inside the building.

Prior to entering the building, monitor around the door. Once this step is completed, open the door only far enough to insert the meter, and take another sample. This should be done at the top, middle, and bottom of the door. After interpreting those readings, enter the building if it appears to be safe to do so. In the same manner that you would conduct a primary search of a building during firefighting operations, work your way through all of the rooms and other open areas.

The monitoring of a building can be very time consuming. Buildings with multiple stories will take even longer to fully cover. Additionally, you may need to rely on several monitoring teams to cover large areas in a timely fashion. Remember—even if you do find a specific source for the leak, it does not rule out the possibility that the substance might have found its way into other areas of the building. Even after the situation has been controlled and/or otherwise ventilated, the entire building will need to be monitored again to deem it safe for the occupants.

Unfortunately, achieving the same flexibility is not feasible with PPE: You cannot wear several different types of garments simultaneously to address multiple threats. Just as no single instrument will detect all chemicals, no single type or configuration of PPE will protect responders against all possible hazards. In general terms, inhalation hazards are the most dangerous, but the easiest to protect against. To that end, the self-contained breathing apparatus (SCBA) offers the best protection. Research regarding the specific hazard and the general situation is critical in choosing the right PPE for a particular incident, while keeping in mind the mission of the monitoring and detection activities.

Not only do personnel and PPE need to be decontaminated after exposure to hazardous materials, but detection and monitoring equipment may also need to be decontaminated. The type of decontamination will depend on the exposure or contamination. To avoid damage to the device during decontamination, refer to the manufacturer's guidelines. In some cases, the equipment will need to be bagged and tagged and then decontaminated away from the scene. The manufacturer may need to be contacted for further recommendations on how and where decontamination should be done.

Detection and Monitoring Unknowns

Monitoring for unknown hazards can be difficult. In most cases, because there is not one type of detection device that will check the atmosphere for everything, several different devices may need to be used. Much valuable information may be missed if only one type of detection device is used or if the person using the instrument does not pay attention to the information provided by the device.

When the atmosphere is completely unknown to the responders, the potential to end up in a dangerous situation is greatly increased, especially when the atmosphere contains flammable gases or vapors. The potential for accidental ignition of these substances creates an extreme hazard for responders. In some cases, you may be able to identify only the most basic characteristics of an unknown atmosphere, such as whether it contains a flammable or corrosive material. The PID, for example, may pick up a reading in the parts per million (ppm) or parts per billion (ppb) range, whereas the flammable gas detector, pH paper, radiation survey meter, and other instruments may not show any readings in similar circumstances. Or perhaps the flammable gas detector shows that the substance is 3 percent LEL/LFL but you don't know what the substance is, so you cannot use any correction factors to ensure the accuracy of the reading. Many variables come into play here, and it is incumbent on responders to place the readings that are obtained into the proper context (which might be understood at only a superficial level) before forming an opinion on the substance(s) that might have been released.

Sometimes you may use detection/monitoring devices to confirm what you already suspect. For example, acetonitrile (a flammable liquid) might have been spilled inside the laboratory at a biotechnology facility, and you might use a device to confirm the presence of that substance and monitor the environment for the LEL/LFL. At other times you may use detection/monitoring devices to ask questions about the atmosphere: Is it flammable, or corrosive, or radioactive? Suppose a suspected methamphetamine laboratory has been discovered, and you are making an initial entry into the facility to determine if any hazard exists. In this case, what is the precise hazard?

VOICES OF EXPERIENCE

We were dispatched to a possible carbon monoxide incident in a duplex structure. When we arrived on the scene we discovered five birds, each in a separate cage on the front lawn. As we got closer, we realized that three of the birds were dead. All of the occupants of the house had already been evacuated. We interviewed each of the residents separately as well as together. There was no variation of the story. We learned that the woman had cooked tortillas for dinner using a gas stove.

We entered the house and obtained a normal reading of 5 parts per million (ppm) of CO. Every now and again we would hit 10 to 15 ppm in dead pockets of air in the house. We checked the stove, furnace, and hot water heater, but we did not obtain any readings above 5 ppm.

We asked the woman to come back in and demonstrate what she had been doing on the stove. She fired up the stove, but no unusual readings were obtained. As she placed the pan on the stove over the fire to heat the tortillas, however, we instantly began to get readings upwards of 125 ppm. I took the pan, cooled it, and read the engraved label. The pan was made with Teflon. Because I was a bird owner, I knew immediately why the birds died: Heated Teflon is deadly to birds. But what had caused the high readings?

After returning to quarters, we did some research, but nothing could be found to explain why we were getting the high CO readings. We contacted the CO detector manufacturer, and it was unaware of why this spike was happening. About 3 months later I was contacted by the manufacturer. It confirmed that the meter will, in fact, detect heated Teflon.

Dominick Iannelli
Fairfax County Fire Department
Fairfax, Virginia

When responders are going into an unknown atmosphere, the following monitoring and detection devices may prove useful (these devices are discussed in the following sections):

- Photo-ionization detector or flame ionization detector
- Combustible gas indicator or multi-gas meter with CO, O_2, and H_2S sensors
- Oxygen monitoring device (when appropriate)
- Carbon monoxide detector (when appropriate)
- pH paper (wet and dry)
- Radiation detection device
- WMD agent detection capability (if suspected)

Types of Detectors and Monitors

There are a wide variety of devices on the market to assist first responders with detection and monitoring activities. In one respect, the diversity of available instruments is beneficial: It allows for a multifaceted approach to detection and monitoring. The informed responder knows that one "box" cannot detect all types of hazardous materials/WMD substances, and that to rely on only one type of technology is a mistake. At the same time, the diverse nature of the instruments on the market today may create confusion. Responders may not completely understand the benefits and limitations of all available instruments, so they may end up purchasing a machine that will not do the job they had intended it to do. The following sections outline the basic operating principles of the various types of instruments in use throughout the hazardous materials response industry.

RESPONDER TIP

All detectors/monitors have limitations. It is important for any responder using instrumentation to fully understand those limitations. Examples of instrument limitations may include length of operation due to battery life; ability to operate in conditions marked by temperature extremes, humidity, or moisture; sensor life; need for user training; lack of proper calibration; software installation; and radio frequency interference.

Photo-Ionization Detector

A photo-ionization detector (PID) is a general survey instrument designed to detect vaporous chemicals at very low levels, often in the ppm range FIGURE 15-6. PIDs operate by using an ultraviolet light lamp to break down the sample gas into electrically charged components (ions) and negatively charged electrons, which produce a current; this current is then amplified and displayed by the instrument. It is common for a PID to detect concentrations as low as 0.1 ppm. In fact, some PIDs are capable of picking up vapors and mists whose concentrations are in the range of parts per billion (ppb).

Unfortunately, PIDs will not identify the material. Instead, they simply alert the user to the presence of something in the air, usually an organic vapor or mist. These vapors or mists are sometimes referred to as volatile organic compounds (VOCs).

PIDs are also capable of detecting many other substances, such as ammonia and hydrogen sulfide. PIDs can help pinpoint the sources of small leaks and identify whether any vaporous chemicals are present in low concentrations.

RESPONDER TIP

Photo-ionization detectors come with different lamps for detection purposes. In order to be read by a PID, the vapor or gas to be sampled must be ionized, which is called ionization potential (IP). The unit of measurement of an IP is electron volts (eV). The gas must have an IP less than the eV rating on the lamp. The most common are 9.8, 10.2, 10.6, and 11.7 electron volts (eV). The most common lamp for emergency responders is 10.6, but keep in mind that some products like methanol (10.85), ethane (11.52) and carbon monoxide (14.01) have higher ionization potentials. Nerve agents can be detected with the PID using the 11.7 lamp.

Combustible Gas Indicator

Combustible gas indicators (CGIs), also known as flammable gas detectors, are employed to detect flammable and potentially explosive atmospheres FIGURE 15-7. Several types of

FIGURE 15-6 Photo-ionization detectors will alert users to the presence of organic vapors or mists in the air. Courtesy of Rob Schnepp

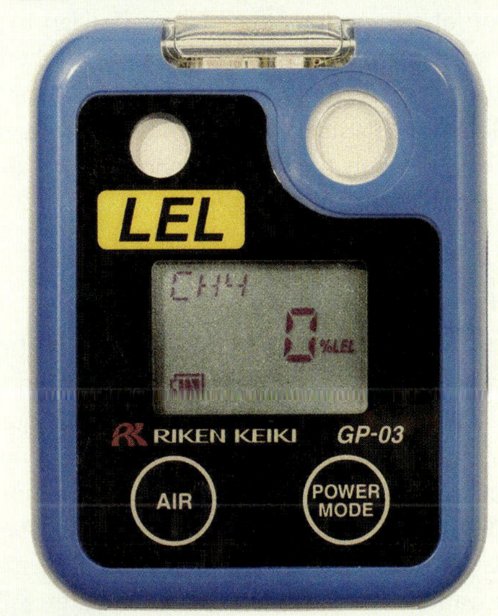

FIGURE 15-7 Combustible gas indicators provide information about flammable or explosive atmospheres. Courtesy of RKI Instruments, Inc.

sensors are used in CGIs, and it is important to understand the advantages and limitations of each because they do not all operate in the same way. Refer to the manufacturer's specifications for a complete operational description of the meter and the sensors on which it relies.

RESPONDER TIP

The CGI needs a minimum of 10 percent oxygen concentration to function properly; however, it is also relatively accurate when the atmospheric oxygen concentration is between 19.5 and 23.5 percent.

CGIs detect flammable atmospheres at or below their LEL/LFL. Most CGIs measure a percentage of the LEL/LFL of the gas they have been calibrated for, not a percentage of the LEL/LFL as it relates to the volume of the atmosphere—an important distinction. Most of these devices are set to alarm at a value of 10 percent of the LEL/LFL of the substance for which they have been calibrated, in increments from 0 to 100 percent of the LEL/LFL. For example, if the device is calibrated for methane and the operator obtains a reading of 5 percent of the LEL/LFL for a flammable atmosphere containing methane, the atmosphere is halfway to the alarm point of 10 percent of the LEL/LFL of methane. If the device is calibrated for methane and the same 5 percent of the LEL/LFL reading is obtained but the chemical is unknown, it is not possible to accurately assess the atmosphere because the appropriate correction factor cannot be used. At this point, the detector is simply "seeing" the presence of a flammable atmosphere.

The highest level of danger occurs when the atmosphere reaches 100 percent of the LEL/LFL. At this point, the operator should interpret the reading as indicating that a sufficient amount of flammable gas or vapor is present in the air to support combustion.

An operations level responder who has been trained under the competencies of this section should always work under the direction of a hazardous materials technician to assist the responder with understanding the nuances of the detection technology on the market. When using a CGI, for example, you must know the flammable range of the gas you are detecting and be able to apply the relative response factor (correction factor) for that instrument.

Gas Chromatography

Gas chromatography (GC) uses a series of techniques to break down sample gases into their various components. The operational principles of GC are complex, so this quite simple description merely touches on the most basic of concepts. Gas chromatography pulls a sample gas into a separation column (also called an analytical column) through which the different compounds move at different rates, allowing the machine to identify the various components of the gas mixture **FIGURE 15-8**. In some cases, GC instruments may be combined with flame ionization detectors (discussed in the next section) to measure the amount of each component. Some larger hazardous materials teams use GC as part of their detection/monitoring strategies; however, same of these devices can be quite expensive and require specialized training to become proficient in their use. Others are compact and designed to be portable for field development.

FIGURE 15-8 Gas chromatography identifies the various components of a gas mixture. Courtesy of SRI Instruments

Flame Ionization Detector

A flame ionization detector (FID) is a versatile instrument that can be used as either a general survey instrument or a qualitative instrument **FIGURE 15-9**. The operational principle underlying an FID is similar to that for a PID: A sample gas is broken down into electrically charged ions, which produce a current that is amplified and displayed by the instrument. Instead of an ultraviolet light, however, FIDs use a tiny hydrogen flame to break down the organic substance into ions. FIDs also detect methane whereas PIDs cannot. When used in conjunction with a GC instrument, the FID can determine the exact amount of each component of the gaseous mixture.

FIDs are sensitive and can read concentrations that fall within the low ppm range. FID and GC instruments are expensive, however, and both require a lot of training to correctly use them and interpret their results. Organized hazardous materials teams most commonly use these types of instruments.

Oxygen Monitoring Device

An oxygen (O_2) monitoring device measures the amount of oxygen in the air **FIGURE 15-10**. Ambient air at sea level contains 20.9 percent oxygen. Therefore, any reading demonstrating an oxygen concentration of less than 19.5 percent is considered an oxygen-deficient atmosphere. Any reading in excess of 23.5 percent is considered an oxygen-enriched atmosphere. Both conditions are problematic for the emergency responder. Too little oxygen creates a health risk; too much oxygen produces an elevated fire risk.

When performing any detection and/or monitoring activities, it is important to operate with these parameters in mind. When monitoring for a combustible gas, both oxygen-deficient and oxygen-enriched atmospheres will affect the performance of the instrument, possibly putting you in a dangerous

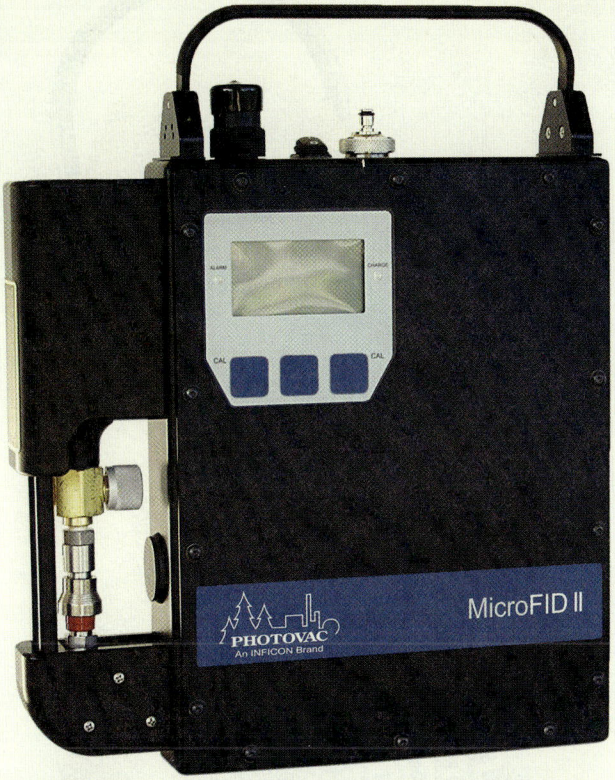

FIGURE 15-9 Flame ionization detectors can determine the identity of a hazardous substance.
Photo courtesy INFICON

FIGURE 15-10 Oxygen monitoring devices are needed to evaluate the amount of oxygen in the air.
Courtesy of BW Technologies by Honeywell

situation. Oxygen monitors typically provide accurate readings when the O_2 concentration is between 0 and 25 percent, and they are commonly employed in confined-space work or other situations where oxygen deficiency may be encountered.

Carbon Monoxide Detector

A carbon monoxide (CO) detector is a designed to identify the presence of carbon monoxide FIGURE 15-11. This colorless, odorless gas is lighter than air and has a flammable range between 12 percent and 74 percent. CO is generated during the combustion process, with its levels reaching as high as 10,000 ppm during the course of a normal working structure fire. Long-term exposure to small amounts of CO can cause chronic health effects such as cardiac disease. Short-term exposures to large amounts of CO can prove rapidly fatal, because they severely compromise the ability of the bloodstream to supply oxygen to the heart, brain, and other vital organs. The immediately dangerous to life and health (IDLH) exposure limit for CO exposure is 1200 ppm for 30 minutes in a healthy nonsmoker with a normal resting respiratory rate.

FIGURE 15-11 Carbon monoxide detectors identify the presence of this odorless and colorless gas.
Courtesy of Rob Schnepp

During the overhaul phase of a structure fire, fire fighters should consider the need for some form of mechanical ventilation, such as an electric smoke ejector. Be wary of using gas-powered blowers because they may pick up exhaust (containing CO) from the motor and blow it into the structure. In atmospheres where the CO concentration is in the vicinity of 35 ppm, all emergency personnel should be wearing SCBA. See the chapter *Fire Smoke: Responder Health and Safety* for more detailed information on CO and other gases at the fire scene.

RESPONDER TIP

Potential Health Effects of a CO Exposure

- Concentration of CO that can be inhaled for 1 hour without appreciable effect: 400–500 ppm
- Concentration of CO causing unpleasant symptoms after 1 hour of exposure: 1000–2000 ppm
- Dangerous concentration of CO for exposure of 1 hour: 1500–2000 ppm
- Lethal concentrations of CO for exposures of less than 1 hour: 4000 ppm or more

Responder Safety Tip

Cigarette smoking subjects the lungs to a CO concentration of approximately 475 ppm for 6 minutes per cigarette. Smokers have residual CO levels of 3–13 percent in their bloodstream.

FIGURE 15-12 Hydrogen sulfide monitors are often used in confined spaces. Courtesy of Sensit Technologies

Hydrogen Sulfide Monitor

A hydrogen sulfide (H_2S) monitor is specific for the gas H_2S and is often used in confined spaces, and used extensively in petroleum manufacturing facilities such as refineries FIGURE 15-12. Hydrogen sulfide is a by-product of decaying organic materials and is often referred to as "sewer gas." This gas is flammable, is heavier than air, and affects the body much like hydrogen cyanide—that is, it blocks the cells from using oxygen. Hydrogen sulfide has a strong pungent odor, at least when a person first smells it; it quickly deadens the sense of smell to the point where the individual may no longer be able to detect the gas by its odor. H_2S monitors are particularly useful because of this property, because the monitor will be able to "smell" the gas long after you cannot! It is common practice to have responders or workers wear personal H_2S monitors while working in a confined space.

FIGURE 15-13 Multi-gas meters monitor for several gases at the same time. © Jones & Bartlett Learning. Photographed by Glen E. Ellman

Multi-gas Meter

A multi-gas meter is capable of detecting several different hazards at the same time FIGURE 15-13. Some agencies refer to these types of instruments as "three-gas" or "four-gas" monitors. A typical arrangement includes sensors for oxygen, carbon monoxide, hydrogen sulfide, and flammable gas. These units are an excellent choice for general gas detection at confined-space incidents and as multipurpose detectors.

Multi- and single-gas detectors usually rely on specific electrochemical sensors—a reliable technology that has been around for quite some time—to detect the presence of target gases FIGURE 15-14. This type of sensor is filled with a chemical solution that, upon contact with the gas it is intended to detect, causes a chemical reaction that generates some amount of electrical current. The current is read by electrodes, which then display the value on the instrument's screen.

Electrochemical sensors have a shelf life of 1 to 1 1/2 years, depending on the use. This defined life expectancy is one reason why calibration and bump testing are so important for multi-gas meters: Over time, the sensors can (and will) lose their ability to function at 100 percent.

VIDEO

On some occasions, other gases may interfere with the target gas, thereby creating the potential for a false reading on a multi-gas meter. For example, a carbon monoxide sensor may pick up the presence of hydrogen sulfide, hydrogen, or even hydrogen cyanide; similarly, a cyanide sensor is subject to interference from H_2S and sulfur dioxide. As stated earlier in the chapter, detection and monitoring is not a "point and shoot" proposition; rather, many subtleties and nuances must be understood to truly appreciate what the machines are telling you.

FIGURE 15-14 A variety of electrochemical sensors ready to be used.

Courtesy of Rob Schnepp

To use a multi-gas meter to provide atmospheric monitoring, follow the steps in SKILL DRILL 15-3:

1. Turn the unit on and let it warm up (usually 5 minutes is sufficient).
2. Ensure the battery has sufficient life for the operational period.
3. Identify the installed sensors (e.g., oxygen, lower explosive limit, hydrogen sulfide, carbon monoxide) and verify they are not expired. (Note: some units have a built in PID function.) Review alarm limits and the audio and visual alarm notifications associated with those limits. Review and understand the types of gases and vapors that could harm or destroy the sensors. Use other methods to check for those substances (e.g., pH paper) to ensure they are not present in the atmosphere to be sampled. Care must be taken to avoid pulling liquids into the device—it is designed to sample air, not liquid! (STEP 1)
4. Perform a test on the pump by occluding the inlet and ensuring the appropriate alarm sounds. (STEP 2)
5. Perform a fresh air calibration and "zero" the unit. (STEP 3)
6. Perform a bump test—expose the unit of a substance or substances that the unit should detect and react to accordingly. In essence, you are making sure the unit will "see" what it is supposed to see before it might be called upon to see it in a real situation. (STEP 4)
7. Allow the device to reset, or return to fresh air calibration state.
8. Review the alarm levels and resetting procedures for addressing sensors that become saturated or are exposed to too much gas or vapor. (STEP 5)
9. Review other device functions such as screen illumination, data logging, if available, and low battery alarm. (STEP 6)
10. Review decontamination procedures.
11. Carry out monitoring and detection. (STEP 7)

SKILL DRILL 15-3

Using a Multi-gas Meter

NFPA 6.7.4.1

VIDEO

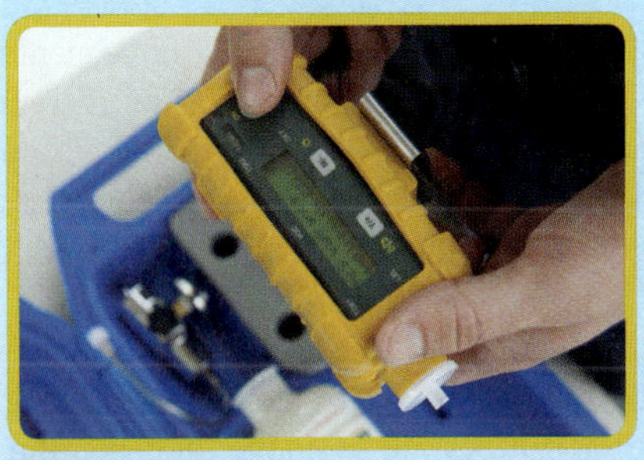

1 Turn the unit on and let it warm up (usually 5 minutes is sufficient).

2 Perform a test on the pump.

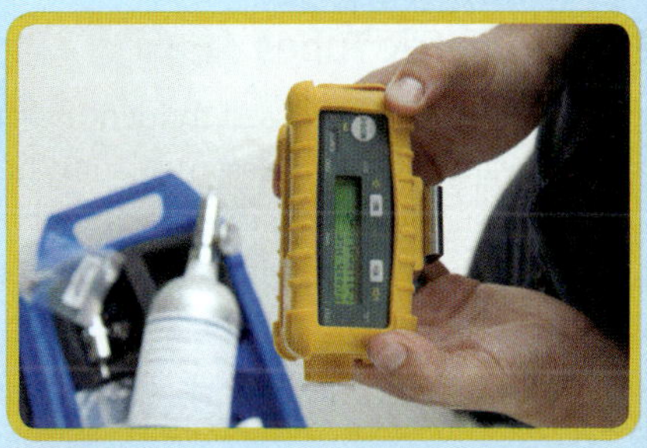

3 Perform a fresh air calibration and zero the unit.

(Continued)

SKILL DRILL 15-3 (CONTINUED)

Using a Multi-gas Meter

NFPA 6.7.4.1

VIDEO

4 Perform a bump test.
© Jones & Bartlett Learning. Photographed by Glem E. Ellman

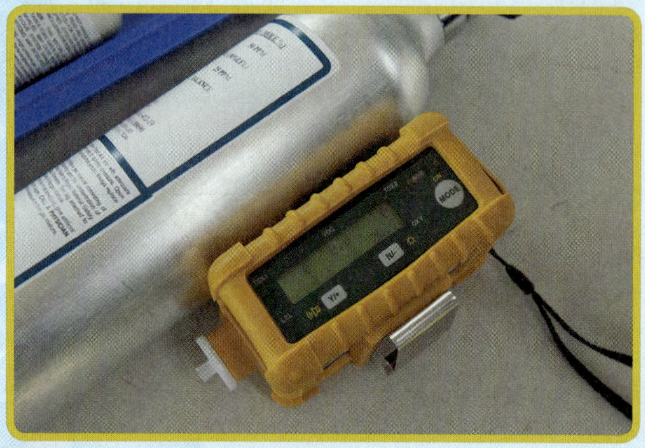

5 Allow the device to reset, or return to fresh air calibration state. © Jones & Bartlett Learning. Photographed by Glem E. Ellman

6 Review other device functions, such as screen illumination, data logging if available, and low battery alarm. © Jones & Bartlett Learning. Photographed by Glem E. Ellman

7 Review decontamination procedures. Carry out monitoring and detection.
© Jones & Bartlett Learning. Photographed by Glem E. Ellman

Colorimetric Tubes

Colorimetric tubes are used to identify known and unknown chemical vapors. These glass tubes are filled with reagents—that is, substances designed to bring about a reaction when they come in contact with another substance **FIGURE 15-15**. Each reagent reacts to a unique substance at a particular concentration. Colorimetric tubes are designed to detect single substances and/or chemical families (groups).

Colorimetric tubes operate by using a manual bellows-type hand or automatic pump to draw an air sample through a glass tube or a series of tubes (with both ends snapped off to allow for air to be drawn through the tube) **FIGURE 15-16**. If the suspected airborne contaminant is encountered, the reagent in the tube progressively changes color, much like the mercury of a thermometer going up in response to an increased temperature. The numerical values on the tube indicate the level of contamination. (Refer to manufacturer guidance to determine correct color change for specific chemicals.)

Colorimetric tubes offer hazardous materials responders the ability to detect a wide variety of substances using a single technology. WMD-specific tubes are available, and combinations of tubes can be used to test for several different substances at the same time. Colorimetric tubes are a solid, reliable method to test for airborne contamination and have been used in hazardous materials response for many years.

FIGURE 15-15 Colorimetric tubes are used to identify chemical gases and vapors using reagents.

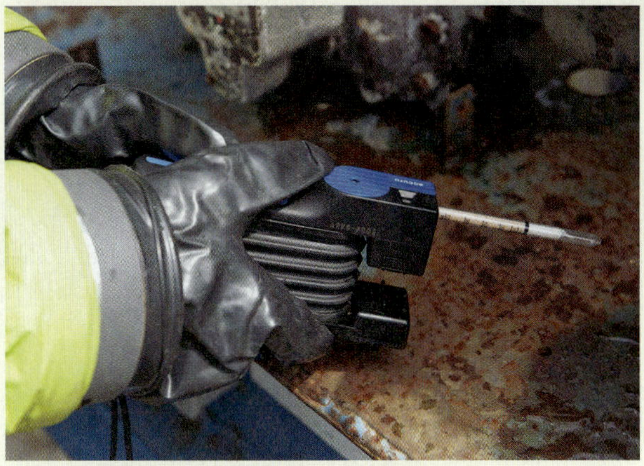

FIGURE 15-16 The bellows-type hand pump draws in the air sample through the tube. Courtesy of Rob Schnepp

To use colorimetric tubes, follow the steps in SKILL DRILL 15-4:

1. Determine the mission and relevance of using colorimetric tubes for air sampling. (STEP 1)
2. Select the appropriate tubes for the mission. Ensure the tubes are within their life expectancy—do not use outdated tubes. (STEP 2)
3. Collect all parts of the tube sampling system including pumps, tubes, extension hoses, etc. (STEP 3)
4. Review the directions for using each tube, including pump stroke count, tube cross sensitivities, and expected reactions and color changes with the selected tubes. (STEP 4)
5. Perform a pump test as specified by the manufacturer to ensure there are no leaks. A typical test is to use an unopened tube to hold the negative pressure inside the pump for a given period of time. (STEP 5)
6. Assemble the tubes and other sampling apparatus prior to entering a contaminated area. (STEP 6)
7. Perform air sampling in accordance with the directions on the tube, using an unreacted tube as a control to view any color changes. (STEP 7)
8. Discard used tubes in accordance with manufacturer's guidelines.

SKILL DRILL 15-4 Using Colorimetric Tubes

1 Determine the need for using colorimetric tubes.

2 Select the appropriate tubes for the mission.

(Continued)

SKILL DRILL 15-4 (CONTINUED) Using Colorimetric Tubes

3 Collect all parts of the tube sampling system, including pumps, tubes, extension hoses, etc.

4 Review the directions for each selected tube. Always take an unopened, unreacted tube to use as a control for tubes used in air sampling.

5 Perform a pump test.

6 Assemble a sampling kit with the needed equipment, including the proper tools to open both ends of the selected tubes.

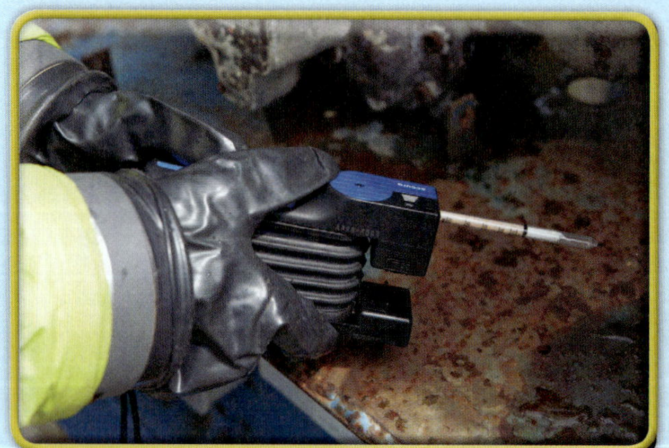

7 Perform air sampling.

pH Paper

pH paper is a chemical paper that allows the user to determine if a liquid, vapor, or solid is an acid or a base FIGURE 15-17. There are a number of technical ways to determine if a particular substance is an acid or a base, but the most common way to define them is by their pH. (If a small amount of the unknown product is mixed with distilled or neutral water, away from the bulk of the product in the event it is water reactive, pH paper can be used on solids to determine if the product is an acid or base.) In simple terms, you can think of pH as the **p**ower of **H**ydrogen. Essentially, it is an expression of the amount of dissolved hydrogen ions (H^+) in a solution. When thinking about pH from a field perspective, it can be viewed as a measurement of corrosive strength, which may be loosely translated to the degree of hazard. We can use pH to judge how aggressive a particular corrosive might be. To assess that hazard, you must understand the pH scale. Refer to the chapter *Hazardous Materials: Properties and Effects* for a review of the pH scale.

Common acids, such as sulfuric, hydrochloric, phosphoric, nitric, and acetic (vinegar) acids, have a predominant amount of hydrogen ions (H^+) in the solution and, therefore, have pH values less than 7. Chemicals that are considered to be bases, such as sodium hydroxide, potassium hydroxide, sodium carbonate, and ammonium hydroxide, have a predominant amount of hydroxide ions (OH^-) in the solution and, therefore, have pH values greater than 7. The middle of the scale (pH = 7) is where a chemical is considered to be "neutral"—it is neither acidic nor basic. At this point, the concentration of hydrogen ions is exactly equal to the concentration of hydroxide ions produced by dissociation of the water. Such a chemical will not harm human tissue. Solvents such as gasoline, toluene, and diesel fuel, for example, will not have high or low pH values because they lack free hydrogen that might otherwise react with the pH paper. That fact may be used to confirm or rule out whether a particular substance is corrosive in nature.

Generally, chemicals with pH values of 2.5 or less, and 12.5 or more, are considered to be "strong." Strong corrosives (either acids or bases) will react more aggressively with metallic substances such as steel and iron; they will cause more damage to unprotected skin; they will react more adversely when contacting other chemicals; and they may react violently with water. The pH scale goes from 0 (strong acid) on the low end of the scale, to 14 (strong base) on the opposing end of the scale.

pH paper is simple to use, and it can serve as a valuable diagnostic and confirmation tool when used in conjunction with other detection technologies. When performing air monitoring where an unknown substance is present, pH paper is used initially to ensure that the airborne environment is not corrosive. This step is taken to protect the sensors of some detection devices. The sensors of many multi-gas meters, for example, may be "poisoned" and rendered useless if assaulted by a heavy concentration of a corrosive vapor.

Chemical Test Strips

A chemical test strip looks somewhat like pH paper, but can perform several tests at once FIGURE 15-18. Each test strip will have multiple areas on it that allow for a wide range of tests to be carried out simultaneously—for example, tests for halogens (chlorine, bromine, iodine); fluoride compounds; acids and bases; and solvents, pesticides, and oxidizers. The test strip is dipped into an unknown liquid, allowing the chemical to come into contact with several small "windows," each of which contains specific reagents designed to identify the presence of different classes of chemicals.

Specialized Detection Devices

A specialized detection device goes beyond the capabilities offered by most of the more common detection and monitoring devices. It is sometimes called a hazardous materials identifier because it can analyze a substance and, in many cases, identify it by chemical name. It is not necessary that you completely understand these sophisticated devices unless one is used in your agency. If so, highly trained responders will undoubtedly run the tests and use these machines, because they require skill and training to operate properly.

FIGURE 15-17 pH paper is used to determine if a substance is an acid or a base. The reading seen here indicates an acid.

FIGURE 15-18 Chemical test strips contain multiple areas that allow for a wide range of tests to be carried out simultaneously.

Courtesy of Rob Schnepp

One of the most commonly used identification technologies is Fourier transform infrared spectroscopy (FTIR) **FIGURE 15-19**. FTIR technology uses infrared radiation to excite the molecules of the sample substance. This excitement creates a unique fingerprint for the substance in question, which is then compared against a library of fingerprints (also called spectra) stored in the device's memory to see if there is a match. Some libraries include tens of thousands of spectra. FTIR technology is used to identify many common industrial chemicals as well as WMD agents, pesticides, and drugs. It works best on pure substances, however; analysis of mixtures can prove somewhat problematic, though the obstacles are not insurmountable when experienced responders use the technology. Some manufacturers allow users to add spectra to the existing library, thereby providing for customization of the FTIR equipment.

A complementary technology to FTIR is Raman spectroscopy. This technology, which relies on the molecular scattering of light, is named after Indian physicist C. V. Raman and has been used in the laboratory setting since the late 1920s. Raman spectroscopy uses a laser as its infrared light source. The light is scattered when it collides with the sample source, instead of being absorbed by it, as occurs with FTIR **FIGURE 15-20**. Additionally, this method of analysis does not destroy the sample source, and it can be used to identify substances still inside clear containers (such as glass and clear plastic containers). In these situations, the laser is pointed at the sample substance to scatter the light in an attempt to identify it. This can be a hazardous proposition, however, because the laser is unprotected and can cause severe eye damage if a person looks into the laser.

Radiation Detection Devices

Radiation detection devices detect and, in some cases, identify which type of radiation is present (alpha, beta, or gamma). These instruments may function as a general survey meter—such as the Ludlum General Purpose Rate Meter **FIGURE 15-21**, looking for any type of radiation—or as a specific meter, looking only for a certain type of radiation.

Recall from the chapter *Hazardous Materials: Properties and Effects* that radioactive isotopes give off energy (alpha, beta, or gamma) from the nucleus of an unstable atom in an attempt to reach a stable state. Each of the forms of energy given off by a radioactive isotope varies in intensity; consequently, the type of radiation determines how responders should attempt to reduce their exposure potential.

Recall also that radioactive *contamination* occurs when radioactive material is deposited on or in an object or a person. Radioactive materials released into the environment can cause air, water, surfaces, soil, plants, buildings, people, or animals to become contaminated. A contaminated person has radioactive materials on or inside his or her body. In some cases, it is possible to use radiation detectors to determine if radioactive contamination exists on humans, perhaps after decontamination has been performed.

As mentioned previously, some radiation detectors now on the market will identify the isotope present **FIGURE 15-22**. These are specialized devices, although they are usually easy to operate and their readings are easy to interpret. In addition,

FIGURE 15-19 Fourier transform infrared spectroscopy can be used to analyze a substance and, in many cases, identify it by chemical name.
Courtesy of Rob Schnepp

FIGURE 15-20 Raman spectroscopy uses a laser as its infrared source; the resulting light is scattered when it collides with the sample source. Courtesy of Rob Schnepp

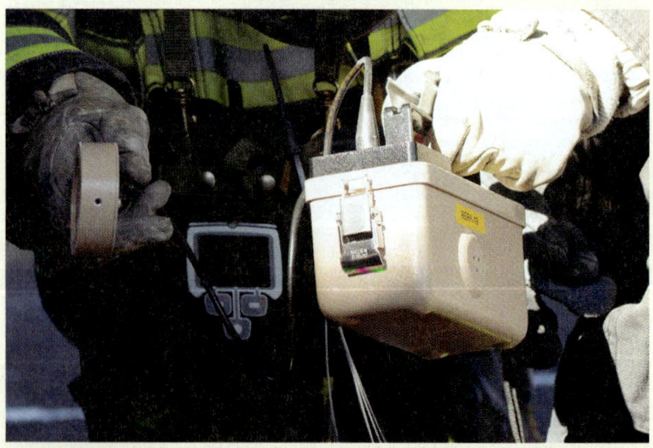

FIGURE 15-21 Ludlum General Purpose Rate Meter. Courtesy of Rob Schnepp

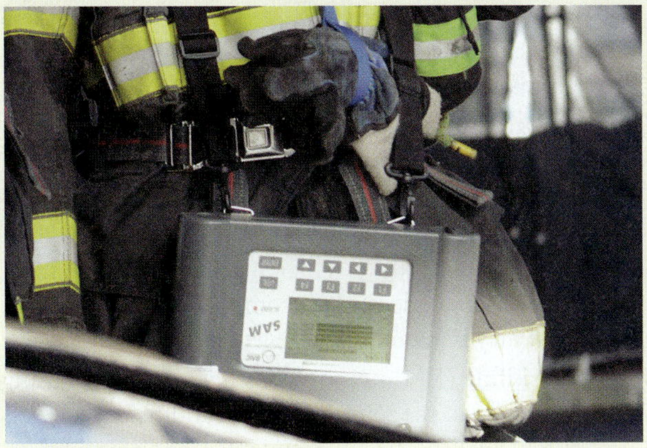

FIGURE 15-22 Some radiation detectors now on the market will identify the isotope present. Courtesy of Rob Schnepp

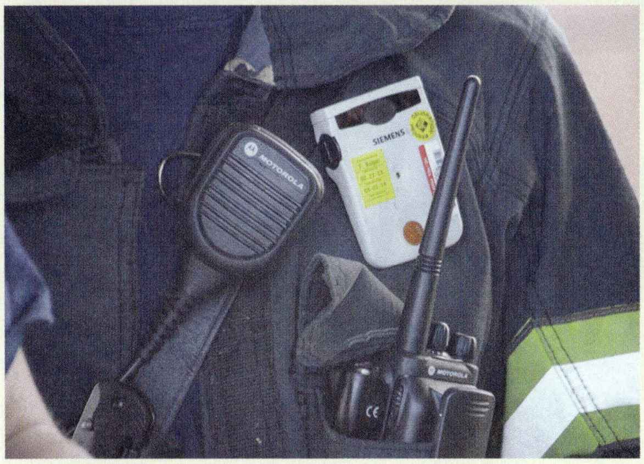

FIGURE 15-23 Small radiation detectors are available that can be worn on turnout gear or on any other type of uniform or civilian clothing. Courtesy of Rob Schnepp

small radiation detectors are available that can be worn on turnout gear or on any other type of uniform or civilian clothing FIGURE 15-23. These detectors sound an alarm when dangerous radiation levels (beyond background radiation) are encountered, thereby alerting the personnel to leave the scene and call for more specialized assistance.

Personal Dosimeters

A personal dosimeter is a small unit, clipped to a front shirt pocket, that is capable of measuring a specific contaminant FIGURE 15-24. A radiation dosimeter, for example, will record the potential dose of radiation the wearer might have received. Other personal dosimeters are designed to measure the amount of a particular chemical contaminant the wearer may have been exposed to over a specified time period. These small detection devices are intended to be worn over a specific time period to determine if, or at what level, an exposure occurred.

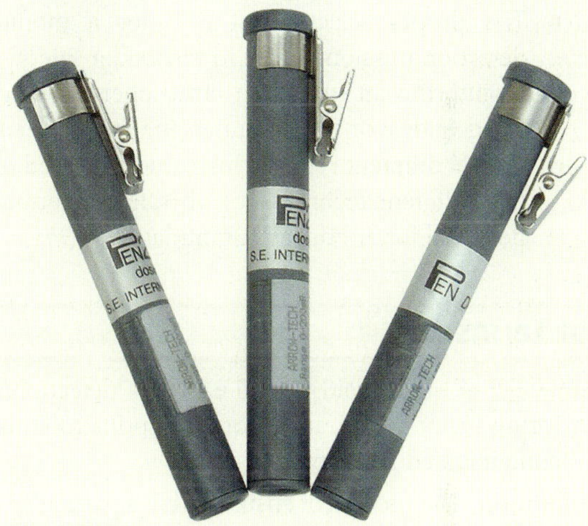

FIGURE 15-24 Dosimeters stay on the responder throughout an incident. Courtesy of S.E. International, Inc. www.seintl.com

Wrap-Up

Chief Concepts

- Detection and monitoring activities should be approached from a multisystem viewpoint. It is important to remember that a particular type of detector or monitoring device serves a unique purpose.
- The nature of the incident, and the intent of the atmospheric monitoring mission, will drive which individual machine or combination of technologies is the most appropriate.
- Emergency responders must understand the operating principles of the detection/monitoring equipment, its limitations, the benefits of using the device, and the way in which the instrument fits into existing response procedures.
- All detection/monitoring instruments require a certain amount of maintenance and upkeep. When considering the purchase of such equipment, it is important to recognize the time, cost, and supplies that the instrument will require on an annual basis.
- All machines have limitations—and responders must know those limitations prior to using any device at the scene of an emergency.

- An accurately calibrated machine ensures that the detection/monitoring device is operating correctly and obtaining the proper readings.
- A bump test is a quick field test carried out to ensure the meter is operating correctly prior to entering a contaminated atmosphere. Bump testing is recommended when doing daily checks, prior to using the device at an incident, and after using the meter.
- The recovery time of a particular device is a function of how much time it takes a detector/monitor to clear itself so that a new reading can be taken. Responders should know the recovery time of any instrument proposed for use at an incident.
- Correction factors or relative response curves may be needed to obtain accurate readings with some detection/monitoring devices.
- The proper selection of PPE for a monitoring/detection mission is vital to responder safety.
- Monitoring an unknown atmosphere is a complicated endeavor that requires the responder to use several different types of instruments, based on several different technologies, to safely determine the potential for airborne contamination.

Hot Terms

Bump test A quick field test to ensure a detection/monitoring meter is operating correctly prior to entering a contaminated atmosphere.

Calibration The process of ensuring that a particular detection/monitoring instrument will respond appropriately to a predetermined concentration of gas. An accurately calibrated machine ensures that the device is detecting the gas or vapor it is intended to detect, *at a given level*.

Carbon monoxide (CO) detector A single-gas device using a specific toxic gas sensor to detect and measure levels of CO in an airborne environment.

Chemical test strip A test strip that gives the responder the ability to test for several different classifications of liquids at the same time. Also known as chemical classifier strips.

Colorimetric tube A reagent-filled tube designed to draw in a sample of air by way of a manual hand-held pump. The reagent will undergo a particular color change when exposed to the contaminant it is intended to detect.

Combustible gas indicator (CGI) A device designed to detect flammable gases and vapors. Also referred to as a flammable gas detector.

Flame ionization detector (FID) A detection/monitoring device that is similar in operational concept to the photo-ionization detector (PID). The key difference is that FIDs can detect methane, whereas PIDs cannot.

Gas chromatography (GC) A sophisticated type of spectroscopy used to identify the components of an air sample that may be a mixture of several different chemicals.

Hydrogen sulfide (H_2S) monitor A single-gas device using a specific toxic gas sensor to detect and measure levels of H_2S in an airborne environment.

Lower explosive limit (LEL) The minimum amount of gaseous fuel that must be present in the air for the air/fuel mixture to be flammable or explosive. Also referred to as the lower flammable limit (LFL).

Multi-gas meter A versatile detection device typically equipped with a combination of toxic gas sensors and the ability to detect flammable gases and vapors.

Oxygen (O_2) monitoring device A single-gas device intended to measure high (more than 23.5 percent) and low (less than 19.5 percent) levels of oxygen in the air.

Personal dosimeter A device that measures the amount of radioactive exposure incurred by an individual.

pH An expression of the amount of dissolved hydrogen ions (H^+) in a solution.

pH paper Also called litmus paper. A type of paper used to measure the pH of corrosive liquids and gases.

Photo-ionization detector (PID) An instrument that provides real-time measurements of airborne contaminant levels of volatile organic compounds.

Radiation detection devices Instruments that employ a variety of technologies to detect the presence of radiation. These devices include personal dosimeters and general survey meters designed to detect alpha, beta, and/or gamma radiation.

Reaction time An expression of the time from when an air sample is drawn into a detection/monitoring meter until the meter processes the sample and provides a reading. Also referred to as response time.

Recovery time The amount of time it takes a detector/monitor to clear itself, so a new reading can be taken.

Relative response curve A curve that accounts for the different types of gases and vapors that might be encountered other than the one used for calibration of a flammable gas detector and/or monitor.

Relative response factor A mathematical computation that must be completed for a detector to correlate the differences between the gas that is used to calibrate the device and the gas that is being detected in the atmosphere.

Situational awareness (SA) A term primarily used in the military, especially the Air Force, that describes the process of observing and understanding the visual cues available, orienting yourself to those inputs relative to your current situation, and making rapid decisions based on those inputs.

Specialized detection device A unique instrument designed to specifically identify a sample substance. These devices are capable of naming the substance in question, sometimes even by its trade name.

Volatile organic compound (VOC) Any organic (carbon-containing) compound that is capable of vaporizing into the atmosphere under normal environmental conditions.

RESPONDER *in action*

You are dispatched for an odor investigation in a 10-story office building. You are met in the lobby of the building by a maintenance supervisor. He states that one of his employees is on the fifth floor, in a small room that contains the phone switching box and a bank of computers. He was called there by the occupant of the office space because of an unusual odor coming from the room. Your crew goes up to the floor in question and begins to assess the situation. You immediately notice a faint odor in the air.

1. Which of the following instruments would be the most appropriate choice to assess this situation?
   - **A.** Personal dosimeter
   - **B.** Hydrogen sulfide detector
   - **C.** Photo-ionization detector
   - **D.** Gas chromatograph
2. Based on the amount of plastics in the room, you are concerned about the amount of VOCs that may be present. What does the acronym VOC stand for?
   - **A.** Volatile activated carbon
   - **B.** Ventilated allowable compound
   - **C.** Volatile organic canister
   - **D.** Volatile organic compound
3. A ____________ is a quick field test to ensure a detection/monitoring meter is operating correctly prior to entering a contaminated atmosphere.
   - **A.** calibration test
   - **B.** field test
   - **C.** recovery time check
   - **D.** bump test
4. You instruct one of your crew members to perform a bump test prior to monitoring the room. What is the recommended amount of time to run the detection/monitoring machine prior to performing an effective bump test?
   - **A.** 15 minutes
   - **B.** 30 minutes
   - **C.** 1 hour
   - **D.** 2 hours

Appendix A: An Extract from NFPA 472, *Standard for Competencies of Responders to Hazardous Materials/Weapons of Mass Destruction Incidents*

Chapter 4 Competencies for Awareness Level Personnel

4.1 General.

4.1.1 Introduction.

4.1.1.1 Awareness level personnel shall be persons who, in the course of their normal duties, could encounter an emergency involving hazardous materials/weapons of mass destruction (WMD) and who are expected to recognize the presence of the hazardous materials/WMD, protect themselves, call for trained personnel, and secure the area.

4.1.1.2 Awareness level personnel shall be trained to meet all competencies of this chapter.

4.1.1.3 Awareness level personnel shall receive additional training to meet applicable governmental occupational health and safety regulations.

4.1.2 Goal.

4.1.2.1 The goal of the competencies in this chapter shall be to provide personnel already on the scene of a hazardous materials/WMD incident with the knowledge and skills to perform the tasks in 4.1.2.2 safely and effectively.

4.1.2.2 When already on the scene of a hazardous materials/WMD incident, the awareness level personnel shall be able to perform the following tasks:

- **(1)** Analyze the incident to determine both the hazardous materials/WMD present and the basic hazard and response information for each hazardous materials/WMD agent by completing the following tasks:
  - **(a)** Detect the presence of hazardous materials/WMD.
  - **(b)** Survey a hazardous materials/WMD incident from a safe location to identify the name, UN/NA identification number, type of placard, or other distinctive marking applied for the hazardous materials/WMD involved.
  - **(c)** Collect hazard information from the current edition of the DOT *Emergency Response Guidebook*.
- **(2)** Implement actions consistent with the authority having jurisdiction (AHJ), and the current edition of the DOT *Emergency Response Guidebook* by completing the following tasks:
  - **(a)** Initiate protective actions
  - **(b)** Initiate the notification process

4.2 Competencies — Analyzing the Incident.

4.2.1* Detecting the Presence of Hazardous Materials/WMD. Given examples of various situations, awareness level personnel shall identify those situations where hazardous materials/WMD are present by completing the following requirements:

- **(1)***Identify the definitions of both *hazardous material* (or *dangerous goods*, in Canada) and *WMD*
- **(2)** Identify the UN/DOT hazard classes and divisions of hazardous materials/WMD and identify common examples of materials in each hazard class or division
- **(3)***Identify the primary hazards associated with each UN/DOT hazard class and division
- **(4)** Identify the difference between hazardous materials/WMD incidents and other emergencies
- **(5)** Identify typical occupancies and locations in the community where hazardous materials/WMD are manufactured, transported, stored, used, or disposed of
- **(6)** Identify typical container shapes that can indicate the presence of hazardous materials/WMD
- **(7)** Identify facility and transportation markings and colors that indicate hazardous materials/WMD, including the following:
  - **(a)** Transportation markings, including UN/NA identification number marks, marine pollutant mark, elevated temperature (HOT) mark, commodity marking, and inhalation hazard mark
  - **(b)** NFPA 704, *Standard System for the Identification of the Hazards of Materials for Emergency Response*, markings
  - **(c)***Military hazardous materials/WMD markings
  - **(d)** Special hazard communication markings for each hazard class
  - **(e)** Pipeline markings
  - **(f)** Container markings
- **(8)** Given an NFPA 704 marking, describe the significance of the colors, numbers, and special symbols
- **(9)** Identify U.S. and Canadian placards and labels that indicate hazardous materials/WMD
- **(10)** Identify the following basic information on material safety data sheets (MSDS) and shipping papers for hazardous materials:
  - **(a)** Identify where to find MSDS
  - **(b)** Identify major sections of an MSDS
  - **(c)** Identify the entries on shipping papers that indicate the presence of hazardous materials
  - **(d)** Match the name of the shipping papers found in transportation (air, highway, rail, and water) with the mode of transportation
  - **(e)** Identify the person responsible for having the shipping papers in each mode of transportation
  - **(f)** Identify where the shipping papers are found in each mode of transportation

(g) Identify where the papers can be found in an emergency in each mode of transportation

(11)*Identify examples of clues (other than occupancy/location, container shape, markings/color, placards/labels, MSDS, and shipping papers) to include sight, sound, and odor of which indicate hazardous materials/WMD

(12) Describe the limitations of using the senses in determining the presence or absence of hazardous materials/WMD

(13)*Identify at least four types of locations that could be targets for criminal or terrorist activity using hazardous materials/WMD

(14)*Describe the difference between a chemical and a biological incident

(15)*Identify at least four indicators of possible criminal or terrorist activity involving chemical agents

(16)*Identify at least four indicators of possible criminal or terrorist activity involving biological agents

(17) Identify at least four indicators of possible criminal or terrorist activity involving radiological agents

(18) Identify at least four indicators of possible criminal or terrorist activity involving illicit laboratories (clandestine laboratories, weapons lab, ricin lab)

(19) Identify at least four indicators of possible criminal or terrorist activity involving explosives

(20)*Identify at least four indicators of secondary devices

4.2.2 **Surveying Hazardous Materials/WMD Incidents.** Given examples of hazardous materials/WMD incidents, awareness level personnel shall, from a safe location, identify the hazardous material(s)/WMD involved in each situation by name, UN/NA identification number, or type placard applied by completing the following requirements:

(1) Identify difficulties encountered in determining the specific names of hazardous materials/WMD at facilities and in transportation

(2) Identify sources for obtaining the names of, UN/NA identification numbers for, or types of placard associated with hazardous materials/WMD in transportation

(3) Identify sources for obtaining the names of hazardous materials/WMD at a facility

4.2.3* **Collecting Hazard Information.** Given the identity of various hazardous materials/WMD (name, UN/NA identification number, or type placard), awareness level personnel shall identify the fire, explosion, and health hazard information for each material by using the current edition of the DOT *Emergency Response Guidebook* by completing the following requirements:

(1)*Identify the three methods for determining the guidebook page for a hazardous material/WMD

(2) Identify the two general types of hazards found on each guidebook page

4.3* **Competencies — Planning the Response. (Reserved)**

4.4 **Competencies — Implementing the Planned Response.**

4.4.1* **Initiating Protective Actions.** Given examples of hazardous materials/WMD incidents, the emergency response plan, the standard operating procedures, and the current edition of the DOT *Emergency Response Guidebook,* awareness level personnel shall be able to identify the actions to be taken to protect themselves and others and to control access to the scene by completing the following requirements:

(1) Identify the location of both the emergency response plan and/or standard operating procedures

(2) Identify the role of the awareness level personnel during hazardous materials/WMD incidents

(3) Identify the following basic precautions to be taken to protect themselves and others in hazardous materials/WMD incidents:

(a) Identify the precautions necessary when providing emergency medical care to victims of hazardous materials/WMD incidents

(b) Identify typical ignition sources found at the scene of hazardous materials/WMD incidents

(c)* Identify the ways hazardous materials/WMD are harmful to people, the environment, and property

(d)*Identify the general routes of entry for human exposure to hazardous materials/WMD

(4)*Given examples of hazardous materials/WMD and the identity of each hazardous material/WMD (name, UN/NA identification number, or type placard), identify the following response information:

(a) Emergency action (fire, spill, or leak and first aid)

(b) Personal protective equipment necessary

(c) Initial isolation and protective action distances

(5) Given the name of a hazardous material, identify the recommended personal protective equipment from the following list:

(a) Street clothing and work uniforms

(b) Structural fire-fighting protective clothing

(c) Positive pressure self-contained breathing apparatus

(d) Chemical-protective clothing and equipment

(6) Identify the definitions for each of the following protective actions:

(a) Isolation of the hazard area and denial of entry

(b) Evacuation

(c)* Shelter-in-place

(7) Identify the size and shape of recommended initial isolation and protective action zones

(8) Describe the difference between small and large spills as found in the Table of Initial Isolation and Protective Action Distances in the DOT *Emergency Response Guidebook*

(9) Identify the circumstances under which the following distances are used at a hazardous materials/WMD incidents:

(a) Table of Initial Isolation and Protective Action Distances

(b) Isolation distances in the numbered guides

(10) Describe the difference between the isolation distances on the orange-bordered guidebook pages and the protective action distances on the green-bordered ERG (*Emergency Response Guidebook*) pages

(11) Identify the techniques used to isolate the hazard area and deny entry to unauthorized persons at hazardous materials/WMD incidents

(12)*Identify at least four specific actions necessary when an incident is suspected to involve criminal or terrorist activity

4.4.2 Initiating the Notification Process. Given scenarios involving hazardous materials/WMD incidents, awareness level personnel shall identify the initial notifications to be made and how to make them, consistent with the AHJ.

4.5* Competencies — Evaluating Progress. (Reserved)

4.6* Competencies — Terminating the Incident. (Reserved)

Chapter 5 Core Competencies for Operations Level Responders

5.1 General.

5.1.1 Introduction.

5.1.1.1* The operations level responder shall be that person who responds to hazardous materials/weapons of mass destruction (WMD) incidents for the purpose of protecting nearby persons, the environment, or property from the effects of the release.

5.1.1.2 The operations level responder shall be trained to meet all competencies at the awareness level (*see Chapter 4*) and the competencies of this chapter.

5.1.1.3* The operations level responder shall receive additional training to meet applicable governmental occupational health and safety regulations.

5.1.2 Goal.

5.1.2.1 The goal of the competencies in this chapter shall be to provide operations level responders with the knowledge and skills to perform the core competencies in 5.1.2.2 safely.

5.1.2.2 When responding to hazardous materials/WMD incidents, operations level responders shall be able to perform the following tasks:

(1) Analyze a hazardous materials/WMD incident to determine the scope of the problem and potential outcomes by completing the following tasks:

(a) Survey a hazardous materials/WMD incident to identify the containers and materials involved, determine whether hazardous materials/WMD have been released, and evaluate the surrounding conditions

(b) Collect hazard and response information from MSDS; CHEMTREC/CANUTEC/SETIQ; local, state, and federal authorities; and shipper/manufacturer contacts

(c) Predict the likely behavior of a hazardous material/WMD and its container

(d) Estimate the potential harm at a hazardous materials/WMD incident

(2) Plan an initial response to a hazardous materials/WMD incident within the capabilities and competencies of available personnel and personal protective equipment by completing the following tasks:

(a) Describe the response objectives for the hazardous materials/WMD incident

(b) Describe the response options available for each objective

(c) Determine whether the personal protective equipment provided is appropriate for implementing each option

(d) Describe emergency decontamination procedures

(e) Develop a plan of action, including safety considerations

(3) Implement the planned response for a hazardous materials/WMD incident to favorably change the outcomes consistent with the emergency response plan and/or standard operating procedures by completing the following tasks:

(a) Establish and enforce scene control procedures, including control zones, emergency decontamination, and communications

(b) Where criminal or terrorist acts are suspected, establish means of evidence preservation

(c) Initiate an incident command system (ICS) for hazardous materials/WMD incidents

(d) Perform tasks assigned as identified in the incident action plan

(e) Demonstrate emergency decontamination

(4) Evaluate the progress of the actions taken at a hazardous materials/WMD incident to ensure that the response objectives are being met safely, effectively, and efficiently by completing the following tasks:

(a) Evaluate the status of the actions taken in accomplishing the response objectives

(b) Communicate the status of the planned response

5.2 Core Competencies — Analyzing the Incident.

5.2.1* Surveying Hazardous Materials/WMD Incidents. Given scenarios involving hazardous materials/WMD incidents, the operations level responder shall collect information about the incident to identify the containers, the materials involved, the surrounding conditions, and whether hazardous materials/WMD have been released by completing the requirements of 5.2.1.1 through 5.2.1.6.

5.2.1.1* Given three examples each of liquid, gas, and solid hazardous material or WMD, including various hazard classes, operations level personnel shall identify the general shapes of containers in which the hazardous materials/WMD are typically found.

5.2.1.1.1 Given examples of the following tank cars, the operations level responder shall identify each tank car by type, as follows:

(1) Cryogenic liquid tank cars

(2) Nonpressure tank cars (general service or low pressure cars)

(3) Pressure tank cars

5.2.1.1.2 Given examples of the following intermodal tanks, the operations level responder shall identify each intermodal tank by type, as follows:

(1) Nonpressure intermodal tanks

(2) Pressure intermodal tanks

(3) Specialized intermodal tanks, including the following:

(a) Cryogenic intermodal tanks

(b) Tube modules

5.2.1.1.3 Given examples of the following cargo tanks, the operations level responder shall identify each cargo tank by type, as follows:

(1) Compressed gas tube trailers

(2) Corrosive liquid tanks

(3) Cryogenic liquid tanks

(4) Dry bulk cargo tanks

(5) High pressure tanks

(6) Low pressure chemical tanks

(7) Nonpressure liquid tanks

5.2.1.1.4 Given examples of the following storage tanks, the operations level responder shall identify each tank by type, as follows:

(1) Cryogenic liquid tank

(2) Nonpressure tank

(3) Pressure tank

5.2.1.1.5 Given examples of the following nonbulk packaging, the operations level responder shall identify each package by type, as follows:

(1) Bags

(2) Carboys

(3) Cylinders

(4) Drums

(5) Dewar flask (cryogenic liquids)

5.2.1.1.6 Given examples of the following packaging, the operations level responder shall identify the characteristics of each container or package by type as follows:

(1) Intermediate bulk container (IBC)

(2) Ton container

5.2.1.1.7* Given examples of the following radioactive material packages, the operations level responder shall identify the characteristics of each container or package by type, as follows:

(1) Excepted

(2) Industrial

(3) Type A

(4) Type B

(5) Type C

5.2.1.2 Given examples of containers, the operations level responder shall identify the markings that differentiate one container from another.

5.2.1.2.1 Given examples of the following marked transport vehicles and their corresponding shipping papers, the operations level responder shall identify the following vehicle or tank identification marking:

(1) Highway transport vehicles, including cargo tanks

(2) Intermodal equipment, including tank containers

(3) Rail transport vehicles, including tank cars

5.2.1.2.2 Given examples of facility containers, the operations level responder shall identify the markings indicating container size, product contained, and/or site identification numbers.

5.2.1.3 Given examples of hazardous materials incidents, the operations level responder shall identify the name(s) of the hazardous material(s) in 5.2.1.3.1 through 5.2.1.3.3.

5.2.1.3.1 The operations level responder shall identify the following information on a pipeline marker:

(1) Emergency telephone number

(2) Owner

(3) Product

5.2.1.3.2 Given a pesticide label, the operations level responder shall identify each of the following pieces of information, then match the piece of information to its significance in surveying hazardous materials incidents:

(1) Active ingredient

(2) Hazard statement

(3) Name of pesticide

(4) Pest control product (PCP) number (in Canada)

(5) Precautionary statement

(6) Signal word

5.2.1.3.3 Given a label for a radioactive material, the operations level responder shall identify the type or category of label, contents, activity, transport index, and criticality safety index as applicable.

5.2.1.4* The operations level responder shall identify and list the surrounding conditions that should be noted when a hazardous materials/WMD incident is surveyed.

5.2.1.5 The operations level responder shall describe ways to verify information obtained from the survey of a hazardous materials/WMD incident.

5.2.1.6* The operations level responder shall identify at least three additional hazards that could be associated with an incident involving terrorist or criminal activities.

5.2.2 Collecting Hazard and Response Information. Given scenarios involving known hazardous materials/WMD, the operations level responder shall collect hazard and response information using MSDS, CHEMTREC/CANUTEC/SETIQ, governmental authorities, and shippers and manufacturers by completing the following requirements:

(1) Match the definitions associated with the UN/DOT hazard classes and divisions of hazardous materials/WMD, including refrigerated liquefied gases and cryogenic liquids, with the class or division

(2) Identify two ways to obtain an MSDS in an emergency

(3) Using an MSDS for a specified material, identify the following hazard and response information:

(a) Physical and chemical characteristics

(b) Physical hazards of the material

(c) Health hazards of the material

(d) Signs and symptoms of exposure

(e) Routes of entry

(f) Permissible exposure limits

(g) Responsible party contact

(h) Precautions for safe handling (including hygiene practices, protective measures, and procedures for cleanup of spills and leaks)

(i) Applicable control measures, including personal protective equipment

(j) Emergency and first-aid procedures

(4) Identify the following:

(a) Type of assistance provided by CHEMTREC/CANUTEC/SETIQ and governmental authorities

(b) Procedure for contacting CHEMTREC/CANUTEC/SETIQ and governmental authorities

(c) Information to be furnished to CHEMTREC/CANUTEC/SETIQ and governmental authorities

(5) Identify two methods of contacting the manufacturer or shipper to obtain hazard and response information

(6) Identify the type of assistance provided by governmental authorities with respect to criminal or terrorist activities

involving the release or potential release of hazardous materials/WMD

(7) Identify the procedure for contacting local, state, and federal authorities as specified in the emergency response plan and/or standard operating procedures

(8)* Describe the properties and characteristics of the following:

(a) Alpha radiation

(b) Beta radiation

(c) Gamma radiation

(d) Neutron radiation

5.2.3* **Predicting the Likely Behavior of a Material and Its Container.** Given scenarios involving hazardous materials/WMD incidents, each with a single hazardous material/WMD, the operations level responder shall describe the likely behavior of the material or agent and its container by completing the following requirements:

(1) Use the hazard and response information obtained from the current edition of the DOT *Emergency Response Guidebook*, MSDS, CHEMTREC/CANUTEC/SETIQ, governmental authorities, and shipper and manufacturer contacts, as follows:

(a) Match the following chemical and physical properties with their significance and impact on the behavior of the container and its contents:

i. Boiling point
ii. Chemical reactivity
iii. Corrosivity (pH)
iv. Flammable (explosive) range [lower explosive limit (LEL) and upper explosive limit (UEL)]
v. Flash point
vi. Ignition (autoignition) temperature
vii. Particle size
viii. Persistence
ix. Physical state (solid, liquid, gas)
x. Radiation (ionizing and non-ionizing)
xi. Specific gravity
xii. Toxic products of combustion
xiii. Vapor density
xiv. Vapor pressure
xv. Water solubility

(b) Identify the differences between the following terms:

i. *Contamination* and *secondary contamination*
ii. *Exposure* and *contamination*
iii. *Exposure* and *hazard*
iv. *Infectious* and *contagious*
v. *Acute effects* and *chronic effects*
vi. *Acute exposures* and *chronic exposures*

(2)* Identify three types of stress that can cause a container system to release its contents

(3)* Identify five ways in which containers can breach

(4)* Identify four ways in which containers can release their contents

(5)* Identify at least four dispersion patterns that can be created upon release of a hazardous material

(6)* Identify the time frames for estimating the duration that hazardous materials/WMD will present an exposure risk

(7)* Identify the health and physical hazards that could cause harm

(8)* Identify the health hazards associated with the following terms:

(a) Alpha, beta, gamma, and neutron radiation

(b) Asphyxiant

(c)* Carcinogen

(d) Convulsant

(e) Corrosive

(f) Highly toxic

(g) Irritant

(h) Sensitizer, allergen

(i) Target organ effects

(j) Toxic

(9)* Given the following, identify the corresponding UN/DOT hazard class and division:

(a) Blood agents

(b) Biological agents and biological toxins

(c) Choking agents

(d) Irritants (riot control agents)

(e) Nerve agents

(f) Radiological materials

(g) Vesicants (blister agents)

5.2.4* **Estimating Potential Harm.** Given scenarios involving hazardous materials/WMD incidents, the operations level responder shall describe the potential harm within the endangered area at each incident by completing the following requirements:

(1)* Identify a resource for determining the size of an endangered area of a hazardous materials/WMD incident

(2) Given the dimensions of the endangered area and the surrounding conditions at a hazardous materials/WMD incident, describe the number and type of exposures within that endangered area

(3) Identify resources available for determining the concentrations of a released hazardous materials/WMD within an endangered area

(4)* Given the concentrations of the released material, describe the factors for determining the extent of physical, health, and safety hazards within the endangered area of a hazardous materials/WMD incident

(5) Describe the impact that time, distance, and shielding have on exposure to radioactive materials specific to the expected dose rate

5.3 **Core Competencies — Planning the Response.**

5.3.1 **Describing Response Objectives.** Given at least two scenarios involving hazardous materials/WMD incidents, the operations level responder shall describe the response objectives for each example by completing the following requirements:

(1) Given an analysis of a hazardous materials/WMD incident and the exposures, describe the number of exposures that could be saved with the resources provided by the AHJ

(2) Given an analysis of a hazardous materials/WMD incident, describe the steps for determining response objectives

(3) Describe how to assess the risk to a responder for each hazard class in rescuing injured persons at a hazardous materials/WMD incident

(4)* Describe the potential for secondary attacks and devices at criminal or terrorist events

5.3.2 Identifying Action Options. Given examples of hazardous materials/WMD incidents (facility and transportation), the operations level responder shall identify the options for each response objective and shall meet the following requirements:

(1) Identify the options to accomplish a given response objective

(2) Describe the prioritization of emergency medical care and removal of victims from the hazard area relative to exposure and contamination concerns

5.3.3 Determining Suitability of Personal Protective Equipment. Given examples of hazardous materials/WMD incidents, including the names of the hazardous materials/WMD involved and the anticipated type of exposure, the operations level responder shall determine whether available personal protective equipment is applicable to performing assigned tasks by completing the following requirements:

(1)* Identify the respiratory protection required for a given response option and the following:

(a) Describe the advantages, limitations, uses, and operational components of the following types of respiratory protection at hazardous materials/WMD incidents:

i. Positive pressure self-contained breathing apparatus (SCBA)
ii. Positive pressure air-line respirator with required escape unit
iii. Closed-circuit SCBA
iv. Powered air-purifying respirator (PAPR)
v. Air-purifying respirator (APR)
vi. Particulate respirator

(b) Identify the required physical capabilities and limitations of personnel working in respiratory protection

(2) Identify the personal protective clothing required for a given option and the following:

(a) Identify skin contact hazards encountered at hazardous materials/WMD incidents

(b) Identify the purpose, advantages, and limitations of the following types of protective clothing at hazardous materials/WMD incidents:

i. Chemical-protective clothing such as liquid splash–protective clothing and vapor-protective clothing
ii. High temperature–protective clothing such as proximity suit and entry suits
iii. Structural fire-fighting protective clothing

5.3.4* Identifying Decontamination Issues. Given scenarios involving hazardous materials/WMD incidents, the operations level responder shall identify when decontamination is needed by completing the following requirements:

(1) Identify ways that people, personal protective equipment, apparatus, tools, and equipment become contaminated

(2) Describe how the potential for secondary contamination determines the need for decontamination

(3) Explain the importance and limitations of decontamination procedures at hazardous materials incidents

(4) Identify the purpose of emergency decontamination procedures at hazardous materials incidents

(5) Identify the methods, advantages, and limitations of emergency decontamination procedures

5.4 Core Competencies — Implementing the Planned Response.

5.4.1 Establishing Scene Control. Given two scenarios involving hazardous materials/WMD incidents, the operations level responder shall explain how to establish and maintain scene control, including control zones and emergency decontamination, and communications between responders and to the public by completing the following requirements:

(1) Identify the procedures for establishing scene control through control zones

(2) Identify the criteria for determining the locations of the control zones at hazardous materials/WMD incidents

(3) Identify the basic techniques for the following protective actions at hazardous materials/WMD incidents:

(a) Evacuation

(b) Shelter-in-place

(4)* Demonstrate the ability to perform emergency decontamination

(5)* Identify the items to be considered in a safety briefing prior to allowing personnel to work at the following:

(a) Hazardous material incidents

(b)* Hazardous materials/WMD incidents involving criminal activities

(6) Identify the procedures for ensuring coordinated communication between responders and to the public

5.4.2* Preserving Evidence. Given two scenarios involving hazardous materials/WMD incidents, the operations level responder shall describe the process to preserve evidence as listed in the emergency response plan and/or standard operating procedures.

5.4.3* Initiating the Incident Command System. Given scenarios involving hazardous materials/WMD incidents, the operations level responder shall implement the incident command system as required by the AHJ by completing the following requirements:

(1) Identify the role of the operations level responder during hazardous materials/WMD incidents as specified in the emergency response plan and/or standard operating procedures

(2) Identify the levels of hazardous materials/WMD incidents as defined in the emergency response plan

(3) Identify the purpose, need, benefits, and elements of the incident command system for hazardous materials/WMD incidents

(4) Identify the duties and responsibilities of the following functions within the incident management system:

(a) Incident safety officer

(b) Hazardous materials branch or group

(5) Identify the considerations for determining the location of the incident command post for a hazardous materials/WMD incident

(6) Identify the procedures for requesting additional resources at a hazardous materials/WMD incident

(7) Describe the role and response objectives of other agencies that respond to hazardous materials/WMD incidents

5.4.4 Using Personal Protective Equipment. Given the personal protective equipment provided by the AHJ, the operations level responder shall describe considerations for the use of personal protective equipment provided by the AHJ by completing the following requirements:

(1) Identify the importance of the buddy system

(2) Identify the importance of the backup personnel

(3) Identify the safety precautions to be observed when approaching and working at hazardous materials/WMD incidents

(4) Identify the signs and symptoms of heat and cold stress and procedures for their control

(5) Identify the capabilities and limitations of personnel working in the personal protective equipment provided by the AHJ

(6) Identify the procedures for cleaning, disinfecting, and inspecting personal protective equipment provided by the AHJ

(7) Describe the maintenance, testing, inspection, and storage procedures for personal protective equipment provided by the AHJ according to the manufacturer's specifications and recommendations

5.5 Core Competencies — Evaluating Progress.

5.5.1 Evaluating the Status of Planned Response. Given two scenarios involving hazardous materials/WMD incidents, including the incident action plan, the operations level responder shall determine the effectiveness of the actions taken in accomplishing the response objectives and shall meet the following requirements:

(1) Identify the considerations for evaluating whether actions taken were effective in accomplishing the objectives

(2) Describe the circumstances under which it would be prudent to withdraw from a hazardous materials/WMD incident

5.5.2 Communicating the Status of Planned Response. Given two scenarios involving hazardous materials/WMD incidents, including the incident action plan, the operations level responder shall report the status of the planned response through the normal chain of command by completing the following requirements:

(1) Identify the procedures for reporting the status of the planned response through the normal chain of command

(2) Identify the methods for immediate notification of the incident commander and other response personnel about critical emergency conditions at the incident

5.6* Competencies — Terminating the Incident. (Reserved)

Chapter 6 Competencies for Operations Level Responders Assigned Mission-Specific Responsibilities

6.1 General.

6.1.1 Introduction.

6.1.1.1* This chapter shall address competencies for the following operations level responders assigned mission-specific responsibilities at hazardous materials/WMD incidents by the AHJ beyond the core competencies at the operations level (*see Chapter 5*):

(1) Operations level responders assigned to use personal protective equipment (*see Section 6.2*)

(2) Operations level responders assigned to perform mass decontamination (*see Section 6.3*)

(3) Operations level responders assigned to perform technical decontamination (*see Section 6.4*)

(4) Operations level responders assigned to perform evidence preservation and sampling (*see Section 6.5*)

(5) Operations level responders assigned to perform product control (*see Section 6.6*)

(6) Operations level responders assigned to perform air monitoring and sampling (*see Section 6.7*)

(7) Operations level responders assigned to perform victim rescue/recovery (*see Section 6.8*)

(8) Operations level responders assigned to respond to illicit laboratory incidents (*see Section 6.9*)

(9) Operational level responders assigned to perform disablement/disruption of improvised explosives devices (IED), improvised WMD dispersal devices, and operations at improvised explosive laboratories. (*see Section 6.10*)

6.1.1.2 The operations level responder who is assigned mission-specific responsibilities at hazardous materials/WMD incidents shall be trained to meet all competencies at the awareness level (*see Chapter 4*), all core competencies at the operations level (*see Chapter 5*), and all competencies for the assigned responsibilities in the applicable section(s) in this chapter.

6.1.1.3* The operations level responder who is assigned mission-specific responsibilities at hazardous materials/WMD incidents shall receive additional training to meet applicable governmental occupational health and safety regulations.

6.1.1.4 The operations level responder who is assigned mission-specific responsibilities at hazardous materials/WMD incidents shall operate under the guidance of a hazardous materials technician, an allied professional, an emergency response plan, or standard operating procedures.

6.1.1.5 The development of assigned mission-specific knowledge and skills shall be based on the tools, equipment, and procedures provided by the AHJ for the mission-specific responsibilities assigned.

6.1.2 Goal. The goal of the competencies in this chapter shall be to provide the operations level responder assigned mission-specific responsibilities at hazardous materials/WMD incidents by the AHJ with the knowledge and skills to perform the assigned mission-specific responsibilities safely and effectively.

6.1.3 Mandating of Competencies. This standard shall not mandate that the response organizations perform mission-specific responsibilities.

6.1.3.1 Operations level responders assigned mission-specific responsibilities at hazardous materials/WMD incidents, operating within the scope of their training in this chapter, shall be able to perform their assigned mission-specific responsibilities.

6.1.3.2 If a response organization desires to train some or all of its operations level responders to perform mission-specific responsibilities at hazardous materials/WMD incidents, the minimum required competencies shall be as set out in this chapter.

6.2 Mission-Specific Competencies: Personal Protective Equipment.

6.2.1 General.

6.2.1.1 Introduction.

6.2.1.1.1 The operations level responder assigned to use personal protective equipment shall be that person, competent at the operations level, who is assigned to use personal protective equipment at hazardous materials/WMD incidents.

6.2.1.1.2 The operations level responder assigned to use personal protective equipment at hazardous materials/WMD incidents shall be trained to meet all competencies at the awareness level (*see Chapter 4*), all core competencies at the operations level (*see Chapter 5*), and all competencies in this section.

6.2.1.1.3 The operations level responder assigned to use personal protective equipment at hazardous materials/WMD incidents shall operate under the guidance of a hazardous materials technician, an allied professional, or standard operating procedures.

6.2.1.1.4* The operations level responder assigned to use personal protective equipment shall receive the additional training necessary to meet specific needs of the jurisdiction.

6.2.1.2 Goal. The goal of the competencies in this section shall be to provide the operations level responder assigned to use personal protective equipment with the knowledge and skills to perform the following tasks safely and effectively:

(1) Plan a response within the capabilities of personal protective equipment provided by the AHJ in order to perform mission-specific tasks assigned

(2) Implement the planned response consistent with the standard operating procedures and site safety and control plan by donning, working in, and doffing personal protective equipment provided by the AHJ

(3) Terminate the incident by completing the reports and documentation pertaining to personal protective equipment

6.2.2 Competencies — Analyzing the Incident. (Reserved)

6.2.3 Competencies — Planning the Response.

6.2.3.1 Selecting Personal Protective Equipment. Given scenarios involving hazardous materials/WMD incidents with known and unknown hazardous materials/WMD and the personal protective equipment provided by the AHJ, the operations level responder assigned to use personal protective equipment shall select the personal protective equipment required to support mission-specific tasks at hazardous materials/WMD incidents based on local procedures by completing the following requirements:

(1)*Describe the types of personal protective equipment that are available for response based on NFPA standards and how these items relate to EPA levels of protection

(2) Describe personal protective equipment options for the following hazards:

- (a) Thermal
- (b) Radiological
- (c) Asphyxiating
- (d) Chemical
- (e) Etiological/biological
- (f) Mechanical

(3) Select personal protective equipment for mission-specific tasks at hazardous materials/WMD incidents based on local procedures

- (a) Describe the following terms and explain their impact and significance on the selection of chemical-protective clothing:
  - i. Degradation
  - ii. Penetration
  - iii. Permeation
- (b) Identify at least three indications of material degradation of chemical-protective clothing
- (c) Identify the different designs of vapor-protective and splash-protective clothing and describe the advantages and disadvantages of each type
- (d)*Identify the relative advantages and disadvantages of the following heat exchange units used for the cooling of personnel operating in personal protective equipment:
  - i. Air cooled
  - ii. Ice cooled
  - iii. Water cooled
  - iv. Phase change cooling technology
- (e) Identify the physiological and psychological stresses that can affect users of personal protective equipment
- (f) Describe local procedures for going through the technical decontamination process

6.2.4 Competencies — Implementing the Planned Response.

6.2.4.1 Using Protective Clothing and Respiratory Protection. Given the personal protective equipment provided by the AHJ, the operations level responder assigned to use personal protective equipment shall demonstrate the ability to don, work in, and doff the equipment provided to support mission-specific tasks by completing the following requirements:

(1) Describe at least three safety procedures for personnel wearing protective clothing

(2) Describe at least three emergency procedures for personnel wearing protective clothing

(3) Demonstrate the ability to don, work in, and doff personal protective equipment provided by the AHJ

(4) Demonstrate local procedures for responders undergoing the technical decontamination process

(5) Describe the maintenance, testing, inspection, storage, and documentation procedures for personal protective equipment provided by the AHJ according to the manufacturer's specifications and recommendations

6.2.5 Competencies — Terminating the Incident.

6.2.5.1 Reporting and Documenting the Incident. Given a scenario involving a hazardous materials/WMD incident, the operations level responder assigned to use personal protective equipment shall document use of the personal protective equipment by completing the documentation requirements of the emergency response plan or standard operating procedures regarding personal protective equipment.

6.3 Mission-Specific Competencies: Mass Decontamination.

6.3.1 General.

6.3.1.1 Introduction.

6.3.1.1.1 The operations level responder assigned to perform mass decontamination at hazardous materials/WMD incidents shall be that person, competent at the operations level, who is assigned to implement mass decontamination operations at hazardous materials/WMD incidents.

6.3.1.1.2 The operations level responder assigned to perform mass decontamination at hazardous materials/WMD incidents shall be trained to meet all competencies at the awareness level (*see Chapter 4*), all core competencies at the operations level (*see Chapter 5*), all mission-specific competencies for personal protective equipment (*see Section 6.2*), and all competencies in this section.

6.3.1.1.3 The operations level responder assigned to perform mass decontamination at hazardous materials/WMD incidents shall operate under the guidance of a hazardous materials technician, an allied professional, or standard operating procedures.

6.3.1.1.4* The operations level responder assigned to perform mass decontamination at hazardous materials/WMD incidents shall receive the additional training necessary to meet specific needs of the jurisdiction.

6.3.1.2 Goal.

6.3.1.2.1 The goal of the competencies in this section shall be to provide the operations level responder assigned to perform mass decontamination at hazardous materials/WMD incidents with the knowledge and skills to perform the tasks in 6.3.1.2.2 safely and effectively.

6.3.1.2.2 When responding to hazardous materials/WMD incidents, the operations level responder assigned to perform mass decontamination shall be able to perform the following tasks:

(1) Plan a response within the capabilities of available personnel, personal protective equipment, and control equipment by selecting a mass decontamination process to minimize the hazard

(2) Implement the planned response to favorably change the outcomes consistent with standard operating procedures and the site safety and control plan by completing the following tasks:

(a) Perform the decontamination duties as assigned

(b) Perform the mass decontamination functions identified in the incident action plan

(3) Evaluate the progress of the planned response by evaluating the effectiveness of the mass decontamination process

(4) Terminate the incident by providing reports and documentation of decontamination operations

6.3.2 Competencies — Analyzing the Incident. (Reserved)

6.3.3 Competencies — Planning the Response.

6.3.3.1 Selecting Personal Protective Equipment. Given an emergency response plan or standard operating procedures and the personal protective equipment provided by the AHJ, the operations level responder assigned to mass decontamination shall select the personal protective equipment required to support mass decontamination at hazardous materials/WMD incidents based on local procedures (*see Section 6.2*).

6.3.3.2 Selecting Decontamination Procedures. Given scenarios involving hazardous materials/WMD incidents, the operations level responder assigned to mass decontamination operations shall select a mass decontamination procedure that will minimize the hazard and spread of contamination, determine the equipment required to implement that procedure, and meet the following requirements:

(1) Identify the advantages and limitations of mass decontamination operations

(2) Describe the advantages and limitations of each of the following mass decontamination methods:

(a) Dilution

(b) Isolation

(c) Washing

(3) Identify sources of information for determining the correct mass decontamination procedure and identify how to access those resources in a hazardous materials/WMD incident

(4) Given resources provided by the AHJ, identify the supplies and equipment required to set up and implement mass decontamination operations

(5) Identify procedures, equipment, and safety precautions for communicating with crowds and crowd management techniques that can be used at incidents where a large number of people might be contaminated

6.3.4 Competencies — Implementing the Planned Response.

6.3.4.1 Performing Incident Management Duties. Given a scenario involving a hazardous materials/WMD incident and the emergency response plan or standard operating procedures, the operations level responder assigned to mass decontamination operations shall demonstrate the mass decontamination duties assigned in the incident action plan by describing the local procedures for the implementation of the mass decontamination function within the incident command system.

6.3.4.2 Performing Decontamination Operations Identified in Incident Action Plan. The operations level responder assigned to mass decontamination operations shall demonstrate the ability to set up and implement mass decontamination operations for ambulatory and nonambulatory victims.

6.3.5 Competencies — Evaluating Progress.

6.3.5.1 Evaluating the Effectiveness of the Mass Decontamination Process. Given examples of contaminated items that have undergone the required decontamination, the operations level responder assigned to mass decontamination operations shall identify procedures for determining whether the items have been fully decontaminated according to the standard operating procedures of the AHJ or the incident action plan.

6.3.6 Competencies — Terminating the Incident.

6.3.6.1 Reporting and Documenting the Incident. Given a scenario involving a hazardous materials/WMD incident, the operations level responder assigned to mass decontamination operations shall document the mass decontamination activities as required by the AHJ by completing the following:

(1) Identify the reports and supporting documentation required by the emergency response plan or standard operating procedures

(2) Describe the importance of personnel exposure records

(3) Identify the steps in keeping an activity log and exposure records

(4) Identify the requirements for filing documents and maintaining records

6.4 Mission-Specific Competencies: Technical Decontamination.

6.4.1 General.

6.4.1.1 Introduction.

6.4.1.1.1 The operations level responder assigned to perform technical decontamination at hazardous materials/WMD incidents shall be that person, competent at the operations level, who is assigned to implement technical decontamination operations at hazardous materials/WMD incidents.

6.4.1.1.2 The operations level responder assigned to perform technical decontamination at hazardous materials/WMD incidents shall be trained to meet all competencies at the awareness level (*see Chapter 4*), all core competencies at the operations level (*see Chapter 5*), all mission-specific competencies for personal protective equipment (*see Section 6.2*), and all competencies in this section.

6.4.1.1.3 The operations level responder assigned to perform technical decontamination at hazardous materials/WMD incidents shall operate under the guidance of a hazardous materials technician, an allied professional, or standard operating procedures.

6.4.1.1.4* The operations level responder assigned to perform technical decontamination at hazardous materials/WMD incidents shall receive the additional training necessary to meet specific needs of the jurisdiction.

6.4.1.2 Goal.

6.4.1.2.1 The goal of the competencies in this section shall be to provide the operations level responder assigned to perform technical decontamination at hazardous materials/WMD incidents with the knowledge and skills to perform the tasks in 6.4.1.2.2 safely and effectively.

6.4.1.2.2 When responding to hazardous materials/WMD incidents, the operations level responder assigned to perform technical decontamination shall be able to perform the following tasks:

(1) Plan a response within the capabilities of available personnel, personal protective equipment, and control equipment by selecting a technical decontamination process to minimize the hazard

(2) Implement the planned response to favorably change the outcomes consistent with standard operating procedures and the site safety and control plan by completing the following tasks:

(a) Perform the technical decontamination duties as assigned

(b) Perform the technical decontamination functions identified in the incident action plan

(3) Evaluate the progress of the planned response by evaluating the effectiveness of the technical decontamination process

(4) Terminate the incident by completing the reports and documentation of decontamination operations

6.4.2 Competencies — Analyzing the Incident. (Reserved)

6.4.3 Competencies — Planning the Response.

6.4.3.1 Selecting Personal Protective Equipment. Given an emergency response plan or standard operating procedures and the personal protective equipment provided by the AHJ, the operations level responder assigned to technical decontamination operations shall select the personal protective equipment required to support technical decontamination at hazardous materials/WMD incidents based on local procedures (*see Section 6.2*).

6.4.3.2 Selecting Decontamination Procedures. Given scenarios involving hazardous materials/WMD incidents, the operations level responder assigned to technical decontamination operations shall select a technical decontamination procedure that will minimize the hazard and spread of contamination and determine the equipment required to implement that procedure by completing the following requirements:

(1) Identify the advantages and limitations of technical decontamination operations

(2) Describe the advantages and limitations of each of the following technical decontamination methods:

(a) Absorption

(b) Adsorption

(c) Chemical degradation

(d) Dilution

(e) Disinfection

(f) Evaporation

(g) Isolation and disposal

(h) Neutralization

(i) Solidification

(j) Sterilization

(k) Vacuuming

(l) Washing

(3) Identify sources of information for determining the correct technical decontamination procedure and identify how to access those resources in a hazardous materials/WMD incident

(4) Given resources provided by the AHJ, identify the supplies and equipment required to set up and implement technical decontamination operations

(5) Identify the procedures, equipment, and safety precautions for processing evidence during technical decontamination operations at hazardous materials/WMD incidents

(6) Identify procedures, equipment, and safety precautions for handling tools, equipment, weapons, criminal suspects, and law enforcement/search canines brought to the decontamination corridor at hazardous materials/WMD incidents

6.4.4 Competencies — Implementing the Planned Response.

6.4.4.1 Performing Incident Management Duties. Given a scenario involving a hazardous materials/WMD incident and the emergency response plan or standard operating procedures, the operations level responder assigned to technical decontamination operations shall demonstrate the technical decontamination duties assigned in the incident action plan by completing the following requirements:

(1) Identify the role of the operations level responder assigned to technical decontamination operations during hazardous materials/WMD incidents

(2) Describe the procedures for implementing technical decontamination operations within the incident command system

6.4.4.2 Performing Decontamination Operations Identified in Incident Action Plan. The responder assigned to technical decontamination operations shall demonstrate the ability to set up and implement the following types of decontamination operations:

(1) Technical decontamination operations in support of entry operations

(2) Technical decontamination operations for ambulatory and nonambulatory victims

6.4.5 Competencies — Evaluating Progress.

6.4.5.1 Evaluating the Effectiveness of the Technical Decontamination Process. Given examples of contaminated items that have undergone the required decontamination, the operations level responder assigned to technical decontamination operations shall identify procedures for determining whether the items have been fully decontaminated according to the standard operating procedures of the AHJ or the incident action plan.

6.4.6 Competencies — Terminating the Incident.

6.4.6.1 Reporting and Documenting the Incident. Given a scenario involving a hazardous materials/WMD incident, the operations level responder assigned to technical decontamination operations shall document the mass decontamination activities as required by the AHJ by completing the following:

(1) Identify the reports and supporting technical documentation required by the emergency response plan or standard operating procedures

(2) Describe the importance of personnel exposure records

(3) Identify the steps in keeping an activity log and exposure records

(4) Identify the requirements for filing documents and maintaining records

6.5 Mission-Specific Competencies: Evidence Preservation and Sampling.

6.5.1 General.

6.5.1.1 Introduction.

6.5.1.1.1 The operations level responder assigned to perform evidence preservation and sampling shall be that person, competent at the operations level, who is assigned to preserve forensic evidence, take samples, and/or seize evidence at hazardous materials/WMD incidents involving potential violations of criminal statutes or governmental regulations.

6.5.1.1.2 The operations level responder assigned to perform evidence preservation and sampling at hazardous materials/WMD incidents shall be trained to meet all competencies at the awareness level (*see Chapter 4*), all core competencies at the operations level (*see Chapter 5*), all mission-specific competencies for personal protective equipment (*see Section 6.2*), and all competencies in this section.

6.5.1.1.3 The operations level responder assigned to perform evidence preservation and sampling at hazardous materials/WMD incidents shall operate under the guidance of a hazardous materials technician, an allied professional, or standard operating procedures.

6.5.1.1.4* The operations level responder assigned to perform evidence preservation and sampling at hazardous materials/WMD incidents shall receive the additional training necessary to meet specific needs of the jurisdiction.

6.5.1.2 Goal.

6.5.1.2.1 The goal of the competencies in this section shall be to provide the operations level responder assigned to evidence preservation and sampling at hazardous materials/WMD incidents with the knowledge and skills to perform the tasks in 6.5.1.2.2 safely and effectively.

6.5.1.2.2 When responding to hazardous materials/WMD incidents involving potential violations of criminal statutes or governmental regulations, the operations level responder assigned to perform evidence preservation and sampling shall be able to perform the following tasks:

(1) Analyze a hazardous materials/WMD incident to determine the complexity of the problem and potential outcomes by completing the following tasks:

(a) Determine if the incident is potentially criminal in nature and identify the law enforcement agency having investigative jurisdiction

(b) Identify unique aspects of criminal hazardous materials/WMD incidents

(2) Plan a response for an incident where there is potential criminal intent involving hazardous materials/WMD within the capabilities and competencies of available personnel, personal protective equipment, and control equipment by completing the following tasks:

(a) Determine the response options to conduct sampling and evidence preservation operations

(b) Describe how the options are within the legal authorities, capabilities, and competencies of available personnel, personal protective equipment, and control equipment

(3) Implement the planned response to a hazardous materials/WMD incident involving potential violations of criminal statutes or governmental regulations by completing the following tasks under the guidance of law enforcement:

(a) Preserve forensic evidence

(b) Take samples

(c) Seize evidence

6.5.2 Competencies — Analyzing the Incident.

6.5.2.1 Determining If the Incident Is Potentially Criminal in Nature and Identifying the Law Enforcement Agency That Has Investigative Jurisdiction. Given examples of hazardous materials/WMD incidents involving potential criminal intent, the operations level responder assigned to evidence preservation and sampling shall describe the potential criminal violation and identify the law enforcement agency having investigative jurisdiction by completing the following requirements:

(1) Given examples of the following hazardous materials/WMD incidents, the operations level responder shall describe products that might be encountered in the incident associated with each situation:

(a) Hazardous materials/WMD suspicious letter

(b) Hazardous materials/WMD suspicious package

(c) Hazardous materials/WMD illicit laboratory

(d) Release/attack with a WMD agent

(e) Environmental crimes

(2) Given examples of the following hazardous materials/WMD incidents, the operations level responder shall identify the agency(cies) with investigative authority and the incident response considerations associated with each situation:

(a) Hazardous materials/WMD suspicious letter

(b) Hazardous materials/WMD suspicious package

(c) Hazardous materials/WMD illicit laboratory

(d) Release/attack with a WMD agent

(e) Environmental crimes

6.5.3 Competencies — Planning the Response.

6.5.3.1 Identifying Unique Aspects of Criminal Hazardous Materials/WMD Incidents. The operations level responder assigned to evidence preservation and sampling shall describe the unique aspects associated with illicit laboratories, hazardous materials/WMD incidents, and environmental crimes by completing the following requirements:

(1) Given an incident involving illicit laboratories, a hazardous materials/WMD incident, or an environmental crime, the operations level responder shall perform the following tasks:

(a) Describe the procedure for securing the scene and characterizing and preserving evidence at the scene

(b) Describe the procedure to document personnel and scene activities associated with the incident

(c) Describe the procedure to determine whether the operations level responders are within their legal authority to perform evidence preservation and sampling tasks

(d) Describe the procedure to notify the agency with investigative authority

(e) Describe the procedure to notify the explosive ordnance disposal (EOD) personnel

(f) Identify potential sample/evidence

(g) Identify the applicable sampling equipment

(h) Describe the procedures to protect samples and evidence from secondary contamination

(i) Describe documentation procedures

(j) Describe evidentiary sampling techniques

(k) Describe field screening protocols for collected samples and evidence

(l) Describe evidence labeling and packaging procedures

(m) Describe evidence decontamination procedures

(n) Describe evidence packaging procedures for evidence transportation

(o) Describe chain-of-custody procedures

(2) Given an example of an illicit laboratory, the operations level responder assigned to evidence preservation and sampling shall be able to perform the following tasks:

(a) Describe the hazards, safety procedures, decontamination, and tactical guidelines for this type of incident

(b) Describe the factors to be evaluated in selecting the personal protective equipment, sampling equipment, detection devices, and sample and evidence packaging and transport containers

(c) Describe the sampling options associated with liquid and solid sample and evidence collection

(d) Describe the field screening protocols for collected samples and evidence

(3) Given an example of an environmental crime, the operations level responder assigned to evidence preservation and sampling shall be able to perform the following tasks:

(a) Describe the hazards, safety procedures, decontamination, and tactical guidelines for this type of incident

(b) Describe the factors to be evaluated in selecting the personal protective equipment, sampling equipment, detection devices, and sample and evidence packaging and transport containers

(c) Describe the sampling options associated with the collection of liquid and solid samples and evidence

(d) Describe the field screening protocols for collected samples and evidence

(4) Given an example of a hazardous materials/WMD suspicious letter, the operations level responder assigned to evidence preservation and sampling shall be able to perform the following tasks:

(a) Describe the hazards, safety procedures, decontamination, and tactical guidelines for this type of incident

(b) Describe the factors to be evaluated in selecting the personal protective equipment, sampling equipment, detection devices, and sample and evidence packaging and transport containers

(c) Describe the sampling options associated with the collection of liquid and solid samples and evidence

(d) Describe the field screening protocols for collected samples and evidence

(5) Given an example of a hazardous materials/WMD suspicious package, the operations level responder assigned to evidence preservation and sampling shall be able to perform the following tasks:

(a) Describe the hazards, safety procedures, decontamination, and tactical guidelines for this type of incident

(b) Describe the factors to be evaluated in selecting the personal protective equipment, sampling equipment, detection devices, and sample and evidence packaging and transport containers

(c) Describe the sampling options associated with liquid and solid sample/evidence collection

(d) Describe the field screening protocols for collected samples and evidence

(6) Given an example of a release/attack involving a hazardous materials/WMD agent, the operations level responder assigned to evidence preservation and sampling shall be able to perform the following tasks:

(a) Describe the hazards, safety procedures, decontamination, and tactical guidelines for this type of incident

(b) Describe the factors to be evaluated in selecting the personal protective equipment, sampling equipment, detection devices, and sample and evidence packaging and transport containers

(c) Describe the sampling options associated with the collection of liquid and solid samples and evidence

(d) Describe the field screening protocols for collected samples and evidence

(7) Given examples of different types of potential criminal hazardous materials/WMD incidents, the operations level responder shall identify and describe the application, use, and limitations of the various types field screening tools that can be utilized for screening the following:

(a) Corrosivity

(b) Flammability

(c) Oxidation

(d) Radioactivity

(e) Volatile organic compounds (VOC)

(8) Describe the potential adverse impact of using destructive field screening techniques

(9) Describe the procedures for maintaining the evidentiary integrity of any item removed from the crime scene

6.5.3.2 Selecting Personal Protective Equipment. Given the personal protective equipment provided by the AHJ, the operations level responder assigned to evidence preservation and sampling shall select the personal protective equipment required to support evidence preservation and sampling at hazardous materials/WMD incidents based on local procedures (*see Section 6.2*).

6.5.4 Competencies — Implementing the Planned Response.

6.5.4.1 Implementing the Planned Response. Given the incident action plan for a criminal incident involving hazardous materials/WMD, the operations level responder assigned to evidence preservation and sampling shall implement selected response actions consistent with the emergency response plan or standard operating procedures by completing the following requirements:

(1) Demonstrate how to secure the scene and characterize and preserve evidence at the scene

(2) Document personnel and scene activities associated with the incident

(3) Determine whether responders are within their legal authority to perform evidence collection and sampling tasks

(4) Describe the procedure to notify the agency with investigative authority

(5) Notify the EOD personnel

(6) Identify potential samples and evidence to be collected

(7) Demonstrate procedures to protect samples and evidence from secondary contamination

(8) Demonstrate correct techniques to collect samples utilizing the equipment provided

(9) Demonstrate documentation procedures

(10) Demonstrate sampling protocols

(11) Demonstrate field screening protocols for samples and evidence collected

(12) Demonstrate evidence/sample labeling and packaging procedures

(13) Demonstrate evidence/sample decontamination procedures

(14) Demonstrate evidence/sample packaging procedures for evidence transportation

(15) Describe chain of custody procedures for evidence/sample preservation

6.5.4.2 The operations level responder assigned to evidence preservation and sampling shall describe local procedures for the technical decontamination process.

6.5.5 Competencies — Implementing the Planned Response. (Reserved)

6.5.6 Competencies — Terminating the Incident. (Reserved)

6.6 Mission-Specific Competencies: Product Control.

6.6.1 General.

6.6.1.1 Introduction.

6.6.1.1.1 The operations level responder assigned to perform product control shall be that person, competent at the operations level, who is assigned to implement product control measures at hazardous materials/WMD incidents.

6.6.1.1.2 The operations level responder assigned to perform product control at hazardous materials/WMD incidents shall be trained to meet all competencies at the awareness level (*see Chapter 4*), all core competencies at the operations level (*see Chapter 5*), all mission-specific competencies for personal protective equipment (*see Section 6.2*), and all competencies in this section.

6.6.1.1.3 The operations level responder assigned to perform product control at hazardous materials/WMD incidents shall operate under the guidance of a hazardous materials technician, an allied professional, or standard operating procedures.

6.6.1.1.4* The operations level responder assigned to perform product control at hazardous materials/WMD incidents shall receive the additional training necessary to meet specific needs of the jurisdiction.

6.6.1.2 Goal.

6.6.1.2.1 The goal of the competencies in this section shall be to provide the operations level responder assigned to product control at hazardous materials/WMD incidents with the knowledge and skills to perform the tasks in 6.6.1.2.2 safely and effectively.

6.6.1.2.2 When responding to hazardous materials/WMD incidents, the operations level responder assigned to perform product control shall be able to perform the following tasks:

(1) Plan an initial response within the capabilities and competencies of available personnel, personal protective equipment, and control equipment and in accordance with the emergency response plan or standard operating procedures by completing the following tasks:

(a) Describe the control options available to the operations level responder

(b) Describe the control options available for flammable liquid and flammable gas incidents

(2) Implement the planned response to a hazardous materials/WMD incident

6.6.2 Competencies — Analyzing the Incident. (Reserved)

6.6.3 Competencies — Planning the Response.

6.6.3.1 Identifying Control Options. Given examples of hazardous materials/WMD incidents, the operations level responder assigned to perform product control shall identify the options for each response objective by completing the following requirements as prescribed by the AHJ:

(1) Identify the options to accomplish a given response objective

(2) Identify the purpose for and the procedures, equipment, and safety precautions associated with each of the following control techniques:

(a) Absorption

(b) Adsorption

(c) Damming

(d) Diking

(e) Dilution

(f) Diversion

(g) Remote valve shutoff

(h) Retention

(i) Vapor dispersion

(j) Vapor suppression

6.6.3.2 Selecting Personal Protective Equipment. Given the personal protective equipment provided by the AHJ, the operations level responder assigned to perform product control shall select the personal protective equipment required to support product control at hazardous materials/WMD incidents based on local procedures (*see Section 6.2*).

6.6.4 Competencies — Implementing the Planned Response.

6.6.4.1 Performing Control Options. Given an incident action plan for a hazardous materials/WMD incident, within the capabilities and equipment provided by the AHJ, the operations level responder assigned to perform product control shall demonstrate control functions set out in the plan by completing the following requirements as prescribed by the AHJ:

(1) Using the type of special purpose or hazard suppressing foams or agents and foam equipment furnished by the AHJ, demonstrate the application of the foam(s) or agent(s) on a spill or fire involving hazardous materials/WMD

(2) Identify the characteristics and applicability of the following Class B foams if supplied by the AHJ:

(a) Aqueous film-forming foam (AFFF)

(b) Alcohol-resistant concentrates

(c) Fluoroprotein

(d) High-expansion foam

(3) Given the required tools and equipment, demonstrate how to perform the following control activities:

(a) Absorption

(b) Adsorption

(c) Damming

(d) Diking

(e) Dilution

(f) Diversion

(g) Retention

(h) Remote valve shutoff

(i) Vapor dispersion

(j) Vapor suppression

(4) Identify the location and describe the use of emergency remote shutoff devices on MC/DOT-306/406, MC/DO-T307/407, and MC-331 cargo tanks containing flammable liquids or gases

(5) Describe the use of emergency remote shutoff devices at fixed facilities

6.6.4.2 The operations level responder assigned to perform product control shall describe local procedures for going through the technical decontamination process.

6.6.5 Competencies — Evaluating Progress. (Reserved)

6.6.6 Competencies — Terminating the Incident. (Reserved)

6.7 Mission-Specific Competencies: Air Monitoring and Sampling.

6.7.1 General.

6.7.1.1 Introduction.

6.7.1.1.1 The operations level responder assigned to perform air monitoring and sampling shall be that person, competent at the operations level, who is assigned to implement air monitoring and sampling operations at hazardous materials/WMD incidents.

6.7.1.1.2 The operations level responder assigned to perform air monitoring and sampling at hazardous materials/WMD incidents shall be trained to meet all competencies at the awareness level (*see Chapter 4*), all core competencies at the operations level (*see Chapter 5*), all mission-specific competencies for personal protective equipment (*see Section 6.2*), and all competencies in this section.

6.7.1.1.3 The operations level responder assigned to perform air monitoring and sampling at hazardous materials/WMD incidents shall operate under the guidance of a hazardous materials technician, an allied professional, or standard operating procedures.

6.7.1.1.4* The operations level responder assigned to perform air monitoring and sampling at hazardous materials/WMD incidents shall receive the additional training necessary to meet specific needs of the jurisdiction.

6.7.1.2 Goal.

6.7.1.2.1 The goal of the competencies in this section shall be to provide the operations level responder assigned to air monitoring and sampling at hazardous materials/WMD incidents with the knowledge and skills to perform the tasks in 6.7.1.2.2 safely and effectively.

6.7.1.2.2 When responding to hazardous materials/WMD incidents, the operations level responder assigned to perform air monitoring and sampling shall be able to perform the following tasks:

(1) Plan the air monitoring and sampling activities within the capabilities and competencies of available personnel, personal protective equipment, and control equipment and in accordance with the emergency response plan or standard operating procedures describe the air monitoring and sampling options available to the operations level responder

(2) Implement the air monitoring and sampling activities as specified in the incident action plan

6.7.2 Competencies — Analyzing the Incident. (Reserved)

6.7.3 Competencies — Planning the Response.

6.7.3.1 Given the air monitoring and sampling equipment provided by the AHJ, the operations level responder assigned to perform air monitoring and sampling shall select the detection or monitoring equipment suitable for detecting or monitoring solid, liquid, or gaseous hazardous materials/WMD.

6.7.3.2 Given detection and monitoring device(s) provided by the AHJ, the operations level responder assigned to perform air monitoring and sampling shall describe the operation, capabilities and limitations, local monitoring procedures, field testing, and maintenance procedures associated with each device.

6.7.3.3 Selecting Personal Protective Equipment (PPE). Given the PPE provided by the AHJ, the operations level responder assigned to perform air monitoring and sampling shall select the personal protective equipment required to support air monitoring and sampling at hazardous materials/WMD incidents based on local procedures (*see Section 6.2*).

6.7.4 Competencies — Implementing the Planned Response.

6.7.4.1 Given a scenario involving hazardous materials/WMD and detection and monitoring devices provided by the AHJ, the operations level responder assigned to perform air monitoring and sampling shall demonstrate the field test and operation of each device and interpret the readings based on local procedures.

6.7.4.2 The operations level responder assigned to perform air monitoring and sampling shall describe local procedures for decontamination of themselves and their detection and monitoring devices upon completion of the air monitoring mission.

6.7.5 Competencies — Evaluating Progress. (Reserved)

6.7.6 Competencies — Terminating the Incident. (Reserved)

6.8 Mission-Specific Competencies: Victim Rescue and Recovery.

6.8.1 General.

6.8.1.1 Introduction.

6.8.1.1.1 The operations level responder assigned to perform victim rescue and recovery shall be that person, competent at the operations level, who is assigned to rescue and recover exposed and contaminated victims at hazardous materials/WMD incidents.

6.8.1.1.2 The operations level responder assigned to perform victim rescue and recovery at hazardous materials/WMD incidents shall be trained to meet all competencies at the awareness level (*see Chapter 4*), all core competencies at the operations level (*see Chapter 5*), all mission-specific competencies for personal protective equipment (*see Section 6.2*), and all competencies in this section.

6.8.1.1.3 The operations level responder assigned to perform victim rescue and recovery at hazardous materials/WMD incidents shall operate under the guidance of a hazardous materials technician, an allied professional, or standard operating procedures.

6.8.1.1.4* The operations level responder assigned to perform victim rescue and recovery at hazardous materials/WMD incidents shall receive the additional training necessary to meet specific needs of the jurisdiction.

6.8.1.2 Goal.

6.8.1.2.1 The goal of the competencies in this section shall be to provide the operations level responder assigned victim rescue and recovery at hazardous materials/WMD incidents with the knowledge and skills to perform the tasks in 6.8.1.2.2 safely and effectively.

6.8.1.2.2 When responding to hazardous materials/WMD incidents, the operations level responder assigned to perform victim rescue and recovery shall be able to perform the following tasks:

(1) Plan a response for victim rescue and recovery operations involving the release of hazardous materials/WMD agent within the capabilities of available personnel and personal protective equipment

(2) Implement the planned response to accomplish victim rescue and recovery operations within the capabilities of available personnel and personal protective equipment

6.8.2 Competencies — Analyzing the Incident. (Reserved)

6.8.3 Competencies — Planning the Response.

6.8.3.1 Given scenarios involving hazardous materials/WMD incidents, the operations level responder assigned to victim rescue and recovery shall determine the feasibility of conducting victim rescue and recovery operations at an incident involving hazardous materials/WMD and shall be able to perform the following tasks:

(1) Determine the feasibility of conducting rescue and recovery operations

(2) Describe the safety procedures, tactical guidelines, and incident response considerations to effect a rescue associated with each of the following situations:

(a) Line-of-sight with ambulatory victims

(b) Line-of-sight with nonambulatory victims

(c) Non-line-of-sight with ambulatory victims

(d) Non-line-of-sight with nonambulatory victims

(e) Victim rescue operations versus victim recovery operations

(3) Determine if the options are within the capabilities of available personnel and personal protective equipment

(4) Describe the procedures for implementing victim rescue and recovery operations within the incident command system

6.8.3.2 Selecting Personal Protective Equipment (PPE). Given the PPE provided by the AHJ, the operations level responder assigned to perform victim rescue and recovery shall select the personal protective equipment required to support victim rescue and recovery at hazardous materials/WMD incidents based on local procedures (*see Section 6.2*).

6.8.4 Competencies — Implementing the Planned Response.

6.8.4.1 Given a scenario involving hazardous materials/WMD, the operations level responder assigned to victim rescue and recovery shall perform the following tasks:

(1) Identify the different team positions and describe their main functions

(2) Select and use specialized rescue equipment and procedures provided by the AHJ to support victim rescue and recovery operations

(3) Demonstrate safe and effective methods for victim rescue and recovery

(4) Demonstrate the ability to triage victims

(5) Describe local procedures for performing decontamination upon completion of the victim rescue and removal mission

6.8.5 Competencies — Evaluating Progress. (Reserved)

6.8.6 Competencies — Terminating the Incident. (Reserved)

6.9 Mission-Specific Competencies: Response to Illicit Laboratory Incidents.

6.9.1 General.

6.9.1.1 Introduction.

6.9.1.1.1 The operations level responder assigned to respond to illicit laboratory incidents shall be that person, competent at the operations level, who, at hazardous materials/WMD incidents involving potential violations of criminal statutes specific to the illegal manufacture of methamphetamines, other drugs, or WMD, is assigned to secure the scene, identify the laboratory or process, and preserve evidence at hazardous materials/WMD incidents involving potential violations of criminal statutes specific to the illegal manufacture of methamphetamines, other drugs, or WMD.

6.9.1.1.2 The operations level responder who responds to illicit laboratory incidents shall be trained to meet all competencies at the awareness level (*see Chapter 4*), all core competencies at the operations level (*see Chapter 5*), all mission-specific competencies for personal protective equipment (*see Section 6.2*), and all competencies in this section.

6.9.1.1.3 The operations level responder who responds to illicit laboratory incidents shall operate under the guidance of a hazardous materials technician, an allied professional, or standard operating procedures.

6.9.1.1.4* The operations level responder who responds to illicit laboratory incidents shall receive the additional training necessary to meet specific needs of the jurisdiction.

6.9.1.2 Goal.

6.9.1.2.1 The goal of the competencies in this section shall be to provide the operations level responder assigned to respond to illicit laboratory incidents with the knowledge and skills to perform the tasks in 6.9.1.2.2 safely and effectively.

6.9.1.2.2 When responding to hazardous materials/WMD incidents, the operations level responder assigned to respond to illicit laboratory incidents shall be able to perform the following tasks:

(1) Analyze a hazardous materials/WMD incident to determine the complexity of the problem and potential outcomes and whether the incident is potentially a criminal illicit laboratory operation

(2) Plan a response for a hazardous materials/WMD incident involving potential illicit laboratory operations in compliance with evidence preservation operations within the capabilities and competencies of available personnel, personal protective equipment, and control equipment after notifying the responsible law enforcement agencies of the problem

(3) Implement the planned response to a hazardous materials/WMD incident involving potential illicit laboratory operations utilizing applicable evidence preservation guidelines

6.9.2 Competencies — Analyzing the Incident.

6.9.2.1 Determining If a Hazardous Materials/WMD Incident Is an Illicit Laboratory Operation. Given examples of hazardous materials/WMD incidents involving illicit laboratory operations, the operations level responder assigned to respond to illicit laboratory incidents shall identify the potential drugs/WMD being manufactured by completing the following related requirements:

(1) Given examples of illicit drug manufacturing methods, describe the operational considerations, hazards, and products involved in the illicit process

(2) Given examples of illicit chemical WMD methods, describe the operational considerations, hazards, and products involved in the illicit process

(3) Given examples of illicit biological WMD methods, describe the operational considerations, hazards, and products involved in the illicit process

(4) Given examples of illicit laboratory operations, describe the potential booby traps that have been encountered by response personnel

(5) Given examples of illicit laboratory operations, describe the agencies that have investigative authority and operational responsibility to support the response

6.9.3 Competencies — Planning the Response.

6.9.3.1 Determining the Response Options. Given an analysis of hazardous materials/WMD incidents involving illicit laboratories, the operations level responder assigned to respond to illicit laboratory incidents shall identify possible response options.

6.9.3.2 Identifying Unique Aspects of Criminal Hazardous Materials/WMD Incidents.

6.9.3.2.1 The operations level responder assigned to respond to illicit laboratory incidents shall identify the unique operational aspects associated with illicit drug manufacturing and illicit WMD manufacturing.

6.9.3.2.2 Given an incident involving illicit drug manufacturing or illicit WMD manufacturing, the operations level responder assigned to illicit laboratory incidents shall describe the following tasks:

(1) Law enforcement securing and preserving the scene

(2) Joint hazardous materials and EOD personnel site reconnaissance and hazard identification

(3) Determining atmospheric hazards through air monitoring and detection

(4) Mitigation of immediate hazards while preserving evidence

(5) Coordinated crime scene operation with the law enforcement agency having investigative authority

(6) Documenting personnel and scene activities associated with incident

6.9.3.3 Identifying the Law Enforcement Agency That Has Investigative Jurisdiction. The operations level responder assigned to respond to illicit laboratory incidents shall identify the law enforcement agency having investigative jurisdiction by completing the following:

(1) Given scenarios involving illicit drug manufacturing or illicit WMD manufacturing, identify the law enforcement agency(s) with investigative authority for the following situations:

(a) Illicit drug manufacturing

(b) Illicit WMD manufacturing

(c) Environmental crimes resulting from illicit laboratory operations

6.9.3.4 Identifying Unique Tasks and Operations at Sites Involving Illicit Laboratories.

6.9.3.4.1 The operations level responder assigned to respond to illicit laboratory incidents shall identify and describe the unique tasks and operations encountered at illicit laboratory scenes.

6.9.3.4.2 Given scenarios involving illicit drug manufacturing or illicit WMD manufacturing, describe the following:

(1) Hazards, safety procedures, and tactical guidelines for this type of emergency

(2) Factors to be evaluated in selection of the appropriate personal protective equipment for each type of tactical operation

(3) Factors to be considered in selection of appropriate decontamination procedures

(4) Factors to be evaluated in the selection of detection devices

(5) Factors to be considered in the development of a remediation plan

6.9.3.5 Selecting Personal Protective Equipment (PPE). Given the PPE provided by the AHJ, the operations level responder assigned to respond to illicit laboratory incidents shall select the personal protective equipment required to respond to illicit laboratory incidents based on local procedures. (*See Section 6.2.*)

6.9.4 Competencies — Implementing the Planned Response.

6.9.4.1 Implementing the Planned Response. Given scenarios involving an illicit drug/WMD laboratory operation involving hazardous materials/WMD, the operations level responder assigned to respond to illicit laboratory incidents shall implement or oversee the implementation of the selected response options safely and effectively.

6.9.4.1.1 Given a simulated illicit drug/WMD laboratory incident, the operations level responder assigned to respond to illicit laboratory incidents shall be able to perform the following tasks:

(1) Describe safe and effective methods for law enforcement to secure the scene

(2) Demonstrate decontamination procedures for tactical law enforcement personnel (SWAT or K-9) securing an illicit laboratory

(3) Describe methods to identify and avoid potential unique safety hazards found at illicit laboratories such as booby traps and releases of hazardous materials

(4) Describe methods to conduct joint hazardous materials/EOD operations to identify safety hazards and implement control procedures

6.9.4.1.2 Given a simulated illicit drug/WMD laboratory entry operation, the operations level responder assigned to respond to illicit laboratory incidents shall describe methods for identifying the following during reconnaissance operations:

(1) Potential manufacture of illicit drugs

(2) Potential manufacture of illicit WMD materials

(3) Potential environmental crimes associated with the manufacture of illicit drugs/WMD materials

6.9.4.1.3 Given a simulated illicit drug/WMD laboratory incident, the operations level responder assigned to respond to illicit laboratory incidents shall describe joint agency crime scene operations, including support to forensic crime scene processing teams.

6.9.4.1.4 Given a simulated illicit drug/WMD laboratory incident, the operations level responder assigned to respond to illicit laboratory incidents shall describe the policy and procedures for post–crime scene processing and site remediation operations.

6.9.4.1.5 The operations level responder assigned to respond to illicit laboratory incidents shall describe local procedures for performing decontamination upon completion of the illicit laboratory mission.

6.9.5 Competencies — Evaluating Progress. (Reserved)

6.9.6 Competencies — Terminating the Incident. (Reserved)

6.10 Mission-Specific Competencies: Disablement/Disruption of Improvised Explosives Devices (IEDs), Improvised WMD Dispersal Devices, and Operations at Improvised Explosives Laboratories.

6.10.1 General.

6.10.1.1 Introduction.

6.10.1.1.1 The operations level responder assigned to perform disablement/disruption of IEDs, improvised WMD dispersal devices, and operations at improvised explosives laboratories shall be that person, competent at the operations level, who is assigned to interrupt the functioning of an IED or an improvised WMD dispersal device or conduct operations at improvised explosives laboratories.

6.10.1.1.2 The operations level responder assigned to perform disablement/disruption of IEDs, improvised WMD dispersal devices, and operations at improvised explosives laboratories shall possess current certification as a Hazardous Device Technician from the FBI Hazardous Devices School, Department of Defense, or equivalent certifying agency as determined by the AHJ and be functioning as a member of a bomb squad or recognized military unit.

6.10.1.1.3 The operations level responder assigned to perform disablement/disruption of IEDs, improvised WMD dispersal devices, and operations at improvised explosives laboratories shall be trained to meet all competencies at the awareness level (*see Chapter 4*), all core competencies at the operations level (*see Chapter 5*), all mission-specific competencies for personal protective equipment (*see Section 6.2*), mission-specific competencies for response to illicit laboratories (*see Section 6.9*), and all competencies in this section.

6.10.1.1.4 The operations level responder assigned to perform disablement/disruption of IEDs, improvised WMD dispersal devices, and operations at improvised explosives laboratories shall operate under the guidance of an allied professional or standard operating procedures.

6.10.1.1.5 The operations level responder assigned to perform disablement/disruption of IEDs, improvised WMD dispersal devices, and operations at improvised explosives laboratories shall receive the additional training necessary to meet the specific needs of the jurisdiction and/or agency.

6.10.1.2 Goal.

6.10.1.2.1 The goal of the competencies in this section shall be to provide the operations level responder assigned to perform disablement/disruption of IEDs, improvised WMD dispersal devices, and operations at improvised explosives laboratories with the knowledge and skills to perform the tasks in 6.10.1.2.2 and 6.10.1.2.3 safely and effectively.

6.10.1.2.2 When responding to hazardous materials/WMD incidents involving a potential IED or improvised WMD dispersal device, the operations level responder assigned to perform disablement/disruption of IEDs, improvised WMD dispersal devices, and operations at improvised explosives laboratories shall be able to perform the following tasks:

(1) Analyze a hazardous materials/WMD incident involving an improvised WMD dispersal device to determine the complexity of the problem and potential outcomes by completing the following tasks:

- **(a)** Determine if an IED or WMD dispersal device is potentially present
- **(b)** Categorize the device by its delivery method

(2) Plan a response for a hazardous materials/WMD incident where there is a potential improvised WMD dispersal device within the capabilities and competencies of available personnel, personal protective equipment, and control equipment by completing the following tasks

- **(a)** Determine if response options can be effectively employed to conduct a disablement/disruption of the device
- **(b)** Describe the actions to be taken and the resources to be requested if the incident exceeds the available capabilities

(3) Implement the planned response to a hazardous materials/WMD incident involving an IED or WMD dispersal device by completing the following tasks under the guidance of the senior hazardous devices technician (HDT) present:

- **(a)** Employ disablement/disruption techniques in accordance with the FBI Hazardous Devices School "logic tree," the current edition of the National Bomb Squad Commanders Advisory Board (NBSCAB) "A Model for Bomb Squad Standard Operating Procedures," established protocol of military units, or the AHJ

6.10.1.2.3 When responding to hazardous materials/WMD incidents involving potential improvised explosives laboratories, the operations level responder assigned to perform disablement/disruption of IEDs, improvised WMD dispersal devices, and operations at improvised explosives laboratories shall be able to perform the following tasks:

(1) Analyze a hazardous materials/WMD incident involving a potential improvised explosives laboratory to determine the complexity of the problem and potential outcomes and whether the incident is potentially an improvised explosives laboratory operation

(2) Plan a response to a hazardous materials/WMD incident involving a potential improvised explosives laboratory in compliance with mitigation techniques and evidence recovery within the capabilities and competencies of available personnel, personal protective equipment, and control equipment, after notifying the responsible investigative agencies of the problem

(3) Implement the planned response to a hazardous materials/WMD incident involving a potential improvised explosives laboratory utilizing applicable standard operating procedures and/or technical advice from qualified allied professionals

6.10.2 Competencies — Analyzing the Incident.

6.10.2.1 Determining If the Incident Involves the Potential Presence of an Improvised WMD Dispersal Device. Given examples of hazardous materials/WMD incidents involving an IED or improvised WMD dispersal device, the operations level responder assigned to perform disablement/disruption of IEDs, improvised WMD dispersal devices, and operations at improvised explosives laboratories shall identify and/or categorize the hazard by completing the following:

(1) Given examples of the following hazardous materials/WMD incidents involving an IED or improvised WMD dispersal device, describe products that might be encountered in the incident associated with each situation:

- **(a)** Letter/package-based improvised dispersal device
- **(b)** Briefcase/backpack-based improvised dispersal device
- **(c)** Transportation-borne WMD dispersal device
- **(d)** Fixed location hazards where an IED has been placed to cause the deliberate release of a material

6.10.2.2 Determining If the Hazardous Materials/WMD Incident Involves an Improvised Explosives Laboratory Operation. Given examples of hazardous materials/WMD incidents involving improvised explosives laboratories, the operations level responder assigned to perform disablement/disruption of IEDs, improvised WMD dispersal devices, and operations at improvised explosives laboratories shall identify the potential explosives/WMD being manufactured by completing the following related requirements:

(1) Given examples of improvised explosives manufacturing methods, describe the operational considerations, hazards, and products involved in the process

(2) Given examples of improvised explosives laboratory operations, describe the potential booby traps that have been encountered by response personnel

(3) Given examples of improvised explosives laboratory operations, describe the agencies that have investigative authority and operational responsibility to support the response

6.10.3 Competencies — Planning the Response.

6.10.3.1 Identifying Unique Aspects of Improvised WMD Dispersal Device Related Hazardous Materials/WMD Incidents. When responding to hazardous materials/WMD incidents, the operations level responder assigned to perform disablement/disruption of IEDs, improvised WMD dispersal devices, and operations at improvised explosives laboratory incidents shall be capable of identifying the unique aspects associated with such incidents by completing the following requirements:

(1) Given an incident involving a nonvehicle based WMD dispersal device, shall be able to perform the following tasks:

(a) Describe the hazards, safety procedures, and tactical guidelines for this type of incident

(b) Describe the factors to be evaluated in selecting the personal protective equipment

(c) Describe the procedure for identifying and obtaining the appropriate emergency response elements to support disablement/disruption activities

(2) Given an incident involving a vehicle-borne WMD dispersal device, shall be able to perform the following tasks:

(a) Describe the hazards, safety procedures, and tactical guidelines for this type of incident

(b) Describe the factors to be evaluated in selecting the personal protective equipment

(c) Describe the procedure for identifying and obtaining the appropriate emergency response elements to support disablement/disruption activities

(3) Given examples of different types of incidents involving an improvised WMD dispersal device, shall identify and describe the application use and limitations of various types of field screening tools that can be utilized for determining the presence of the following materials:

(a) Gamma and neutron radiation

(b) Explosive materials [commercial and home made explosives (HME)]

6.10.3.2 Identifying Unique Aspects of Improvised Explosives Laboratory Related Hazardous Materials/WMD Incidents. When responding to conduct mitigation procedures on energetic materials at an improvised explosive laboratory, the operations level responder assigned to perform disablement/disruption of IEDs, improvised WMD dispersal devices, and operations at improvised explosives shall be capable of identifying the unique aspects associated with such incidents by completing the following requirements:

(1) Given a scenario involving an improvised explosive laboratory and detection devices provided by the AHJ, complete the following:

(a) Describe the hazards, safety procedures, and tactical guidelines for this type of incident

(b) Describe the factors to be evaluated in selecting the personal protective equipment

(c) Describe the application, use, and limitations of various types of field screening tools that can be utilized for determining the presence of the following materials:

i. Radioactive materials that emit alpha, beta, gamma, or neutron radiation, including radionuclide identification of gamma emitting radioactive materials

ii. Explosive materials (commercial and HME)

(d) Demonstrate the field test and operation of each detection device and interpret the readings based on local procedures

(e) Describe local procedures for decontamination of themselves and their detection devices upon completion of the material detection mission

(f) Describe the procedure for identifying and obtaining the appropriate emergency response elements to support disablement/disruption or mitigation activities

6.10.3.3 Identifying Potential Response Options.

6.10.3.3.1 Given scenarios involving a potential IED or improvised WMD materials dispersal device, the operations level responder assigned to perform disablement/disruption of IEDs, improvised WMD dispersal devices, and operations at improvised explosives laboratories incident shall identify possible response options.

6.10.3.3.2 Given scenarios involving a potential improvised explosives laboratories, the operations level responder assigned to perform disablement/disruption of IEDs, improvised WMD dispersal devices, and operations at improvised explosives laboratories incident shall identify possible response options.

6.10.3.4 Selecting Personal Protective Equipment. Given the personal protective equipment provided by the AHJ, the operations level responder assigned to perform disablement/disruption of IEDs, improvised WMD dispersal devices, and operations at improvised explosives laboratories incident, shall select the personal protective equipment required to support such operations at hazardous materials/WMD incidents based on the National Guidelines for Bomb Technicians adopted by the National Bomb Squad Commanders Advisory Board (NBSCAB) (*see Section 6.2*).

6.10.4 Competencies — Implementing the Planned Response.

6.10.4.1 Given scenarios involving a potential IED or improvised WMD dispersal device, the operations level responder assigned to perform disablement/disruption of IEDs, improvised WMD dispersal devices, and operations at improvised explosives laboratory incident shall be able to complete the following tasks:

(1) Using detection and monitoring devices provided by the AHJ, demonstrate the field test and operation of each device and interpret the readings based on local or agency procedures

(2) Perform diagnostics based on procedures instructed by a nationally accredited hazardous devices school or program

(3) Perform disablement/disruption techniques in accordance with the FBI Hazardous Devices School "logic tree," the NBSCAB *A Model for Bomb Squad Standard Operating Procedures*, established protocol for military units, or established protocol of the AHJ

(4) Assist in planning the air monitoring and sampling activities within the capabilities and competencies of available personnel, personal protective equipment, and control equipment; and in accordance with the AHJ, describe the air monitoring and sampling options available

(5) Given the air monitoring and sampling equipment provided by the AHJ, shall complete the following:

(a) Select the detection or monitoring equipment suitable for detecting or monitoring of the IED or improvised WMD dispersal device

(b) Describe the operation, capabilities, limitations, local monitoring procedures, field-testing, and maintenance procedures associated with each device provided by the AHJ

(c) Describe local procedures for decontamination of the detection and monitoring devices upon completion of the mission

6.10.4.2 Given a simulated improvised explosives laboratory incident, the operations level responder assigned to perform disablement/disruption of IEDs, improvised WMD dispersal devices, and operations at improvised explosives laboratory incident shall be able to perform the following tasks:

(1) Describe the safe and effective methods for law enforcement to secure the scene

(2) Demonstrate methods to identify and avoid unique safety hazards at improvised explosives laboratories such as booby traps, releases of hazardous materials, and initiating components

(3) Using detection and monitoring devices provided by the AHJ, demonstrate the field test and operation of each device and interpret the readings based on local or agency procedures

(4) Describe the methods that could be utilized to mitigate the hazards identified

6.10.4.3 The operations level responder assigned to perform disablement/disruption of IEDs, improvised WMD dispersal devices, and operations at improvised explosives laboratories shall demonstrate the ability to wear an appropriate combination of chemical protective clothing, respiratory protection, and ballistic protection for the hazards identified in 6.10.2.1 and 6.10.2.2.

6.10.4.4 The operations level responder assigned to perform disablement/disruption of IEDs, improvised WMD dispersal devices, and operations at improvised explosives laboratories shall describe the local procedures for the technical decontamination process.

6.10.5 Competencies — Evaluating Progress. (Reserved)

6.10.6 Competencies — Terminating the Incident (Reserved)

Appendix B: An Extract from NFPA 473, *Standard for Competencies for EMS Personnel Responding to Hazardous Materials/Weapons of Mass Destruction Incidents*

Chapter 4 Competencies for Hazardous Materials/WMD Basic Life Support (BLS) Responder

4.1 **General.**

4.1.1 **Introduction.** All emergency medical services (EMS) personnel at the hazardous materials/WMD basic life support (BLS) responder level, in addition to their BLS certification, shall be trained to meet at least the core competencies of the operations level responders as defined in Chapter 5 of NFPA 472, *Standard for Competence of Responders to Hazardous Materials/Weapons of Mass Destruction Incidents*, and all competencies of this chapter.

4.1.2 **Goal.** The goal of the competencies at the BLS responder level shall be to provide the individual with the knowledge and skills necessary to safely deliver BLS at hazardous materials/WMD incidents, function within the established IMS/ICS, and perform the following duties:

(1) Analyze a hazardous materials/WMD incident to determine the potential health hazards encountered by the BLS responder, other responders, and anticipated and actual patients by completing the following tasks:

(a) Survey an incident where hazardous materials/WMD have been released and evaluate suspected and identified patients for signs and symptoms of exposure

(b) Collect hazard and response information from available technical resources to determine the nature of the problem and potential health effects of the substances involved

(2) Plan to deliver BLS to any exposed patient within the scope of practice by completing the following tasks:

(a) Identify preplans of high-risk areas and occupancies to identify potential locations where significant human exposures can occur

(b) Identify the capabilities of the hospital network to accept exposed patients and perform emergency decontamination if required

(c) Identify the medical components of the communication plan

(d) Describe the role of the BLS level responder as it relates to the local emergency response plan and established IMS/ICS

(3) Implement a prehospital treatment plan within the scope of practice by completing the following tasks:

(a) Determine the nature of the hazardous materials/ WMD incident as it relates to anticipated or actual patient exposures and subsequent medical treatment

(b) Identify the need for and the effectiveness of decontamination efforts

(c) Determine if the available medical resources will meet or exceed patient care needs

(d) Describe evidence preservation issues associated with patient care

(e) Develop and implement a medical monitoring plan for responders

(f) Report and document the actions taken by the BLS responder at the incident scene

4.2 **Competencies — Analyzing the Incident.**

4.2.1 **Surveying Hazardous Materials/WMD Incidents.** Given scenarios of hazardous materials/WMD incidents, the BLS responder shall assess the nature and severity of the incident as it relates to anticipated or actual EMS responsibilities at the scene.

4.2.1.1 Given examples of the following types of containers, the BLS responder shall identify the potential mechanisms of injury/harm and possible treatment modalities:

(1) Pressure

(2) Nonpressure

(3) Cryogenic

(4) Radioactive

4.2.1.2 Given examples of the nine U.S. Department of Transportation (DOT) hazard classes, the BLS responder shall identify possible treatment modalities associated with each hazard class.

4.2.1.3 Given examples of various hazardous materials/WMD incidents at fixed facilities, the BLS responder shall identify the following available health-related resource personnel:

(1) Environmental heath and safety representatives

(2) Radiation safety officers (RSO)

(3) Occupational physicians and nurses

(4) Site emergency response teams

(5) Product or container specialists

4.2.1.4 Given various scenarios of hazardous materials/WMD incidents, the BLS responder, working within an IMS/ICS, shall evaluate the off-site consequences of the release based on the physical and chemical nature of the released substance and the prevailing environmental factors, to determine the need to evacuate or to shelter-in-place affected persons.

4.2.1.5 Given the following biological agents, the BLS responder shall describe the signs and symptoms of exposure and/or illness and the likely means of dissemination:

(1) Variola major virus (smallpox)

(2) *Clostridium botulinum*

(3) Coliforms (e.g., *E. coli* O157:H7)

(4) Ricin toxin

(5) *Bacillus anthracis* (anthrax)

(6) Venezuelan equine encephalitis virus

(7) *Rickettsia*

(8) *Yersinia pestis* (plague)

(9) *Francisella tularensis* (tularemia)

(10) Viral hemorrhagic fever

(11) Other CDC Category A, B, or C–listed organism

4.2.1.6 Given examples of various types of hazardous materials/WMD incidents involving toxic industrial chemicals (TICs) and toxic industrial materials (TIMs) (e.g., corrosives, reproductive hazards, carcinogens, nerve agents, flammable and/or explosive hazards, blister agents, blood agents, choking agents, and irritants), the BLS responder shall determine the general health risks to patients exposed to those substances in the case of any release with the following:

(1) Visible cloud

(2) Liquid pooling

(3) Solid dispersion

4.2.1.7 **Determining If a Hazardous Materials/WMD Incident Is an Illicit Laboratory Operation.** Given examples of hazardous materials/WMD incidents involving illicit laboratory operations, BLS responders assigned to respond to illicit laboratory incidents shall identify the potential drugs/WMD being manufactured and shall meet the following related requirements:

(1)*Given examples of illicit drug manufacturing methods, describe the operational considerations, hazards, and products involved in the illicit process

(2) Given examples of illicit chemical WMD methods, describe the operational considerations, hazards, and products involved in the illicit process

(3) Given examples of illicit biological WMD methods, describe the operational considerations, hazards, and products involved in the illicit process

(4) Given examples of illicit laboratory operations, describe the potential booby traps that have been encountered by response personnel

(5) Given examples of illicit laboratory operations, describe the agencies that have investigative authority and operational responsibility to support the response

4.2.1.8 Given a scenario involving radioactive materials, including an accidental release, a radiological dispersion device (RDD), and a radiological exposure device (RED), the BLS responder shall determine the pre-hospital care based upon the probable health risks and potential patient outcomes by completing the following:

(1) Determine the most likely exposure pathways for a given radiation exposure, including inhalation, ingestion, injection, and direct skin exposure

(2) Identify the difference between radiation exposure and radioactive contamination and the health concerns associated with each

4.2.1.9 Given three examples of pesticide labels and labeling, the BLS responder shall use the following information to determine the associated health risks:

(1) Hazard statement

(2) Precautionary statement

(3) Signal word

(4) Pesticide name

4.2.2 **Collecting and Interpreting Hazard and Response Information.** The BLS responder shall obtain information from the following sources to determine the nature of the medical problem and potential health effects:

(1) Hazardous materials databases

(2) Clinical monitoring

(3) Reference materials (e.g., MSDS and ERG)

(4)*Technical information centers (e.g., CHEMTREC, CANUTEC, and SETIQ) and local, state, and federal authorities

(5) Allied professionals

(6) Regional poison control centers

4.2.3 **Establishing and Enforcing Scene Control Procedures.** Given two scenarios involving hazardous materials/WMD incidents, the BLS responder shall identify how to establish and enforce scene control, including control zones and emergency decontamination, and communications between responders and to the public and shall meet the following requirements:

(1) Identify the procedures for establishing scene control through control zones

(2) Identify the criteria for determining the locations of the control zones at hazardous materials/WMD incidents

(3) Identify the basic techniques for the following protective actions at hazardous materials/WMD incidents:

(a) Evacuation

(b) Shelter-in-place protection

(c) Isolation of the hazard area and denial of entry

(4) Demonstrate the ability to perform emergency decontamination

(5) Identify the items to be considered in a safety briefing prior to allowing personnel to work at the following:

(a) Hazardous materials incidents

(b) Hazardous materials/WMD incidents involving criminal activities

(6) Identify the procedures for ensuring coordinated communication between responders and to the public

4.3 **Competencies — Planning the Response.**

4.3.1 **Identifying High Risk Areas for Potential Exposures.**

4.3.1.1 The BLS responder, given an events calendar and pre-incident plans, which can include the local emergency planning committee plan, as well as the agency's emergency response plan and standard operating procedures (SOPs), shall identify the venues for mass gatherings, industrial facilities, potential targets for terrorism, and any other location where an accidental or intentional release of a harmful substance can pose a health risk to any person in the local geographical area as determined by the AHJ and shall identify the following:

(1) Locations where hazardous materials/WMD are used, stored, or transported

(2) Areas and locations that present a potential for a high loss of life or rate of injury in the event of an accidental or intentional release of hazardous materials/WMD

(3)*External factors that may complicate a hazardous materials/WMD incident

4.3.2 **Determining the Capabilities of the Local Hospital Network.**

4.3.2.1 The BLS responder shall identify the following methods and vehicles available to transport hazardous materials patients and shall determine the location and potential routes of travel to the medically appropriate local and regional hospitals, based on the patients' needs:

(1) Adult trauma centers

(2) Pediatric trauma centers

(3) Adult burn centers

(4) Pediatric burn centers

(5) Hyperbaric chambers

(6) Field hospitals

(7) Dialysis centers

(8) Supportive care facilities

(9) Forward deployable assets

(10) Other specialty hospitals or medical centers

4.3.2.2 Given a list of receiving hospitals in the region, the BLS responder shall describe the location, availability, and capability of hospital-based decontamination facilities.

4.3.2.3 The BLS responder shall describe the BLS protocols for pre-hospital care and SOPs for a mass casualty incident (MCI), including triage, treatment, and transport at a hazardous materials/WMD incident where exposures have occurred as developed by the AHJ and the prescribed role of medical control and poison control centers.

4.3.2.4 The BLS responder shall identify the formal and informal mutual aid resources (hospital and nonhospital based) for the field management of multicasualty incidents, as follows:

(1) Mass-casualty trailers with medical supplies

(2) Mass-decedent capabilities

(3) Regional decontamination units

(4) Replenishment of medical supplies during long-term incidents

(5) Rehabilitation units for the EMS responders

(6) Replacement transport units for vehicles lost to mechanical trouble, collision, theft, and contamination

4.3.2.5 The BLS responder shall identify the special hazards associated with inbound and outbound air transportation of patients exposed to hazardous materials/WMD.

4.3.3 Identifying Incident Communications.

4.3.3.1 Given an incident communications plan, the BLS responder shall identify the following:

(1) Medical components of the communications plan

(2) Ability to communicate with other responders, transport units, and receiving facilities

4.3.3.2 Given examples of various patient exposure scenarios, the BLS responder shall describe the following information to be transmitted to the medical or poison control center or the receiving hospital prior to arrival:

(1) Name of the substance(s) involved

(2) Physical and chemical properties of the substance(s) involved

(3) Number of victims being transported

(4) Age and sex of transported patient

(5) Patient condition and chief complaint

(6) Medical history

(7) Circumstances and history of the exposure, such as duration of exposure and primary route of exposure

(8) Vital signs, initial and current

(9) Symptoms described by the patient, initial and current

(10) Presence of associated injuries, such as burns and trauma

(11) Decontamination status

(12) Treatment rendered or in progress

(13) Patient response to treatment(s)

(14) Estimated time of arrival

4.3.4 Identifying the Role of the BLS Responder.

4.3.4.1 Given scenarios involving hazardous materials/WMD, the BLS responder shall identify his or her role during hazardous materials/WMD incidents as specified in the emergency response plan and SOPs developed by the AHJ, as follows:

(1) Describe the purpose, benefits, and elements of the incident command system as it relates to the BLS responder

(2) Describe the typical incident command structure, for the emergency medical component of a hazardous materials/WMD incident as specified in the emergency response plan and SOPs, as developed by the AHJ

(3) Demonstrate the ability of the BLS responder to function within the incident command system

(4) Demonstrate the ability to implement an incident command system for a hazardous materials/WMD incident where an ICS does not currently exist

(5) Identify the procedures for requesting additional resources at a hazardous materials/WMD incident

4.3.4.2 The hazardous materials/WMD BLS responder shall describe his or her role within the hazardous materials response plan developed by the AHJ or identified in the local emergency response plan, as follows:

(1) Determine the toxic effect of hazardous materials/WMD

(2) Estimate the number of patients

(3) Recognize and assess the presence and severity of symptoms

(4) Take and record vital signs

(5) Determine resource maximization and assessment

(6) Assess the impact on the health care system

(7) Perform appropriate patient monitoring

(8) Communicate pertinent information

4.4 Competencies — Implementing the Planned Response.

4.4.1 Determining the Nature of the Incident/Providing Medical Care. The BLS responder shall demonstrate the ability to identify the mechanisms of injury or harm and the clinical implications and provide emergency medical care to those patients exposed to hazardous materials/WMD agent by completing the following tasks:

(1) Determine the physical state of the released substance, in addition to the environmental influences surrounding the release, as follows:

(a) Solid

(b) Liquid

(c) Gas

(d) Vapor

(e) Dust

(f) Mist

(g) Aerosol

(2) Identify potential routes of exposure and correlate those routes of exposure to the physical state of the released substance, to determine the origin of the illness or injury, as follows:

(a) Inhalation

(b) Absorption

(c) Ingestion

(d) Injection

(3)*Describe the potential routes of entry into the body, the common signs and symptoms of exposure, and the BLS treatment options approved by the AHJ for exposure(s) to the following classification of substances:

(a) Corrosives

(b) Pesticides

(c) Chemical asphyxiants

(d) Simple asphyxiants

(e) Organic solvents

(f) Nerve agents

(g) Vesicants and blister agents

(h) Blood agents

(i) Choking agents

(j) Irritants

(k) Biological agents and toxins

(l) Incapacitating agents

(m) Radioactive materials

(n) Nitrogen compounds

(o) Opiate compounds

(p) Fluorine compounds

(q) Phenolic compounds

(4) Describe the basic toxicological principles relative to assessment and treatment of persons exposed to hazardous materials, including the following:

(a) Acute and chronic effects

(b) Local and systemic effects

(c) Dose–response relationship

(5) Given examples of various hazardous materials/WMD, define the basic toxicological terms as applied to patient care:

(a) Threshold limit value — time-weighted average (TLVTWA)

(b) Permissible exposure limit (PEL)

(c) Threshold limit value — short-term exposure limit (TLV-STEL)

(d) Immediately dangerous to life and health (IDLH)

(e) Threshold limit value — ceiling (TLV-C)

(f) Parts per million/parts per billion/parts per trillion (ppm/ppb/ppt)

(6) Given examples of hazardous materials/WMD incidents with exposed patients, evaluate the progress and effectiveness of the medical care provided at a hazardous materials/WMD incident to ensure that the overall incident response objectives, along with patient care goals, are being met by completing the following tasks:

(a) Locate and track all exposed patients at a hazardous materials/WMD incident, from triage and treatment to transport to a medically appropriate facility

(b) Review the incident objectives at periodic intervals to ensure that patient care is being carried out within the overall incident action plan

(c) Ensure that the required incident command system forms are completed, along with the patient care forms, during the course of the incident

(d) Evaluate the need for trained and qualified EMS personnel, medical equipment, transport units, and other supplies based on the scope and duration of the incident

4.4.2 **Decontamination.** Given the emergency response plan and SOPs developed by the AHJ, the BLS responder shall do the following:

(1) Determine if patient decontamination activities were performed prior to accepting responsibility and transferring care of exposed patients

(2) Determine the need and location for patient decontamination, including mass casualty decontamination, in the event none has been performed prior to arrival of EMS personnel and complete the following tasks:

(a) Given the emergency response plan and SOPs developed by the AHJ, identify sources of information for determining the appropriate decontamination procedure and identify how to access those resources in a hazardous materials/WMD incident

(b) Given the emergency response plan and SOPs developed by the AHJ, identify (within the plan) the supplies and equipment required to set up and implement the following:

i. Emergency decontamination operations for ambulatory and nonambulatory patients

ii. Mass decontamination operations for ambulatory and nonambulatory patients

(c) Identify procedures, equipment, and safety precautions for the treatment and handling of emergency service animals brought to the decontamination corridor at hazardous materials/WMD incidents

(d) Identify procedures, equipment, and safety precautions for communicating with critical, urgent, and potentially exposed patients and identify population prioritization as it relates to decontamination purposes

(e) Identify procedures, equipment, and safety precautions for preventing cross contamination

4.4.3 **Determining the Ongoing Need for Medical Supplies.**

4.4.3.1 Given examples of single-patient and multicasualty hazardous materials/WMD incidents, the BLS responder shall determine the following:

(1) If the available medical equipment will meet or exceed patient care needs throughout the duration of the incident

(2) If the available transport units will meet or exceed patient care needs throughout the duration of the incident

4.4.4 **Preserving Evidence.** Given examples of hazardous materials/WMD incidents where criminal acts are suspected, the BLS responder shall make every attempt to preserve evidence during the course of delivering patient care by completing the following tasks:

(1) Determine if the incident is potentially criminal in nature and cooperate with the law enforcement agency having investigative jurisdiction

(2) Identify the unique aspects of criminal hazardous materials/WMD incidents, including crime scene preservation and evidence preservation, to avoid the destruction of potential

evidence on medical patients during the decontamination process

(3) Identify, within the emergency response plan and SOPs developed by the AHJ, procedures, equipment, and safety precautions for securing evidence during decontamination operations at hazardous materials/WMD incidents

(4) Ensure that any information regarding suspects, sequence of events during a potentially criminal act, and observations made based on patient presentation or during patient assessment are documented and communicated to the law enforcement agency having investigative jurisdiction

4.4.5 Medical Support at Hazardous Materials/WMD Incidents. Given examples of hazardous materials/WMD incident, the BLS responder shall describe the procedures of the AHJ for performing medical monitoring and support of hazardous materials incident response personnel and shall complete the following tasks:

(1) Given examples of various hazardous materials/WMD incidents requiring the use of chemical protective ensembles, the BLS responder shall complete the following tasks:

(a) Demonstrate the ability to set up and operate a medical monitoring station

(b) Demonstrate the ability to recognize the signs and symptoms of heat stress, cold stress, heat exhaustion, and heat stroke

(c) Determine the BLS needs for responders exhibiting the effects of heat stress, cold stress, and heat exhaustion

(d) Describe the medical significance of heat stroke and the importance of rapid transport to an appropriate medical receiving facility

(e) Given a simulated hazardous materials incident, demonstrate the appropriate documentation of medical monitoring activities

(2) The BLS responder responsible for pre-entry medical monitoring shall obtain hazard and toxicity information on the hazardous materials/WMD from the designated hazardous materials technical reference resource or other sources of information at the scene.

(3) The following information shall be conveyed to the entry team, incident safety officer, hazardous materials officer, other EMS personnel at the scene, and any other responders responsible for the health and well-being of those personnel operating at the scene:

(a) Chemical name

(b) Hazard class

(c) Multiple hazards and toxicity information

(d) Applicable decontamination methods and procedures

(e) Potential for cross contamination

(f) Procedure for transfer of patients from the constraints of the incident to the EMS

(g) Prehospital management of medical emergencies and exposures

(4) The BLS responder shall evaluate the pre-entry health status of responders to hazardous materials/WMD incidents as per the AHJ policies and procedures prior to their donning personal protective equipment (PPE) by performing the following tasks (consideration shall be given to excluding responders if they do not meet criteria specified by the AHJ prior to working in chemical protective clothing):

(a) Record vital signs

(b) Body weight measurements to address hydration considerations

(c) General health observations

(d) Body temperature: hypothermia/hyperthermia

(e) Blood pressure: hypotension/hypertension

(f) Pulse rate: bradycardia/tachycardia as defined

(g) Respiratory rate: bradypnea/tachypnea

(5) The BLS responder shall determine how the following factors influence heat stress on hazardous materials/WMD response personnel:

(a) Baseline level of hydration

(b) Underlying physical fitness

(c) Environmental factors

(d) Activity levels during the entry

(e) Level of PPE worn

(f) Duration of entry

(g) Cold stress

(6) The BLS responder shall medically evaluate all team members after decontamination and PPE removal, using the following criteria:

(a) Pulse rate determined within the first minute

(b) Pulse rate determined 3 minutes after initial evaluation

(c) Temperature

(d) Body weight

(e) Blood pressure

(f) Respiratory rate

(7) The BLS responder shall recommend that any hazardous materials team member be prohibited from redonning chemical protective clothing if any of the following criteria is exhibited:

(a) Signs or symptoms of heat stress or heat exhaustion

(b) Abnormal vital signs

(c) Abnormal core body temperature

(d) Abnormal heart rate and/or rhythm

(e) Abnormal blood pressure

(f)*Significant acute body weight loss

(8) Any team member exhibiting the signs or symptoms of extreme heat exhaustion or heat stroke shall be transported to the medical facility.

(9) The BLS responder responsible for medical monitoring and support shall immediately notify the persons designated by the incident action plan that a team member required significant medical treatment or transport. Transportation shall be arranged through the designee identified in the emergency response plan.

4.5 Reporting and Documenting the Incident. Given a scenario involving a hazardous materials/WMD incident, the responder assigned to use PPE shall complete the reporting and documentation requirements consistent with the emergency response plan or SOPs and identify the reports and supporting documentation required by the emergency response plan or SOPs.

4.6 Compiling Incident Reports. The BLS responder shall describe his or her role in compiling incident reports that meet federal, state, local, and organizational requirements, as follows:

(1) List the information to be gathered regarding the exposure of all patient(s) and describe the reporting procedures, including the following:

(a) Detailed information on the substances released

(b) Pertinent information on each patient treated and transported

(c) Routes, extent, and duration of exposures

(d) Actions taken to limit exposure

(e) Decontamination activities

(2) At the conclusion of the hazardous materials/WMD incident, identify the methods used by the AHJ to evaluate transport units that might have been contaminated and the process and locations available to decontaminate those units

Chapter 5 Competencies for Hazardous Materials/WMD Advanced Life Support (ALS) Responder

5.1 General.

5.1.1 Introduction. All emergency medical services (EMS) personnel at the hazardous materials/WMD advanced life support (ALS) responder level, in addition to their ALS certification, shall be trained to meet at least the core competencies of the operations level responders as defined in Chapter 5 of NFPA 472, *Standard for Competence of Responders to Hazardous Materials/Weapons of Mass Destruction Incidents*, and all competencies of this chapter.

5.1.2 Goal. The goal of the competencies at the ALS responder shall be to provide the individual with the knowledge and skills necessary to safely deliver ALS at hazardous materials/WMD incidents and to function within the established IMS/ICS, as follows:

(1) Analyze a hazardous materials/WMD incident to determine the potential health risks to the ALS provider, other responders, and anticipated/actual patients by completing the following tasks:

(a) Survey a hazardous materials/WMD incident to determine whether harmful substances have been released and to evaluate suspected and identified patients for telltale signs of exposure

(b) Collect hazard and response information from reference sources and allied professionals on the scene to determine the nature of the problem and potential health effects of the substances involved *(See Annex C for a list of informational references.)*

(c) Survey the hazardous materials/WMD scene for the presence of secondary devices and other potential hazards

(2) Plan to deliver ALS to exposed patients, within the scope of practice and training competencies established by the AHJ, by completing the following tasks:

(a) Evaluate high-risk areas/occupancies within the AHJ to identify potential locations where significant human exposures can occur

(b) Identify the capabilities of the hospital network within the AHJ to accept exposed patients and to perform emergency decontamination if required

(c) Evaluate the components of the incident communication plan within the AHJ

(d) Describe the role of the ALS responder as it relates to the local emergency response plan and established IMS/ICS

(e) Identify supplemental regional and national medical resources, including but not limited to assets of the strategic national stockpile (SNS) and the metropolitan medical response system (MMRS)

(3) Implement a prehospital treatment plan for exposed patients, within the scope of practice and training competencies established by the AHJ, by completing the following tasks:

(a) Determine the nature of the hazardous materials/WMD incident as it relates to anticipated or actual patient exposures and subsequent medical treatment

(b) Determine the need or effectiveness of decontamination prior to accepting an exposed patient

(c) Determine if the available medical equipment, transport units, and other supplies, including antidotes and therapeutic modalities, will meet patient care needs

(d) Describe the process of evidence preservation where criminal or terrorist acts are suspected or confirmed

(e) Develop and implement a medical monitoring plan for those responders operating in chemical protective clothing at a hazardous materials/WMD incident

(f) Evaluate the need to administer antidotes to affected patients

(4) Participate in the termination of the incident by completing the following tasks:

(a) Participate in an incident debriefing

(b) Participate in an incident critique with the appropriate agencies

(c) Report and document the actions taken by the ALS level responder at the scene of the incident

5.2 Competencies — Analyzing the Hazardous Materials Incident.

5.2.1 Surveying Hazardous Materials/WMD Incidents. Given scenarios of hazardous materials/WMD incidents, the ALS responder shall assess the nature and severity of the incident as it relates to anticipated or actual EMS responsibilities at the scene.

5.2.1.1 Given examples of the following marked transport vehicles (and their corresponding shipping papers or identification systems) that can be involved in hazardous materials/WMD incidents, the ALS responder shall evaluate the general health risks based on the physical and chemical properties of the anticipated contents:

(1) Highway transport vehicles, including cargo tanks

(2) Intermodal equipment, including tank containers

(3) Rail transport vehicles, including tank cars

5.2.1.2 Given examples of various hazardous materials/WMD incidents at fixed facilities, the ALS responder shall demonstrate the ability to perform the following tasks:

(1) Identify a variety of containers and their markings, including bulk and nonbulk packages and containers, drums, underground and aboveground storage tanks, specialized storage tanks, or any other specialized containers found in the AHJ's geographic area, and evaluate the general health risks based on the physical and chemical properties of the anticipated contents

(2) Identify the following job functions of health-related resource personnel available at fixed facility hazardous materials/WMD incidents:

(a) Environmental heath and safety representatives

(b) Radiation safety officers

(c) Occupational physicians and nurses

(d) Site emergency response teams

(e) Specialized experts

5.2.1.3 The ALS responder shall identify two ways to obtain a material safety data sheet (MSDS) at a hazardous materials/WMD incident and shall demonstrate the ability to identify the following health-related information:

(1) Proper chemical name or synonyms

(2) Physical and chemical properties

(3) Health hazards of the material

(4) Signs and symptoms of exposure

(5) Routes of entry

(6) Permissible exposure limits

(7) Emergency medical procedures or recommendations

(8) Responsible party contact

5.2.1.4 Given scenarios at various fixed facilities, transportation incidents, pipeline release scenarios, maritime incidents, or any other unexpected hazardous materials/WMD incident, the ALS responder, working within an incident command system must evaluate the off-site consequences of the release, based on the physical and chemical nature of the released substance, and the prevailing environmental factors to determine the need to evacuate or shelter-in-place affected persons.

5.2.1.5* Given examples of the following biological threat agents, the ALS responder shall define the various types of biological threat agents, including the signs and symptoms of exposure, mechanism of toxicity, incubation periods, possible disease patterns, and likely means of dissemination:

(1) Variola major virus (smallpox)

(2) *Clostridium botulinum* (botulism)

(3) Coliforms (e.g., *E. coli* O157:H7)

(4) Ricin toxin

(5) *Bacillus anthracis* (anthrax)

(6) Venezuelan equine encephalitis virus

(7) *Rickettsia*

(8) *Yersinia pestis* (plague)

(9) *Francisella tularensis* (tularemia)

(10) Viral hemorrhagic fever

(11) Other CDC Category A, B, or C–listed organism

5.2.1.6* Given examples of various types of hazardous materials/WMD incidents involving toxic industrial chemicals, toxic industrial materials, blister agents, blood agents, nerve agents, choking agents, and irritants, the ALS responder shall determine the general health risks to patients exposed to those substances and identify those patients who may be candidates for antidotes.

5.2.1.7* Given examples of hazardous materials/WMD found at illicit laboratories, the ALS responder shall identify general health hazards associated with the chemical substances that are expected to be encountered.

5.2.1.8* Given examples of a hazardous materials/WMD incident involving radioactive materials, including radiation dispersion devices (RDD), radiation exposure devices (RED), and improvised nuclear devices (IND), the ALS responder shall determine the probable health risks and potential patient outcomes by completing the following tasks:

(1) Determine the types of radiation (alpha, beta, gamma, and neutron), the isotope if possible, and potential health effects of each

(2) Determine the most likely exposure pathways for a given radiation exposure, including inhalation, ingestion, injection, and direct skin exposure

(3) Describe the difference between radioactive contamination and radiation exposure

(4) Identify priorities for decontamination in scenarios involving radioactive materials

(5) Describe the manner in which acute medical illness or traumatic injury can influence decisions about decontamination and patient transport

5.2.1.9 Given examples of typical labels found on pesticide containers, the ALS responder shall define the following terms:

(1) Pesticide name

(2) Pesticide classification (e.g., insecticide, rodenticide, organophosphate, carbamate, organochlorine)

(3) Environmental Protection Agency (EPA) registration number

(4) Manufacturer name

(5) Ingredients broken down by percentage

(6) Cautionary statement (e.g., Danger, Warning, Caution, Keep from Waterways)

(7) Strength and concentration

(8) Treatment information

5.2.2 Collecting and Interpreting Hazard and Response Information. The ALS responder shall demonstrate the ability to utilize various reference sources at a hazardous materials/ WMD incident, including the following:

(1) DOT *Emergency Response Guidebook*

(2) MSDS

(3) CHEMTREC/CANUTEC/SETIQ

(4) Regional poison control centers

(5) NFPA 704, *Standard System for the Identification of the Hazards of Materials for Emergency Response*, identification system

(6) Hazardous materials information system (HMIS)

(7) Local, state, federal, tribal, and provincial authorities

(8) Shipper/manufacturer contacts

(9) Agency for Toxic Substances and Disease Registry (ATSDR) medical management guidelines

(10) Allied professionals

(11) Electronic databases

(12) Radiation safety officer (RSO)

5.2.2.1 Identifying Secondary Devices. Given scenarios involving hazardous materials/WMD, the ALS responders shall describe the importance of evaluating the scene for secondary devices prior to rendering patient care, including the following safety points:

(1) Evaluate the scene for likely areas where secondary devices can be placed

(2) Visually scan operating areas for a secondary device before providing patient care

(3) Avoid touching or moving anything that can conceal an explosive device

(4) Designate and enforce scene control zones

(5) Evacuate victims, other responders, and nonessential personnel as quickly and safely as possible

5.3 Competencies — Planning the Response.

5.3.1 Identifying High-Risk Areas for Potential Exposures.

5.3.1.1 The ALS responder, given an events calendar and pre-incident plans, which can include the local emergency planning committee plan as well as the agency's emergency response plan and SOPs, shall identify the venues for mass gatherings, industrial facilities, potential targets for terrorism, or any other locations where an accidental or intentional release of a harmful substance can pose an unreasonable health risk to any person within the local geographical area as determined by the AHJ and shall do the following:

(1) Identify locations where hazardous materials/WMD are used, stored, or transported

(2) Identify areas and locations presenting a potential for a high loss of life or rate of injury in the event of an accidental/intentional release of a hazardous materials/ WMD substance

(3) Evaluate the geographic and environmental factors that can complicate a hazardous materials/WMD incident, including prevailing winds, water supply, vehicle and pedestrian traffic flow, ventilation systems, and other natural or man-made influences, including air and rail corridors

5.3.2 Determining the Capabilities of the Local Hospital Network.

5.3.2.1 The ALS responder shall identify the methods and vehicles available to transport hazardous materials patients and shall determine the location and potential routes of travel to the following appropriate local and regional hospitals, based on patient need:

(1) Adult trauma centers

(2) Pediatric trauma centers

(3) Adult burn centers

(4) Pediatric burn centers

(5) Hyperbaric chambers

(6) Field hospitals

(7) Other specialty hospitals or medical centers

5.3.2.2 Given a list of local receiving hospitals in the AHJ's geographic area, the ALS responder shall describe the location and availability of hospital-based decontamination facilities.

5.3.2.3 The ALS responder shall describe the ALS protocols and SOPs developed by the AHJ and the prescribed role of medical control and poison control centers during mass casualty incidents, at hazardous materials/WMD incidents where exposures have occurred, and in the event of disrupted radio communications.

5.3.2.4 The ALS responder shall identify the following mutual aid resources (hospital and nonhospital based) identified by the AHJ for the field management of multicasualty incidents:

(1) Mass-casualty trailers with medical supplies

(2) Mass-decedent capability

(3) Regional decontamination units

(4) Replenishment of medical supplies during long-term incidents

(5) Locations and availability of mass-casualty antidotes for selected exposures, including but not limited to the following:

(a) Nerve agents and organophosphate pesticides

(b) Biological agents and other toxins

(c) Asphyxiants

(d) Opiate exposures

(e) Radiation exposures or contamination events

(6) Rehabilitation units for the EMS responders

(7) Replacement transport units for those vehicles lost to mechanical trouble, collision, theft, and contamination

5.3.2.5 The ALS responder shall identify the special hazards associated with inbound and outbound air transportation of patients exposed to hazardous materials/WMD.

5.3.2.6 The ALS responder shall describe the available medical information resources concerning hazardous materials toxicology and response.

5.3.3 Identifying Incident Communications.

5.3.3.1 The ALS responder shall identify the components of the communication plan within the AHJ geographic area and determine that the EMS providers have the ability to communicate with other responders on the scene, with transport units, and with local hospitals.

5.3.3.2 Given examples of various patient exposure scenarios, the ALS responder shall describe the following information to be transmitted to the medical control or poison control center or the receiving hospital prior to arrival:

(1) The exact name of the substance(s) involved

(2) The physical and chemical properties of the substance(s) involved

(3) Number of victims being transported

(4) Age and sex of transported patients

(5) Patient condition and chief complaint

(6) Medical history

(7) Circumstances and history of the exposure, such as duration of exposure and primary route of exposure

(8) Vital signs, initial and current

(9) Symptoms described by the patient, initial and current

(10) Presence of associated injuries, such as burns and trauma

(11) Decontamination status

(12) Treatment rendered or in progress, including the effectiveness of antidotes administered

(13) Estimated time of arrival

5.3.4 Identifying the Role of the ALS Level Responder.

5.3.4.1 Given scenarios involving hazardous materials/WMD, the ALS responder shall identify his or her role during hazardous materials/ WMD incidents as specified in the emergency response plan and SOPs developed by the AHJ, as follows:

(1) Describe the purpose, benefits, and elements of the incident command system at it relates to the ALS responder

(2) Describe the typical incident command structure for the emergency medical component of a hazardous materials/WMD incident as specified in the emergency response plan and SOPs developed by the AHJ

(3) Describe the role of the ALS responder within the incident command system

(4) Describe the role of the ALS responder within the incident command system during a hazardous materials/WMD incident when an ICS does not currently exist

(5) Identify the procedures for requesting additional resources at a hazardous materials/WMD incident

5.3.4.2 Describe the hazardous materials/WMD ALS responder's role in the hazardous materials/WMD response plan developed by the AHJ or identified in the local emergency response plan as follows:

(1) Determine the toxic effect of hazardous materials/WMD

(2) Estimate the number of patients

(3) Recognize and assess the presence and severity of symptoms

(4) Assess the impact on the health care system

(5) Perform appropriate patient monitoring

(6) Communicate pertinent information

(7) Estimate pharmacological need

(8) Address threat potential for clinical latency

(9) Estimate dosage — exposure

(10) Estimate dosage — treatment

(11) Train in appropriate monitoring

5.3.5 **Supplemental Medical Resources.** Given scenarios of various hazardous materials/WMD mass casualty incidents, the ALS responder shall identify the supplemental medical resources available to the AHJ, including the following:

(1) Describe the strategic national stockpile (SNS) program, including the following components:

(a) Intent and goals of the SNS program

(b) Procedures and requirements for deploying the SNS to a local jurisdiction

(c) Typical supplies contained in 12-hour push package

(d) Role of the technical advisory response unit (TARU)

(2) Describe the metropolitan medical response system (MMRS) including the following components:

(a) Scope, intent, and goals of the MMRS

(b) Capabilities and resources of the MMRS

(c) Eight capability focus areas of the MMRS

5.4 **Competencies — Implementing the Planned Response.**

5.4.1 **Determining the Nature of the Incident and Providing Medical Care.** The ALS responder shall describe the ability to provide emergency medical care to those patients exposed to hazardous materials/WMD by completing the following tasks:

(1) The ALS responder shall determine the physical state of the released substance and the environmental influences surrounding the release, as follows:

(a) Solid

(b) Liquid

(c) Gas, vapor, dust, mist, aerosol

(2)* The ALS responder shall identify potential routes of exposure, and correlate those routes of exposure to the physical state of the released substance, to determine the origin of the illness or injury, as follows:

(a) Inhalation

(b) Absorption

(c) Ingestion

(d) Injection

(3) The ALS responder shall describe the potential routes of entry into the body, the common signs and symptoms of exposure, and the ALS treatment options approved by the AHJ (e.g., advanced airway management, drug therapy), including antidote administration where appropriate, for exposure(s) to the following classification of substances:

(a) Corrosives

(b) Pesticides

(c) Chemical asphyxiants

(d) Simple asphyxiants

(e) Organic solvents

(f) Nerve agents

(g) Vesicants

(h) Irritants (riot control agents)

(i) Biological agents and toxins

(j) Incapacitating agents

(k) Radioactive materials

(l) Nitrogen compounds

(m) Opiate compounds

(n) Fluorine compounds

(o) Phenolic compounds

(4) The ALS responder shall describe the basic toxicological principles relative to assessment and treatment of persons exposed to hazardous materials, including the following:

(a) Acute and chronic effects

(b) Local and systemic effects

(c) Dose–response relationship

(5) Given examples of various hazardous substances, the ALS responder shall define the basic toxicological terms as they relate to the treatment of an exposed patient, as follows:

(a) *Threshold limit value — time weighted average* (TLV-TWA)

(b) *Lethal doses* and *lethal* concentrations, as follows:

i. LD_{lo}

ii. LD_{50}

iii. LD_{hi}

iv. LC_{lo}

v. LC_{50}

vi. LC_{hi}

(c) *Parts per million/parts per billion/parts per trillion* (ppm/ppb/ppt)

(d) *Immediately dangerous to life and health* (IDLH)

(e) *Permissible exposure limit* (PEL)

(f) *Recommended exposure limit* (REL)

(g) *Threshold limit value — short-term exposure limit* (TLVSTEL)

(h) *Threshold limit value — ceiling* (TLV-C)

(i) *Solubility*

(j) *Poison* — a substance that causes injury, illness, or death

(k) *Toxic* — harmful nature related to amount and concentration

(6) Given examples of hazardous materials/WMD incidents with exposed patients, the ALS responder shall evaluate the progress and effectiveness of the medical care provided at a hazardous materials/WMD incident, to ensure that the overall incident response objectives, along with patient care goals, are being met by completing the following tasks:

(a) Locate and track all exposed patients at a hazardous materials/WMD incident, from triage and treatment to transport to the appropriate hospital

(b) Review the incident objectives at periodic intervals to ensure that patient care is being carried out within the overall incident response plan

(c) Ensure that the incident command system forms are completed, along with the patient care forms required by the AHJ, during the course of the incident

(d) Evaluate the need for trained and qualified EMS personnel, medical equipment, transport units, and other supplies, including antidotes based on the scope and duration of the incident

5.4.2* Decontaminating Exposed Patients. Given the emergency response plan and SOPs developed by the AHJ and given examples of hazardous materials/WMD incidents with exposed patients, the ALS responder shall do as follows:

(1) Given the emergency response plan and SOPs developed by the AHJ, identify and evaluate the patient decontamination activities performed prior to accepting responsibility for and transferring care of exposed patients

(2) Determine the need and location for patient decontamination, including mass-casualty decontamination, in the event none has been performed prior to arrival of EMS personnel, and complete the following tasks:

(a) Given the emergency response plan and SOPs developed by the AHJ, identify and evaluate the patient decontamination activities performed prior to accepting responsibility for and transferring care of exposed patients; identify sources of information for determining the appropriate decontamination procedure and how to access those resources in a hazardous materials/WMD incident

(b) Given the emergency response plan and SOPs developed by the AHJ, identify and evaluate the patient decontamination activities performed prior to accepting responsibility for and transferring care of exposed patients

(c) Given the emergency response plan and SOPs provided by the AHJ, identify the supplies and equipment required to set up and implement technical or mass-casualty decontamination operations for ambulatory and nonambulatory patients

(d) Given the emergency response plan and SOPs developed by the AHJ, identify the procedures, equipment, and safety precautions for securing evidence during decontamination operations at hazardous materials/WMD incidents

(e) Identify procedures, equipment, and safety precautions for handling tools, equipment, weapons, and law enforcement and K-9 search dogs brought to the decontamination corridor at hazardous materials/WMD incidents

(f) Identify procedures, equipment, and safety precautions for communicating with critically, urgently, and potentially exposed patients, and population prioritization and management techniques

(g) Determine the threat of cross contamination to all responders and patients by completing the following tasks:

i. Identify hazardous materials/WMD with a high risk of cross contamination

ii. Identify hazardous materials/WMD agents with a low risk of cross contamination

iii. Describe how the physical state of the hazardous materials/WMD provides clues to its potential for secondary contamination, when the exact identity of the hazardous materials/WMD is not known

5.4.3 Evaluating the Need for Medical Supplies. Given examples of single-patient and multicasualty hazardous materials/WMD incidents, the ALS responder shall determine if the available medical equipment, transport units, and other supplies, including antidotes, will meet or exceed expected patient care needs throughout the duration of the incident.

5.4.4 Evidence Preservation. Given examples of hazardous materials/WMD incidents where criminal acts are suspected, the ALS responder shall make every attempt to preserve evidence during the course of delivering patient care by completing the following tasks:

(1) Determine if the incident is potentially criminal in nature and cooperate with the law enforcement agency having investigative jurisdiction

(2) Identify the unique aspects of criminal hazardous materials/WMD incidents, including crime scene preservation, evidence preservation, and destruction of potential evidence found on medical patients, and/or the destruction of evidence during the decontamination process

(3) Ensure that any information regarding suspects, sequence of events during a potential criminal act, or observations made based on patient presentation or during patient assessment are documented and communicated and passed on to the law enforcement agency having investigative jurisdiction

5.4.5 Medical Support at Hazardous Materials/WMD Incidents. Given examples of hazardous materials/WMD incidents, the ALS responder shall describe the procedures of the AHJ for performing medical monitoring and support of hazardous materials incident response personnel, and shall complete the following tasks:

(1) The ALS responder responsible for pre-entry medical monitoring shall obtain hazard and toxicity information on the released substance from the designated hazardous materials technical reference resource or other reliable sources of information at the scene. The following information shall be conveyed to the entry team, incident safety officer, hazardous materials officer, other EMS personnel at the scene, and any other responders responsible for the health and well-being of those personnel operating at the scene:

(a) Chemical name

(b) Hazard class

(c) Hazard and toxicity information

(d) Applicable decontamination methods and procedures

(e) Potential for secondary contamination

(f) Procedure for transfer of patients from the constraints of the incident to the emergency medical system

(g) Prehospital management of medical emergencies and exposures, including antidote administration

(2) The ALS responder shall evaluate the pre-entry health status of hazardous materials/WMD responders prior to donning PPE by performing the following tasks:

(a) Record vital signs

(b) Record body weight measurements

(c) Record general health observations

(3) The ALS responder shall determine the medical fitness of those personnel charged with donning chemical protective clothing, using the criteria set forth in the emergency action plan (EAP) and the SOP developed by the AHJ. Consideration shall be given to excluding responders from working in personal protective equipment if they exhibit any significant abnormalities in the following areas:

(a) Body temperature (taking temperature or skin temperature does not correlate to body temperature)

(b) Vital signs

(4) The ALS responder shall determine how the following factors influence heat stress on hazardous materials/WMD response personnel:

(a) Baseline level of hydration

(b) Underlying physical fitness

(c) Environmental factors

(d) Activity levels during the entry

(e) Level of PPE worn

(f) Duration of entry

(g) Cold stress

(5) Given examples of various hazardous materials/WMD incidents requiring the use of chemical protective ensembles, the ALS responder shall complete the following tasks:

(a) Demonstrate the ability to set up and operate a medical monitoring station

(b) Demonstrate the ability to recognize the signs and symptoms of heat stress, heat exhaustion, and heat stroke

(c) Determine the ALS needs for responders exhibiting the effects of heat stress, cold stress, and heat exhaustion

(d) Describe the medical significance of heat stroke and the importance of rapid transport to an appropriate medical receiving facility

(6) Given a simulated hazardous materials/WMD incident, the ALS responder shall demonstrate documentation of medical monitoring activities.

(7) The ALS responder shall evaluate all team members after decontamination and PPE removal, using the following criteria:

(a) Pulse rate — done within the first minute

(b) Pulse rate — 3 minutes after initial evaluation

(c) Temperature

(d) Body weight

(e) Blood pressure

(f) Respiratory rate

(8) The ALS responder shall recommend that any hazardous materials team member exhibiting any of the following signs be prohibited from redonning chemical protective clothing:

(a) Signs or symptoms of heat stress or heat exhaustion

(b) Abnormal vital signs

(c) Abnormal core body temperature

(d) Abnormal heart rate or rhythm

(e) Significant acute body weight loss

(9) The ALS responder shall notify immediately the appropriate persons designated by the emergency response plan if a team member requires significant medical treatment or transport (arranged through the appropriate designee identified by the emergency response plan).

5.5 **Competencies — Terminating the Incident.** Upon termination of the hazardous materials/WMD incident, the ALS responder shall complete the reporting, documentation, and EMS termination activities as required by the local emergency response plan or the organization's SOPs and shall meet the following requirements:

(1) Identify the reports and supporting documentation required by the emergency response plan or SOPs

(2) Demonstrate completion of the reports required by the emergency response plan or SOPs

(3) Describe the importance of personnel exposure records

(4) Describe the importance of debriefing records

(5) Describe the importance of critique records

(6) Identify the steps in keeping an activity log and exposure records

(7) Identify the steps to be taken in compiling incident reports that meet federal, state, local, and organizational requirements

(8) Identify the requirements for compiling personal protective equipment logs

(9) Identify the requirements for filing documents and maintaining records, as follows:

(a) List the information to be gathered regarding the exposure of all patient(s) and describe the reporting procedures, including the following:

i. Detailed information on the substances released

ii. Pertinent information on each patient treated or transported

iii. Routes, extent, and duration of exposures

iv. Actions taken to limit exposure

v. Decontamination activities

(b) Identify the methods used by the AHJ to evaluate transport units for potential contamination and the process and locations available to decontaminate those units

Appendix C: OSHA 1910.120, Hazardous Waste Operations and Emergency Response

© Greg Henry/ShutterStock, Inc.

Courtesy of Occupational Safety & Health Administration (OSHA), Standard Number 1910.120.

1910.120(a) **Scope, application, and definitions.**

1910.120(a)(1) Scope. This section covers the following operations, unless the employer can demonstrate that the operation does not involve employee exposure or the reasonable possibility for employee exposure to safety or health hazards:

1910.120(a)(1)(i) Clean-up operations required by a governmental body, whether Federal, state local or other involving hazardous substances that are conducted at uncontrolled hazardous waste sites (including, but not limited to, the EPA's National Priority Site List (NPL), state priority site lists, sites recommended for the EPA NPL, and initial investigations of government identified sites which are conducted before the presence or absence of hazardous substances has been ascertained);

1910.120(a)(1)(ii) Corrective actions involving clean-up operations at sites covered by the Resource Conservation and Recovery Act of 1976 (RCRA) as amended (42 U.S.C. 6901 **et seq**);

1910.120(a)(1)(iii) Voluntary clean-up operations at sites recognized by Federal, state, local or other governmental bodies as uncontrolled hazardous waste sites;

1910.120(a)(1)(iv) Operations involving hazardous waste that are conducted at treatment, storage, disposal (TSD) facilities regulated by 40 CFR Parts 264 and 265 pursuant to RCRA; or by agencies under agreement with U.S.E.P.A. to implement RCRA regulations; and

1910.120(a)(1)(v) Emergency response operations for releases of, or substantial threats of releases of, hazardous substances without regard to the location of the hazard.

1910.120(a)(2) **Application.**

1910.120(a)(2)(i) All requirements of Part 1910 and Part 1926 of Title 29 of the Code of Federal Regulations apply pursuant to their terms to hazardous waste and emergency response operations whether covered by this section or not. If there is a conflict or overlap, the provision more protective of employee safety and health shall apply without regard to 29 CFR 1910.5(c)(1).

1910.120(a)(2)(ii) Hazardous substance clean-up operations within the scope of paragraphs (a)(1)(i) through (a)(1)(iii) of this section must comply with all paragraphs of this section except paragraphs (p) and (q).

1910.120(a)(2)(iii) Operations within the scope of paragraph (a)(1)(iv) of this section must comply only with the requirements of paragraph (p) of this section.

Notes and Exceptions:

1910.120(a)(2)(iii)(A) All provisions of paragraph (p) of this section cover any treatment, storage or disposal (TSD) operation regulated by 40 CFR parts 264 and 265 or by state law authorized under RCRA, and required to have a permit or interim status from EPA pursuant to 40 CFR 270.1 or from a state agency pursuant to RCRA.

1910.120(a)(2)(iii)(B) Employers who are not required to have a permit or interim status because they are conditionally exempt small quantity generators under 40 CFR 261.5 or are generators who qualify under 40 CFR 262.34 for exemptions from regulation under 40 CFR parts 264, 265 and 270 ("excepted employers") are not covered by paragraphs (p)(1) through (p)(7) of this section. Excepted employers who are required by the EPA or state agency to have their employees engage in emergency response or who direct their employees to engage in emergency response are covered by paragraph (p)(8) of this section, and cannot be exempted by (p)(8)(i) of this section.

1910.120(a)(2)(iii)(C) If an area is used primarily for treatment, storage or disposal, any emergency response operations in that area shall comply with paragraph (p) (8) of this section. In other areas not used primarily for treatment, storage, or disposal, any emergency response operations shall comply with paragraph (q) of this section. Compliance with the requirements of paragraph (q) of this section shall be deemed to be in compliance with the requirements of paragraph (p)(8) of this section.

1910.120(a)(2)(iv) Emergency response operations for releases of, or substantial threats of releases of, hazardous substances which are not covered by paragraphs (a)(1)(i) through (a)(1)(iv) of this section must only comply with the requirements of paragraph (q) of this section.

1910.120(a)(3) **Definitions**

Buddy system means a system of organizing employees into work groups in such a manner that each employee of the work group is designated to be observed by at least one other employee in the work group. The purpose of the buddy system is to provide rapid assistance to employees in the event of an emergency.

Clean-up operation means an operation where hazardous substances are removed, contained, incinerated, neutralized, stabilized, cleared-up, or in any other manner processed or handled with the ultimate goal of making the site safer for people or the environment.

Decontamination means the removal of hazardous substances from employees and their equipment to the extent necessary to preclude the occurrence of foreseeable adverse health effects.

Emergency response or responding to emergencies means a response effort by employees from outside the immediate release area or by other designated responders (i.e., mutual aid groups, local fire departments, etc.) to an occurrence which results, or is likely to result, in an uncontrolled release of a hazardous substance. Responses to incidental releases of hazardous substances where the substance can be absorbed, neutralized, or otherwise controlled at the time of release by employees in the immediate release area, or by maintenance personnel are not considered to be emergency responses within the scope of this standard. Responses to releases of hazardous substances where there is no potential safety or health hazard (i.e., fire, explosion, or chemical exposure) are not considered to be emergency responses.

Facility means (A) any building, structure, installation, equipment, pipe or pipeline (including any pipe into a sewer or publicly owned treatment works), well, pit, pond, lagoon, impoundment, ditch, storage container, motor vehicle, rolling stock, or aircraft, or (B) any site or area where a hazardous substance has been deposited, stored, disposed of, or placed, or otherwise come to be located; but does not include any consumer product in consumer use or any water-borne vessel.

Hazardous materials response (HAZMAT) team means an organized group of employees, designated by the employer, who are expected to perform work to handle and control actual or potential leaks or spills of hazardous substances requiring possible close approach to the substance. The team members perform responses to releases or potential releases of hazardous substances for the purpose of control or stabilization of the incident. A HAZMAT team is not a fire brigade nor is a typical fire brigade a HAZMAT team. A HAZMAT team, however, may be a separate component of a fire brigade or fire department.

Hazardous substance means any substance designated or listed under (A) through (D) of this definition, exposure to which results or may result in adverse effects on the health or safety of employees:

[A] Any substance defined under section 101(14) of CERCLA;

[B] Any biologic agent and other disease causing agent which after release into the environment and upon exposure, ingestion, inhalation, or assimilation into any person, either directly from the environment or indirectly by ingestion through food chains, will or may reasonably be anticipated to cause death, disease, behavioral abnormalities, cancer, genetic mutation, physiological malfunctions (including malfunctions in reproduction) or physical deformations in such persons or their offspring.

[C] Any substance listed by the U.S. Department of Transportation as hazardous materials under 49 CFR 172.101 and appendices; and

[D] Hazardous waste as herein defined.

Hazardous waste means—

[A] A waste or combination of wastes as defined in 40 CFR 261.3, or

[B] Those substances defined as hazardous wastes in 49 CFR 171.8.

Hazardous waste operation means any operation conducted within the scope of this standard.

Hazardous waste site or Site means any facility or location within the scope of this standard at which hazardous waste operations take place.

Health hazard means a chemical, mixture of chemicals or a pathogen for which there is statistically significant evidence based on at least one study conducted in accordance with established scientific principles that acute or chronic health effects may occur in exposed employees. The term "health hazard" includes chemicals which are carcinogens, toxic or highly toxic agents, reproductive toxins, irritants, corrosives, sensitizers, hepatotoxins, nephrotoxins, neurotoxins, agents which act on the hematopoietic system, and agents which damage the lungs, skin, eyes, or mucous membranes. It also includes stress due to temperature extremes. Further definition of the terms used above can be found in Appendix A to 29 CFR 1910.1200.

IDLH or Immediately dangerous to life or health means an atmospheric concentration of any toxic, corrosive or asphyxiant substance that poses an immediate threat to life or would interfere with an individual's ability to escape from a dangerous atmosphere.

Oxygen deficiency means that concentration of oxygen by volume below which atmosphere supplying respiratory protection must be provided. It exists in atmospheres where the percentage of oxygen by volume is less than 19.5 percent oxygen.

Permissible exposure limit means the exposure, inhalation or dermal permissible exposure limit specified in 29 CFR Part 1910, Subparts G and Z.

Published exposure level means the exposure limits published in "NIOSH Recommendations for Occupational Health Standards" dated 1986, which is incorporated by reference as specified in § 1910.6, or if none is specified, the exposure limits published in the standards specified by the American Conference of Governmental Industrial Hygienists in their publication "Threshold Limit Values and Biological Exposure Indices for 1987-88" dated 1987, which is incorporated by reference as specified in § 1910.6.

Post emergency response means that portion of an emergency response performed after the immediate threat of a release has been stabilized or eliminated and clean-up of the site has begun. If post emergency response is performed by an employer's own employees who were part of the initial emergency response, it is considered to be part of the initial response and not post emergency response. However, if a group of an employer's own employees, separate from the group providing initial response, performs the clean-up operation, then the separate group of employees would be considered to be performing post-emergency response and subject to paragraph (q)(11) of this section.

Qualified person means a person with specific training, knowledge and experience in the area for which the person has the responsibility and the authority to control.

Site safety and health supervisor (or official) means the individual located on a hazardous waste site who is responsible to the employer and has the authority and knowledge necessary to implement the site safety and health plan and verify compliance with applicable safety and health requirements.

Small quantity generator means a generator of hazardous wastes who in any calendar month generates no more than 1,000 kilograms (2,205) pounds of hazardous waste in that month.

Uncontrolled hazardous waste site means an area identified as an uncontrolled hazardous waste site by a governmental body, whether Federal, state, local or other where an accumulation of hazardous substances creates a threat to the health and safety of individuals or the environment or both. Some sites are found on public lands such as those created by former municipal, county or state landfills where illegal or poorly managed waste disposal has taken place. Other sites are found on private property, often belonging to generators or former generators of hazardous substance wastes. Examples of such sites include, but are not limited to, surface impoundments, landfills, dumps, and tank or drum farms. Normal operations at TSD sites are not covered by this definition.

1910.120(b) Safety and health program.

NOTE TO (b): Safety and health programs developed and implemented to meet other federal, state, or local regulations are considered acceptable in meeting this requirement if they cover or are modified to cover the topics required in this paragraph. An additional or separate safety and health program is not required by this paragraph.

1910.120(b)(1) General.

1910.120(b)(1)(i) Employers shall develop and implement a written safety and health program for their employees involved in hazardous waste operations. The program shall be designed to identify, evaluate, and control safety and health hazards, and provide for emergency response for hazardous waste operations.

1910.120(b)(1)(ii) The written safety and health program shall incorporate the following:

1910.120(b)(1)(ii)(A) An organizational structure;

1910.120(b)(1)(ii)(B) A comprehensive workplan;

1910.120(b)(1)(ii)(C) A site-specific safety and health plan which need not repeat the employer's standard operating procedures required in paragraph (b)(1)(ii)(F) of this section;

1910.120(b)(1)(ii)(D) The safety and health training program;

1910.120(b)(1)(ii)(E) The medical surveillance program;

1910.120(b)(1)(ii)(F) The employer's standard operating procedures for safety and health; and

1910.120(b)(1)(ii)(G) Any necessary interface between general program and site specific activities.

1910.120(b)(1)(iii) **Site excavation.** Site excavations created during initial site preparation or during hazardous waste operations shall be shored or sloped as appropriate to prevent accidental collapse in accordance with Subpart P of 29 CFR Part 1926.

1910.120(b)(1)(iv) **Contractors and sub-contractors.** An employer who retains contractor or sub-contractor services for work in hazardous waste operations shall inform those contractors, sub-contractors, or their representatives of the site emergency response procedures and any potential fire, explosion, health, safety or other hazards of the hazardous waste operation that have been identified by the employer's information program.

1910.120(b)(1)(v) **Program availability.** The written safety and health program shall be made available to any contractor or subcontractor or their representative who will be involved with the hazardous waste operation; to employees; to employee designated representatives; to OSHA personnel, and to personnel of other Federal, state, or local agencies with regulatory authority over the site.

1910.120(b)(2) **Organizational structure part of the site program.**

1910.120(b)(2)(i) The organizational structure part of the program shall establish the specific chain of command and specify the overall responsibilities of supervisors and employees. It shall include, at a minimum, the following elements:

1910.120(b)(2)(i)(A) A general supervisor who has the responsibility and authority to direct all hazardous waste operations.

1910.120(b)(2)(i)(B) A site safety and health supervisor who has the responsibility and authority to develop and implement the site safety and health plan and verify compliance.

1910.120(b)(2)(i)(C) All other personnel needed for hazardous waste site operations and emergency response and their general functions and responsibilities.

1910.120(b)(2)(i)(D) The lines of authority, responsibility, and communication.

1910.120(b)(2)(ii) The organizational structure shall be reviewed and updated as necessary to reflect the current status of waste site operations.

1910.120(b)(3) **Comprehensive workplan part of the site program.** The comprehensive workplan part of the program shall address the tasks and objectives of the site operations and the logistics and resources required to reach those tasks and objectives.

1910.120(b)(3)(i) The comprehensive workplan shall address anticipated clean-up activities as well as normal operating procedures which need not repeat the employer's procedures available elsewhere.

1910.120(b)(3)(ii) The comprehensive workplan shall define work tasks and objectives and identify the methods for accomplishing those tasks and objectives.

1910.120(b)(3)(iii) The comprehensive workplan shall establish personnel requirements for implementing the plan.

1910.120(b)(3)(iv) The comprehensive workplan shall provide for the implementation of the training required in paragraph (e) of this section.

1910.120(b)(3)(v) The comprehensive workplan shall provide for the implementation of the required informational programs required in paragraph (i) of this section.

1910.120(b)(3)(vi) The comprehensive workplan shall provide for the implementation of the medical surveillance program described in paragraph (f) of this section.

1910.120(b)(4) **Site-specific safety and health plan part of the program.**

1910.120(b)(4)(i) **General.** The site safety and health plan, which must be kept on site, shall address the safety and health hazards of each phase of site operation and include the requirements and procedures for employee protection.

1910.120(b)(4)(ii) **Elements.** The site safety and health plan, as a minimum, shall address the following:

1910.120(b)(4)(ii)(A) A safety and health risk or hazard analysis for each site task and operation found in the workplan.

1910.120(b)(4)(ii)(B) Employee training assignments to assure compliance with paragraph (e) of this section.

1910.120(b)(4)(ii)(C) Personal protective equipment to be used by employees for each of the site tasks and operations being conducted as required by the personal protective equipment program in paragraph (g)(5) of this section.

1910.120(b)(4)(ii)(D) Medical surveillance requirements in accordance with the program in paragraph (f) of this section.

1910.120(b)(4)(ii)(E) Frequency and types of air monitoring, personnel monitoring, and environmental sampling techniques and instrumentation to be used, including methods of maintenance and calibration of monitoring and sampling equipment to be used.

1910.120(b)(4)(ii)(F) Site control measures in accordance with the site control program required in paragraph (d) of this section.

1910.120(b)(4)(ii)(G) Decontamination procedures in accordance with paragraph (k) of this section.

1910.120(b)(4)(ii)(H) An emergency response plan meeting the requirements of paragraph (l) of this section for safe and effective responses to emergencies, including the necessary PPE and other equipment.

1910.120(b)(4)(ii)(I) Confined space entry procedures.

1910.120(b)(4)(ii)(J) A spill containment program meeting the requirements of paragraph (j) of this section.

1910.120(b)(4)(iii) **Pre-entry briefing.** The site specific safety and health plan shall provide for pre-entry briefings to be held prior to initiating any site activity, and at such other times as necessary to ensure that employees are apprised of the site safety and health plan and that this plan is being followed. The information and data obtained from site characterization and analysis work required in paragraph (c) of this section shall be used to prepare and update the site safety and health plan.

1910.120(b)(4)(iv) **Effectiveness of site safety and health plan.** Inspections shall be conducted by the site safety and health supervisor or, in the absence of that individual, another individual who is knowledgeable in occupational safety and health, acting on behalf of the employer as necessary to determine the effectiveness of the site safety and health plan. Any deficiencies in the effectiveness of the site safety and health plan shall be corrected by the employer.

1910.120(c) Site characterization and analysis

1910.120(c)(1) General. Hazardous waste sites shall be evaluated in accordance with this paragraph to identify specific site hazards and to determine the appropriate safety and health control procedures needed to protect employees from the identified hazards.

1910.120(c)(2) Preliminary evaluation. A preliminary evaluation of a site's characteristics shall be performed prior to site entry by a qualified person in order to aid in the selection of appropriate employee protection methods prior to site entry. Immediately after initial site entry, a more detailed evaluation of the site's specific characteristics shall be performed by a qualified person in order to further identify existing site hazards and to further aid in the selection of the appropriate engineering controls and personal protective equipment for the tasks to be performed.

1910.120(c)(3) Hazard identification. All suspected conditions that may pose inhalation or skin absorption hazards that are immediately dangerous to life or health (IDLH) or other conditions that may cause death or serious harm shall be identified during the preliminary survey and evaluated during the detailed survey. Examples of such hazards include, but are not limited to, confined space entry, potentially explosive or flammable situations, visible vapor clouds, or areas where biological indicators such as dead animals or vegetation are located.

1910.120(c)(4) Required information. The following information to the extent available shall be obtained by the employer prior to allowing employees to enter a site:

1910.120(c)(4)(i) Location and approximate size of the site.

1910.120(c)(4)(ii) Description of the response activity and/or the job task to be performed.

1910.120(c)(4)(iii) Duration of the planned employee activity.

1910.120(c)(4)(iv) Site topography and accessibility by air and roads.

1910.120(c)(4)(v) Safety and health hazards expected at the site.

1910.120(c)(4)(vi) Pathways for hazardous substance dispersion.

1910.120(c)(4)(vii) Present status and capabilities of emergency response teams that would provide assistance to on-site employees at the time of an emergency.

1910.120(c)(4)(viii) Hazardous substances and health hazards involved or expected at the site and their chemical and physical properties.

1910.120(c)(5) Personal protective equipment. Personal protective equipment (PPE) shall be provided and used during initial site entry in accordance with the following requirements:

1910.120(c)(5)(i) Based upon the results of the preliminary site evaluation, an ensemble of PPE shall be selected and used during initial site entry which will provide protection to a level of exposure below permissible exposure limits and published exposure levels for known or suspected hazardous substances and health hazards and which will provide protection against other known and suspected hazards identified during the preliminary site evaluation. If there is no permissible exposure limit or published exposure level, the employer may use other published studies and information as a guide to appropriate personal protective equipment.

1910.120(c)(5)(ii) If positive-pressure self-contained breathing apparatus is not used as part of the entry ensemble, and if respiratory protection is warranted by the potential hazards identified during the preliminary site evaluation, an escape self-contained breathing apparatus of at least five minutes' duration shall be carried by employees during initial site entry.

1910.120(c)(5)(iii) If the preliminary site evaluation does not produce sufficient information to identify the hazards or suspected hazards of the site an ensemble providing equivalent to Level B PPE shall be provided as minimum protection, and direct reading instruments shall be used as appropriate for identifying IDLH conditions. (See Appendix B for guidelines on Level B protective equipment.)

1910.120(c)(5)(iv) Once the hazards of the site have been identified, the appropriate PPE shall be selected and used in accordance with paragraph (g) of this section.

1910.120(c)(6) Monitoring. The following monitoring shall be conducted during initial site entry when the site evaluation produces information which shows the potential for ionizing radiation or IDLH conditions, or when the site information is not sufficient reasonably to eliminate these possible conditions:

1910.120(c)(6)(i) Monitoring with direct reading instruments for hazardous levels of ionizing radiation.

1910.120(c)(6)(ii) Monitoring the air with appropriate direct reading test equipment (i.e., combustible gas meters, detector tubes) for IDLH and other conditions that may cause death or serious harm (combustible or explosive atmospheres, oxygen deficiency, toxic substances.)

1910.120(c)(6)(iii) Visually observing for signs of actual or potential IDLH or other dangerous conditions.

1910.120(c)(6)(iv) An ongoing air monitoring program in accordance with paragraph (h) of this section shall be implemented after site characterization has determined the site is safe for the start-up of operations.

1910.120(c)(7) Risk identification. Once the presence and concentrations of specific hazardous substances and health hazards have been established, the risks associated with these substances shall be identified. Employees who will be working on the site shall be informed of any risks that have been identified. In situations covered by the Hazard Communication Standard, 29 CFR 1910.1200, training required by that standard need not be duplicated.

NOTE TO PARAGRAPH (c)(7).—Risks to consider include, but are not limited to:

[a] Exposures exceeding the permissible exposure limits and published exposure levels.

[b] IDLH Concentrations.

[c] Potential Skin Absorption and Irritation Sources.

[d] Potential Eye Irritation Sources.

[e] Explosion Sensitivity and Flammability Ranges.

[f] Oxygen deficiency.

1910.120(c)(8) Employee notification. Any information concerning the chemical, physical, and toxicologic properties of each substance known or expected to be present on site that is available to the employer and relevant to the duties an employee is expected to perform shall be made available to the affected employees prior to the commencement of their work activities. The employer may utilize information developed for the hazard communication standard for this purpose.

1910.120(d) Site control.

1910.120(d)(1) General. Appropriate site control procedures shall be implemented to control employee exposure to hazardous substances before clean-up work begins.

1910.120(d)(2) Site control program. A site control program for protecting employees which is part of the employer's site safety and health program required in paragraph (b) of this section shall be developed during the planning stages of a hazardous waste clean-up operation and modified as necessary as new information becomes available.

1910.120(d)(3) Elements of the site control program. The site control program shall, as a minimum, include: A site map; site work zones; the use of a "buddy system"; site communications including alerting means for emergencies; the standard operating procedures or safe work practices; and, identification of the nearest medical assistance. Where these requirements are covered elsewhere they need not be repeated.

1910.120(e) Training.

1910.120(e)(1) General.

1910.120(e)(1)(i) All employees working on site (such as but not limited to equipment operators, general laborers and others) exposed to hazardous substances, health hazards, or safety hazards and their supervisors and management responsible for the site shall receive training meeting the requirements of this paragraph before they are permitted to engage in hazardous waste operations that could expose them to hazardous substances, safety, or health hazards, and they shall receive review training as specified in this paragraph.

1910.120(e)(1)(ii) Employees shall not be permitted to participate in or supervise field activities until they have been trained to a level required by their job function and responsibility.

1910.120(e)(2) Elements to be covered. The training shall thoroughly cover the following:

1910.120(e)(2)(i) Names of personnel and alternates responsible for site safety and health;

1910.120(e)(2)(ii) Safety, health and other hazards present on the site;

1910.120(e)(2)(iii) Use of personal protective equipment;

1910.120(e)(2)(iv) Work practices by which the employee can minimize risks from hazards;

1910.120(e)(2)(v) Safe use of engineering controls and equipment on the site;

1910.120(e)(2)(vi) Medical surveillance requirements including recognition of symptoms and signs which might indicate over exposure to hazards; and

1910.120(e)(2)(vii) The contents of paragraphs (G) through (J) of the site safety and health plan set forth in paragraph (b)(4)(ii) of this section.

1910.120(e)(3) Initial training.

1910.120(e)(3)(i) General site workers (such as equipment operators, general laborers and supervisory personnel) engaged in hazardous substance removal or other activities which expose or potentially expose workers to hazardous substances and health hazards shall receive a minimum of 40 hours of instruction off the site, and a minimum of three days actual field experience under the direct supervision of a trained experienced supervisor.

1910.120(e)(3)(ii) Workers on site only occasionally for a specific limited task (such as, but not limited to, ground water monitoring, land surveying, or geophysical surveying) and who are unlikely to be exposed over permissible exposure limits and published exposure limits shall receive a minimum of 24 hours of instruction off the site, and the minimum of one day actual field experience under the direct supervision of a trained, experienced supervisor.

1910.120(e)(3)(iii) Workers regularly on site who work in areas which have been monitored and fully characterized indicating that exposures are under permissible exposure limits and published exposure limits where respirators are not necessary, and the characterization indicates that there are no health hazards or the possibility of an emergency developing, shall receive a minimum of 24 hours of instruction off the site, and the minimum of one day actual field experience under the direct supervision of a trained, experienced supervisor.

1910.120(e)(3)(iv) Workers with 24 hours of training who are covered by paragraphs (e)(3)(ii) and (e)(3)(iii) of this section, and who become general site workers or who are required to wear respirators, shall have the additional 16 hours and two days of training necessary to total the training specified in paragraph (e)(3)(i).

1910.120(e)(4) Management and supervisor training. On-site management and supervisors directly responsible for, or who supervise employees engaged in, hazardous waste operations shall receive 40 hours initial training, and three days of supervised field experience (the training may be reduced to 24 hours and one day if the only area of their responsibility is employees covered by paragraphs (e)(3)(ii) and (e)(3)(iii)) and at least eight additional hours of specialized training at the time of job assignment on such topics as, but not limited to, the employer's safety and health program and the associated employee training program, personal protective equipment program, spill containment program, and health hazard monitoring procedure and techniques.

1910.120(e)(5) Qualifications for trainers. Trainers shall be qualified to instruct employees about the subject matter that is being presented in training. Such trainers shall have satisfactorily completed a training program for teaching the subjects they are expected to teach, or they shall have the academic credentials and instructional experience necessary for teaching the subjects. Instructors shall demonstrate competent instructional skills and knowledge of the applicable subject matter.

1910.120(e)(6) Training certification. Employees and supervisors that have received and successfully completed the training and field experience specified in paragraphs (e)(1) through (e)(4) of this section shall be certified by their instructor or the head instructor and trained supervisor as having completed the necessary training. A written certificate shall be given to each person so certified. Any person who has not been so certified or who does not meet the requirements of paragraph (e)(9) of this section shall be prohibited from engaging in hazardous waste operations.

1910.120(e)(7) Emergency response. Employees who are engaged in responding to hazardous emergency situations at hazardous waste clean-up sites that may expose them to hazardous substances shall be trained in how to respond to such expected emergencies.

1910.120(e)(8) Refresher training. Employees specified in paragraph (e)(1) of this section, and managers and supervisors specified in paragraph (e)(4) of this section, shall receive eight hours of refresher training annually on the items specified in paragraph (e)(2) and/or (e)(4) of this section, any critique of incidents that have occurred in the past year that can serve as training examples of related work, and other relevant topics.

1910.120(e)(9) Equivalent training. Employers who can show by documentation or certification that an employee's work experience and/or training has resulted in training equivalent to that training required in paragraphs (e)(1) through (e)(4) of this section shall not be required to provide the initial training requirements of those paragraphs to such employees and shall provide a copy of the certification or documentation to the employee upon request. However, certified employees or employees with equivalent training new to a site shall receive appropriate, site specific training before site entry and have appropriate supervised field experience at the new site. Equivalent training includes any academic training or the training that existing employees might have already received from actual hazardous waste site experience.

1910.120(f) Medical surveillance

1910.120(f)(1) General. Employees engaged in operations specified in paragraphs (a)(1)(i) through (a)(1)(iv) of this section and not covered by (a)(2)(iii) exceptions and employers of employees specified in paragraph (q)(9) shall institute a medical surveillance program in accordance with this paragraph.

1910.120(f)(2) Employees covered. The medical surveillance program shall be instituted by the employer for the following employees:

1910.120(f)(2)(i) All employees who are or may be exposed to hazardous substances or health hazards at or above the established permissible exposure limit, above the published exposure levels for these substances, without regard to the use of respirators, for 30 days or more a year;

1910.120(f)(2)(ii) All employees who wear a respirator for 30 days or more a year or as required by 1910.134;

1910.120(f)(2)(iii) All employees who are injured, become ill or develop signs or symptoms due to possible overexposure involving hazardous substances or health hazards from an emergency response or hazardous waste operation; and

1910.120(f)(2)(iv) Members of HAZMAT teams.

1910.120(f)(3) Frequency of medical examinations and consultations. Medical examinations and consultations shall be made available by the employer to each employee covered under paragraph (f)(2) of this section on the following schedules:

1910.120(f)(3)(i) For employees covered under paragraphs (f)(2)(i), (f)(2)(ii), and (f)(2)(iv);

1910.120(f)(3)(i)(A) Prior to assignment;

1910.120(f)(3)(i)(B) At least once every twelve months for each employee covered unless the attending physician believes a longer interval (not greater than biennially) is appropriate;

1910.120(f)(3)(i)(C) At termination of employment or reassignment to an area where the employee would not be covered if the employee has not had an examination within the last six months.

1910.120(f)(3)(i)(D) As soon as possible upon notification by an employee that the employee has developed signs or symptoms indicating possible overexposure to hazardous substances or health hazards, or that the employee has been injured or exposed above the permissible exposure limits or published exposure levels in an emergency situation;

1910.120(f)(3)(i)(E) At more frequent times, if the examining physician determines that an increased frequency of examination is medically necessary.

1910.120(f)(3)(ii) For employees covered under paragraph (f)(2)(iii) and for all employees including of employers covered by paragraph (a)(1)(iv) who may have been injured, received a health impairment, developed signs or symptoms which may have resulted from exposure to hazardous substances resulting from an emergency incident, or exposed during an emergency incident to hazardous substances at concentrations above the permissible exposure limits or the published exposure levels without the necessary personal protective equipment being used:

1910.120(f)(3)(ii)(A) As soon as possible following the emergency incident or development of signs or symptoms;

1910.120(f)(3)(ii)(B) At additional times, if the examining physician determines that follow-up examinations or consultations are medically necessary.

1910.120(f)(4) Content of medical examinations and consultations.

1910.120(f)(4)(i) Medical examinations required by paragraph (f)(3) of this section shall include a medical and work history (or updated history if one is in the employee's file) with special emphasis on symptoms related to the handling of hazardous substances and health hazards, and to fitness for duty including the ability to wear any required PPE under conditions (i.e., temperature extremes) that may be expected at the work site.

1910.120(f)(4)(ii) The content of medical examinations or consultations made available to employees pursuant to paragraph (f) shall be determined by the attending physician. The guidelines in the **Occupational Safety and Health Guidance Manual for Hazardous Waste Site Activities** (See Appendix D, reference # 10) should be consulted.

1910.120(f)(5) Examination by a physician and costs. All medical examinations and procedures shall be performed by or under the supervision of a licensed physician, preferably one knowledgeable in occupational medicine, and shall be provided without cost to the employee, without loss of pay, and at a reasonable time and place.

1910.120(f)(6) Information provided to the physician. The employer shall provide one copy of this standard and its appendices to the attending physician and in addition the following for each employee:

1910.120(f)(6)(i) A description of the employee's duties as they relate to the employee's exposures,

1910.120(f)(6)(ii) The employee's exposure levels or anticipated exposure levels.

1910.120(f)(6)(iii) A description of any personal protective equipment used or to be used.

1910.120(f)(6)(iv) Information from previous medical examinations of the employee which is not readily available to the examining physician.

1910.120(f)(6)(v) Information required by §1910.134.

1910.120(f)(7) Physician's written opinion.

1910.120(f)(7)(i) The employer shall obtain and furnish the employee with a copy of a written opinion from the examining physician containing the following:

1910.120(f)(7)(i)(A) The physician's opinion as to whether the employee has any detected medical conditions which would place the employee at increased risk of material impairment of the employee's health from work in hazardous waste operations or emergency response, or from respirator use.

1910.120(f)(7)(i)(B) The physician's recommended limitations upon the employees assigned work.

1910.120(f)(7)(i)(C) The results of the medical examination and tests if requested by the employee.

1910.120(f)(7)(i)(D) A statement that the employee has been informed by the physician of the results of the medical examination and any medical conditions which require further examination or treatment.

1910.120(f)(7)(ii) The written opinion obtained by the employer shall not reveal specific findings or diagnoses unrelated to occupational exposure.

1910.120(f)(8) Recordkeeping.

1910.120(f)(8)(i) An accurate record of the medical surveillance required by paragraph (f) of this section shall be retained. This record shall be retained for the period specified and meet the criteria of 29 CFR 1910.1020.

1910.120(f)(8)(ii) The record required in paragraph (f)(8)(i) of this section shall include at least the following information:

1910.120(f)(8)(ii)(A) The name and social security number of the employee;

1910.120(f)(8)(ii)(B) Physicians' written opinions, recommended limitations and results of examinations and tests;

1910.120(f)(8)(ii)(C) Any employee medical complaints related to exposure to hazardous substances;

1910.120(f)(8)(ii)(D) A copy of the information provided to the examining physician by the employer, with the exception of the standard and its appendices.

1910.120(g) Engineering controls, work practices, and personal protective equipment for employee protection. Engineering controls, work practices and PPE for substances regulated in Subpart Z. (i) Engineering controls, work practices, personal protective equipment, or a combination of these shall be implemented in accordance with this paragraph to protect employees from exposure to hazardous substances and safety and health hazards.

1910.120(g)(1) Engineering controls, work practices and PPE for substances regulated in Subparts G and Z.

1910.120(g)(1)(i) Engineering controls and work practices shall be instituted to reduce and maintain employee exposure to or below the permissible exposure limits for substances regulated by 29 CFR Part 1910, to the extent required by Subpart Z, except to the extent that such controls and practices are not feasible.

NOTE TO PARAGRAPH (g)(1)(i): Engineering controls which may be feasible include the use of pressurized cabs or control booths on equipment, and/or the use of remotely operated material handling equipment. Work practices which may be feasible are removing all non-essential employees from potential exposure during opening of drums, wetting down dusty operations and locating employees upwind of possible hazards.

1910.120(g)(1)(ii) Whenever engineering controls and work practices are not feasible, or not required, any reasonable combination of engineering controls, work practices and PPE shall be used to reduce and maintain to or below the permissible exposure limits or dose limits for substances regulated by 29 CFR Part 1910, Subpart Z.

1910.120(g)(1)(iii) The employer shall not implement a schedule of employee rotation as a means of compliance with permissible exposure limits or dose limits except when there is no other feasible way of complying with the airborne or dermal dose limits for ionizing radiation.

1910.120(g)(1)(iv) The provisions of 29 CFR, subpart G, shall be followed.

1910.120(g)(2) Engineering controls, work practices, and PPE for substances not regulated in Subparts G and Z. An appropriate combination of engineering controls, work practices, and personal protective equipment shall be used to reduce and maintain employee exposure to or below published exposure levels for hazardous substances and health hazards not regulated by 29 CFR Part 1910, Subparts G and Z. The employer may use the published literature and MSDS as a guide in making the employer's determination as to what level of protection the employer believes is appropriate for hazardous substances and health hazards for which there is no permissible exposure limit or published exposure limit.

1910.120(g)(3) Personal protective equipment selection.

1910.120(g)(3)(i) Personal protective equipment (PPE) shall be selected and used which will protect employees from the hazards and potential hazards they are likely to encounter as identified during the site characterization and analysis.

1910.120(g)(3)(ii) Personal protective equipment selection shall be based on an evaluation of the performance characteristics of the PPE relative to the requirements and limitations of the site, the task-specific conditions and duration, and the hazards and potential hazards identified at the site.

1910.120(g)(3)(iii) Positive pressure self-contained breathing apparatus, or positive pressure air-line respirators equipped with an escape air supply shall be used when chemical exposure levels present will create a substantial possibility of immediate death, immediate serious illness or injury, or impair the ability to escape.

1910.120(g)(3)(iv) Totally-encapsulating chemical protective suits (protection equivalent to Level A protection as recommended in Appendix B) shall be used in conditions where skin absorption of a hazardous substance may result in a substantial possibility of immediate death, immediate serious illness or injury, or impair the ability to escape.

1910.120(g)(3)(v) The level of protection provided by PPE selection shall be increased when additional information or site conditions show that increased protection is necessary to reduce employee exposures below permissible exposure limits and published exposure levels for hazardous substances and health hazards. (See Appendix B for guidance on selecting PPE ensembles.)

NOTE TO PARAGRAPH (g)(3): The level of employee protection provided may be decreased when additional information or site conditions show that decreased protection will not result in hazardous exposures to employees.

1910.120(g)(3)(vi) Personal protective equipment shall be selected and used to meet the requirements of 29 CFR Part 1910, Subpart I, and additional requirements specified in this section.

1910.120(g)(4) Totally-encapsulating chemical protective suits.

1910.120(g)(4)(i) Totally-encapsulating suits shall protect employees from the particular hazards which are identified during site characterization and analysis.

1910.120(g)(4)(ii) Totally-encapsulating suits shall be capable of maintaining positive air pressure. (See Appendix A for a test method which may be used to evaluate this requirement.)

1910.120(g)(4)(iii) Totally-encapsulating suits shall be capable of preventing inward test gas leakage of more than 0.5 percent. (See Appendix A for a test method which may be used to evaluate this requirement.)

1910.120(g)(5) Personal protective equipment (PPE) program. A personal protective equipment program, which is part of the employer's safety and health program required in paragraph (b) of this section or required in paragraph (p)(1) of this section and which is also a part of the site-specific safety and health plan shall be established. The PPE program shall address the elements listed below. When elements, such as donning and doffing procedures, are provided by the manufacturer of a piece of equipment and are attached to the plan, they need not be rewritten into the plan as long as they adequately address the procedure or element.

1910.120(g)(5)(i) PPE selection based upon site hazards,

1910.120(g)(5)(ii) PPE use and limitations of the equipment,

1910.120(g)(5)(iii) Work mission duration,

1910.120(g)(5)(iv) PPE maintenance and storage,

1910.120(g)(5)(v) PPE decontamination and disposal,

1910.120(g)(5)(vi) PPE training and proper fitting,

1910.120(g)(5)(vii) PPE donning and doffing procedures,

1910.120(g)(5)(viii) PPE inspection procedures prior to, during, and after use,

1910.120(g)(5)(ix) Evaluation of the effectiveness of the PPE program, and

1910.120(g)(5)(x) Limitations during temperature extremes, heat stress, and other appropriate medical considerations.

1910.120(h) Monitoring.

1910.120(h)(1) General.

1910.120(h)(1)(i) Monitoring shall be performed in accordance with this paragraph where there may be a question of employee exposure to hazardous concentrations of hazardous substances in order to assure proper selection of engineering controls, work practices and personal protective equipment so that employees are not exposed to levels which exceed permissible exposure limits, or published exposure levels if there are no permissible exposure limits, for hazardous substances.

1910.120(h)(1)(ii) Air monitoring shall be used to identify and quantify airborne levels of hazardous substances and safety and health hazards in order to determine the appropriate level of employee protection needed on site.

1910.120(h)(2) Initial entry. Upon initial entry, representative air monitoring shall be conducted to identify any IDLH condition, exposure over permissible exposure limits or published exposure levels, exposure over a radioactive material's dose limits or other dangerous condition such as the presence of flammable atmospheres, oxygen-deficient environments.

1910.120(h)(3) Periodic monitoring. Periodic monitoring shall be conducted when the possibility of an IDLH condition or flammable atmosphere has developed or when there is indication that exposures may have risen over permissible exposure limits or published exposure levels since prior monitoring. Situations where it shall be considered whether the possibility that exposures have risen are as follows:

1910.120(h)(3)(i) When work begins on a different portion of the site.

1910.120(h)(3)(ii) When contaminants other than those previously identified are being handled.

1910.120(h)(3)(iii) When a different type of operation is initiated (e.g., drum opening as opposed to exploratory well drilling.)

1910.120(h)(3)(iv) When employees are handling leaking drums or containers or working in areas with obvious liquid contamination (e.g., a spill or lagoon.)

1910.120(h)(4) Monitoring of high-risk employees. After the actual clean-up phase of any hazardous waste operation commences; for example, when soil, surface water or containers are moved or disturbed; the employer shall monitor those employees likely to have the highest exposures to those hazardous substances and health hazards likely to be present above permissible exposure limits or published exposure levels by using personal sampling frequently enough to characterize employee exposures. The employer may utilize a representative sampling approach by documenting that the employees and chemicals chosen for monitoring are based on the criteria stated in the first sentence of this paragraph. If the employees likely to have the highest exposure are over permissible exposure limits or published exposure limits, then monitoring shall continue to determine all employees likely to be above those limits. The employer may utilize a representative sampling approach by documenting that the employees and chemicals chosen for monitoring are based on the criteria stated above.

NOTE TO PARAGRAPH (h): It is not required to monitor employees engaged in site characterization operations covered by paragraph (c) of this section.

1910.120(i) Informational programs. Employers shall develop and implement a program which is part of the employer's safety and health program required in paragraph (b) of this section to inform employees, contractors, and subcontractors (or their representative) actually engaged in hazardous waste operations of the nature, level and degree of exposure likely as a result of participation in such hazardous waste operations. Employees, contractors and subcontractors working outside of the operations part of a site are not covered by this standard.

1910.120(j) Handling drums and containers

1910.120(j)(1) General.

1910.120(j)(1)(i) Hazardous substances and contaminated liquids and other residues shall be handled, transported, labeled, and disposed of in accordance with this paragraph.

1910.120(j)(1)(ii) Drums and containers used during the clean-up shall meet the appropriate DOT, OSHA, and EPA regulations for the wastes that they contain.

1910.120(j)(1)(iii) When practical, drums and containers shall be inspected and their integrity shall be assured prior to being moved. Drums or containers that cannot be inspected before being moved because of storage conditions (i.e., buried beneath the earth, stacked behind other drums, stacked several tiers high in a pile, etc.) shall be moved to an accessible location and inspected prior to further handling.

1910.120(j)(1)(iv) Unlabeled drums and containers shall be considered to contain hazardous substances and handled accordingly until the contents are positively identified and labeled.

1910.120(j)(1)(v) Site operations shall be organized to minimize the amount of drum or container movement.

1910.120(j)(1)(vi) Prior to movement of drums or containers, all employees exposed to the transfer operation shall be warned of the potential hazards associated with the contents of the drums or containers.

1910.120(j)(1)(vii) U.S. Department of Transportation specified salvage drums or containers and suitable quantities of proper absorbent shall be kept available and used in areas where spills, leaks, or ruptures may occur.

1910.120(j)(1)(viii) Where major spills may occur, a spill containment program, which is part of the employer's safety and health program required in paragraph (b) of this section, shall be implemented to contain and isolate the entire volume of the hazardous substance being transferred.

1910.120(j)(1)(ix) Drums and containers that cannot be moved without rupture, leakage, or spillage shall be emptied into a sound container using a device classified for the material being transferred.

1910.120(j)(1)(x) A ground-penetrating system or other type of detection system or device shall be used to estimate the location and depth of buried drums or containers.

1910.120(j)(1)(xi) Soil or covering material shall be removed with caution to prevent drum or container rupture.

1910.120(j)(1)(xii) Fire extinguishing equipment meeting the requirements of 29 CFR Part 1910, Subpart L, shall be on hand and ready for use to control incipient fires.

1910.120(j)(2) **Opening drums and containers.** The following procedures shall be followed in areas where drums or containers are being opened:

1910.120(j)(2)(i) Where an airline respirator system is used, connections to the source of air supply shall be protected from contamination and the entire system shall be protected from physical damage.

1910.120(j)(2)(ii) Employees not actually involved in opening drums or containers shall be kept a safe distance from the drums or containers being opened.

1910.120(j)(2)(iii) If employees must work near or adjacent to drums or containers being opened, a suitable shield that does not interfere with the work operation shall be placed between the employee and the drums or containers being opened to protect the employee in case of accidental explosion.

1910.120(j)(2)(iv) Controls for drum or container opening equipment, monitoring equipment, and fire suppression equipment shall be located behind the explosion-resistant barrier.

1910.120(j)(2)(v) When there is a reasonable possibility of flammable atmospheres being present, material handling equipment and hand tools shall be of the type to prevent sources of ignition.

1910.120(j)(2)(vi) Drums and containers shall be opened in such a manner that excess interior pressure will be safely relieved. If pressure cannot be relieved from a remote location, appropriate shielding shall be placed between the employee and the drums or containers to reduce the risk of employee injury.

1910.120(j)(2)(vii) Employees shall not stand upon or work from drums or containers.

1910.120(j)(3) **Material handling equipment.** Material handling equipment used to transfer drums and containers shall be selected, positioned and operated to minimize sources of ignition related to the equipment from igniting vapors released from ruptured drums or containers.

1910.120(j)(4) **Radioactive wastes.** Drums and containers containing radioactive wastes shall not be handled until such time as their hazard to employees is properly assessed.

1910.120(j)(5) **Shock sensitive wastes.** As a minimum, the following special precautions shall be taken when drums and containers containing or suspected of containing shock-sensitive wastes are handled:

1910.120(j)(5)(i) All non-essential employees shall be evacuated from the area of transfer.

1910.120(j)(5)(ii) Material handling equipment shall be provided with explosive containment devices or protective shields to protect equipment operators from exploding containers.

1910.120(j)(5)(iii) An employee alarm system capable of being perceived above surrounding light and noise conditions shall be used to signal the commencement and completion of explosive waste handling activities.

1910.120(j)(5)(iv) Continuous communications (i.e., portable radios, hand signals, telephones, as appropriate) shall be maintained between the employee-in-charge of the immediate handling area and both the site safety and health supervisor and the command post until such time as the handling operation is completed. Communication equipment or methods that could cause shock sensitive materials to explode shall not be used.

1910.120(j)(5)(v) Drums and containers under pressure, as evidenced by bulging or swelling, shall not be moved until such time as the cause for excess pressure is determined and appropriate containment procedures have been implemented to protect employees from explosive relief of the drum.

1910.120(j)(5)(vi) Drums and containers containing packaged laboratory wastes shall be considered to contain shock-sensitive or explosive materials until they have been characterized.

Caution: Shipping of shock sensitive wastes may be prohibited under U.S. Department of Transportation regulations. Employers and their shippers should refer to 49 CFR 173.21 and 173.50.

1910.120(j)(6) **Laboratory waste packs.** In addition to the requirements of paragraph (j)(5) of this section, the following precautions shall be taken, as a minimum, in handling laboratory waste packs (lab packs):

1910.120(j)(6)(i) Lab packs shall be opened only when necessary and then only by an individual knowledgeable in the inspection, classification, and segregation of the containers within the pack according to the hazards of the wastes.

1910.120(j)(6)(ii) If crystalline material is noted on any container, the contents shall be handled as a shock-sensitive waste until the contents are identified.

1910.120(j)(7) **Sampling of drum and container contents.** Sampling of containers and drums shall be done in accordance with a sampling procedure which is part of the site safety and health plan developed for and available to employees and others at the specific worksite.

1910.120(j)(8) **Shipping and transport.**

1910.120(j)(8)(i) Drums and containers shall be identified and classified prior to packaging for shipment.

1910.120(j)(8)(ii) Drum or container staging areas shall be kept to the minimum number necessary to safely identify and classify materials and prepare them for transport.

1910.120(j)(8)(iii) Staging areas shall be provided with adequate access and egress routes.

1910.120(j)(8)(iv) Bulking of hazardous wastes shall be permitted only after a thorough characterization of the materials has been completed.

1910.120(j)(9) **Tank and vault procedures.**

1910.120(j)(9)(i) Tanks and vaults containing hazardous substances shall be handled in a manner similar to that for drums and containers, taking into consideration the size of the tank or vault.

1910.120(j)(9)(ii) Appropriate tank or vault entry procedures as described in the employer's safety and health plan shall be followed whenever employees must enter a tank or vault.

1910.120(k) **Decontamination**

1910.120(k)(1) **General.** Procedures for all phases of decontamination shall be developed and implemented in accordance with this paragraph.

1910.120(k)(2) **Decontamination procedures.**

1910.120(k)(2)(i) A decontamination procedure shall be developed, communicated to employees and implemented before any employees or equipment may enter areas on site where potential for exposure to hazardous substances exists.

1910.120(k)(2)(ii) Standard operating procedures shall be developed to minimize employee contact with hazardous substances or with equipment that has contacted hazardous substances.

1910.120(k)(2)(iii) All employees leaving a contaminated area shall be appropriately decontaminated; all contaminated clothing and equipment leaving a contaminated area shall be appropriately disposed of or decontaminated.

1910.120(k)(2)(iv) Decontamination procedures shall be monitored by the site safety and health supervisor to determine their effectiveness. When such procedures are found to be ineffective, appropriate steps shall be taken to correct any deficiencies.

1910.120(k)(3) Location. Decontamination shall be performed in geographical areas that will minimize the exposure of uncontaminated employees or equipment to contaminated employees or equipment.

1910.120(k)(4) Equipment and solvents. All equipment and solvents used for decontamination shall be decontaminated or disposed of properly.

1910.120(k)(5) Personal protective clothing and equipment.

1910.120(k)(5)(i) Protective clothing and equipment shall be decontaminated, cleaned, laundered, maintained or replaced as needed to maintain their effectiveness.

1910.120(k)(5)(ii) Employees whose non-impermeable clothing becomes wetted with hazardous substances shall immediately remove that clothing and proceed to shower. The clothing shall be disposed of or decontaminated before it is removed from the work zone.

1910.120(k)(6) Unauthorized employees. Unauthorized employees shall not remove protective clothing or equipment from change rooms.

1910.120(k)(7) Commercial laundries or cleaning establishments. Commercial laundries or cleaning establishments that decontaminate protective clothing or equipment shall be informed of the potentially harmful effects of exposures to hazardous substances.

1910.120(k)(8) Showers and change rooms. Where the decontamination procedure indicates a need for regular showers and change rooms outside of a contaminated area, they shall be provided and meet the requirements of 29 CFR 1910.141. If temperature conditions prevent the effective use of water, then other effective means for cleansing shall be provided and used.

1910.120(l) Emergency response by employees at uncontrolled hazardous waste sites

1910.120(l)(1) Emergency response plan.

1910.120(l)(1)(i) An emergency response plan shall be developed and implemented by all employers within the scope of paragraphs (a)(1)(i) through (ii) of this section to handle anticipated emergencies prior to the commencement of hazardous waste operations. The plan shall be in writing and available for inspection and copying by employees, their representatives, OSHA personnel and other governmental agencies with relevant responsibilities.

1910.120(l)(1)(ii) Employers who will evacuate their employees from the danger area when an emergency occurs, and who do not permit any of their employees to assist in handling the emergency, are exempt from the requirements of this paragraph if they provide an emergency action plan complying with 29 CFR 1910.38.

1910.120(l)(2) Elements of an emergency response plan. The employer shall develop an emergency response plan for emergencies which shall address, as a minimum, the following:

1910.120(l)(2)(i) Pre-emergency planning.

1910.120(l)(2)(ii) Personnel roles, lines of authority, training, and communication.

1910.120(l)(2)(iii) Emergency recognition and prevention.

1910.120(l)(2)(iv) Safe distances and places of refuge.

1910.120(l)(2)(v) Site security and control.

1910.120(l)(2)(vi) Evacuation routes and procedures.

1910.120(l)(2)(vii) Decontamination procedures which are not covered by the site safety and health plan.

1910.120(l)(2)(viii) Emergency medical treatment and first aid.

1910.120(l)(2)(ix) Emergency alerting and response procedures.

1910.120(l)(2)(x) Critique of response and follow-up.

1910.120(l)(2)(xi) PPE and emergency equipment.

1910.120(l)(3) Procedures for handling emergency incidents.

1910.120(l)(3)(i) In addition to the elements for the emergency response plan required in paragraph (l)(2) of this section, the following elements shall be included for emergency response plans:

1910.120(l)(3)(i)(A) Site topography, layout, and prevailing weather conditions.

1910.120(l)(3)(i)(B) Procedures for reporting incidents to local, state, and federal governmental agencies.

1910.120(l)(3)(ii) The emergency response plan shall be a separate section of the Site Safety and Health Plan.

1910.120(l)(3)(iii) The emergency response plan shall be compatible and integrated with the disaster, fire and/or emergency response plans of local, state, and federal agencies.

1910.120(l)(3)(iv) The emergency response plan shall be rehearsed regularly as part of the overall training program for site operations.

1910.120(l)(3)(v) The site emergency response plan shall be reviewed periodically and, as necessary, be amended to keep it current with new or changing site conditions or information.

1910.120(l)(3)(vi) An employee alarm system shall be installed in accordance with 29 CFR 1910.165 to notify employees of an emergency situation, to stop work activities if necessary, to lower background noise in order to speed communication, and to begin emergency procedures.

1910.120(l)(3)(vii) Based upon the information available at time of the emergency, the employer shall evaluate the incident and the site response capabilities and proceed with the appropriate steps to implement the site emergency response plan.

1910.120(m) Illumination. Areas accessible to employees shall be lighted to not less than the minimum illumination intensities listed in the following Table H-120.1 while any work is in progress:

Table H-120.1	Minimum Illumination Intensities in Foot-Candles
Foot-candles	**Area or operations**
5	General site areas.
3	Excavation and waste areas, accessways, active storage areas, loading platforms, refueling, and field maintenance areas.
5	Indoors: warehouses, corridors, hallways, and exitways.
5	Tunnels, shafts, and general underground work areas; (Exception: minimum of 10 foot-candles is required at tunnel and shaft heading during drilling, mucking, and scaling. Mine Safety and Health Administration approved cap lights shall be acceptable for use in the tunnel heading.
10	General shops (e.g., mechanical and electrical equipment rooms, active storerooms, barracks or living quarters, locker or dressing rooms, dining areas, and indoor toilets and workrooms.
30	First aid stations, infirmaries, and offices.

1910.120(n) **Sanitation at temporary workplaces**

1910.120(n)(1) **Potable water.**

1910.120(n)(1)(i) An adequate supply of potable water shall be provided on the site.

1910.120(n)(1)(ii) Portable containers used to dispense drinking water shall be capable of being tightly closed, and equipped with a tap. Water shall not be dipped from containers.

1910.120(n)(1)(iii) Any container used to distribute drinking water shall be clearly marked as to the nature of its contents and not used for any other purpose.

1910.120(n)(1)(iv) Where single service cups (to be used but once) are supplied, both a sanitary container for the unused cups and a receptacle for disposing of the used cups shall be provided.

1910.120(n)(2) **Nonpotable water.**

1910.120(n)(2)(i) Outlets for nonpotable water, such as water for firefighting purposes shall be identified to indicate clearly that the water is unsafe and is not to be used for drinking, washing, or cooking purposes.

1910.120(n)(2)(ii) There shall be no cross-connection, open or potential, between a system furnishing potable water and a system furnishing nonpotable water.

1910.120(n)(3) **Toilet facilities.**

1910.120(n)(3)(i) Toilets shall be provided for employees according to Table H-120.2.

Table H-120.2 Toilet Facilities

Number of employees	Minimum number of facilities
20 or fewer	One
More than 20 fewer than 200	One toilet seat and 1 urinal per 40 employees
More than 200	One toilet seat and 1 urinal per 50 employees

1910.120(n)(3)(ii) Under temporary field conditions, provisions shall be made to assure not less than one toilet facility is available.

1910.120(n)(3)(iii) Hazardous waste sites, not provided with a sanitary sewer, shall be provided with the following toilet facilities unless prohibited by local codes:

1910.120(n)(3)(iii)(A) Chemical toilets;

1910.120(n)(3)(iii)(B) Recirculating toilets;

1910.120(n)(3)(iii)(C) Combustion toilets; or

1910.120(n)(3)(iii)(D) Flush toilets.

1910.120(n)(3)(iv) The requirements of this paragraph for sanitation facilities shall not apply to mobile crews having transportation readily available to nearby toilet facilities.

1910.120(n)(3)(v) Doors entering toilet facilities shall be provided with entrance locks controlled from inside the facility.

1910.120(n)(4) **Food handling.** All food service facilities and operations for employees shall meet the applicable laws, ordinances, and regulations of the jurisdictions in which they are located.

1910.120(n)(5) **Temporary sleeping quarters.** When temporary sleeping quarters are provided, they shall be heated, ventilated, and lighted.

1910.120(n)(6) **Washing facilities.** The employer shall provide adequate washing facilities for employees engaged in operations where hazardous substances may be harmful to employees. Such facilities shall be in near proximity to the worksite; in areas where exposures are below permissible exposure limits and which are under the controls of the employer; and shall be so equipped as to enable employees to remove hazardous substances from themselves.

1910.120(n)(7) **Showers and change rooms.** When hazardous waste clean-up or removal operations commence on a site and the duration of the work will require six months or greater time to complete, the employer shall provide showers and change rooms for all employees exposed to hazardous substances and health hazards involved in hazardous waste clean-up or removal operations.

1910.120(n)(7)(i) Showers shall be provided and shall meet the requirements of 29 CFR 1910.141(d)(3).

1910.120(n)(7)(ii) Change rooms shall be provided and shall meet the requirements of 29 CFR 1910.141(e). Change rooms shall consist of two separate change areas separated by the shower area required in paragraph (n)(7)(i) of this section. One change area, with an exit leading off the worksite, shall provide employees with an area where they can put on, remove and store work clothing and personal protective equipment.

1910.120(n)(7)(iii) Showers and change rooms shall be located in areas where exposures are below the permissible exposure limits and published exposure levels. If this cannot be accomplished, then a ventilation system shall be provided that will supply air that is below the permissible exposure limits and published exposure levels.

1910.120(n)(7)(iv) Employers shall assure that employees shower at the end of their work shift and when leaving the hazardous waste site.

1910.120(o) **New technology programs.**

1910.120(o)(1) The employer shall develop and implement procedures for the introduction of effective new technologies and equipment developed for the improved protection of employees working with hazardous waste clean-up operations, and the same shall be implemented as part of the site safety and health program to assure that employee protection is being maintained.

1910.120(o)(2) New technologies, equipment or control measures available to the industry, such as the use of foams, absorbents, absorbents, neutralizers, or other means to suppress the level of air contaminants while excavating the site or for spill control, shall be evaluated by employers or their representatives. Such an evaluation shall be done to determine the effectiveness of the new methods, materials, or equipment before implementing their use on a large scale for enhancing employee protection. Information and data from manufacturers or suppliers may be used as part of the employer's evaluation effort. Such evaluations shall be made available to OSHA upon request.

1910.120(p) **Certain Operations Conducted Under the Resource Conservation and Recovery Act of 1976 (RCRA).** Employers conducting operations at treatment, storage and disposal (TSD) facilities specified in paragraph (a)(1)(iv) of this section shall provide and implement the programs specified in this paragraph. See the "Notes and Exceptions" to paragraph (a)(2)(iii) of this section for employers not covered.

1910.120(p)(1) **Safety and health program.** The employer shall develop and implement a written safety and health program for employees involved in hazardous waste operations that shall be available for inspection by employees, their representatives and OSHA personnel. The program shall be designed to identify, evaluate and control safety and health hazards in their facilities for the purpose of employee protection, to provide for emergency response meeting the requirements of paragraph (p)(8) of this section and to address as appropriate site analysis, engineering controls, maximum exposure limits, hazardous waste handling procedures and uses of new technologies.

1910.120(p)(2) **Hazard communication program.** The employer shall implement a hazard communication program meeting the requirements of 29 CFR 1910.1200 as part of the employer's safety and program.

NOTE TO §1910.120— The exemption for hazardous waste provided in 1910.1200 is applicable to this section.

1910.120(p)(3) **Medical surveillance program.** The employer shall develop and implement a medical surveillance program meeting the requirements of paragraph (f) of this section.

1910.120(p)(4) **Decontamination program.** The employer shall develop and implement a decontamination procedure meeting the requirements of paragraph (k) of this section.

1910.120(p)(5) **New technology program.** The employer shall develop and implement procedures meeting the requirements of paragraph (o) of this section for introducing new and innovative equipment into the workplace.

1910.120(p)(6) **Material handling program.** Where employees will be handling drums or containers, the employer shall develop and implement procedures meeting the requirements of paragraphs (j)(1)(ii) through (viii) and (xi) of this section, as well as (j)(3) and (j)(8) of this section prior to starting such work.

1910.120(p)(7) **Training program**

1910.120(p)(7)(i) **New employees.** The employer shall develop and implement a training program which is part of the employer's safety and health program, for employees exposed to health hazards or hazardous substances at TSD operations to enable the employees to perform their assigned duties and functions in a safe and healthful manner so as not to endanger themselves or other employees. The initial training shall be for 24 hours and refresher training shall be for eight hours annually. Employees who have received the initial training required by this paragraph shall be given a written certificate attesting that they have successfully completed the necessary training.

1910.120(p)(7)(II) **Current employees.** Employers who can show by an employee's previous work experience and/or training that the employee has had training equivalent to the initial training required by this paragraph, shall be considered as meeting the initial training requirements of this paragraph as to that employee. Equivalent training includes the training that existing employees might have already received from actual site work experience. Current employees shall receive eight hours of refresher training annually.

1910.120(p)(7)(iii) **Trainers.** Trainers who teach initial training shall have satisfactorily completed a training course for teaching the subjects they are expected to teach or they shall have the academic credentials and instruction experience necessary to demonstrate a good command of the subject matter of the courses and competent instructional skills.

1910.120(p)(8) **Emergency response program**

1910.120(p)(8)(i) Emergency response plan. An emergency response plan shall be developed and implemented by all employers. Such plans need not duplicate any of the subjects fully addressed in the employer's contingency planning required by permits, such as those issued by the U.S. Environmental Protection Agency, provided that the contingency plan is made part of the emergency response plan. The emergency response plan shall be a written portion of the employer's safety and health program required in paragraph (p)(1) of this section. Employers who will evacuate their employees from the worksite location when an emergency occurs and who do not permit any of their employees to assist in handling the emergency are exempt from the requirements of paragraph (p)(8) if they provide an emergency action plan complying with 29 CFR 1910.38.

1910.120(p)(8)(ii) Elements of an emergency response plan. The employer shall develop an emergency response plan for emergencies which shall address, as a minimum, the following areas to the extent that they are not addressed in any specific program required in this paragraph:

1910.120(p)(8)(ii)(A) Pre-emergency planning and coordination with outside parties.

1910.120(p)(8)(ii)(B) Personnel roles, lines of authority, training, and communication.

1910.120(p)(8)(ii)(C) Emergency recognition and prevention.

1910.120(p)(8)(ii)(D) Safe distances and places of refuge.

1910.120(p)(8)(ii)(E) Site security and control.

1910.120(p)(8)(ii)(F) Evacuation routes and procedures.

1910.120(p)(8)(ii)(G) Decontamination procedures.

1910.120(p)(8)(ii)(H) Emergency medical treatment and first aid.

1910.120(p)(8)(ii)(I) Emergency alerting and response procedures.

1910.120(p)(8)(ii)(J) Critique of response and follow-up.

1910.120(p)(8)(ii)(K) PPE and emergency equipment.

1910.120(p)(8)(iii) **Training.**

1910.120(p)(8)(iii)(A) Training for emergency response employees shall be completed before they are called upon to perform in real emergencies. Such training shall include the elements of the emergency response plan, standard operating procedures the employer has established for the job, the personal protective equipment to be worn and procedures for handling emergency incidents.

Exception #1: An employer need not train all employees to the degree specified if the employer divides the work force in a manner such that a sufficient number of employees who have responsibility to control emergencies have the training specified, and all other employees, who may first respond to an emergency incident, have sufficient awareness training to recognize that an emergency response situation exists and that they are instructed in that case to summon the fully trained employees and not attempt to control activities for which they are not trained.

Exception #2: An employer need not train all employees to the degree specified if arrangements have been made in advance for an outside fully-trained emergency response team to respond in a reasonable period and all employees, who may come to the incident first, have sufficient awareness training to recognize that an emergency response situation exists and they have been instructed to call the designated outside fully-trained emergency response team for assistance.

1910.120(p)(8)(iii)(B) Employee members of TSD facility emergency response organizations shall be trained to a level of competence in the recognition of health and safety hazards to protect themselves and other employees. This would include training in the methods used to minimize the risk from safety and health hazards; in the safe use of control equipment; in the selection and use of appropriate personal protective equipment; in the safe operating procedures to be used at the incident scene; in the techniques of coordination with other employees to minimize risks; in the appropriate response to over exposure from health hazards or injury to themselves and other employees; and in the recognition of subsequent symptoms which may result from over exposures.

1910.120(p)(8)(iii)(C) The employer shall certify that each covered employee has attended and successfully completed the training required in paragraph (p)(8)(iii) of this section, or shall certify the employee's competency for certification of training shall be recorded and maintained by the employer.

1910.120(p)(8)(iv) Procedures for handling emergency incidents.

1910.120(p)(8)(iv)(A) In addition to the elements for the emergency response plan required in paragraph (p)(8)(ii) of this section, the following elements shall be included for emergency response plans to the extent that they do not repeat any information already contained in the emergency response plan:

1910.120(p)(8)(iv)(A)(1) Site topography, layout, and prevailing weather conditions.

1910.120(p)(8)(iv)(A)(2) Procedures for reporting incidents to local, state, and federal governmental agencies.

1910.120(p)(8)(iv)(B) The emergency response plan shall be compatible and integrated with the disaster, fire and/or emergency response plans of local, state, and federal agencies.

1910.120(p)(8)(iv)(C) The emergency response plan shall be rehearsed regularly as part of the overall training program for site operations.

1910.120(p)(8)(iv)(D) The site emergency response plan shall be reviewed periodically and, as necessary, be amended to keep it current with new or changing site conditions or information.

1910.120(p)(8)(iv)(E) An employee alarm system shall be installed in accordance with 29 CFR 1910.165 to notify employees of an emergency situation, to stop work activities if necessary, to lower background noise in order to speed communication; and to begin emergency procedures.

1910.120(p)(8)(iv)(F) Based upon the information available at time of the emergency, the employer shall evaluate the incident and the site response capabilities and proceed with the appropriate steps to implement the site emergency response plan.

1910.120(q) Emergency response program to hazardous substance releases. This paragraph covers employers whose employees are engaged in emergency response no matter where it occurs except that it does not cover employees engaged in operations specified in paragraphs (a)(1)(i) through (a)(1)(iv) of this section. Those emergency response organizations who have developed and implemented programs equivalent to this paragraph for handling releases of hazardous substances pursuant to section 303 of the Superfund Amendments and Reauthorization Act of 1986 (Emergency Planning and Community Right-to-Know Act of 1986, 42 U.S.C. 11003) shall be deemed to have met the requirements of this paragraph.

1910.120(q)(1) Emergency response plan. An emergency response plan shall be developed and implemented to handle anticipated emergencies prior to the commencement of emergency response operations. The plan shall be in writing and available for inspection and copying by employees, their representatives and OSHA personnel. Employers who will evacuate their employees from the danger area when an emergency occurs, and who do not permit any of their employees to assist in handling the emergency, are exempt from the requirements of this paragraph if they provide an emergency action plan in accordance with 29 CFR 1910.38.

1910.120(q)(2) Elements of an emergency response plan. The employer shall develop an emergency response plan for emergencies which shall address, as a minimum, the following areas to the extent that they are not addressed in any specific program required in this paragraph:

1910.120(q)(2)(i) Pre-emergency planning and coordination with outside parties.

1910.120(q)(2)(ii) Personnel roles, lines of authority, training, and communication.

1910.120(q)(2)(iii) Emergency recognition and prevention.

1910.120(q)(2)(iv) Safe distances and places of refuge.

1910.120(q)(2)(v) Site security and control.

1910.120(q)(2)(vi) Evacuation routes and procedures.

1910.120(q)(2)(vii) Decontamination.

1910.120(q)(2)(viii) Emergency medical treatment and first aid.

1910.120(q)(2)(ix) Emergency alerting and response procedures.

1910.120(q)(2)(x) Critique of response and follow-up.

1910.120(q)(2)(xi) PPE and emergency equipment.

1910.120(q)(2)(xii) Emergency response organizations may use the local emergency response plan or the state emergency response plan or both, as part of their emergency response plan to avoid duplication. Those items of the emergency response plan that are being properly addressed by the SARA Title III plans may be substituted into their emergency plan or otherwise kept together for the employer and employee's use.

1910.120(q)(3) Procedures for handling emergency response.

1910.120(q)(3)(i) The senior emergency response official responding to an emergency shall become the individual in charge of a site-specific Incident Command System (ICS). All emergency responders and their communications shall be coordinated and controlled through the individual in charge of the ICS assisted by the senior official present for each employer.

NOTE TO PARAGRAPH (q)(3)(i).— The "senior official" at an emergency response is the most senior official on the site who has the responsibility for controlling the operations at the site. Initially it is the senior officer on the first-due piece of responding emergency apparatus to arrive on the incident scene. As more senior officers arrive (i.e., battalion chief, fire chief, state law enforcement official, site coordinator, etc.) the position is passed up the line of authority which has been previously established.

1910.120(q)(3)(ii) The individual in charge of the ICS shall identify, to the extent possible, all hazardous substances or conditions present and shall address as appropriate site analysis, use of engineering controls, maximum exposure limits, hazardous substance handling procedures, and use of any new technologies.

1910.120(q)(3)(iii) Based on the hazardous substances and/or conditions present, the individual in charge of the ICS shall implement appropriate emergency operations, and assure that the personal protective equipment worn is appropriate for the hazards to be encountered. However, personal protective equipment shall meet, at a minimum, the criteria contained in 29 CFR 1910.156(e) when worn while performing fire fighting operations beyond the incipient stage for any incident.

1910.120(q)(3)(iv) Employees engaged in emergency response and exposed to hazardous substances presenting an inhalation hazard or potential inhalation hazard shall wear positive pressure self-contained breathing apparatus while engaged in emergency response, until such time that the individual in charge of the ICS determines through the use of air monitoring that a decreased level of respiratory protection will not result in hazardous exposures to employees.

1910.120(q)(3)(v) The individual in charge of the ICS shall limit the number of emergency response personnel at the emergency site, in those areas of potential or actual exposure to incident or site hazards, to those who are actively performing emergency operations. However, operations in hazardous areas shall be performed using the buddy system in groups of two or more.

1910.120(q)(3)(vi) Back-up personnel shall be standing by with equipment ready to provide assistance or rescue. Qualified basic life support personnel, as a minimum, shall also be standing by with medical equipment and transportation capability.

1910.120(q)(3)(vii) The individual in charge of the ICS shall designate a safety officer, who is knowledgeable in the operations being implemented at the emergency response site, with specific responsibility to identify and evaluate hazards and to provide direction with respect to the safety of operations for the emergency at hand.

1910.120(q)(3)(viii) When activities are judged by the safety officer to be an IDLH and/or to involve an imminent danger condition, the safety officer shall have the authority to alter, suspend, or terminate those activities. The safety official shall immediately inform the individual in charge of the ICS of any actions needed to be taken to correct these hazards at the emergency scene.

1910.120(q)(3)(ix) After emergency operations have terminated, the individual in charge of the ICS shall implement appropriate decontamination procedures.

1910.120(q)(3)(x) When deemed necessary for meeting the tasks at hand, approved self-contained compressed air breathing apparatus may be used with approved cylinders from other approved self-contained compressed air breathing apparatus provided that such cylinders are of the same capacity and pressure rating. All compressed air cylinders used with self-contained breathing apparatus shall meet U.S. Department of Transportation and National Institute for Occupational Safety and Health criteria.

1910.120(q)(4) Skilled support personnel. Personnel, not necessarily an employer's own employees, who are skilled in the operation of certain equipment, such as mechanized earth moving or digging equipment or crane and hoisting equipment, and who are needed temporarily to perform immediate emergency support work that cannot reasonably be performed in a timely fashion by an employer's own employees, and who will be or may be exposed to the hazards at an emergency response scene, are not required to meet the training required in this paragraph for the employer's regular employees. However, these personnel shall be given an initial briefing at the site prior to their participation in any emergency response. The initial briefing shall include instruction in the wearing of appropriate personal protective equipment, what chemical hazards are involved, and what duties are to be performed. All other appropriate safety and health precautions provided to the employer's own employees shall be used to assure the safety and health of these personnel.

1910.120(q)(5) Specialist employees. Employees who, in the course of their regular job duties, work with and are trained in the hazards of specific hazardous substances, and who will be called upon to provide technical advice or assistance at a hazardous substance release incident to the individual in charge, shall receive training or demonstrate competency in the area of their specialization annually.

1910.120(q)(6) Training. Training shall be based on the duties and function to be performed by each responder of an emergency response organization. The skill and knowledge levels required for all new responders, those hired after the effective date of this standard, shall be conveyed to them through training before they are permitted to take part in actual emergency operations on an incident. Employees who participate, or are expected to participate, in emergency response, shall be given training in accordance with the following paragraphs:

1910.120(q)(6)(i) First responder awareness level. First responders at the awareness level are individuals who are likely to witness or discover a hazardous substance release and who have been trained to initiate an emergency response sequence by notifying the proper authorities of the release. They would take no further action beyond notifying the authorities of the release. First responders at the awareness level shall have sufficient training or have had sufficient experience to objectively demonstrate competency in the following areas:

1910.120(q)(6)(i)(A) An understanding of what hazardous substances are, and the risks associated with them in an incident.

1910.120(q)(6)(i)(B) An understanding of the potential outcomes associated with an emergency created when hazardous substances are present.

1910.120(q)(6)(i)(C) The ability to recognize the presence of hazardous substances in an emergency.

1910.120(q)(6)(i)(D) The ability to identify the hazardous substances, if possible.

1910.120(q)(6)(i)(E) An understanding of the role of the first responder awareness individual in the employer's emergency response plan including site security and control and the U.S. Department of Transportation's *Emergency Response Guidebook*.

1910.120(q)(6)(i)(F) The ability to realize the need for additional resources, and to make appropriate notifications to the communication center.

1910.120(q)(6)(ii) First responder operations level. First responders at the operations level are individuals who respond to releases or potential releases of hazardous substances as part of the initial response to the site for the purpose of protecting nearby persons, property, or the environment from the effects of the release. They are trained to respond in a defensive fashion without actually trying to stop the release. Their function is to contain the release from a safe distance, keep it from spreading, and prevent exposures. First responders at the operational level shall have received at least eight hours of training or have had sufficient experience to objectively demonstrate competency in the following areas in addition to those listed for the awareness level and the employer shall so certify:

1910.120(q)(6)(ii)(A) Knowledge of the basic hazard and risk assessment techniques.

1910.120(q)(6)(ii)(B) Know how to select and use proper personal protective equipment provided to the first responder operational level.

1910.120(q)(6)(ii)(C) An understanding of basic hazardous materials terms.

1910.120(q)(6)(ii)(D) Know how to perform basic control, containment and/or confinement operations within the capabilities of the resources and personal protective equipment available with their unit.

1910.120(q)(6)(ii)(E) Know how to implement basic decontamination procedures.

1910.120(q)(6)(ii)(F) An understanding of the relevant standard operating procedures and termination procedures.

1910.120(q)(6)(iii) Hazardous materials technician. Hazardous materials technicians are individuals who respond to releases or potential releases for the purpose of stopping the release. They assume a more aggressive role than a first responder at the operations level in that they will approach the point of release in order to plug, patch or otherwise stop the release of a hazardous substance. Hazardous materials technicians shall have received at least 24 hours of training equal to the first responder operations level and in addition have competency in the following areas and the employer shall so certify:

1910.120(q)(6)(iii)(A) Know how to implement the employer's emergency response plan.

1910.120(q)(6)(iii)(B) Know the classification, identification and verification of known and unknown materials by using field survey instruments and equipment.

1910.120(q)(6)(iii)(C) Be able to function within an assigned role in the Incident Command System.

1910.120(q)(6)(iii)(D) Know how to select and use proper specialized chemical personal protective equipment provided to the hazardous materials technician.

1910.120(q)(6)(iii)(E) Understand hazard and risk assessment techniques.

1910.120(q)(6)(iii)(F) Be able to perform advance control, containment, and/or confinement operations within the capabilities of the resources and personal protective equipment available with the unit.

1910.120(q)(6)(iii)(G) Understand and implement decontamination procedures.

1910.120(q)(6)(iii)(H) Understand termination procedures.

1910.120(q)(6)(iii)(I) Understand basic chemical and toxicological terminology and behavior.

1910.120(q)(6)(iv) Hazardous materials specialist. Hazardous materials specialists are individuals who respond with and provide support to hazardous materials technicians. Their duties parallel those of the hazardous materials technician, however, those duties require a more directed or specific knowledge of the various substances they may be called upon to contain. The hazardous materials specialist would also act as the site liaison with Federal, state, local and other government authorities in regards to site activities. Hazardous materials specialists shall have received at least 24 hours of training equal to the technician level and in addition have competency in the following areas and the employer shall so certify:

1910.120(q)(6)(iv)(A) Know how to implement the local emergency response plan.

1910.120(q)(6)(iv)(B) Understand classification, identification and verification of known and unknown materials by using advanced survey instruments and equipment.

1910.120(q)(6)(iv)(C) Know the state emergency response plan.

1910.120(q)(6)(iv)(D) Be able to select and use proper specialized chemical personal protective equipment provided to the hazardous materials specialist.

1910.120(q)(6)(iv)(E) Understand in-depth hazard and risk techniques.

1910.120(q)(6)(iv)(F) Be able to perform specialized control, containment, and/or confinement operations within the capabilities of the resources and personal protective equipment available.

1910.120(q)(6)(iv)(G) Be able to determine and implement decontamination procedures.

1910.120(q)(6)(iv)(H) Have the ability to develop a site safety and control plan.

1910.120(q)(6)(iv)(I) Understand chemical, radiological and toxicological terminology and behavior.

1910.120(q)(6)(v) On scene incident commander. Incident commanders, who will assume control of the incident scene beyond the first responder awareness level, shall receive at least 24 hours of training equal to the first responder operations level and in addition have competency in the following areas and the employer shall so certify:

1910.120(q)(6)(v)(A) Know and be able to implement the employer's incident command system.

1910.120(q)(6)(v)(B) Know how to implement the employer's emergency response plan.

1910.120(q)(6)(v)(C) Know and understand the hazards and risks associated with employees working in chemical protective clothing.

1910.120(q)(6)(v)(D) Know how to implement the local emergency response plan.

1910.120(q)(6)(v)(E) Know of the state emergency response plan and of the Federal Regional Response Team.

1910.120(q)(6)(v)(F) Know and understand the importance of decontamination procedures.

1910.120(q)(7) Trainers. Trainers who teach any of the above training subjects shall have satisfactorily completed a training course for teaching the subjects they are expected to teach, such as the courses offered by the U.S. National Fire Academy, or they shall have the training and/or academic credentials and instructional experience necessary to demonstrate competent instructional skills and a good command of the subject matter of the courses they are to teach.

1910.120(q)(8) Refresher training.

1910.120(q)(8)(i) Those employees who are trained in accordance with paragraph (q)(6) of this section shall receive annual refresher training of sufficient content and duration to maintain their competencies, or shall demonstrate competency in those areas at least yearly.

1910.120(q)(8)(ii) A statement shall be made of the training or competency, and if a statement of competency is made, the employer shall keep a record of the methodology used to demonstrate competency.

1910.120(q)(9) Medical surveillance and consultation.

1910.120(q)(9)(i) Members of an organized and designated HAZMAT team and hazardous materials specialist shall receive a baseline physical examination and be provided with medical surveillance as required in paragraph (f) of this section.

1910.120(q)(9)(ii) Any emergency response employees who exhibit signs or symptoms which may have resulted from exposure to hazardous substances during the course of an emergency incident either immediately or subsequently, shall be provided with medical consultation as required in paragraph (f)(3)(ii) of this section.

1910.120(q)(10) Chemical protective clothing. Chemical protective clothing and equipment to be used by organized and designated HAZMAT team members, or to be used by hazardous materials specialists, shall meet the requirements of paragraphs (g)(3) through (5) of this section.

1910.120(q)(11) Post-emergency response operations. Upon completion of the emergency response, if it is determined that it is necessary to remove hazardous substances, health hazards and materials contaminated with them (such as contaminated soil or other elements of the natural environment) from the site of the incident, the employer conducting the clean-up shall comply with one of the following:

1910.120(q)(11)(i) Meet all the requirements of paragraphs (b) through (o) of this section; or

1910.120(q)(11)(ii) Where the clean-up is done on plant property using plant or workplace employees, such employees shall have completed the training requirements of the following: 29 CFR 1910.38, 1910.134, 1910.1200, and other appropriate safety and health training made necessary by the tasks they are expected to perform such as personal protective equipment and decontamination procedures.

APPENDICES TO §1910.120—Hazardous Waste Operations and Emergency Response

NOTE: The following appendices serve as non-mandatory guidelines to assist employees and employers in complying with the appropriate requirements of this section. However paragraph 1910.120(g) makes mandatory in certain circumstances the use of Level A and Level B PPE protection.

1910.120 App A—Personal protective equipment test methods.

This appendix sets forth the non-mandatory examples of tests which may be used to evaluate compliance with paragraphs 1910.120(g)(4) (ii) and (iii). Other tests and other challenge agents may be used to evaluate compliance.

A. Totally-Encapsulating chemical protective suit pressure test

1.0—Scope

1.1 This practice measures the ability of a gas tight totally encapsulating chemical protective suit material, seams, and closures to maintain a fixed positive pressure. The results of this practice allow the gas tight integrity of a total-encapsulating chemical protective suit to be evaluated.

1.2 Resistance of the suit materials to permeation, penetration, and degradation by specific hazardous substances is not determined by this test method.

2.0—Description of Terms

2.1 "Totally-encapsulated chemical protective suit (TECP suit)" means a full body garment which is constructed of protective clothing materials; covers the wearer's torso, head, arms, legs and respirator; may cover the wearer's hands and feet with tightly attached gloves and boots; completely encloses the wearer and respirator by itself or in combination with the wearer's gloves and boots.

2.2 "Protective clothing material" means any material or combination of materials used in an item of clothing for the purpose of isolating parts of the body from direct contact with a potentially hazardous liquid or gaseous chemicals.

2.3 "Gas tight" means, for the purpose of the test method, the limited flow of a gas under pressure from the inside of a TECP suit to atmosphere at a prescribed pressure and time interval.

3.0 Summary of test method

3.1 The TECP suit is visually inspected and modified for the test. The test apparatus is attached to the suit to permit inflation to the pre-test suit expansion pressure for removal of suit wrinkles and creases. The pressure is lowered to the test pressure and monitored for three minutes. If the pressure drop is excessive, the TECP suit fails the test and is removed from service. The test is repeated after leak location and repair.

4.0—Required Supplies

4.1 Source of compressed air.

4.2 Test apparatus for suit testing including a pressure measurement device with a sensitivity of at least 1/4 inch water gauge.

4.3 Vent valve closure plugs or sealing tape.

4.4 Soapy water solution and soft brush.

4.5 Stop watch or appropriate timing device.

5.0—Safety Precautions

5.1 Care shall be taken to provide the correct pressure safety devices required for the source of compressed air used.

6.0—Test Procedure

6.1 Prior to each test, the tester shall perform a visual inspection of the suit. Check the suit for seam integrity by visually examining the seams and gently pulling on the seams. Ensure that all air supply lines, fittings, visor, zippers, and valves are secure and show no signs of deterioration.

6.1.1 Seal off the vent valves along with any other normal inlet or exhaust points (such as umbilical air line fittings or face piece opening) with tape or other appropriate means (caps, plugs, fixture, etc.). Care should be exercised in the sealing process not to damage any of the suit components.

6.1.2 Close all closure assemblies.

6.1.3 Prepare the suit for inflation by providing an improvised connection point on the suit for connecting an airline. Attach the pressure test apparatus to the suit to permit suit inflation from a compressed air source equipped with a pressure indicating regulator. The leak tightness of the pressure test apparatus should be tested before and after each test by closing off the end of the tubing attached to the suit and assuring a pressure of three inches water gauge for three minutes can be maintained. If a component is removed for the test, that component shall be replaced and a second test conducted with another component removed to permit a complete test of the ensemble.

6.1.4 The pre-test expansion pressure (A) and the suit test pressure (B) shall be supplied by the suit manufacturer, but in no case shall they be less than: (A) = 3 inches water gauge and (B) = 2 inches water gauge. The ending suit pressure (C) shall be no less than 80 percent of the test pressure (B); i.e., the pressure drop shall not exceed 20 percent of the test pressure (B).

6.1.5 Inflate the suit until the pressure inside is equal to pressure (A), the pre-test expansion suit pressure. Allow at least one minute to fill out the wrinkles in the suit. Release sufficient air to reduce the suit pressure to pressure (B), the suit test pressure. Begin timing. At the end of three minutes, record the suit pressure as pressure (C), the ending suit pressure. The difference between the suit test pressure and the ending suit test pressure (B–C) shall be defined as the suit pressure drop.

6.1.6 If the suit pressure drop is more than 20 percent of the suit test pressure (B) during the three minute test period, the suit fails the test and shall be removed from service.

7.0—Retest Procedure

7.1 If the suit fails the test check for leaks by inflating the suit to pressure (A) and brushing or wiping the entire suit (including seams, closures, lens gaskets, glove-to-sleeve joints, etc.) with a mild soap and water solution. Observe the suit for the formation of soap bubbles, which is an indication of a leak. Repair all identified leaks.

7.2 Retest the TECP suit as outlined in Test procedure 6.0.

8.0—Report

8.1 Each TECP suit tested by this practice shall have the following information recorded.

8.1.1 Unique identification number, identifying brand name, date of purchase, material of construction, and unique fit features; e.g., special breathing apparatus.

8.1.2 The actual values for test pressures (A), (B), and (C) shall be recorded along with the specific observation times. If the ending pressure (C) is less than 80 percent of the test pressure (B), the suit shall be identified as failing the test. When possible, the specific leak location shall be identified in the test records. Retest pressure data shall be recorded as an additional test.

8.1.3 The source of the test apparatus used shall be identified and the sensitivity of the pressure gauge shall be recorded.

8.1.4 Records shall be kept for each pressure test even if repairs are being made at the test location.

Caution

Visually inspect all parts of the suit to be sure they are positioned correctly and secured tightly before putting the suit back into service. Special care should be taken to examine each exhaust valve to make sure it is not blocked.

Care should also be exercised to assure that the inside and outside of the suit is completely dry before it is put into storage.

B. Totally-encapsulated chemical protective suit qualitative leak test

1.0—Scope

1.1 This practice semi-qualitatively tests gas tight totally-encapsulating chemical protective suit integrity by detecting inward leakage of ammonia vapor. Since no modifications are made to the suit to carry out this test, the results from this practice provide a realistic test for the integrity of the entire suit.

1.2 Resistance of the suit materials to permeation, penetration, and degradation is not determined by this test method. ASTM test methods are available to test suit materials for these characteristics and the tests are usually conducted by the manufacturers of the suits.

2.0—Description of Terms

2.1 "Totally-encapsulated chemical protective suit (TECP suit)" means a full body garment which is constructed of protective clothing materials; covers the wearer's torso, head, arms, legs and respirator; may cover the wearer's hands and feet with tightly attached gloves and boots; completely encloses the wearer and respirator by itself or in combination with the wearer's gloves, and boots.

2.2 "Protective clothing material" means any material or combination of materials used in an item of clothing for the purpose of isolating parts of the body from direct contact with a potentially hazardous liquid or gaseous chemicals.

2.3 "Gas tight" means, for the purpose of this practice the limited flow of a gas under pressure from the inside of a TECP suit to atmosphere at a prescribed pressure and time interval.

2.4 "Intrusion Coefficient" means a number expressing the level of protection provided by a gas tight totally-encapsulating chemical protective suit. The intrusion coefficient is calculated by dividing the test room challenge agent concentration by the concentration of challenge agent found inside the suit. The accuracy of the intrusion coefficient is dependent on the challenge agent monitoring methods. The larger the intrusion coefficient the greater the protection provided by the TECP suit.

3.0—Summary of recommended practice

3.1 The volume of concentrated aqueous ammonia solution (ammonia hydroxide, NH(4) OH) required to generate the test atmosphere is determined using the directions outlined in 6.1. The suit is donned by a person wearing the appropriate respiratory equipment (either a self-contained breathing apparatus or a supplied air respirator) and worn inside the enclosed test room. The concentrated aqueous ammonia solution is taken by the suited individual into the test room and poured into an open plastic pan. A two-minute evaporation period is observed before the test room concentration is measured using a high range ammonia length of stain detector tube. When the ammonia vapor reaches a concentration of between 1000 and 1200 ppm, the suited individual starts a standardized exercise protocol to stress and flex the suit. After this protocol is completed the test room concentration is measured again. The suited individual exits the test room and his stand-by person measures the ammonia concentration inside the suit using a low range ammonia length of stain detector tube or other more sensitive ammonia detector. A stand-by person is required to observe the test individual during the test procedure, aid the person in donning and doffing the TECP suit; and monitor the suit interior. The intrusion coefficient of the suit can be calculated by dividing the average test area concentration by the interior suit concentration. A colorimetric indicator strip of bromophenol blue is placed on the inside of the suit face piece lens so that the suited individual is able to detect a color change and know if the suit has a significant leak. If a color change is observed the individual should leave the test room immediately.

4.0—Required supplies

4.1 A supply of concentrated aqueous ammonium hydroxide (58 percent by weight).

4.2 A supply of bromophenol/blue indicating paper, sensitive to 5-10 ppm ammonia or greater over a two-minute period of exposure. [pH 3.0(yellow) to pH 4.6(blue)]

4.3 A supply of high range (0.5–10 volume percent) and low range (5–700 ppm) detector tubes for ammonia and the corresponding sampling pump. More sensitive ammonia detectors can be substituted for the low range detector tubes to improve the sensitivity of this practice.

4.4 A plastic pan (PVC) at least 12":14":1" and a half pint plastic container (PVC) with tightly closing lid.

4.5 A graduated cylinder or other volumetric measuring device of at least 50 milliliters in volume with an accuracy of at least + or –1 milliliters.

5.0—Safety precautions

5.1 Concentrated aqueous ammonium hydroxide, NH(4)OH, is a corrosive volatile liquid requiring eye, skin, and respiratory protection. The person conducting test shall review the MSDS for aqueous ammonia.

5.2 Since the established permissible exposure limit for ammonia is 35 ppm as a 15 minute STEL, only persons wearing a positive pressure self-contained breathing apparatus or a supplied air respirator shall be in the chamber. Normally only the person wearing the total-encapsulating suit will be inside the chamber. A stand-by person shall have a positive pressure self-contained breathing apparatus, or a supplied air respirator, available to enter the test area should the suited individual need assistance.

5.3 A method to monitor the suited individual must be used during this test. Visual contact is the simplest but other methods using communication devices are acceptable.

5.4 The test room shall be large enough to allow the exercise protocol to be carried out and and then to be ventilated to allow for easy exhaust of the ammonia test atmosphere after the test(s) are completed.

5.5 Individuals shall be medically screened for the use of respiratory protection and checked for allergies to ammonia before participating in this test procedure.

6.0—Test procedure

6.1.1 Measure the test area to the nearest foot and calculate its volume in cubic feet. Multiply the test area volume by 0.2 milliliters of concentrated aqueous ammonia solution per cubic foot of test area volume to determine the approximate volume of concentrated aqueous ammonia required to generate 1000 ppm in the test area.

6.1.2 Measure this volume from the supply of concentrated ammonia and place it into a closed plastic container.

6.1.3 Place the container, several high range ammonia detector tubes, and the pump in the clean test pan and locate it near the test area entry door so that the suited individual has easy access to these supplies.

6.2.1 In a non-contaminated atmosphere, open a pre-sealed ammonia indicator strip and fasten one end of the strip to the inside of suit face shield lens where it can be seen by the wearer. Moisten the indicator strip with distilled water. Care shall be taken not to contaminate the detector part of the indicator paper by touching it. A small piece of masking tape or equivalent should be used to attach the indicator strip to the interior of the suit face shield.

6.2.2 If problems are encountered with this method of attachment, the indicator strip can be attached to the outside of the respirator face piece being used during the test.

6.3 Don the respiratory protective device normally used with the suit, and then don the TECP suit to be tested. Check to be sure all openings which are intended to be sealed (zippers, gloves, etc.) are completely sealed. DO NOT, however, plug off any venting valves.

6.4 Step into the enclosed test room such as a closet, bathroom, or test booth, equipped with an exhaust fan. No air should be exhausted from the chamber during the test because this will dilute the ammonia challenge concentrations.

6.5 Open the container with the pre-measured volume of concentrated aqueous ammonia within the enclosed test room, and pour the liquid into the empty plastic test pan. Wait two minutes to allow for adequate volatilization of the concentrated aqueous ammonia. A small mixing fan can be used near the evaporation pan to increase the evaporation rate of ammonia solution.

6.6 After two minutes a determination of the ammonia concentration within the chamber should be made using the high range colorimetric detector tube. A concentration of 1000 ppm ammonia or greater shall be generated before the exercises are started.

6.7 To test the integrity of the suit the following four minute exercise protocol should be followed:

6.7.1 Raising the arms above the head with at least 15 raising motions completed in one minute.

6.7.2 Walking in place for one minute with at least 15 raising motions of each leg in a one-minute period.

6.7.3 Touching the toes with a least 10 complete motions of the arms from above the head to touching of the toes in a one-minute period.

6.7.4 Knee bends with at least 10 complete standing and squatting motions in a one-minute period.

6.8 If at any time during the test the colorimetric indicating paper should change colors, the test should be stopped and section 6.10 and 6.12 initiated (See 4.2).

6.9 After completion of the test exercise, the test area concentration should be measured again using the high range colorimetric detector tube.

6.10 Exit the test area.

6.11 The opening created by the suit zipper or other appropriate suit penetration should be used to determine the ammonia concentration in the suit with the low range length of stain detector tube or other ammonia monitor. The internal TECP suit air should be sampled far enough from the enclosed test area to prevent a false ammonia reading.

6.12 After completion of the measurement of the suit interior ammonia concentration the test is concluded and the suit is doffed and the respirator removed.

6.13 The ventilating fan for the test room should be turned on and allowed to run for enough time to remove the ammonia gas. The fan shall be vented to the outside of the building.

6.14 Any detectable ammonia in the suit interior (five ppm (NH(3)) or more for the length of stain detector tube) indicates the suit has failed the test. When other ammonia detectors are used a lower level of detection is possible, and it should be specified as the pass/fail criteria.

6.15 By following this test method, an intrusion coefficient of approximately 200 or more can be measured with the suit in a completely operational condition. If the coefficient is 200 or more, then the suit is suitable for emergency response and field use.

7.0—Retest procedures

7.1 If the suit fails this test, check for leaks by following the pressure test in test A above.

7.2 Retest the TECP suit as outlined in the test procedure 6.0.

8.0—Report

8.1 Each gas tight totally-encapsulating chemical protective suit tested by this practice shall have the following information recorded.

8.1.1 Unique identification number identifying brand name, date of purchase, material of construction, and unique suit features; e.g., special breathing apparatus.

8.1.2 General description of test room used for test.

8.1.3 Brand name and purchase date of ammonia detector strips and color change date.

8.1.4 Brand name, sampling range, and expiration date of the length of stain ammonia detector tubes. The brand name and model of the sampling pump should also be recorded. If another type of ammonia detector is used, it should be identified along with its minimum detection limit for ammonia.

8.1.5 Actual test results shall list the two test area concentrations, their average, the interior suit concentration, and the calculated intrusion coefficient. Retest data shall be recorded as an additional test.

8.2 The evaluation of the data shall be specified as "suit passed" or "suit failed," and the date of the test. Any detectable ammonia (five ppm or greater for the length of stain detector tube) in the suit interior indicates the suit has failed this test. When other ammonia detectors are used, a lower level of detection is possible and it should be specified as the pass fail criteria.

Caution Visually inspect all parts of the suit to be sure they are positioned correctly and secured tightly before putting the suit back into service. Special care should be taken to examine each exhaust valve to make sure it is not blocked.

Care should also be exercised to assure that the inside and outside of the suit is completely dry before it is put into storage.

1910.120 App B General description and discussion of the levels of protection and protective gear

This appendix sets forth information about personal protective equipment (PPE) protection levels which may be used to assist employers in complying with the PPE requirements of this section.

As required by the standard, PPE must be selected which will protect employees from the specific hazards which they are likely to encounter during their work on-site.

Selection of the appropriate PPE is a complex process which should take into consideration a variety of factors. Key factors involved in this process are identification of the hazards, or suspected hazards; their routes of potential hazard to employees (inhalation, skin absorption, ingestion, and eye or skin contact); and the performance of the PPE materials (and seams) in providing a barrier to these hazards. The amount of protection provided by PPE is material-hazard specific. That is, protective equipment materials will protect well against some hazardous substances and poorly, or not at all, against others. In many instances, protective equipment materials cannot be found which will provide continuous protection from the particular hazardous substance. In these cases the breakthrough time of the protective material should exceed the work durations. (end of sentence deleted—FR 14074, Apr 13, 1990)

Other factors in this selection process to be considered are matching the PPE to the employee's work requirements and task-specific conditions. The durability of PPE materials, such as tear strength and seam strength, should be considered in relation to the employee's tasks. The effects of PPE in relation to heat stress and task duration are a factor in selecting and using PPE. In some cases layers of PPE may be necessary to provide sufficient protection, or to protect expensive PPE inner garments, suits or equipment.

The more that is known about the hazards at the site, the easier the job of PPE selection becomes. As more information about the hazards and conditions at the site becomes available, the site supervisor can make decisions to up-grade or down-grade the level of PPE protection to match the tasks at hand.

The following are guidelines which an employer can use to begin the selection of the appropriate PPE. As noted above, the site information may suggest the use of combinations of PPE selected from the different protection levels (i.e., A, B, C, or D) as being more suitable to the hazards of the work. It should be cautioned that the listing below does not fully address the performance of the specific PPE material in relation to the specific hazards at the job site, and that PPE selection, evaluation and re-selection is an ongoing process until sufficient information about the hazards and PPE performance is obtained.

Part A. Personal protective equipment is divided into four categories based on the degree of protection afforded. (See Part B of this appendix for further explanation of Levels A, B, C, and D hazards.)

I. Level A—To be selected when the greatest level of skin, respiratory, and eye protection is required.

The following constitute Level A equipment; it may be used as appropriate;

1. Positive pressure, full face-piece self-contained breathing apparatus (SCBA), or positive pressure supplied air respirator with escape SCBA, approved by the National Institute for Occupational Safety and Health (NIOSH).
2. Totally-encapsulating chemical-protective suit.
3. Coveralls.(1)
4. Long underwear.(1)
5. Gloves, outer, chemical-resistant.
6. Gloves, inner, chemical-resistant.
7. Boots, chemical-resistant, steel toe and shank.
8. Hard hat (under suit).(1)
9. Disposable protective suit, gloves and boots (depending on suit construction, may be worn over totally-encapsulating suit).

Footnote(1) Optional, as applicable.

II. Level B—The highest level of respiratory protection is necessary but a lesser level of skin protection is needed.

The following constitute Level B equipment; it may be used as appropriate.

1. Positive pressure, full-facepiece self-contained breathing apparatus (SCBA), or positive pressure supplied air respirator with escape SCBA (NIOSH approved).
2. Hooded chemical-resistant clothing (overalls and long-sleeved jacket; coveralls; one or two-piece chemical-splash suit; disposable chemical-resistant overalls).
3. Coveralls.(1)
4. Gloves, outer, chemical-resistant.
5. Gloves, inner, chemical-resistant.
6. Boots, outer, chemical-resistant steel toe and shank.
7. Boot-covers, outer, chemical-resistant (disposable).(1)
8. Hard hat.(1)
9. [Reserved]
10. Face shield.(1)

Footnote(1) Optional, as applicable.

III. Level C—The concentration(s) and type(s) of airborne substance(s) is known and the criteria for using air purifying respirators are met.

The following constitute Level C equipment; it may be used as appropriate.

1. Full-face or half-mask, air purifying respirators (NIOSH approved).
2. Hooded chemical-resistant clothing (overalls; two-piece chemical-splash suit; disposable chemical-resistant overalls).
3. Coveralls.(1)
4. Gloves, outer, chemical-resistant.
5. Gloves, inner, chemical-resistant.
6. Boots (outer), chemical-resistant steel toe and shank.(1)
7. Boot-covers, outer, chemical-resistant (disposable).(1)
8. Hard hat.(1)
9. Escape mask.(1)
10. Face shield.(1)

Footnote(1) Optional, as applicable.

IV. Level D—A work uniform affording minimal protection: used for nuisance contamination only.

The following constitute Level D equipment; it may be used as appropriate:

1. Coveralls.
2. Gloves.(1)
3. Boots/shoes, chemical-resistant steel toe and shank.
4. Boots, outer, chemical-resistant (disposable).(1)
5. Safety glasses or chemical splash goggles.(1)
6. Hard hat.(1)
7. Escape mask.(1)
8. Face shield.(1)

Footnote(1) Optional, as applicable.

Part B. The types of hazards for which levels A, B, C, and D protection are appropriate are described below:

I. Level A—Level A protection should be used when:

1. The hazardous substance has been identified and requires the highest level of protection for skin, eyes, and the respiratory system based on either the measured (or potential for) high concentration of atmospheric vapors, gases, or particulates; or the site operations and work functions involve a high potential for splash, immersion, or exposure to unexpected vapors, gases, or particulates of materials that are harmful to skin or capable of being absorbed through the skin,
2. Substances with a high degree of hazard to the skin are known or suspected to be present, and skin contact is possible; or
3. Operations must be conducted in confined, poorly ventilated areas, and the absence of conditions requiring Level A have not yet been determined.

II. Level B protection should be used when:

1. The type and atmospheric concentration of substances have been identified and require a high level of respiratory protection, but less skin protection.
2. The atmosphere contains less than 19.5 percent oxygen; or
3. The presence of incompletely identified vapors or gases is indicated by a direct-reading organic vapor detection instrument, but vapors and gases are not suspected of containing high levels of chemicals harmful to skin or capable of being absorbed through the skin.

Note: This involves atmospheres with IDLH concentrations of specific substances that present severe inhalation hazards and that do not represent a severe skin hazard; or that do not meet the criteria for use of air-purifying respirators.

III. Level C—Level C protection should be used when:

1. The atmospheric contaminants, liquid splashes, or other direct contact will not adversely affect or be absorbed through any exposed skin;
2. The types of air contaminants have been identified, concentrations measured, and an air-purifying respirator is available that can remove the contaminants; and
3. All criteria for the use of air-purifying respirators are met.

IV. Level D—Level D protection should be used when:

1. The atmosphere contains no known hazard; and
2. Work functions preclude splashes, immersion, or the potential for unexpected inhalation of or contact with hazardous levels of any chemicals.

Note: As stated before, combinations of personal protective equipment other than those described for Levels A, B, C, and D protection may be more appropriate and may be used to provide the proper level of protection.

As an aid in selecting suitable chemical protective clothing, it should be noted that the National Fire Protection Association (NFPA) has developed standards on chemical protective clothing. The standards that have been adopted include:

NFPA 1991—Standard on Vapor-Protective Suits for Hazardous Chemical Emergencies (EPA Level A Protective Clothing)

NFPA 1992—Standard on Liquid Splash-Protective Suits for Hazardous Chemical Emergencies (EPA Level B Protective Clothing)

NFPA 1993—Standard on Liquid Splash-Protective Suits for Non-emergency, Non-flammable Hazardous Chemical Situations (EPA Level B Protective Clothing)

These standards apply documentation and performance requirements to the manufacture of chemical protective suits. Chemical protective suits meeting these requirements are labeled as compliant with the appropriate standard. It is recommended that chemical protective suits that meet these standards be used.

1910.120 App C Compliance guidelines

1. Occupational Safety and Health Program. Each hazardous waste site clean-up effort will require a site specific occupational safety and health program headed by the site coordinator or the employer's representative. The purpose of the program will be the protection of employees at the site and will be an extension of the employer's overall safety and health program work. The program will need to be developed before work begins on the site and implemented as work proceeds as stated in paragraph (b). The program is to facilitate coordination and communication of safety and health issues among personnel responsible for the various activities which will take place at the site. It will provide the overall means for planning and implementing the needed safety and health training and job orientation of employees who will be working at the site. The program will provide the means for identifying and controlling worksite hazards and the means for monitoring program effectiveness. The program will need to cover the responsibilities and authority of the site coordinator for the safety and health of employees at the site, and the relationships with contractors or support services as to what each employer's safety and health responsibilities are for their employees on the site. Each contractor on the site needs to have its own safety and health program so structured that it will smoothly interface with the program of the site coordinator or principal contractor.

Also those employers involved with treating, storing or disposal of hazardous waste as covered in paragraph (p) must have implemented a safety and health program for their employees. This program is to include the hazard communication program required in paragraph (p)(1) and the training required in paragraphs (p)(7) and (p)(8) as parts of the employers' comprehensive overall safety and health program. This program is to be in writing.

Each site safety and health program will need to include the following: (1) Policy statements of the line of authority and accountability for implementing the program, the objectives of the program and the role of the site safety and health officer or manager and staff; (2) means or methods for the development of procedures for identifying and controlling workplace hazards at the site; (3) means or methods for the development and communication to employees of the various plans, work rules, standard operating procedures and practices that pertain to individual employees and supervisors; (4) means for the training of supervisors and employees to develop the needed skills and knowledge to perform their work in a safe and healthful manner; (5) means to anticipate and prepare for emergency situations and; (6) means for obtaining information feedback to aid in evaluating the program and for improving the effectiveness of the program. The management and employees should be trying continually to improve the effectiveness of the program thereby enhancing the protection being afforded those working on the site.

Accidents on the site or workplace should be investigated to provide information on how such occurrences can be avoided in the future. When injuries or illnesses occur on the site or workplace, they will need to be investigated to determine what needs to be done to prevent this incident from occurring again. Such information will need to be used as feedback on the effectiveness of the program and the information turned into positive steps to prevent any reoccurrence. Receipt of employee suggestions or complaints relating to safety and health issues involved with site activities is also a feedback mechanism that can be used effectively to improve the program and may serve in part as an evaluative tool(s).

For the development and implementation of the program to be the most effective, professional safety and health personnel should be used. Certified Safety Professionals, Board Certified Industrial Hygienists or Registered Professional Safety Engineers are good examples of professional stature for safety and health managers who will administer the employer's program.

2. Training. The training programs for employees subject to the requirements of paragraph (e) of this standard should address: the safety and health hazards employees should expect to find on hazardous waste clean-up sites; what control measures or techniques are effective for those hazards; what monitoring procedures are effective in characterizing exposure levels; what makes an effective employer's safety and health program; what a site safety and health plan should include; hands on training with personal protective equipment and clothing they may be expected to use; the contents of the OSHA standard relevant to the employee's duties and function; and employee's responsibilities under OSHA and other regulations. Supervisors will need training in their responsibilities under the safety and health program and its subject areas such as the spill containment program, the personal protective equipment program, the medical surveillance program, the emergency response plan and other areas.

The training programs for employees subject to the requirements of paragraph (p) of this standard should address: the employer's safety and health program elements impacting employees; the hazard communication program; the hazards and the controls for such hazards that employees need to know for their job duties and functions. All require annual refresher training.

The training programs for employees covered by the requirements of paragraph (q) of this standard should address those competencies required for the various levels of response such as: the hazards associated with hazardous substances; hazard identification and awareness; notification of appropriate persons; the need for and use of personal protective equipment including respirators; the decontamination procedures to be used; preplanning activities for hazardous substance incidents including the emergency response plan; company standard operating procedures for hazardous substance emergency responses; the use of the incident command system and other subjects. Hands-on training should be stressed whenever possible. Critiques done after an incident which include an evaluation of what worked and what did not and how could the incident be better handled the next time may be counted as training time.

For hazardous materials specialists (usually members of hazardous materials teams), the training should address the care, use and/or testing of chemical protective clothing including totally encapsulating suits, the medical surveillance program, the standard operating procedures for the hazardous materials team including the use of plugging and patching equipment and other subject areas.

Officers and leaders who may be expected to be in charge at an incident should be fully knowledgeable of their company's incident command system. They should know where and how to obtain additional assistance and be familiar with the local district's emergency response plan and the state emergency response plan.

Specialist employees such as technical experts, medical experts or environmental experts that work with hazardous materials in their regular jobs, who may be sent to the incident scene by the shipper, manufacturer or governmental agency to advise and assist the person in charge of the incident should have training on an annual basis. Their training should include the care and use of personal protective equipment including respirators; knowledge of the incident command system and how they are to relate to it; and those areas needed to keep them current in their respective field as it relates to safety and health involving specific hazardous substances.

Those skilled support personnel, such as employees who work for public works departments or equipment operators who operate bulldozers, sand trucks, backhoes, etc., who may be called to the incident scene to provide emergency support assistance, should have at least a safety and health briefing before entering the area of potential or actual exposure. These skilled support personnel, who have not been a part of the emergency response plan and do not meet the training requirements, should be made aware of the hazards they face and should be provided all necessary protective clothing and equipment required for their tasks.

There are two National Fire Protection Association standards. NFPA 472—"Standard for Professional Competence of Responders to Hazardous Material Incidents" and NFPA 471—"Recommended Practice for Responding to Hazardous Material Incidents", which are excellent resource documents to aid fire departments and other emergency response organizations in developing their training program materials. NFPA 472 provides guidance on the skills and knowledge needed for first responder awareness level, first responder operations level, hazmat technicians, and hazmat specialist. It also offers guidance for the officer corp who will be in charge of hazardous substance incidents.

3. Decontamination. Decontamination procedures should be tailored to the specific hazards of the site and will vary in complexity and number of steps, depending on the level of hazard and the employee's exposure to the hazard. Decontamination procedures and PPE decontamination methods will vary depending upon the specific substance, since one procedure or method will not work for all substances. Evaluation of decontamination methods and procedures should be performed, as necessary, to assure that employees are not exposed to hazards by reusing PPE. References in Appendix D may be used for guidance in establishing an effective decontamination program. In addition, the U.S. Coast Guard's Manual, "Policy Guidance for Response to Hazardous Chemical Releases," U.S. Department of Transportation, Washington, DC (COMDTINST M16465.30) is a good reference for establishing an effective decontamination program.

4. Emergency response plans. States, along with designated districts within the states, will be developing or have developed emergency response plans. These state and district plans should be utilized in the emergency response plans called for in the standard. Each employer should assure that its emergency response plan is compatible with the local plan. The major reference being used to aid in developing the state and local district plans is the Hazardous Materials Emergency Planning Guide, NRT—1. The current *Emergency Response Guidebook* from the U.S. Department of Transportation, CMA's CHEMTREC and the Fire Service Emergency Management Handbook may also be used as resources.

Employers involved with treatment, storage, and disposal facilities for hazardous waste, which have the required contingency plan called for by their permit, would not need to duplicate the same planning elements. Those items of the emergency response plan may be substituted into the emergency response plan required in 1910.120 or otherwise kept together for employer and employee use.

5. Personal protective equipment programs. The purpose of personal protective clothing and equipment (PPE) is to shield or isolate individuals from the chemical, physical, and biologic hazards that may be encountered at a hazardous substance site.

   As discussed in Appendix B, no single combination of protective equipment and clothing is capable of protecting against all hazards. Thus PPE should be used in conjunction with other protective methods and its effectiveness evaluated periodically.

   The use of PPE can itself create significant worker hazards, such as heat stress, physical and psychological stress, and impaired vision, mobility and communication. For any given situation, equipment and clothing should be selected that provide an adequate level of protection. However, over-protection, as well as under-protection, can be hazardous and should be avoided where possible. Two basic objectives of any PPE program should be to protect the wearer from safety and health hazards, and to prevent injury to the wearer from incorrect use and/or malfunction of the PPE. To accomplish these goals, a comprehensive PPE program should include hazard identification, medical monitoring, environmental surveillance, selection, use, maintenance, and decontamination of PPE and its associated training.

   The written PPE program should include policy statements, procedures, and guidelines. Copies should be made available to all employees, and a reference copy should be made available at the worksite. Technical data on equipment, maintenance manuals, relevant regulations, and other essential information should also be collected and maintained.

6. Incident command system (ICS). Paragraph 1910.120(q)(3)(ii) requires the implementation of an ICS. The ICS is an organized approach to effectively control and manage operations at an emergency incident. The individual in charge of the ICS is the senior official responding to the incident. The ICS is not much different than the "command post" approach used for many years by the fire service. During large complex fires involving several companies and many pieces of apparatus, a command post would be established. This enabled one individual to be in charge of managing the incident, rather than having several officers from different companies making separate, and sometimes conflicting, decisions. The individual in charge of the command post would delegate responsibility for performing various tasks to subordinate officers. Additionally, all communications were routed through the command post to reduce the number of radio transmissions and eliminate confusion. However, strategy, tactics, and all decisions were made by one individual.

   The ICS is a very similar system, except it is implemented for emergency response to all incidents, both large and small, that involve hazardous substances.

   For a small incident, the individual in charge of the ICS may perform many tasks of the ICS. There may not be any, or little, delegation of tasks to subordinates. For example, in response to a small incident, the individual in charge of the ICS, in addition to normal command activities, may become the safety officer and may designate only one employee (with proper equipment) as a backup to provide assistance if needed. OSHA does recommend, however, that at least two employees be designated as back-up personnel since the assistance needed may include rescue.

   To illustrate the operation of the ICS, the following scenario might develop during a small incident, such as an overturned tank truck with a small leak of flammable liquid.

   The first responding senior officer would implement and take command of the ICS. That person would size-up the incident and determine if additional personnel and apparatus were necessary; would determine what actions to take to control the leak; and determine the proper level of personal protective equipment. If additional assistance is not needed, the individual in charge of the ICS would implement actions to stop and control the leak using the fewest number of personnel that can effectively accomplish the tasks. The individual in charge of the ICS then would designate himself as the safety officer and two other employees as a back-up in case rescue may become necessary. In this scenario, decontamination procedures would not be necessary.

   A large complex incident may require many employees and difficult, time-consuming efforts to control. In these situations, the individual in charge of the ICS will want to delegate different tasks to subordinates in order to maintain a span of control that will keep the number of subordinates, that are reporting, to a manageable level.

   Delegation of task at large incidents may be by location, where the incident scene is divided into sectors, and subordinate officers coordinate activities within the sector that they have been assigned.

   Delegation of tasks can also be by function. Some of the functions that the individual in charge of the ICS may want to delegate at a large incident are: medical services; evacuation; water supply; resources (equipment, apparatus); media relations; safety; and, site control (integrate activities with police for crowd and traffic control). Also for a large incident, the individual in charge of the ICS will designate several employees as back-up personnel; and a number of safety officers to monitor conditions and recommend safety precautions.

   Therefore, no matter what size or complexity an incident may be, by implementing an ICS there will be one individual in charge who makes the decisions and gives directions; and, all actions, and communications are coordinated through one central point of command. Such a system should reduce confusion, improve safety, organize and coordinate actions, and should facilitate effective management of the incident.

7. Site Safety and Control Plans. The safety and security of response personnel and others in the area of an emergency response incident site should be of primary concern to the incident commander. The use of a site safety and control plan could greatly assist those in charge of assuring the safety and health of employees on the site.

   A comprehensive site safety and control plan should include the following: summary analysis of hazards on the site and a risk analysis of those hazards; site map or sketch; site work zones (clean zone, transition or decontamination zone, work or hot zone); use of the buddy system; site communications; command post or command center; standard operating procedures and safe work practices; medical assistance and triage area; hazard monitoring plan (air contaminate monitoring, etc.); decontamination procedures and area; and other relevant areas. This plan should be a part of the employer's emergency response plan or an extension of it to the specific site.

8. Medical surveillance programs. Workers handling hazardous substances may be exposed to toxic chemicals, safety hazards, biologic hazards, and radiation. Therefore, a medical surveillance program is essential to assess and monitor workers' health and fitness for employment in hazardous waste operations and during the course

of work; to provide emergency and other treatment as needed; and to keep accurate records for future reference.

The Occupational Safety and Health Guidance Manual for Hazardous Waste Site Activities developed by the National Institute for Occupational Safety and Health (NIOSH), the Occupational Safety and Health Administration (OSHA), the U.S. Coast Guard (USCG), and the Environmental Protection Agency (EPA); October 1985 provides an excellent example of the types of medical testing that should be done as part of a medical surveillance program.

9. New Technology and Spill Containment Programs. Where hazardous substances may be released by spilling from a container that will expose employees to the hazards of the materials, the employer will need to implement a program to contain and control the spilled material. Diking and ditching, as well as use of absorbents like diatomaceous earth, are traditional techniques which have proven to be effective over the years. However, in recent years new products have come into the marketplace, the use of which complement and increase the effectiveness of these traditional methods. These new products also provide emergency responders and others with additional tools or agents to use to reduce the hazards of spilled materials.

These agents can be rapidly applied over a large area and can be uniformly applied or otherwise can be used to build a small dam, thus improving the workers' ability to control spilled material. These application techniques enhance the intimate contact between the agent and the spilled material allowing for the quickest effect by the agent or quickest control of the spilled material. Agents are available to solidify liquid spilled materials, to suppress vapor generation from spilled materials, and to do both. Some special agents, which when applied as recommended by the manufacturer, will react in a controlled manner with the spilled material to neutralize acids or caustics, or greatly reduce the level of hazard of the spilled material.

There are several modern methods and devices for use by emergency response personnel or others involved with spill control efforts to safely apply spill control agents to control spilled material hazards. These include portable pressurized applicators similar to hand-held portable fire extinguishing devices, and nozzle and hose systems similar to portable fire fighting foam systems which allow the operator to apply the agent without having to come into contact with the spilled material. The operator is able to apply the agent to the spilled material from a remote position.

The solidification of liquids provides for rapid containment and isolation of hazardous substance spills. By directing the agent at run-off points or at the edges of the spill, the reactant solid will automatically create a barrier to slow or stop the spread of the material. Clean-up of hazardous substances is greatly improved when solidifying agents, acid or caustic neutralizers, or activated carbon absorbents are used. Properly applied, these agents can totally solidify liquid hazardous substances or neutralize or absorb them, which results in materials which are less hazardous and easier to handle, transport, and dispose of. The concept of spill treatment, to create less hazardous substances, will improve the safety and level of protection of employees working at spill cleanup operations or emergency response operations to spills of hazardous substances.

The use of vapor suppression agents for volatile hazardous substances, such as flammable liquids and those substances which present an inhalation hazard, is important for protecting workers. The rapid and uniform distribution of the agent over the surface of the spilled material can provide quick vapor knockdown. There are temporary and long-term foam-type agents which are effective on vapors and dusts, and activated carbon adsorption agents which are effective for vapor control and soaking-up of the liquid. The proper use of hose lines or hand-held portable pressurized applicators provides good mobility and permits the worker to deliver the agent from a safe distance without having to step into the untreated spilled material. Some of these systems can be recharged in the field to provide coverage of larger spill areas than the design limits of a single charged applicator unit. Some of the more effective agents can solidify the liquid flammable hazardous substances and at the same time elevate the flashpoint above 140 degrees F so the resulting substance may be handled as a nonhazardous waste material if it meets the U.S. Environmental Protection Agency's 40 CFR part 261 requirements (See particularly 261.21).

All workers performing hazardous substance spill control work are expected to wear the proper protective clothing and equipment for the materials present and to follow the employer's established standard operating procedures for spill control. All involved workers need to be trained in the established operating procedures; in the use and care of spill control equipment; and in the associated hazards and control of such hazards of spill containment work.

These new tools and agents are the things that employers will want to evaluate as part of their new technology program. The treatment of spills of hazardous substances or wastes at an emergency incident as part of the immediate spill containment and control efforts is sometimes acceptable to EPA and a permit exception is described in 40 CFR 264.1(g)(8) and 265.1(c)(11).

1910.120 App D References

The following references may be consulted for further information on the subject of this standard:

1. OSHA Instruction DFO CPL 2.70—January 29, 1986, Special Emphasis Program: Hazardous Waste Sites.
2. OSHA Instruction DFO CPL 2-2.37A—January 29, 1986, Technical Assistance and Guidelines for Superfund and Other Hazardous Waste Site Activities.
3. OSHA Instruction DTS CPL 2.74—January 29, 1986, Hazardous Waste Activity Form, OSHA 175.
4. Hazardous Waste Inspections Reference Manual, U.S. Department of Labor, Occupational Safety and Health Administration, 1986.
5. Memorandum of Understanding Among the National Institute for Occupational Safety and Health, the Occupational Safety and Health Administration, the United States Coast Guard, and the United States Environmental Protection Agency, Guidance for Worker Protection During Hazardous Waste Site Investigations and Clean-up and Hazardous Substance Emergencies. December 18, 1980.
6. National Priorities List, 1st Edition, October 1984; U.S. Environmental Protection Agency, Revised periodically.
7. The Decontamination of Response Personnel, Field Standard Operating Procedures (F.S.O.P.) 7; U.S. Environmental Protection Agency, Office of Emergency and Remedial Response, Hazardous Response Support Division, December 1984.
8. Preparation of a Site Safety Plan, Field Standard Operating Procedures (F.S.O.P.) 9; U.S. Environmental Protection Agency, Office of Emergency and Remedial Response, Hazardous Response Support Division, April 1985.
9. Standard Operating Safety Guidelines; U.S. Environmental Protection Agency, Office of Emergency and Remedial Response, Hazardous Response Support Division, Environmental Response Team; November 1984.

10. Occupational Safety and Health Guidance Manual for Hazardous Waste Site Activities, National Institute for Occupational Safety and Health (NIOSH), Occupational Safety and Health Administration (OSHA), U.S. Coast Guard (USCG), and Environmental Protection Agency (EPA); October 1985.

11. Protecting Health and Safety at Hazardous Waste Sites: An Overview, U.S. Environmental Protection Agency, EPA/625/9-85/006; September 1985.

12. Hazardous Waste Sites and Hazardous Substance Emergencies, NIOSH Worker Bulletin, U.S. Department of Health and Human Services, Public Health Service, Centers for Disease Control, National Institute for Occupational Safety and Health; December 1982.

13. Personal Protective Equipment for Hazardous Materials Incidents: A Selection Guide; U.S. Department of Health and Human Services, Public Health Service, Centers for Disease Control, National Institute for Occupational Safety and Health; October 1984.

14. Fire Service Emergency Management Handbook, Federal Emergency Management Agency, Washington, DC, January 1985.

15. Emergency Response Guidebook, U.S. Department of Transportation, Washington, DC, 1987.

16. Report to the Congress on Hazardous Materials Training. Planning and Preparedness, Federal Emergency Management Agency, Washington, DC, July 1986.

17. Workbook for Fire Command, Alan V. Brunacini and J. David Beageron, National Fire Protection Association, Batterymarch Park, Quincy, MA 02269, 1985.

18. Fire Command, Alan B. Brunacini, National Fire Protection Association, Batterymarch Park, Quincy, MA 02269, 1985.

19. Incident Command System, Fire Protection Publications, Oklahoma State University, Stillwater, OK 74078, 1983.

20. Site Emergency Response Planning, Chemical Manufacturers Association, Washington, DC 20037, 1986.

21. Hazardous Materials Emergency Planning Guide, NRT-1, Environmental Protection Agency, Washington, DC, March 1987.

22. Community Teamwork: Working Together to Promote Hazardous Materials Transportation Safety. U.S. Department of Transportation, Washington, DC, May 1983.

23. Disaster Planning Guide for Business and Industry, Federal Emergency Management Agency, Publication No. FEMA 141, August 1987.

(The Office of Management and Budget has approved the information collection requirements in this section under control number 1218–0139)

1910.120 App E Training Curriculum Guidelines (Non-mandatory)

The following non-mandatory general criteria may be used for assistance in developing site-specific training curriculum used to meet the training requirements of 29 CFR 1910.120(e); 29 CFR 1910.120(p)(7), (p)(8)(iii); and 29 CFR 1910.120(q)(6), (q)(7), and (q)(8). These are generic guidelines and they are not presented as a complete training curriculum for any specific employer. Site-specific training programs must be developed on the basis of a needs assessment of the hazardous waste site, RCRA/TSDF, or emergency response operation in accordance with 29 CFR 1910.120.

It is noted that the legal requirements are set forth in the regulatory text of Sec. 1910.120. The guidance set forth here presents a highly effective program that in the areas covered would meet or exceed the regulatory requirements. In addition, other approaches could meet the regulatory requirements.

Suggested General Criteria

Definitions:

"Competent" means possessing the skills, knowledge, experience, and judgment to perform assigned tasks or activities satisfactorily as determined by the employer.

"Demonstration" means the showing by actual use of equipment or procedures.

"Hands-on training" means training in a simulated work environment that permits each student to have experience performing tasks, making decisions, or using equipment appropriate to the job assignment for which the training is being conducted.

"Initial training" means training required prior to beginning work.

"Lecture" means an interactive discourse with a class led by an instructor.

"Proficient" means meeting a stated level of achievement.

"Site-specific" means individual training directed to the operations of a specific job site.

"Training hours" means the number of hours devoted to lecture, learning activities, small group work sessions, demonstration, evaluations, or hands-on experience.

Suggested core criteria:

1. Training facility. The training facility should have available sufficient resources, equipment, and site locations to perform didactic and hands-on training when appropriate. Training facilities should have sufficient organization, support staff, and services to conduct training in each of the courses offered.

2. Training Director. Each training program should be under the direction of a training director who is responsible for the program. The Training Director should have a minimum of two years of employee education experience.

3. Instructors. Instructors should be deem competent on the basis of previous documented experience in their area of instruction, successful completion of a "train-the-trainer" program specific to the topics they will teach, and an evaluation of instructional competence by the Training Director.

   Instructors should be required to maintain professional competency by participating in continuing education or professional development programs or by completing successfully an annual refresher course and having an annual review by the Training Director.

   The annual review by the Training Director should include observation of an instructor's delivery, a review of those observations with the trainer, and an analysis of any instructor or class evaluations completed by the students during the previous year.

4. Course materials. The Training Director should approve all course materials to be used by the training provider. Course materials should be reviewed and updated at least annually. Materials and equipment should be in good working order and maintained properly.

   All written and audio-visual materials in training curricula should be peer reviewed by technically competent outside reviewers or by a standing advisory committee.

   Reviews should possess expertise in the following disciplines where applicable: occupational health, industrial hygiene and safety, chemical/environmental engineering, employee education, or emergency response. One or more of the peer reviewers should be a employee experienced in the work activities to which the training is directed.

5. Students. The program for accepting students should include:
   a. Assurance that the student is or will be involved in work where chemical exposures are likely and that the student possesses the skills necessary to perform the work.
   b. A policy on the necessary medical clearance.
6. Ratios. Student-instructor ratios should not exceed 30 students per instructor. Hands-on activity requiring the use of personal protective equipment should have the following student–instructor ratios. For Level C or Level D personal protective equipment the ratio should be 10 students per instructor. For Level A or Level B personal protective equipment the ratio should be 5 students per instructor.
7. Proficiency assessment. Proficiency should be evaluated and documented by the use of a written assessment and a skill demonstration selected and developed by the Training Director and training staff. The assessment and demonstration should evaluate the knowledge and individual skills developed in the course of training. The level of minimum achievement necessary for proficiency shall be specified in writing by the Training Director.

   If a written test is used, there should be a minimum of 50 questions. If a written test is used in combination with a skills demonstration, a minimum of 25 questions should be used. If a skills demonstration is used, the tasks chosen and the means to rate successful completion should be fully documented by the Training Director.

   The content of the written test or of the skill demonstration shall be relevant to the objectives of the course. The written test and skill demonstration should be updated as necessary to reflect changes in the curriculum and any update should be approved by the Training Director.

   The proficiency assessment methods, regardless of the approach or combination of approaches used, should be justified, documented and approved by the Training Director.

   The proficiency of those taking the additional courses for supervisors should be evaluated and documented by using proficiency assessment methods acceptable to the Training Director. These proficiency assessment methods must reflect the additional responsibilities borne by supervisory personnel in hazardous waste operations or emergency response.
8. Course certificate. Written documentation should be provided to each student who satisfactorily completes the training course.

   The documentation should include:
   a. Student's name.
   b. Course title.
   c. Course date.
   d. Statement that the student has successfully completed the course.
   e. Name and address of the training provider.
   f. An individual identification number for the certificate.
   g. List of the levels of personal protective equipment used by the student to complete the course.

   This documentation may include a certificate and an appropriate wallet-sized laminated card with a photograph of the student and the above information. When such course certificate cards are used, the individual identification number for the training certificate should be shown on the card.
9. Recordkeeping. Training providers should maintain records listing the dates courses were presented, the names of the individual course attenders, the names of those students successfully completing each course, and the number of training certificates issued to each successful student. These records should be maintained for a minimum of five years after the date an individual participated in a training program offered by the training provider. These records should be available and provided upon the student's request or as mandated by law.
10. Program quality control. The Training Director should conduct or direct an annual written audit of the training program. Program modifications to address deficiencies, if any, should be documented, approved, and implemented by the training provider. The audit and the program modification documents should be maintained at the training facility.

    Suggested Program Quality Control Criteria

    Factors listed here are suggested criteria for determining the quality and appropriateness of employee health and safety training for hazardous waste operations and emergency response.

A. Training Plan.

   Adequacy and appropriateness of the training program's curriculum development, instructor training, distribution of course materials, and direct student training should be considered, including:
   1. The duration of training, course content, and course schedules/agendas;
   2. The different training requirements of the various target populations, as specified in the appropriate generic training curriculum;
   3. The process for the development of curriculum, which includes appropriate technical input, outside review, evaluation, program pretesting.
   4. The adequate and appropriate inclusion of hands-on, demonstration, and instruction methods;
   5. Adequate monitoring of student safety, progress, and performance during the training.

B. Program management, Training Director, staff, and consultants.

   Adequacy and appropriateness of staff performance and delivering an effective training program should be considered, including:
   1. Demonstration of the training director's leadership in assuring quality of health and safety training.
   2. Demonstration of the competency of the staff to meet the demands of delivering high quality hazardous waste employee health and safety training.
   3. Organization charts establishing clear lines of authority.
   4. Clearly defined staff duties including the relationship of the training staff to the overall program.
   5. Evidence that the training organizational structure suits the needs of the training program.
   6. Appropriateness and adequacy of the training methods used by the instructors.
   7. Sufficiency of the time committed by the training director and staff to the training program.
   8. Adequacy of the ratio of training staff to students.
   9. Availability and commitment of the training program of adequate human and equipment resources in the areas of:
      a. Health effects,
      b. Safety,
      c. Personal protective equipment (PPE),
      d. Operational procedures
      e. Employee protection practices/procedures.
   10. Appropriateness of management controls.

11. Adequacy of the organization and appropriate resources assigned to assure appropriate training.
12. In the case of multiple-site training programs, adequacy of satellite centers management.

C. Training facilities and resources.

Adequacy and appropriateness of the facilities and resources for supporting the training program should be considered, including:

1. Space and equipment to conduct the training.
2. Facilities for representative hands-on training.
3. In the case of multiple-site programs, equipment and facilities at the satellite centers.
4. Adequacy and appropriateness of the quality control and evaluations program to account for instructor performance.
5. Adequacy and appropriateness of the quality control and evaluation program to ensure appropriate course evaluation, feedback, updating, and corrective action.
6. Adequacy and appropriateness of disciplines and expertise being used within the quality control and evaluation program.
7. Adequacy and appropriateness of the role of student evaluations to provide feedback for training program improvement.

D. Quality control and evaluation.

Adequacy and appropriateness of quality control and evaluation plans for training programs should be considered, including:

1. A balanced advisory committee and/or competent outside reviewers to give overall policy guidance;
2. Clear and adequate definition of the composition and active programmatic role of the advisory committee or outside reviewers.
3. Adequacy of the minutes or reports of the advisory committee or outside reviewers' meetings or written communication.
4. Adequacy and appropriateness of the quality control and evaluations program to account for instructor performance.
5. Adequacy and appropriateness of the quality control and evaluation program to ensure appropriate course evaluation, feedback, updating, and corrective action.
6. Adequacy and appropriateness of disciplines and expertise being used within the quality control and evaluation program.
7. Adequacy and appropriateness of the role of student evaluations to provide feedback for training program improvement.

E. Students

Adequacy and appropriateness of the program for accepting students should be considered, including:

1. Assurance that the student already possess the necessary skills for their job, including necessary documentation.
2. Appropriateness of methods the program uses to ensure that recruits are capable of satisfactorily completing training.
3. Review and compliance with any medical clearance policy.

F. Institutional Environment and Administrative Support

The adequacy and appropriateness of the institutional environment and administrative support system for the training program should be considered, including:

1. Adequacy of the institutional commitment to the employee training program.

2. Adequacy and appropriateness of the administrative structure and administrative support.

G. Summary of Evaluation Questions

Key questions for evaluating the quality and appropriateness of an overall training program should include the following:

1. Are the program objectives clearly stated?
2. Is the program accomplishing its objectives?
3. Are appropriate facilities and staff available?
4. Is there an appropriate mix of classroom, demonstration, and hands-on training?
5. Is the program providing quality employee health and safety training that fully meets the intent of regulatory requirements?
6. What are the program's main strengths?
7. What are the program's main weaknesses?
8. What is recommended to improve the program?
9. Are instructors instructing according to their training outlines?
10. Is the evaluation tool current and appropriate for the program content?
11. Is the course material current and relevant to the target group?

Suggested Training Curriculum Guidelines

The following training curriculum guidelines are for those operations specifically identified in 29 CFR 1910.120 as requiring training. Issues such as qualifications of instructors, training certification, and similar criteria appropriate to all categories of operations addressed in 1910.120 have been covered in the preceding section and are not re-addressed in each of the generic guidelines. Basic core requirements for training programs that are addressed include:

1. General Hazardous Waste Operations
2. RCRA operations—Treatment, storage, and disposal facilities.
3. Emergency Response.

A. General Hazardous Waste Operations and Site-specific Training

1. Off-site training. Training course content for hazardous waste operations, required by 29 CFR 1910.120(e), should include the following topics or procedures:
   a. Regulatory knowledge.
      (1) A review of 29 CFR 1910.120 and the core elements of an occupational safety and health program.
      (2) The content of a medical surveillance program as outlined in 29 CFR 1910.120(f).
      (3) The content of an effective site safety and health plan consistent with the requirements of 29 CFR 1910.120(b)(4)(ii).
      (4) Emergency response plan and procedures as outlined in 29 CFR 1910.38 and 29 CFR 1910.120(l).
      (5) Adequate illumination.
      (6) Sanitation recommendation and equipment.
      (7) Review and explanation of OSHA's hazard-communication standard (29 CFR 1910.1200) and lock-out-tag-out standard (29 CFR 1910.147).
      (8) Review of other applicable standards including but not limited to those in the construction standards (29 CFR Part 1926).
      (9) Rights and responsibilities of employers and employees under applicable OSHA and EPA laws.
   b. Technical knowledge.
      (1) Type of potential exposures to chemical, biological, and radiological hazards; types of human responses to

these hazards and recognition of those responses; principles of toxicology and information about acute and chronic hazards; health and safety considerations of new technology.

(2) Fundamentals of chemical hazards including but not limited to vapor pressure, boiling points, flash points, ph, other physical and chemical properties.

(3) Fire and explosion hazards of chemicals.

(4) General safety hazards such as but not limited to electrical hazards, powered equipment hazards, motor vehicle hazards, walking–working surface hazards, excavation hazards, and hazards associated with working in hot and cold temperature extremes.

(5) Review and knowledge of confined space entry procedures in 29 CFR 1910.146.

(6) Work practices to minimize employee risk from site hazards.

(7) Safe use of engineering controls, equipment, and any new relevant safety technology or safety procedures.

(8) Review and demonstration of competency with air sampling and monitoring equipment that may be used in a site monitoring program.

(9) Container sampling procedures and safeguarding; general drum and container handling procedures including special requirement for laboratory waste packs, shock-sensitive wastes, and radioactive wastes.

(10) The elements of a spill control program.

(11) Proper use and limitations of material handling equipment.

(12) Procedures for safe and healthful preparation of containers for shipping and transport.

(13) Methods of communication including those used while wearing respiratory protection.

c. Technical skills.

(1) Selection, use maintenance, and limitations of personal protective equipment including the components and procedures for carrying out a respirator program to comply with 29 CFR 1910.134.

(2) Instruction in decontamination programs including personnel, equipment, and hardware; hands-on training including level A, B, and C ensembles and appropriate decontamination lines; field activities including the donning and doffing of protective equipment to a level commensurate with the employee's anticipated job function and responsibility and to the degree required by potential hazards.

(3) Sources for additional hazard information; exercises using relevant manuals and hazard coding systems.

d. Additional suggested items.

(1) A laminated, dated card or certificate with photo, denoting limitations and level of protection for which the employee is trained should be issued to those students successfully completing a course.

(2) Attendance should be required at all training modules, with successful completion of exercises and a final written or oral examination with at least 50 questions.

(3) A minimum of one-third of the program should be devoted to hands-on exercises.

(4) A curriculum should be established for the 8-hour refresher training required by 29 CFR 1910.120(e)(8), with delivery of such courses directed toward those areas of previous training that need improvement or reemphasis.

(5) A curriculum should be established for the required 8-hour training for supervisors. Demonstrated competency in the skills and knowledge provided in a 40-hour course should be a prerequisite for supervisor training.

2. Refresher training.

The 8-hour annual refresher training required in 29 CFR 1910.120(e)(8) should be conducted by qualified training providers. Refresher training should include at a minimum the following topics and procedures:

(a) Review of and retraining on relevant topics covered in the 40-hour program, as appropriate, using reports by the students on their work experiences.

(b) Update on developments with respect to material covered in the 40-hour course.

(c) Review of changes to pertinent provisions of EPA or OSHA standards or laws.

(d) Introduction of additional subject areas as appropriate.

(e) Hands-on review of new or altered PPE or decontamination equipment or procedures. Review of new developments in personal protective equipment.

(f) Review of newly developed air and contaminant monitoring equipment.

3. On-site training.

a. The employer should provide employees engaged in hazardous waste site activities with information and training prior to initial assignment into their work area, as follows:

(1) The requirements of the hazard communication program including the location and availability of the written program, required lists of hazardous chemicals, and material safety data sheets.

(2) Activities and locations in their work area where hazardous substance may be present.

(3) Methods and observations that may be used to detect the presence or release of a hazardous chemical in the work area such as monitoring conducted by the employer, continuous monitoring devices, visual appearances, or other evidence (sight, sound or smell) of hazardous chemicals being released, and applicable alarms from monitoring devices that record chemical releases.

(4) The physical and health hazards of substances known or potentially present in the work area.

(5) The measures employees can take to help protect themselves from work-site hazards, including specific procedures the employer has implemented.

(6) An explanation of the labeling system and material safety data sheets and how employees can obtain and use appropriate hazard information.

(7) The elements of the confined space program including special PPE, permits, monitoring requirements, communication procedures, emergency response, and applicable lock-out procedures.

b. The employer should provide hazardous waste employees information and training and should provide a review and access to the site safety and plan as follows:

(1) Names of personnel and alternate responsible for site safety and health.

(2) Safety and health hazards present on the site.

(3) Selection, use, maintenance, and limitations of personal protective equipment specific to the site.

(4) Work practices by which the employee can minimize risks from hazards.

(5) Safe use of engineering controls and equipment available on site.

(6) Safe decontamination procedures established to minimize employee contact with hazardous substances, including:

(A) Employee decontamination,

(B) Clothing decontamination, and

(C) Equipment decontamination.

(7) Elements of the site emergency response plan, including:

(A) Pre-emergency planning.

(B) Personnel roles and lines of authority and communication.

(C) Emergency recognition and prevention.

(D) Safe distances and places of refuge.

(E) Site security and control.

(F) Evacuation routes and procedures.

(G) Decontamination procedures not covered by the site safety and health plan.

(H) Emergency medical treatment and first aid.

(I) Emergency equipment and procedures for handling emergency incidents.

c. The employer should provide hazardous waste employees information and training on personal protective equipment used at the site, such as the following:

(1) PPE to be used based upon known or anticipated site hazards.

(2) PPE limitations of materials and construction; limitations during temperature extremes, heat stress, and other appropriate medical considerations; use and limitations of respirator equipment as well as documentation procedures as outlined in 29 CFR 1910.134.

(3) PPE inspection procedures prior to, during, and after use.

(4) PPE donning and doffing procedures.

(5) PPE decontamination and disposal procedures.

(6) PPE maintenance and storage.

(7) Task duration as related to PPE limitations.

d. The employer should instruct the employee about the site medical surveillance program relative to the particular site, including:

(1) Specific medical surveillance programs that have been adapted for the site.

(2) Specific signs and symptoms related to exposure to hazardous materials on the site.

(3) The frequency and extent of periodic medical examinations that will be used on the site.

(4) Maintenance and availability of records.

(5) Personnel to be contacted and procedures to be followed when signs and symptoms of exposures are recognized.

e. The employees will review and discuss the site safety plan as part of the training program. The location of the site safety plan and all written programs should be discussed with employees including a discussion of the mechanisms for access, review, and references described.

B. RCRA Operations Training for Treatment, Storage and Disposal Facilities.

1. As a minimum, the training course required in 29 CFR 1910.120 (p) should include the following topics:

(a) Review of the applicable paragraphs of 29 CFR 1910.120 and the elements of the employer's occupational safety and health plan.

(b) Review of relevant hazards such as, but not limited to, chemical, biological, and radiological exposures; fire and explosion hazards; thermal extremes; and physical hazards.

(c) General safety hazards including those associated with electrical hazards, powered equipment hazards, lock-out-tag-out procedures, motor vehicle hazards and walking-working surface hazards.

(d) Confined-space hazards and procedures.

(e) Work practices to minimize employee risk from workplace hazards.

(f) Emergency response plan and procedures including first aid meeting the requirements of paragraph (p)(8).

(g) A review of procedures to minimize exposure to hazardous waste and various type of waste streams, including the materials handling program and spill containment program.

(h) A review of hazard communication programs meeting the requirements of 29 CFR 1910.1200.

(i) A review of medical surveillance programs meeting the requirements of 29 CFR 1910.120(p)(3) including the recognition of signs and symptoms of overexposure to hazardous substance including known synergistic interactions.

(j) A review of decontamination programs and procedures meeting the requirements of 29 CFR 1910.120(p)(4).

(k) A review of an employer's requirements to implement a training program and its elements.

(l) A review of the criteria and programs for proper selection and use of personal protective equipment, including respirators.

(m) A review of the applicable appendices to 29 CFR 1910.120.

(n) Principles of toxicology and biological monitoring as they pertain to occupational health.

(o) Rights and responsibilities of employees and employers under applicable OSHA and EPA laws.

(p) Hands-on exercises and demonstrations of competency with equipment to illustrate the basic equipment principles that may be used during the performance of work duties, including the donning and doffing of PPE.

(q) Sources of reference, efficient use of relevant manuals, and knowledge of hazard coding systems to include information contained in hazardous waste manifests.

(r) At least 8 hours of hands-on training.

(s) Training in the job skills required for an employee's job function and responsibility before they are permitted to participate in or supervise field activities.

2. The individual employer should provide hazardous waste employees with information and training prior to an employee's initial assignment into a work area. The training and information should cover the following topics:
   (a) The Emergency response plan and procedures including first aid.
   (b) A review of the employer's hazardous waste handling procedures including the materials handling program and elements of the spill containment program, location of spill response kits or equipment, and the names of those trained to respond to releases.
   (c) The hazardous communication program meeting the requirements of 29 CFR 1910.1200.
   (d) A review of the employer's medical surveillance program including the recognition of signs and symptoms of exposure to relevant hazardous substance including known synergistic interactions.
   (e) A review of the employer's decontamination program and procedures.
   (f) An review of the employer's training program and the parties responsible for that program.
   (g) A review of the employer's personal protective equipment program including the proper selection and use of PPE based upon specific site hazards.
   (h) All relevant site-specific procedures addressing potential safety and health hazards. This may include, as appropriate, biological and radiological exposures, fire and explosion hazards, thermal hazards, and physical hazards such as electrical hazards, powered equipment hazards, lock-out-tag-out hazards, motor vehicle hazards, and walking-working surface hazards.
   (i) Safe use engineering controls and equipment on site.
   (j) Names of personnel and alternates responsible for safety and health.

C. Emergency response training.

Federal OSHA standards in 29 CFR 1910.120(q) are directed toward private sector emergency responders. Therefore, the guidelines provided in this portion of the appendix are directed toward that employee population. However, they also impact indirectly through State OSHA or USEPA regulations some public sector emergency responders. Therefore, the guidelines provided in this portion of the appendix may be applied to both employee populations.

States with OSHA state plans must cover their employees with regulations at least as effective as the Federal OSHA standards. Public employees in states without approved state OSHA programs covering hazardous waste operations and emergency response are covered by the U.S. EPA under 40 CFR 311, a regulation virtually identical to Sec. 1910.120.

Since this is a non-mandatory appendix and therefore not an enforceable standard, OSHA recommends that those employers, employees or volunteers in public sector emergency response organizations outside Federal OSHA jurisdiction consider the following criteria in developing their own training programs. A unified approach to training at the community level between emergency response organizations covered by Federal OSHA and those not covered directly by Federal OSHA can help ensure an effective community response to the release or potential release of hazardous substances in the community.

a. General considerations.

Emergency response organizations are required to consider the topics listed in Sec. 1910.120(q)(6). Emergency response organizations may use some or all of the following topics to supplement those mandatory topics when developing their response training programs. Many of the topics would require an interaction between the response provider and the individuals responsible for the site where the response would be expected.

(1) Hazard recognition, including:
   (A) Nature of hazardous substances present,
   (B) Practical applications of hazard recognition, including presentations on biology, chemistry, and physics.
(2) Principles of toxicology, biological monitoring, and risk assessment.
(3) Safe work practices and general site safety.
(4) Engineering controls and hazardous waste operations.
(5) Site safety plans and standard operating procedures.
(6) Decontamination procedures and practices.
(7) Emergency procedures, first aid, and self-rescue.
(8) Safe use of field equipment.
(9) Storage, handling, use and transportation of hazardous substances.
(10) Use, care, and limitations of personal protective equipment.
(11) Safe sampling techniques.
(12) Rights and responsibilities of employees under OSHA and other related laws concerning right-to-know, safety and health, compensations and liability.
(13) Medical monitoring requirements.
(14) Community relations.

b. Suggested criteria for specific courses.

(1) First responder awareness level.
   (A) Review of and demonstration of competency in performing the applicable skills of 29 CFR 1910.120(q).
   (B) Hands-on experience with the U.S. Department of Transportation's *Emergency Response Guidebook* (*ERG*) and familiarization with OSHA standard 29 CFR 1910.1201.
   (C) Review of the principles and practices for analyzing an incident to determine both the hazardous substances present and the basic hazard and response information for each hazardous substance present.
   (D) Review of procedures for implementing actions consistent with the local emergency response plan, the organization's standard operating procedures, and the current edition of DOT's *ERG* including emergency notification procedures and follow-up communications.
   (E) Review of the expected hazards including fire and explosions hazards, confined space hazards, electrical hazards, powered equipment hazards, motor vehicle hazards, and walking-working surface hazards.
   (F) Awareness and knowledge of the competencies for the First Responder at the Awareness Level covered in the National Fire Protection Association's Standard No. 472, Professional Competence of Responders to Hazardous Materials Incidents.

(2) First responder operations level.

- **(A)** Review of and demonstration of competency in performing the applicable skills of 29 CFR 1910.120(q).
- **(B)** Hands-on experience with the U.S. Department of Transportation's *Emergency Response Guidebook* (*ERG*), manufacturer material safety data sheets, CHEMTREC/CANUTEC, shipper or manufacturer contacts, and other relevant sources of information addressing hazardous substance releases. Familiarization with OSHA standard 29 CFR 1910.1201.
- **(C)** Review of the principles and practices for analyzing an incident to determine the hazardous substances present, the likely behavior of the hazardous substance and its container, the types of hazardous substance transportation containers and vehicles, the types and selection of the appropriate defensive strategy for containing the release.
- **(D)** Review of procedures for implementing continuing response actions consistent with the local emergency response plan, the organization's standard operating procedures, and the current edition of DOT's *ERG* including extended emergency notification procedures and follow-up communications.
- **(E)** Review of the principles and practice for proper selection and use of personal protective equipment.
- **(F)** Review of the principles and practice of personnel and equipment decontamination.
- **(G)** Review of the expected hazards including fire and explosions hazards, confined space hazards, electrical hazards, powered equipment hazards, motor vehicle hazards, and walking-working surface hazards.
- **(H)** Awareness and knowledge of the competencies for the First Responder at the Operations Level covered in the National Fire Protection Association's Standard No. 472, Professional Competence of Responders to Hazardous Materials Incidents.

(3) Hazardous materials technician.

- **(A)** Review of and demonstration of competency in performing the applicable skills of 29 CFR 1910.120(q).
- **(B)** Hands-on experience with written and electronic information relative to response decision making including but not limited to the U.S. Department of Transportation's *Emergency Response Guidebook* (*ERG*), manufacturer material safety data sheets, CHEMTREC/CANUTEC, shipper or manufacturer contacts, computer data bases and response models, and other relevant sources of information addressing hazardous substance releases. Familiarization with OSHA standard 29 CFR 1910.1201.
- **(C)** Review of the principles and practices for analyzing an incident to determine the hazardous substances present, their physical and chemical properties, the likely behavior of the hazardous substance and its container, the types of hazardous substance transportation containers and vehicles involved in the release, the appropriate strategy for approaching release sites and containing the release.
- **(D)** Review of procedures for implementing continuing response actions consistent with the local emergency response plan, the organization's standard operating procedures, and the current edition of DOT's *ERG* including extended emergency notification procedures and follow-up communications.
- **(E)** Review of the principles and practice for proper selection and use of personal protective equipment.
- **(F)** Review of the principles and practices of establishing exposure zones, proper decontamination and medical surveillance stations and procedures.
- **(G)** Review of the expected hazards including fire and explosions hazards, confined space hazards, electrical hazards, powered equipment hazards, motor vehicle hazards, and walking-working surface hazards.
- **(H)** Awareness and knowledge of the competencies for the Hazardous Materials Technician covered in the National Fire Protection Association's Standard No. 472, Professional Competence of Responders to Hazardous Materials Incidents.

(4) Hazardous materials specialist.

- **(A)** Review of and demonstration of competency in performing the applicable skills of 29 CFR 1910.120(q).
- **(B)** Hands-on experience with retrieval and use of written and electronic information relative to response decision making including but not limited to the U.S. Department of Transportation's *Emergency Response Guidebook* (*ERG*), manufacturer material safety data sheets, CHEMTREC/CANUTEC, shipper or manufacturer contacts, computer data bases and response models, and other relevant sources of information addressing hazardous substance releases. Familiarization with OSHA standard 29 CFR 1910.1201.
- **(C)** Review of the principles and practices for analyzing an incident to determine the hazardous substances present, their physical and chemical properties, and the likely behavior of the hazardous substance and its container, vessel, or vehicle.
- **(D)** Review of the principles and practices for identification of the types of hazardous substance transportation containers, vessels and vehicles involved in the release; selecting and using the various types of equipment available for plugging or patching transportation containers, vessels or vehicles; organizing and directing the use of multiple teams of hazardous material technicians and selecting the appropriate strategy for approaching release sites and containing or stopping the release.
- **(E)** Review of procedures for implementing continuing response actions consistent with the local emergency response plan, the organization's standard operating procedures, including knowledge of the available public and private response resources, establishment of an incident command post, direction of hazardous material technician teams, and extended emergency notification procedures and follow-up communications.
- **(F)** Review of the principles and practice for proper selection and use of personal protective equipment.
- **(G)** Review of the principles and practices of establishing exposure zones and proper decontamination, monitoring and medical surveillance stations and procedures.
- **(H)** Review of the expected hazards including fire and explosions hazards, confined space hazards, electrical hazards, powered equipment hazards, motor vehicle hazards, and walking-working surface hazards.

(I) Awareness and knowledge of the competencies for the Off-site Specialist Employee covered in the National Fire Protection Association's Standard No. 472, Professional Competence of Responders to Hazardous Materials Incidents.

(5) Incident commander.

The incident commander is the individual who, at any one time, is responsible for and in control of the response effort. This individual is the person responsible for the direction and coordination of the response effort. An incident commander's position should be occupied by the most senior, appropriately trained individual present at the response site. Yet, as necessary and appropriate by the level of response provided, the position may be occupied by many individuals during a particular response as the need for greater authority, responsibility, or training increases. It is possible for the first responder at the awareness level to assume the duties of incident commander until a more senior and appropriately trained individual arrives at the response site.

Therefore, any emergency responder expected to perform as an incident commander should be trained to fulfill the obligations of the position at the level of response they will be providing including the following:

(A) Ability to analyze a hazardous substance incident to determine the magnitude of the response problem.

(B) Ability to plan and implement an appropriate response plan within the capabilities of available personnel and equipment.

(C) Ability to implement a response to favorably change the outcome of the incident in a manner consistent with the local emergency response plan and the organization's standard operating procedures.

(D) Ability to evaluate the progress of the emergency response to ensure that the response objectives are being met safely, effectively, and efficiently.

(E) Ability to adjust the response plan to the conditions of the response and to notify higher levels of response when required by the changes to the response plan.

Appendix D: Pro Board Assessment Methodology Matrices for NFPA 472

Competencies for Awareness Level Personnel

NFPA 472 - HazMat Awareness - 2013 Edition

IMPORTANT: The language from the standard on the AMMs is truncated. When completing the AMMs, the agency must refer to the NFPA standards for the complete text and a comprehensive statement of each Job Performance Requirement (JPR), Requisite Knowledge (RK), Requisite Skill (RS), and any applicable annex or explanatory information.

INSTRUCTIONS: Please review the instructions for filling out this Assessment Methodology Matrix at http://theproboard.org/AMM.htm. The submission of this form by an agency is affirmation that it is filled out in accordance with the instructions listed above.

AGENCY NAME:					DATE COMPILED:	
OBJECTIVE / JPR, RK, RS		COGNITIVE WRITTEN TEST	MANIPULATIVE SKILLS STATION	PORTFOLIO	PROJECTS	HAZARDOUS MATERIALS AWARENESS AND OPERATIONS, SECOND EDITION
SECTION	ABBREVIATED TEXT					
4.2.1	Detecting the Presence of Hazardous Materials					Chapter 3
4.2.1(1)	definition of hazardous materials and WMD					Chapter 3, p 46–47
4.2.1(2)	hazard classes					Chapter 3, p 64–66
4.2.1(3)	primary hazards					Chapter 3, p 64–66
4.2.1(4)	difference between hazardous materials incidents and other emergencies					Chapter 3, p 47
4.2.1(5)	typical occupancies and locations					Chapter 3, p 47
4.2.1(6)	typical container shapes					Chapter 3, p 48–53
4.2.1(7)	facility and transportation markings and colors					Chapter 3, p 58
4.2.1(8)	NFPA 704					Chapter 3, p 60–61
4.2.1(9)	placards and labels					Chapter 3, p 58–60
4.2.1(10)	information on material safety data sheets					Chapter 3, p 66–71
4.2.1(11)	examples of clues					Chapter 3, p 47
4.2.1(12)	limitations of using the senses					Chapter 3, p 47–48
4.2.1(13)	four types of locations					Chapter 3, p 73–76
4.2.1(14)	difference between a chemical and a biological incident					Chapter 3, p 73–74
4.2.1(15)	indicators of possible criminal or terrorist activity involving chemical agents					Chapter 3, p 73–76
4.2.1(16)	indicators of possible criminal or terrorist activity involving biological agents					Chapter 3, p 73–76
4.2.1(17)	indicators of possible criminal or terrorist activity involving radiological agents					Chapter 3, p 73–76

NFPA 472 - HazMat Awareness - 2013 Edition (*Continued*)

IMPORTANT: The language from the standard on the AMMs is truncated. When completing the AMMs, the agency must refer to the NFPA standards for the complete text and a comprehensive statement of each Job Performance Requirement (JPR), Requisite Knowledge (RK), Requisite Skill (RS), and any applicable annex or explanatory information.

INSTRUCTIONS: Please review the instructions for filling out this Assessment Methodology Matrix at http://theproboard.org/AMM.htm. The submission of this form by an agency is affirmation that it is filled out in accordance with the instructions listed above.

AGENCY NAME:					DATE COMPILED:	
OBJECTIVE / JPR, RK, RS						
SECTION	ABBREVIATED TEXT	COGNITIVE WRITTEN TEST	MANIPULATIVE SKILLS STATION	PORTFOLIO	PROJECTS	HAZARDOUS MATERIALS AWARENESS AND OPERATIONS, SECOND EDITION
4.2.1(18)	indicators of possible criminal or terrorist activity involving illicit laboratories					Chapter 3, p 76–77
4.2.1(19)	indicators of possible criminal or terrorist activity involving explosives					Chapter 3, p 76–77
4.2.1(20)	indicators of secondary device					Chapter 3, p 76–77
4.2.2	Surveying the Hazardous Materials Incident from a safe location.					Chapter 3, p 47–48
4.2.2(1)	determining the specific names					Chapter 3, p 47–60
4.2.2(2)	obtaining the names of, UN/NA identification numbers					Chapter 3, p 58–60
4.2.2(3)	obtaining the names of hazardous materials					Chapter 3, p 58–60
4.2.3	Collecting Hazard Information.					Chapter 3, p 63–66
4.2.3(1)	determining the appropriate guide page					Chapter 3, p 63–66
4.2.3(2)	general types of hazards					Chapter 3, p 63–66
4.4.1	Initiating Protective Actions					Chapters 2, 3, 4
4.4.1(1)	location of both the local emergency response plan and the organization's					Chapter 1, p 5
4.4.1(2)	role of the first responder					Chapter 2, p 31–32
4.4.1(3)	basic precautions					Chapter 2, p 22–39
4.4.1(4)	identification of response information					Chapter 3, p 58–60
4.4.1(5)	identify the recommended personal protective equipment					Chapter 4, p 96–100
4.4.1(6)	identify the definitions					Chapter 4, p 89–90
4.4.1(7)	identify the shapes of recommended initial isolation and protective action zones					Chapter 4, p 89–90
4.4.1(8)	describe the difference between small and large spills					Chapter 4, p 88
4.4.1(9)	identify circumstances under which distances are used at a haz mat incidents					Chapter 4, p 86–91
4.4.1(10)	describe the difference between the isolation distances and protective action distances					Chapter 4, p 89–91
4.4.1(11)	identify the techniques used to isolate the hazard area and deny entry					Chapter 4, p 89–90
4.4.1(12)	identify at least four specific actions necessary when an incident is suspected					Chapter 4, p 92–93
4.4.2	Initiating the Notification Process					Chapter 5, p 111

Competencies for Operations Levels Responders

NFPA 472 - HazMat Operational - Core Competencies - 2013 Edition

IMPORTANT: The language from the standard on the AMMs is truncated. When completing the AMMs, the agency must refer to the NFPA standards for the complete text and a comprehensive statement of each Job Performance Requirement (JPR), Requisite Knowledge (RK), Requisite Skill (RS), and any applicable annex or explanatory information.

INSTRUCTIONS: Please review the instructions for filling out this Assessment Methodology Matrix at http://theproboard.org/AMM.htm. The submission of this form by an agency is affirmation that it is filled out in accordance with the instructions listed above.

AGENCY NAME:					DATE COMPILED:	
OBJECTIVE / JPR, RK, RS						
SECTION	ABBREVIATED TEXT	COGNITIVE WRITTEN TEST	MANIPULATIVE SKILLS STATION	PORTFOLIO	PROJECTS	HAZARDOUS MATERIALS AWARENESS AND OPERATIONS, SECOND EDITION
5.1.1.2	Awareness level	**Prerequisite requirements should be met according to the approved agency policy**				
5.2.1	survey the incident					
5.2.1.1	identify shapes of containers					Chapter 3, p 48–53
5.2.1.1.1	identify each type of tank car					Chapter 3, p 54–57
5.2.1.1.2	identify each type of intermodal tank					Chapter 3, p 49–50
5.2.1.1.3	identify each type of cargo tank					Chapter 3, p 54–57
5.2.1.1.4	identify each type of storage tank					Chapter 3, p 48–57
5.2.1.1.5	identify each package by type					Chapter 3, p 50–54
5.2.1.1.6	identify each container/package by type					Chapter 3, p 48–54
5.2.1.1.7	identify each container/package by type					Chapter 3, p 75–76
5.2.1.2	identify the markings that differentiate one container from another					Chapter 3, p 48–53
5.2.1.2.1	identify the vehicle or tank identification markings					Chapter 3, p 66–71
5.2.1.2.2	identify the markings indicating container size, product contained, and/or site identification numbers					Chapter 3, p 48–62
5.2.1.3	identify the name(s) of haz mats					Chapter 3
5.2.1.3.1	identify pipeline marker information					Chapter 3, p 57
5.2.1.3.2	identify pesticide label information					Chapter 3, p 51, 74–76
5.2.1.3.3	identify radioactive material label information					Chapter 3, p 51, 74–76
5.2.1.4	identify conditions surrounding a haz mat incident					Chapter 3, p 46–48
5.2.1.5	ways to verify information					Chapter 3, p 57–73
5.2.1.6	additional hazards associated with criminal or terrorist activities					Chapter 3, p 73–76
5.2.2	collect hazard and response information					Chapter 3, p 66–73
5.2.2(1)	definitions associated with the DOT hazard classes					Chapter 3, p 66–73
5.2.2(2)	ways to obtain a material safety data sheet					Chapter 3, p 66–73
5.2.2(3)	identify hazard and response information					Chapter 3, p 66–73
5.2.2(4)	Identify type, procedures, and information regarding CHEMTREC					Chapter 3, p 66–73

NFPA 472 - HazMat Operational - Core Competencies - 2013 Edition (*Continued*)

IMPORTANT: The language from the standard on the AMMs is truncated. When completing the AMMs, the agency must refer to the NFPA standards for the complete text and a comprehensive statement of each Job Performance Requirement (JPR), Requisite Knowledge (RK), Requisite Skill (RS), and any applicable annex or explanatory information.

INSTRUCTIONS: Please review the instructions for filling out this Assessment Methodology Matrix at http://theproboard.org/AMM.htm. The submission of this form by an agency is affirmation that it is filled out in accordance with the instructions listed above.

AGENCY NAME:					DATE COMPILED:	
OBJECTIVE / JPR, RK, RS						
SECTION	ABBREVIATED TEXT	COGNITIVE WRITTEN TEST	MANIPULATIVE SKILLS STATION	PORTFOLIO	PROJECTS	HAZARDOUS MATERIALS AWARENESS AND OPERATIONS, SECOND EDITION
5.2.2(5)	methods of contacting the manufacturer					Chapter 3, p 66–73
5.2.2(6)	identify type of assistance available with respect to criminal or terrorist activities					Chapter 3, p 66–73
5.2.2(7)	identify procedure for contacting local, state, and federal authorities					Chapter 3, p 66–73
5.2.2(8)	describe properties or radioactive materials					Chapter 2, p 29–31
5.2.3	Predicting the Behavior of a Material and its Container					Chapter 2, Chapter 3
5.2.3(1)	interpret the hazard and response information					Chapter 2, p 20–38, Chapter 3, p 73
5.2.3(1)(a)	significance of chemical and physical properties on behavior of the container and/or its contents					Chapter 2, p 20–37
5.2.3(1)(b)	differences between terms					Chapter 2, p 31–38
5.2.3(2)	types of stress					Chapter 2, p 20–22
5.2.3(3)	ways in which containers can breach					Chapter 2, p 20–22
5.2.3(4)	ways in which containers can release their contents					Chapter 2, p 20–22
5.2.3(5)	dispersion patterns					Chapter 2, p 20–22
5.2.3(6)	time frames for predicting the length of time that exposures					Chapter 2, p 37–39
5.2.3(7)	health and physical hazards					Chapter 2, p 35–37
5.2.3(8)	health hazards					Chapter 2, p 29–38
5.2.3(9)	warfare agents					Chapter 3, p 66
5.2.4	Estimating the Potential Harm					Chapter 4, p 84–92
5.2.4(1)	resource for determining the size of an endangered area					Chapter 4, p 86–89
5.2.4(2)	number and type of exposures					Chapter 4, p 84–92
5.2.4(3)	resources available for determining the concentrations					Chapter 4, p 84–92
5.2.4(4)	factors for determining the extent of physical, health, and safety hazards					Chapter 4, p 84–92
5.2.4(5)	impact of time, distance and shielding					Chapter 4, p 84–92
5.3.1	Describing Response Objectives for Hazardous Materials Incidents					Chapter 4, p 89–94

(*Continued*)

NFPA 472 - HazMat Operational - Core Competencies - 2013 Edition (*Continued*)

IMPORTANT: The language from the standard on the AMMs is truncated. When completing the AMMs, the agency must refer to the NFPA standards for the complete text and a comprehensive statement of each Job Performance Requirement (JPR), Requisite Knowledge (RK), Requisite Skill (RS), and any applicable annex or explanatory information.

INSTRUCTIONS: Please review the instructions for filling out this Assessment Methodology Matrix at http://theproboard.org/AMM.htm. The submission of this form by an agency is affirmation that it is filled out in accordance with the instructions listed above.

AGENCY NAME:					DATE COMPILED:	
OBJECTIVE / JPR, RK, RS						
SECTION	ABBREVIATED TEXT	COGNITIVE WRITTEN TEST	MANIPULATIVE SKILLS STATION	PORTFOLIO	PROJECTS	HAZARDOUS MATERIALS AWARENESS AND OPERATIONS, SECOND EDITION
5.3.1(1)	steps for determining the number of exposures					Chapter 4, p 89–90
5.3.1(2)	steps for determining defensive response objectives					Chapter 4, p 92–94
5.3.1(3)	assess risk to responder					Chapter 4, p 92–94
5.3.1(4)	assess potential for secondary attack					Chapter 4, p 93–94
5.3.2	Identifying Action Options					Chapter 4, p 92–104
5.3.2(1)	options to accomplish objectives					Chapter 4, p 92–104
5.3.2(2)	prioritization of emergency medical care					Chapter 4, p 84–85
5.3.3	Determining Suitability of Personal Protective Equipment					Chapter 4, p 94
5.3.3(1)	appropriate respiratory protection					Chapter 4, p 100–103
5.3.3(2)	appropriate personal protective clothing					Chapter 4, p 94–103
5.3.4	Identifying Decontamination Issues					Chapter 4, p 103–104
5.3.4(1)	ways that personnel, personal protective equipment, apparatus, and tools and equipment become contaminated					Chapter 4, p 103–104
5.3.4(2)	potential for secondary contamination					Chapter 4, p 103–104
5.3.4(3)	importance and limitations of decontamination procedures					Chapter 4, p 103–104
5.3.4(4)	purpose of emergency decontamination procedures					Chapter 4, p 103–104
5.3.4(5)	factors to be considered in emergency decontamination					Chapter 4, p 103–104
5.4.1	Establishing and Enforcing Scene Control Procedures					Chapter 5, p 112–114
5.4.1(1)	procedures for establishing scene control					Chapter 5, p 112–114
5.4.1(2)	criteria for determining the locations					Chapter 5, p 112–114
5.4.1(3)	basic techniques for the following protective actions					Chapter 5, p 114–118
5.4.1(4)	demonstrate emergency decontamination					Chapter 5, p 114
5.4.1(5)	identify safety items					Chapter 5, p 112–114
5.4.1(6)	procedures for ensuring coordinated communication					Chapter 5, p 119

NFPA 472 - HazMat Operational - Core Competencies - 2013 Edition (*Continued*)

IMPORTANT: The language from the standard on the AMMs is truncated. When completing the AMMs, the agency must refer to the NFPA standards for the complete text and a comprehensive statement of each Job Performance Requirement (JPR), Requisite Knowledge (RK), Requisite Skill (RS), and any applicable annex or explanatory information.

INSTRUCTIONS: Please review the instructions for filling out this Assessment Methodology Matrix at http://theproboard.org/AMM.htm. The submission of this form by an agency is affirmation that it is filled out in accordance with the instructions listed above.

AGENCY NAME:					DATE COMPILED:	
OBJECTIVE / JPR, RK, RS						
SECTION	ABBREVIATED TEXT	COGNITIVE WRITTEN TEST	MANIPULATIVE SKILLS STATION	PORTFOLIO	PROJECTS	HAZARDOUS MATERIALS AWARENESS AND OPERATIONS, SECOND EDITION
5.4.2	Preserving Evidence					Chapter 5, p 119
5.4.3	Initiating ICS					Chapter 5, p 121–127
5.4.3(1)	role of the first responder					Chapter 5, p 124–127
5.4.3(2)	levels of hazardous materials incidents					Chapter 5, p 127
5.4.3(3)	purpose, need, benefits, and elements of an incident management system					Chapter 5, p 121–127
5.4.3(4)	identify duties and responsibilities					Chapter 5, p 124–127
5.4.3(5)	considerations for determining location of command post					Chapter 5, p 123
5.4.3(6)	procedures for requesting additional resources					Chapter 5, p 111
5.4.3(7)	Role and objectives of other agencies					Chapter 5, p 124–127
5.4.4	Using Personal Protective Equipment					Chapter 5, p 114–122
5.4.4(1)	importance of the buddy system					Chapter 5, p 115–116
5.4.4(2)	importance of the backup personnel					Chapter 5, p 115–116
5.4.4(3)	identify safety precautions					Chapter 5, p 114–118
5.4.4(4)	symptoms of heat and cold stress					Chapter 5, p 114–121
5.4.4(5)	physical capabilities required for, and the limitations of, personnel working in the personal protective equipment					Chapter 5, p 114–121
5.4.4(6)	procedures for cleaning, disinfecting, and inspecting personal protective equipment					Chapter 4, p 94–103
5.4.4(7)	maintenance, testing, inspections, storage of PPE					Chapter 4, p 94–103
5.5.1	Evaluating the Status of Planned Response					Chapter 5, p 118–119
5.5.1(1)	considerations for evaluating whether actions were effective					Chapter 5, p 118–119
5.5.1(2)	circumstances under which it would be prudent to withdraw					Chapter 5, p 118–119
5.5.2	Communicating the Status of the Planned Response					Chapter 5, p 118–119
5.5.2(1)	methods for communicating the status of the planned response					Chapter 5, p 118–119
5.5.2(2)	methods for immediate notification					Chapter 5, p 118–119

Competencies for Operations Level Responders Assigned Mission-Specific Responsibilities

NFPA 472 - HazMat Operational w/Personal Protective Equipment Mission Specific Competencies - 2013 Edition

IMPORTANT: The language from the standard on the AMMs is truncated. When completing the AMMs, the agency must refer to the NFPA standards for the complete text and a comprehensive statement of each Job Performance Requirement (JPR), Requisite Knowledge (RK), Requisite Skill (RS), and any applicable annex or explanatory information.

INSTRUCTIONS: Please review the instructions for filling out this Assessment Methodology Matrix at http://theproboard.org/AMM.htm. The submission of this form by an agency is affirmation that it is filled out in accordance with the instructions listed above.

AGENCY NAME:					DATE COMPILED:	
OBJECTIVE / JPR, RK, RS						
SECTION	ABBREVIATED TEXT	COGNITIVE WRITTEN TEST	MANIPULATIVE SKILLS STATION	PORTFOLIO	PROJECTS	HAZARDOUS MATERIALS AWARENESS AND OPERATIONS, SECOND EDITION
6.2.1.1.2	awareness level (Chapter 4), all core competencies at the operations level (Chapter 5), and all competencies in this section.	**Prerequisite requirements should be met according to the approved agency policy**				
6.2.3.1	select PPE					Chapter 8, p 178–183
6.2.3.1(1)	Describe the types of protective clothing and equipment					Chapter 8, p 179–183
6.2.3.1(2)	Describe personal protective equipment options for hazards					Chapter 8, p 179–183
6.2.3.1(3)	Select personal protective equipment for mission-specific tasks at hazardous materials					Chapter 8, p 179–181
6.2.3.1(3)(a)	Describe terms and explain their impact and significance					Chapter 8, p 181
6.2.3.1(3)(b)	identify three indications of material degradation					Chapter 8, p 181
6.2.3.1(3)(c)	identify the different designs of vapor-protective and splash-protective clothing					Chapter 8, p 183
6.2.3.1(3)(d)	identify the relative advantages and disadvantages of heat exchange units					Chapter 8, p 197–199
6.2.3.1(3)(e)	identify physiological and psychological stresses					Chapter 8, p 196
6.2.3.1(3)(f)	describe local procedures for technical decontamination					Chapter 8, p 178
6.2.4.1	Using Protective Clothing and Respiratory Protection					Chapter 8, p 186–195
6.2.4.1(1)	Describe at least three safety procedures					Chapter 8, p 183–185, 193–199
6.2.4.1(2)	Describe at least three emergency procedures					Chapter 8, p 183–185, 193–199
6.2.4.1(3)	Demonstrate the ability to don, work in, and doff personal protective equipment					Chapter 8, p 185–195
6.2.4.1(4)	Demonstrate local procedures					Chapter 8, p 178–179
6.2.4.1(5)	Describe the maintenance, testing, inspection, storage, and documentation procedures					Chapter 8, p 178–181
6.2.5.1	Reporting and Documenting the Incident					Chapter 8, p 200

NFPA 472 - HazMat Operational w/Mass Decontamination Mission Specific Competencies - 2013 Edition

IMPORTANT: The language from the standard on the AMMs is truncated. When completing the AMMs, the agency must refer to the NFPA standards for the complete text and a comprehensive statement of each Job Performance Requirement (JPR), Requisite Knowledge (RK), Requisite Skill (RS), and any applicable annex or explanatory information.

INSTRUCTIONS: Please review the instructions for filling out this Assessment Methodology Matrix at http://theproboard.org/AMM.htm. The submission of this form by an agency is affirmation that it is filled out in accordance with the instructions listed above.

AGENCY NAME:					DATE COMPILED:	
OBJECTIVE / JPR, RK, RS						
SECTION	ABBREVIATED TEXT	COGNITIVE WRITTEN TEST	MANIPULATIVE SKILLS STATION	PORTFOLIO	PROJECTS	HAZARDOUS MATERIALS AWARENESS AND OPERATIONS, SECOND EDITION
6.3.1.1.2	awareness level (Chapter 4), all core competencies at the operations level (Chapter 5), all mission-specific competencies for personal protective equipment (Section 6.2), and all competencies in this section.	**Prerequisite requirements should be met according to the approved agency policy**				
6.3.3.1	select personal protective equipment					Chapter 10, p 220
6.3.3.2	select decontamination procedures					Chapter 10, p 220–231
6.3.3.2(1)	Identify the advantages and limitations of mass decontamination operations					Chapter 10, p 220–222
6.3.3.2(2)	Describe the advantages and limitations of types of mass decontamination methods					Chapter 10, p 226–228
6.3.3.2(3)	Identify sources of information					Chapter 10, p 228–231
6.3.3.2(4)	identify the supplies and equipment					Chapter 10, p 222
6.3.3.2(5)	Identify procedures, equipment, and safety precautions for communicating					Chapter 10, p 231
6.3.4.1	Perform Incident Management Duties					Chapter 10, p 222–223
6.3.4.2	Perform Decontamination Operations					Chapter 10, p 222–227
6.3.5.1	Evaluate Effectiveness of Mass Decon					Chapter 10, p 224–225
6.3.6.1	Report and document the Incident					Chapter 10, p 231–232
6.3.6.1(1)	Identify the reports and supporting documentation					Chapter 10, p 231–232
6.3.6.1(2)	importance of personnel exposure records					Chapter 10, p 231–232
6.3.6.1(3)	steps in keeping an activity log					Chapter 10, p 231–232
6.3.6.1(4)	filing and record maintenance requirements					Chapter 10, p 231–232

NFPA 472 - HazMat Operational w/Technical Decontamination Mission Specific Competencies - 2013 Edition

IMPORTANT: The language from the standard on the AMMs is truncated. When completing the AMMs, the agency must refer to the NFPA standards for the complete text and a comprehensive statement of each Job Performance Requirement (JPR), Requisite Knowledge (RK), Requisite Skill (RS), and any applicable annex or explanatory information.

INSTRUCTIONS: Please review the instructions for filling out this Assessment Methodology Matrix at http://theproboard.org/AMM.htm. The submission of this form by an agency is affirmation that it is filled out in accordance with the instructions listed above.

AGENCY NAME:					DATE COMPILED:	
OBJECTIVE / JPR, RK, RS						
SECTION	ABBREVIATED TEXT	COGNITIVE WRITTEN TEST	MANIPULATIVE SKILLS STATION	PORTFOLIO	PROJECTS	HAZARDOUS MATERIALS AWARENESS AND OPERATIONS, SECOND EDITION
6.4.1.1.2	awareness level (Chapter 4), all core competencies at the operations level (Chapter 5), all mission-specific competencies for personal protective equipment (Section 6.2), and all competencies in this section.	**Prerequisite requirements should be met according to the approved agency policy**				

(Continued)

NFPA 472 - HazMat Operational w/Technical Decontamination Mission Specific Competencies - 2013 Edition (*Continued*)

IMPORTANT: The language from the standard on the AMMs is truncated. When completing the AMMs, the agency must refer to the NFPA standards for the complete text and a comprehensive statement of each Job Performance Requirement (JPR), Requisite Knowledge (RK), Requisite Skill (RS), and any applicable annex or explanatory information.

INSTRUCTIONS: Please review the instructions for filling out this Assessment Methodology Matrix at http://theproboard.org/AMM.htm. The submission of this form by an agency is affirmation that it is filled out in accordance with the instructions listed above.

AGENCY NAME:					DATE COMPILED:	
OBJECTIVE / JPR, RK, RS						
SECTION	ABBREVIATED TEXT	COGNITIVE WRITTEN TEST	MANIPULATIVE SKILLS STATION	PORTFOLIO	PROJECTS	HAZARDOUS MATERIALS AWARENESS AND OPERATIONS, SECOND EDITION
6.4.3.1	selecting personal protective equipment					Chapter 9, p 213
6.4.3.2	selecting decontamination procedures					Chapter 9, p 206–211
6.4.3.2(1)	Identify the advantages and limitations of technical decontamination operations					Chapter 9, p 208–211
6.4.3.2(2)	Describe the advantages and limitations of technical decontamination methods					Chapter 9, p 208–211
6.4.3.2(3)	Identify sources of information					Chapter 9, p 206–213
6.4.3.2(4)	identify the supplies and equipment					Chapter 9, p 213
6.4.3.2(5)	Identify the procedures, equipment, and safety precautions for processing evidence					Chatper 9, p 206–211
6.4.3.2(6)	Identify procedures, equipment, and safety precautions for handling tools					Chapter 9, p 206–207
6.4.4.1	perform incident management duties					Chapter 9, p 206–213
6.4.4.1(1)	Identify the role of the operations level responder					Chapter 9, p 206–215
6.4.4.1(2)	Describe the procedures for implementing technical decontamination operations					Chapter 9, p 211–213
6.4.4.2	perform decontamination operations identified in IAP					Chapter 9, p 213
6.4.4.2(1)	Technical decontamination operations in support of entry operations					Chapter 9, p 213
6.4.4.2(2)	Technical decontamination operations for ambulatory and nonambulatory victims					Chapter 9, p 213
6.4.5.1	evaluate effectiveness of technical decontamination process					Chapter 9, p 215
6.4.6.1	report and document					Chapter 9, p 215
6.4.6.1(1)	Identify the reports and supporting technical documentation					Chapter 9, p 215
6.4.6.1(2)	Describe the importance of personnel exposure records					Chapter 9, p 215
6.4.6.1(3)	Identify the steps in keeping an activity log					Chapter 9, p 215
6.4.6.1(4)	Identify the requirements for filing					Chapter 9, p 215

NFPA 472 - HazMat Operational w/Evidence Preservation and Sampling Mission Specific Competencies - 2013 Edition

IMPORTANT: The language from the standard on the AMMs is truncated. When completing the AMMs, the agency must refer to the NFPA standards for the complete text and a comprehensive statement of each Job Performance Requirement (JPR), Requisite Knowledge (RK), Requisite Skill (RS), and any applicable annex or explanatory information.

INSTRUCTIONS: Please review the instructions for filling out this Assessment Methodology Matrix at http://theproboard.org/AMM.htm. The submission of this form by an agency is affirmation that it is filled out in accordance with the instructions listed above.

AGENCY NAME:						DATE COMPILED:	
OBJECTIVE / JPR, RK, RS							
SECTION	ABBREVIATED TEXT	COGNITIVE WRITTEN TEST	MANIPULATIVE SKILLS STATION	PORTFOLIO	PROJECTS	HAZARDOUS MATERIALS AWARENESS AND OPERATIONS, SECOND EDITION	
6.5.1.1.2	awareness level (Chapter 4), all core competencies at the operations level (Chapter 5), all mission-specific competencies for personal protective equipment (Section 6.2), and all competencies in this section.	Prerequisite requirements should be met according to the approved agency policy					
6.5.2.1	Determining If the Incident Is Potentially Criminal					Chapter 11, p 237–238	
6.5.2.1(1)	describe products that might be encountered in the incident					Chapter 11, p 237–238	
6.5.2.1(2)	identify the agency(s) with investigative authority					Chapter 11, p 237–238	
6.5.3.1	Identify unique aspects of incident					Chapter 11, p 240–248	
6.5.3.1(1)	perform tasks associated illicit laboratories					Chapter 11, p 242–248	
6.5.3.1(2)	perform evidence preservation tasks					Chapter 11, p 237–238	
6.5.3.1(3)	perform evidence preservation tasks					Chapter 11, p 240–249	
6.5.3.1(4)	perform specified tasks associated w/suspicious letter					Chapter 11, p 240–249	
6.5.3.1(5)	perform specified tasks associated w/suspicious package					Chapter 11, p 240–249	
6.5.3.1(6)	perform specified tasks associated we/agents					Chapter 11, p 240–249	
6.5.3.1(7)	identify and describe the application, use, and limitations					Chapter 11, p 246	
6.5.3.1(8)	Describe the potential adverse impact					Chapter 11, p 246	
6.5.3.1(9)	Describe the procedures for maintaining the evidentiary integrity					Chapter 11, p 246–249	
6.5.3.2	select personal protective equipment					Chapter 11, p 238	
6.5.4.1	implement a planned response					Chapter 11, p 246	
6.5.4.2	describe local procedures for the technical decontamination process					Chapter 11, p 237–249	

NFPA 472 - HazMat Operational w/Product Control Mission Specific Competencies - 2013 Edition

IMPORTANT: The language from the standard on the AMMs is truncated. When completing the AMMs, the agency must refer to the NFPA standards for the complete text and a comprehensive statement of each Job Performance Requirement (JPR), Requisite Knowledge (RK), Requisite Skill (RS), and any applicable annex or explanatory information.

INSTRUCTIONS: Please review the instructions for filling out this Assessment Methodology Matrix at http://theproboard.org/AMM.htm. The submission of this form by an agency is affirmation that it is filled out in accordance with the instructions listed above.

AGENCY NAME:					DATE COMPILED:	
OBJECTIVE / JPR, RK, RS						
SECTION	ABBREVIATED TEXT	COGNITIVE WRITTEN TEST	MANIPULATIVE SKILLS STATION	PORTFOLIO	PROJECTS	HAZARDOUS MATERIALS AWARENESS AND OPERATIONS, SECOND EDITION
6.6.1.1.2	awareness level (Chapter 4), all core competencies at the operations level (Chapter 5), all mission-specific competencies for personal protective equipment (Section 6.2), and all competencies in this section.	**Prerequisite requirements should be met according to the approved agency policy**				
6.6.3.1	identify control options					Chapter 12, p 254–255
6.6.3.1(1)	Identify the options to accomplish a given response objective.					Chapter 12, p 254–255
6.6.3.1(2)	Identify the purpose for and the procedures					Chapter 12, p 255–267
6.6.3.2	select personal protective equipment					Chapter 12, p 255–267
6.6.4.1	perform control options					Chapter 12, p 265
6.6.4.1(1)	Using special purpose or hazard suppressing foams or agents					Chapter 12, p 265
6.6.4.1(2)	Identify the characteristics and applicability of Class B foams					Chapter 12, p 265
6.6.4.1(3)	demonstrate how to perform control activities					Chapter 12, p 255–265
6.6.4.1(4)	Identify location and describe use of emergency remote shutoff devices					Chapter 12, p 262–263
6.6.4.1(5)	Describe the use of emergency remote shutoff devices					Chapter 12, p 262–263
6.6.4.2	describe local procedures for going through the technical decontamination process					Chapter 12, p 267–268

NFPA 472 - HazMat Operational w/Air Monitoring and Sampling Specific Competencies - 2013 Edition

IMPORTANT: The language from the standard on the AMMs is truncated. When completing the AMMs, the agency must refer to the NFPA standards for the complete text and a comprehensive statement of each Job Performance Requirement (JPR), Requisite Knowledge (RK), Requisite Skill (RS), and any applicable annex or explanatory information.

INSTRUCTIONS: Please review the instructions for filling out this Assessment Methodology Matrix at http://theproboard.org/AMM.htm. The submission of this form by an agency is affirmation that it is filled out in accordance with the instructions listed above.

AGENCY NAME:					DATE COMPILED:	
OBJECTIVE / JPR, RK, RS						
SECTION	ABBREVIATED TEXT	COGNITIVE WRITTEN TEST	MANIPULATIVE SKILLS STATION	PORTFOLIO	PROJECTS	HAZARDOUS MATERIALS AWARENESS AND OPERATIONS, SECOND EDITION
6.7.1.1.2	awareness level (Chapter 4), all core competencies at the operations level (Chapter 5), all mission-specific competencies for personal protective equipment (Section 6.2), and all competencies in this section.	**Prerequisite requirements should be met according to the approved agency policy**				

NFPA 472 - HazMat Operational w/Air Monitoring and Sampling Specific Competencies - 2013 Edition (*Continued*)

IMPORTANT: The language from the standard on the AMMs is truncated. When completing the AMMs, the agency must refer to the NFPA standards for the complete text and a comprehensive statement of each Job Performance Requirement (JPR), Requisite Knowledge (RK), Requisite Skill (RS), and any applicable annex or explanatory information.

INSTRUCTIONS: Please review the instructions for filling out this Assessment Methodology Matrix at http://theproboard.org/AMM.htm. The submission of this form by an agency is affirmation that it is filled out in accordance with the instructions listed above.

AGENCY NAME:					DATE COMPILED:	
OBJECTIVE / JPR, RK, RS						
SECTION	ABBREVIATED TEXT	COGNITIVE WRITTEN TEST	MANIPULATIVE SKILLS STATION	PORTFOLIO	PROJECTS	HAZARDOUS MATERIALS AWARENESS AND OPERATIONS, SECOND EDITION
6.7.3.1	select the detection or monitoring equipment					Chapter 15, p 314–315
6.7.3.2	describe operation, capabilities and limitations					Chapter 15, p 321–325
6.7.3.3	select personal protective equipment					Chapter 15, p 319
6.7.4.1	demonstrate the field test and operation					Chapter 15, p 316
6.7.4.2	describe local procedures for decontamination of themselves and their devices					Chapter 15, p 318–319

NFPA 472 - HazMat Operational w/Victim Rescue and Recovery Specific Competencies - 2013 Edition

IMPORTANT: The language from the standard on the AMMs is truncated. When completing the AMMs, the agency must refer to the NFPA standards for the complete text and a comprehensive statement of each Job Performance Requirement (JPR), Requisite Knowledge (RK), Requisite Skill (RS), and any applicable annex or explanatory information.

INSTRUCTIONS: Please review the instructions for filling out this Assessment Methodology Matrix at http://theproboard.org/AMM.htm. The submission of this form by an agency is affirmation that it is filled out in accordance with the instructions listed above.

AGENCY NAME:					DATE COMPILED:	
OBJECTIVE / JPR, RK, RS						
SECTION	ABBREVIATED TEXT	COGNITIVE WRITTEN TEST	MANIPULATIVE SKILLS STATION	PORTFOLIO	PROJECTS	HAZARDOUS MATERIALS AWARENESS AND OPERATIONS, SECOND EDITION
6.8.1.1.2	awareness level (Chapter 4), all core competencies at the operations level (Chapter 5), all mission-specific competencies for personal protective equipment (Section 6.2), and all competencies in this section.	**Prerequisite requirements should be met according to the approved agency policy**				
6.8.3.1	determine the feasibility of conducting victim rescue and recovery					Chapter 13, p 274–279
6.8.3.1(1)	Determine feasibility					Chapter 13, p 274–275
6.8.3.1(2)	Describe the safety procedures, tactical guidelines					Chapter 13, p 275–278
6.8.3.1(3)	Determine if the options are within the capabilities					Chapter 13, p 274–279
6.8.3.1(4)	Describe the procedures for implementing victim rescue and recovery operations					Chapter 13, p 274–279
6.8.3.2	select personal protective equipment					Chapter 13, p 274–275
6.8.4.1(1)	Identify the different team positions					Chapter 13, p 275–279
6.8.4.1(2)	Select and use specialized rescue equipment					Chapter 13, p 281–282
6.8.4.1(3)	Demonstrate safe and effective methods					Chapter 13, p 282–297
6.8.4.1(4)	Demonstrate the ability to triage					Chapter 13, p 276–277
6.8.4.1(5)	Describe local procedures for performing decontamination					Chapter 13, p 279

NFPA 472 - HazMat Operational w/Response to Illicit Laboratory Incidents Specific Competencies - 2013 Edition

IMPORTANT: The language from the standard on the AMMs is truncated. When completing the AMMs, the agency must refer to the NFPA standards for the complete text and a comprehensive statement of each Job Performance Requirement (JPR), Requisite Knowledge (RK), Requisite Skill (RS), and any applicable annex or explanatory information.

INSTRUCTIONS: Please review the instructions for filling out this Assessment Methodology Matrix at http://theproboard.org/AMM.htm. The submission of this form by an agency is affirmation that it is filled out in accordance with the instructions listed above.

AGENCY NAME:					DATE COMPILED:	
OBJECTIVE / JPR, RK, RS						
SECTION	ABBREVIATED TEXT	COGNITIVE WRITTEN TEST	MANIPULATIVE SKILLS STATION	PORTFOLIO	PROJECTS	HAZARDOUS MATERIALS AWARENESS AND OPERATIONS, SECOND EDITION
6.9.1.1.2	awareness level (Chapter 4), all core competencies at the operations level (Chapter 5), all mission-specific competencies for personal protective equipment (Section 6.2), and all competencies in this section.	**Prerequisite requirements should be met according to the approved agency policy**				
6.9.2.1	determine if incident is illicit lab operation					Chapter 14, p 305–308
6.9.2.1(1)	Given examples of illicit drug manufacturing methods, describe the operational considerations, hazards, and products					Chapter 14, p 305–308
6.9.2.1(2)	Given examples of illicit chemical WMD methods describe the operational considerations, hazards, and products					Chapter 14, p 306
6.9.2.1(3)	Given examples of illicit biological WMD methods, describe the operational considerations					Chapter 14, p 306
6.9.2.1(4)	Given examples of illicit laboratory operations, describe the potential booby traps					Chapter 14, p 305–308
6.9.2.1(5)	describe law enforcement agencies that have investigative authority					Chapter 14, p 308–309
6.9.3.1	identify possible response options					Chapter 14, p 309–310
6.9.3.2.1	identify the unique operational aspects associated with illicit drug manufacturing					Chapter 14, p 302–305
6.9.3.2.2(1)	describe law enforcement securing and preserving the scene					Chapter 14, p 308–309
6.9.3.2.2(2)	describe joint hazardous materials and EOD personnel site reconnaissance					Chapter 14, p 308
6.9.3.2.2(3)	describe atmospheric hazards through air monitoring and detection					Chapter 14, p 308–309
6.9.3.2.2(4)	describe mitigation of immediate hazards					Chapter 14, p 308–309
6.9.3.2.2(5)	describe coordinated crime scene operation					Chapter 14, p 305–308
6.9.3.2.2(6)	describe documenting personnel and scene activities					Chapter 14, p 309
6.9.3.3(1)	identify the law enforcement agency(s)					Chapter 14, p 308
6.9.3.4.1	identify and describe the unique tasks and operations					Chapter 14, p 306

NFPA 472 - HazMat Operational w/Response to Illicit Laboratory Incidents Specific Competencies - 2013 Edition (*Continued*)

IMPORTANT: The language from the standard on the AMMs is truncated. When completing the AMMs, the agency must refer to the NFPA standards for the complete text and a comprehensive statement of each Job Performance Requirement (JPR), Requisite Knowledge (RK), Requisite Skill (RS), and any applicable annex or explanatory information.

INSTRUCTIONS: Please review the instructions for filling out this Assessment Methodology Matrix at http://theproboard.org/AMM.htm. The submission of this form by an agency is affirmation that it is filled out in accordance with the instructions listed above.

AGENCY NAME:					DATE COMPILED:	
OBJECTIVE / JPR, RK, RS						
SECTION	ABBREVIATED TEXT	COGNITIVE WRITTEN TEST	MANIPULATIVE SKILLS STATION	PORTFOLIO	PROJECTS	HAZARDOUS MATERIALS AWARENESS AND OPERATIONS, SECOND EDITION
6.9.3.4.2(1)	describe hazards, safety procedures, and tactical guidelines					Chapter 14, p 305–306
6.9.3.4.2(2)	describe factors to be evaluated in selection of the appropriate personal protective equipment					Chapter 14, p 305–306
6.9.3.4.2(3)	describe factors to be considered in selection of appropriate decontamination					Chapter 14, p 305–306
6.9.3.4.2(4)	describe factors to be evaluated in the selection of detection devices					Chapter 14, p 305–306
6.9.3.4.2(5)	describe factors to be considered in the development of a remediation plan					Chapter 14, p 305–306
6.9.3.5	select personal protective equipment					Chapter 14, p 309
6.9.4.1.1(1)	Describe safe and effective methods for law enforcement to secure the scene.					Chapter 14, p 306–308
6.9.4.1.1(2)	Demonstrate decontamination procedures for tactical law enforcement personnel					Chapter 14, p 309–310
6.9.4.1.1(3)	Demonstrate methods to identify and avoid potential unique safety hazards					Chapter 14, p 306–308
6.9.4.1.1(4)	Demonstrate methods to conduct joint hazardous materials/EOD operations					Chapter 14, p 305–306
6.9.4.1.2	demonstrate methods of identifying various manufacturing operations					Chapter 14, p 303–305
6.9.4.1.3	describe joint agency crime scene operations					Chapter 14, p 305–306
6.9.4.1.4	describe the policy and procedures for post–crime scene processing					Chapter 14, p 306–310
6.9.4.1.5	describe local procedures for performing decontamination upon					Chapter 14, p 306–310

Glossary

Aboveground storage tank (AST) A tank that can hold anywhere from a few hundred gallons to several million gallons of product. ASTs are usually made of aluminum, steel, or plastic.

Absorption The process by which substances travel through body tissues until they reach the bloodstream.

Absorption (as it relates to product control) The process of applying a material that will soak up and hold a hazardous material in a sponge-like manner, for collection and subsequent disposal.

Acid A material with a pH value less than 7.

Acute exposure A "right now" exposure that produces observable signs such as eye irritation, coughing, dizziness, and skin burns.

Acute health effects Health problems caused by relatively short exposure periods to a harmful substance that produces observable conditions such as eye irritation, coughing, dizziness, and skin burns.

Adsorption The process in which a contaminant adheres to the surface of an added material—such as silica or activated carbon—rather than combining with it (as in absorption).

Aerobic metabolism The creation of energy through the breakdown of nutrients in the presence of oxygen. The byproducts are carbon dioxide and water, which the body disposes of by breathing and sweating.

Agroterrorism The intentional act of using chemical or biological agents against the agricultural industry or food supply.

Air bill The shipping papers on an airplane.

Air-purifying respirator (APR) A device worn to filter particulates and contaminants from the air before it is inhaled. Selection of the filter cartridge for an APR is based on the expected contaminants.

Alcohol-resistant concentrate A foam concentrate with properties similar to those of aqueous film-forming foam, except that the concentrate is formulated so that alcohols and other polar solvents will not break down the foam.

Alpha particle A type of radiation that quickly loses energy and can travel only 1 or 2 inches from its source. Clothing or a sheet of paper can stop this type of energy. Alpha particles are not dangerous to plants, animals, or people unless the alpha-emitting substance has entered the body.

Ambulatory victim A victim who can walk on his or her own and can usually perform a self-rescue with direction and guidance from rescuers.

Ammonium nitrate fertilizer and fuel oil (ANFO) An explosive made of commonly available materials.

Anaerobic metabolism The creation of energy through the breakdown of glucose. Without oxygen, this metabolic process results in the production of lactic acid.

Anthrax An infectious disease spread by the bacterium *Bacillus anthracis*; typically found around farms, infecting livestock.

Aqueous film-forming foam (AFFF) A water-based extinguishing agent used on Class B fires that forms a foam layer over the liquid and stops the production of flammable vapors.

Asphyxiant A material that causes the victim to suffocate.

Authority having jurisdiction (AHJ) The governing body that sets operational policy and procedures for the jurisdiction you operate in.

Awareness level personnel Persons who, in the course of their normal duties, could encounter an emergency involving hazardous materials/weapons of mass destruction (WMD) and who are expected to recognize the presence of hazardous materials/WMD, protect themselves, call for trained personnel, and secure the area.

Backup team Individuals who function as a stand-by rescue crew of relief for those entering the hot zone (entry team). Also referred to as backup personnel.

Base A material with a pH value greater than 7.

Beta particle A type of radiation that is capable of travelling 10 to 15 feet from its source. Heavier materials, such as metal and glass, can stop this type of energy.

Bill of lading The shipping papers used for transport of chemicals over roads and highways. Also referred to as a *freight bill*.

Biological agents Disease-causing bacteria, viruses, and other agents that attack the human body.

BLEVE Boiling liquid/expanding vapor explosion; an explosion that occurs when pressurized liquefied materials (propane or butane, for example) inside a closed vessel are exposed to a source of high heat.

Blister agent A chemical that causes the skin to blister; also known as a vesicant.

Blood agent A chemical that interferes with the utilization of oxygen by the cells of the body. Cyanide is an example of a blood agent.

Boiling point The temperature at which a liquid will continually give off vapors in sustained amounts and, if held at that temperature long enough, will eventually turn completely into a gas.

Buddy system A system in which two responders always work as a team for safety purposes.

Bulk storage container A large-volume container that has an internal volume greater than 119 gallons for liquids and a capacity greater than 882 pounds for solids and greater than 882 pounds for gases.

Bump test A quick field test to ensure a detection/monitoring meter is operating correctly prior to entering a contaminated atmosphere.

Bung A small opening in a closed-head drum.

Calibration The process of ensuring that a particular detection/monitoring instrument will respond appropriately to a predetermined concentration of gas. An accurately calibrated machine ensures that the device is detecting the gas or vapor it is intended to detect, *at a given level*.

Canadian Transport Emergency Centre (CANUTEC) A national call center located in Ottawa, Canada. This organization serves Canadian responders (in French and English) in much the same way CHEMTREC serves responders in the United States.

Carbon monoxide (CO) detector A single-gas device using a specific toxic gas sensor to detect and measure levels of CO in an airborne environment.

Carboy A glass, plastic, or steel nonbulk storage container, ranging in volume from 5 to 15 gallons.

Carcinogen A cancer-causing agent.

Cargo tank Bulk packaging that is permanently attached to or forms a part of a motor vehicle, or is not permanently attached to any motor vehicle, and that, because of its size, construction, or attachment to a motor vehicle, is loaded or unloaded without being removed from the motor vehicle.

CBRN Chemical, biological, radiological, and nuclear.

Chain of custody A record documenting the identities of any personnel who handled the evidence, the date and time that contact occurred or the evidence was transferred from one person to another, and the reason for doing so.

Chemical Abstracts Service (CAS) A division of the American Chemical Society. This resource provides hazardous materials responders with access to an enormous collection of chemical substance information—the CAS Registry.

Chemical and physical properties Measurable characteristics of a chemical, such as its vapor density, flammability, corrosivity, and water reactivity.

Chemical degradation A natural or artificial process that causes the breakdown of a chemical substance.

Chemical reactivity The ability of a chemical to undergo an alteration in its chemical make-up, usually accompanied by a release of some form of energy.

Chemical test strip A test strip that gives the responder the ability to test for several different classifications of liquids at the same time. Also known as chemical classifier strips.

Chemical Transportation Emergency Center (CHEMTREC) A U.S. national call center that provides basic chemical information. It is operated by the American Chemistry Council.

Chemical-resistant materials Clothing (suit fabrics) specifically designed to inhibit or resist the passage of chemicals into and through the material by the processes of penetration, permeation, or degradation.

Chlorine A yellowish gas that is approximately 2.5 times heavier than air and slightly water soluble. Chlorine has many industrial uses. It damages the lungs when inhaled; it is a choking agent.

Choking agent A chemical designed to inhibit breathing and typically intended to incapacitate rather than kill its victims.

Chronic exposure Long-term exposure, occurring over the course of many months or years.

Chronic health hazard An adverse health effect occurring after a long-term exposure to a substance.

Circumstantial evidence Information that can be used to prove a theory based on facts that were observed firsthand.

Clandestine drug laboratory An illicit operation consisting of a sufficient combination of apparatus and chemicals that either has been or could be used in the manufacture or synthesis of controlled substances.

Closed-circuit self-contained breathing apparatus Self-contained breathing apparatus designed to recycle the user's exhaled air. This system removes carbon dioxide and generates fresh oxygen.

Code of Federal Regulations (CFR) A collection of permanent rules published in the *Federal Register* by the executive departments and agencies of the U.S. federal government. Its 50 titles represent broad areas of interest that are governed by federal regulation. Each volume of the CFR is updated annually and issued on a quarterly basis.

Cold zone A safe area at a hazardous materials incident for those personnel involved in the operations. The incident commander, the command post, EMS providers, and other support functions necessary to control the incident should be located in the cold zone. Also referred to as the clean zone or the support zone.

Colorimetric tube A reagent-filled tube designed to draw in a sample of air by way of a manual hand-held pump. The reagent will undergo a particular color change when exposed to the contaminant it is intended to detect.

Combustible gas indicator (CGI) A device designed to detect flammable gases and vapors. Also referred to as a flammable gas detector.

Combustible The process of burning from a gaseous or vaporous state.

Command staff Staff positions that assume responsibility for key activities in the incident command system. Individuals at this level report directly to the incident commander. Command staff members include the safety officer, public information officer, and liaison officer.

Confinement The process of keeping a hazardous material within the immediate area of the release.

Consist A list of the contents of every car on a train. Also called a train list.

Contagious Capable of transmitting a disease.

Container Any vessel or receptacle that holds material, including storage vessels, pipelines, and packaging.

Containment Actions relating to stopping a leak from a hazardous materials container, such as patching, plugging, or righting the container.

Contamination The process of transferring a hazardous material from its source to people, animals, the environment, or equipment, all of which may act as carriers for the material.

Control zones Areas at a hazardous materials incident that are designated as hot, warm, or cold, based on their safety and the degree of hazard found there.

Convulsant A chemical capable of causing convulsions or seizures when absorbed by the body. Chemicals that fall into this category include nerve agents such as sarin, soman, tabun, VX, and the organophosphate and carbamate classes of pesticides.

Corrosivity The ability of a material to cause damage (on contact) to skin, eyes, or other parts of the body.

Cryogenic liquid (cryogen) A gaseous substance that has been chilled to the point where it has liquefied; a liquid having a boiling point lower than –150°F (–101°C) at 14.7 psi (an absolute pressure of 101 kPa).

Cyanide A highly toxic chemical agent that prevents cells from using oxygen.

Cyanide compound Blood agent that prevents the body from using the available oxygen effectively.

Cyberterrorism The intentional act of electronically attacking government or private computer systems.

Cylinder A portable, nonbulk, compressed gas container used to hold liquids and gases. Uninsulated compressed gas cylinders are used to store substances such as nitrogen, argon, helium, and oxygen. They have a range of sizes and internal pressures.

Cytochrome oxidase Found in the mitochondria, this is important in cell respiration as an agent of electron transfer from certain cytochrome molecules to oxygen molecules.

Damming The product-control process used when liquid is flowing in a natural channel or depression, and its progress can be stopped by constructing a barrier to block the flow.

Dangerous cargo manifest The shipping papers on a marine vessel, generally located in a tube-like container.

Decontamination The physical and/or chemical process of reducing and preventing the spread of contaminants from people, animals, the environment, or equipment involved at hazardous materials/weapons of mass destruction incidents.

Decontamination corridor A controlled area within the warm zone where decontamination takes place.

Decontamination team The team responsible for reducing and preventing the spread of contaminants from persons and equipment used at a hazardous materials incident. Members of this team establish the decontamination corridor and conduct all phases of decontamination.

Degradation The physical destruction or decomposition of a clothing material owing to chemical exposure, general use, or ambient conditions (such as storage in sunlight). Materials can also be tested for weight changes, loss of fabric tensile strength, and other properties to measure degradation.

Dehydration An excessive loss of body water. Signs and symptoms of dehydration may include increasing thirst, dry mouth, weakness or dizziness, and a darkening of the urine or a decrease in the frequency of urination.

Demonstrative evidence Materials used to demonstrate a theory or explain an event.

Denial of entry A policy under which, once the perimeter around a release site has been identified and marked out, responders limit access to all but essential personnel.

Department of Transportation (DOT) The U.S. government agency that publicizes and enforces rules and regulations that relate to the transportation of many hazardous materials.

Dewar container A container designed to preserve the temperature of the cold liquid held inside.

Diking The placement of materials to form a barrier that will keep a hazardous material in liquid form from entering an area or that will hold the material in an area.

Dilution The process of adding a substance—usually water—in an attempt to weaken the concentration of another substance.

Direct evidence Statements made by suspects, victims, and witnesses.

Disinfection The process used to destroy recognized disease-carrying (pathogenic) microorganisms.

Diversion The process of redirecting spilled or leaking material to an area where it will have less impact.

Division An organizational level within the incident command system that divides an incident in one location into geographic areas of operational responsibility.

Doffing The process of taking off an ensemble of PPE.

Donning The process of putting on an ensemble of PPE.

Drug Enforcement Administration (DEA) Established in 1973, the federal agency charged with enforcing controlled substances laws and regulations in the United States.

Drum A barrel-like nonbulk storage vessel used to store a wide variety of substances, including food-grade materials, corrosives, flammable liquids, and grease. Drums may be constructed of low-carbon steel, polyethylene, cardboard, stainless steel, nickel, or other materials.

Dry bulk cargo tank A tank designed to carry dry bulk goods such as powders, pellets, fertilizers, or grain. Such tanks are generally *V*-shaped with rounded sides that funnel toward the bottom.

Ecoterrorism Terrorism directed against causes that radical environmentalists think would damage the earth or its creatures.

Emergency decontamination The process of removing the bulk of contaminants off a victim without regard for containment. It is used in potentially life-threatening situations, without the formal establishment of a decontamination corridor.

Emergency Planning and Community Right to Know Act (EPCRA) Legislation that requires a business that handles chemicals to report on those chemicals' type, quantity, and storage methods to the fire department and the local emergency planning committee.

Emergency Response Guidebook (ERG) A preliminary action guide for first responders operating at a hazardous materials incident in coordination with the U.S. Department of Transportation's (DOT's) labels and placards marking system. The DOT and the Secretariat of Communications and Transportation of Mexico (SCT), along with Transport Canada, jointly developed the *Emergency Response Guidebook*.

Emergency Transportation System for the Chemical Industry, Mexico (SETIQ) A national response center that is the Mexican equivalent of CHEMTREC and CANUTEC.

Entry team A team of fully qualified and equipped responders who are assigned to enter the designated hot zone.

Environmental Protection Agency (EPA) Established in 1970, the U.S. federal agency that ensures safe manufacturing, use, transportation, and disposal of hazardous substances.

Evacuation The removal or relocation of those individuals who may be affected by an approaching release of a hazardous material.

Evaporation A natural form of chemical degradation in which a liquid material becomes a gas, allowing for dissipation of a liquid spill. It is sometimes used as a safe, noninvasive way to allow a chemical substance to stabilize without human intervention.

Evidence Information that is gathered and used by an investigator in determining the cause of an incident. When evidence is used in or suitable to courts to answer questions of interest, it is referred to as forensic evidence.

Evidence preservation The process of protecting potential evidence until it can be documented, sampled, and collected appropriately.

Evidence sampling The process of collecting portions of a hazardous material/WMD for the purposes of field screening, laboratory testing, and, ultimately, criminal prosecution.

Excepted packaging Packaging used to transport materials that meets only general design requirements for any hazardous material. Low-level radioactive substances are commonly shipped in excepted packages, which may be constructed out of heavy cardboard.

Expansion ratio A description of the volume increase that occurs when a liquid changes to a gas.

Explosive ordnance disposal (EOD) personnel Personnel trained to detect, identify, evaluate, render safe, recover, and dispose of unexploded explosive devices.

Exposure The process by which people, animals, the environment, and equipment are subjected to or come into contact with a hazardous material.

Federal Bureau of Investigation (FBI) Established in 1908, the federal agency charged with defending and protecting the United States against terrorist and foreign intelligence threats and with enforcing criminal laws including acts of terrorism.

Finance/administration section The command-level section of the incident command system responsible for all costs and financial aspects of the incident, as well as any legal issues that arise.

Fire point The temperature at which sustained combustion will occur. The fire point is usually only slightly higher than the flash point for most materials.

Flame ionization detector (FID) A detection/monitoring device that is similar in operational concept to the photoionization detector (PID). The key difference is that FIDs can detect methane, whereas PIDs cannot.

Flammable range An expression of a fuel/air mixture, defined by upper and lower limits, that reflects an amount of flammable vapor mixed with a given volume of air.

Flash point The minimum temperature at which a liquid or a solid releases sufficient vapor to form an ignitable mixture with air.

Fluoroprotein foam A blended organic and synthetic foam that is made from animal by-products and synthetic surfactants.

Forward staging area A strategically placed area, close to the incident site, where personnel and equipment can be held in readiness for rapid response to an emergency event.

Freight bill The shipping papers used for transport of chemicals along roads and highways. Also referred to as a *bill of lading*.

Gamma radiation A type of radiation that can travel significant distances, penetrating most materials and passing through the body. Gamma radiation is the most destructive type of radiation to the human body.

Gas chromatography (GC) A sophisticated type of spectroscopy used to identify the components of an air sample that may be a mixture of several different chemicals.

Gross decontamination A technique for significantly reducing the amount of surface contaminant by application of a continuous shower of water prior to the removal of outer clothing. It differs from emergency decontamination in that gross decontamination is controlled through the decontamination corridor.

Hazard A material capable of posing an unreasonable risk to health, safety, or the environment; a material capable of causing harm.

Hazardous device personnel (bomb squad) Personnel trained to detect, identify, evaluate, render safe, recover, and dispose of unexploded explosive devices. Also known as a hazardous device technician.

Hazardous material Any substance or material that is capable of posing an unreasonable risk to human health, safety, or the environment when transported in commerce, used incorrectly, or not properly contained or stored.

Hazardous materials branch A unit consisting of some or all of the following positions as needed for the safe control of a hazardous materials incident: a second safety officer known as the hazardous materials safety officer (or assistant safety officer), who reports directly to the incident safety officer; a hazardous materials group supervisor; the entry team; the decontamination team; and the technical reference team.

Hazardous materials group A group often established when companies and crews are working on the same task or objective, albeit not necessarily in the same location. A group is specific because it applies to the incident command system and is assembled to relieve span of control issues.

Hazardous Materials Information System (HMIS) A color-coded marking system by which employers give their personnel the necessary information to work safely around chemicals. The Workplace Hazardous Materials Information System (WHMIS) is the Canadian hazard communication standard.

Hazardous materials safety officer A second safety officer dedicated to the safety needs of the hazardous materials branch. Also referred to as the assistant safety officer.

Hazardous waste A substance that remains after a process or manufacturing plant has used some of the material and the substance is no longer pure.

HAZWOPER (HAZardous Waste OPerations and Emergency Response) The federal OSHA regulation that governs hazardous materials waste site and response training. Specifics to emergency response can be found in 29 CFR 1910.120(q).

Heat exhaustion A mild form of shock that occurs when the circulatory system begins to fail as a result of the body's inadequate effort to give off excessive heat.

Heat stroke A severe, sometimes fatal condition resulting from the failure of the body's temperature-regulating capacity. Reduction or cessation of sweating is an early symptom; body temperature of 105°F (40.5°C) or higher, rapid pulse, hot and dry skin, headache, confusion, unconsciousness, and convulsions may occur as well.

High-expansion foam A foam created by pumping large volumes of air through a small screen coated with a foam solution. Some high-expansion foams have expansion ratios ranging from 200:1 to approximately 1000:1.

High-temperature–protective equipment A type of personal protective equipment that shields the wearer during short-term exposures to high temperatures. Sometimes referred to as a *proximity suit*, this type of equipment allows the properly trained fire fighter to work in extreme fire conditions. It is not designed to protect against hazardous materials or weapons of mass destruction.

Hot zone The area immediately surrounding a hazardous materials spill/incident site that is directly dangerous to life and health. All personnel working in the hot zone must wear complete, appropriate protective clothing and equipment.

Hydrogen sulfide (H_2S) monitor A single-gas device using a specific toxic gas sensor to detect and measure levels of H_2S in an airborne environment.

Hypoxia A deficiency of oxygen supply to the tissues in the human body.

Ignition (autoignition) temperature The minimum temperature at which a fuel, when heated, will ignite in air and continue to burn. Also called the *autoignition temperature*.

Illicit laboratory Any unlicensed or illegal structure, vehicle, facility, or physical location that may be used to manufacture, process, culture, or synthesize an illegal drug, hazardous material/WMD device, or agent.

Immediately dangerous to life and health (IDLH) The atmospheric concentration of any toxic, corrosive, or asphyxiant substance such that it poses an immediate threat to life or could cause irreversible or delayed adverse health effects.

Improvised explosive device (IED) An explosive or incendiary device that is fabricated in an improvised manner.

Incident command post (ICP) The location in the cold zone where the incident command is located; it is where the command, coordination, control, and communications functions are centralized.

Incident command system (ICS) The combination of facilities, equipment, personnel, procedures, and communications under a standard organizational structure organized so as to manage assigned resources and effectively accomplish stated objectives for an incident.

Incident commander (IC) A level of training intended for those assuming command of a hazardous materials incident beyond the operations level. Individuals trained as incident commanders should have at least operations level training and additional training specific to commanding a hazardous materials incident.

Incident safety officer The position within the incident command system responsible for identifying and evaluating hazardous or unsafe conditions at the scene of an incident. Safety officers have the authority to stop any activity that is deemed unsafe. Also referred to as the safety officer.

Incubation period The time period between the initial infection by an organism and the development of symptoms by a victim.

Industrial packaging Packaging used to transport materials that present a limited hazard to the public or the environment. Contaminated equipment is an example of such material, because it contains a non-life-endangering amount of radioactivity. Industrial packaging is classified into three categories, based on the strength of the packaging.

Infectious Capable of causing an illness by entry of a pathogenic microorganism.

Ingestion Exposure to a hazardous material by swallowing the substance.

Inhalation Exposure to a hazardous material by breathing the substance into the lungs.

Injection Exposure to a hazardous material by the substance entering cuts or other breaches in the skin.

Intermodal tank A bulk container that serves as both a shipping and storage vessel. Such tanks hold between 5000 and 6000 gallons of product and can be either pressurized or nonpressurized. Intermodal tanks can be shipped by all modes of transportation—air, sea, or land.

Investigative authority The agency that has the legal jurisdiction to enforce a local, state, or federal law or regulation. It is the most appropriate law enforcement organization to ensure the successful investigation and prosecution of a case.

Ionizing radiation Electromagnetic waves of such intensity that chemical bonds at the atomic level can be broken (creating an ion).

Irritant A substance (such as mace) that can be dispersed to briefly incapacitate a person or groups of people. Irritants cause pain and a burning sensation to exposed skin, eyes, and mucous membranes.

Isolation and disposal A two-step removal process for items that cannot be properly decontaminated. First, the contaminated article is removed and isolated in a designated area. Second, it is packaged in a suitable container and transported to an approved facility, where it is either incinerated or buried in a hazardous waste landfill.

Isolation of the hazard area Steps taken to identify a perimeter around a contaminated atmosphere. Isolating an area is driven largely by the nature of the released chemicals and the environmental conditions that exist at the time of the release.

Label A smaller version of a placard (4-inch, diamond-shaped markings). Labels are placed on all four sides of individual boxes and smaller packages that are being transported.

Lethal concentration (LC) The concentration of a material in air that, on the basis of laboratory tests (inhalation route), is expected to kill a specified number of the group of test animals when administered over a specified period of time.

Lethal dose (LD) A single dose that causes the death of a specified number of the group of test animals exposed by any route other than inhalation.

Level A ensemble Personal protective equipment that provides protection against vapors, gases, mists, and even dusts. The highest level of protection, Level A requires a totally encapsulating suit that includes a self-contained breathing apparatus.

Level B ensemble Personal protective equipment that is used when the type and atmospheric concentration of substances require a high level of respiratory protection but less skin protection. The kinds of gloves and boots worn depend on the identified chemical.

Level C ensemble Personal protective equipment that is used when the type of airborne substance is known, the concentration is measured, the criteria for using an air-purifying respirator are met, and skin and eye exposure is unlikely. A Level C ensemble consists of standard work clothing with the addition of chemical-protective clothing, chemically resistant gloves, and a form of respiratory protection.

Level D ensemble Personal protective equipment that is used when the atmosphere contains no known hazard, and work functions preclude splashes, immersion, or the potential for unexpected inhalation of or contact with hazardous levels of chemicals. A Level D ensemble is primarily a work uniform that includes coveralls and affords minimal protection.

Lewisite A blister-forming agent that is an oily, colorless-to-dark-brown liquid with an odor of geraniums.

Liaison officer The position within the incident command system that establishes a point of contact with outside agency representatives.

Liquid splash–protective clothing Clothing designed to protect the wearer from chemical splashes. It does not provide total body protection from gases or vapors and should not be used for incidents involving liquids that emit vapors known to affect or be absorbed through the skin. NFPA 1992 is the performance document pertaining to liquid-splash garments and ensembles.

Local emergency planning committee (LEPC) A committee made up of members of industry, transportation, the public at large, media, and fire and police agencies that gathers and disseminates information on hazardous materials stored in the community and ensures that there are adequate local resources to respond to a chemical event in the community.

Logistics section chief The general staff position responsible for directing the logistics function. It is generally assigned on complex, resource-intensive, or long duration incidents.

Logistics section The section within the incident command system responsible for providing facilities, services, and materials for the incident.

Lower explosive limit (LEL) The minimum amount of gaseous fuel that must be present in the air for the air/fuel mixture to be flammable or explosive. Also referred to as the lower flammable limit (LFL).

Mass decontamination The physical process of reducing or removing surface contaminants from large numbers of victims in potentially life-threatening situations in the fastest time possible.

Material safety data sheet (MSDS) *See* Safety data sheet (SDS).

MC-306/DOT 406 cargo tank Such a vehicle typically carries between 6000 and 10,000 gallons of a product such as gasoline or other flammable and combustible materials. The tank is nonpressurized.

MC-307/DOT 407 chemical hauler A tank with a rounded or horseshoe-shaped tank capable of holding 6000 to 7000 gallons of flammable liquid, mild corrosives, and poisons. The tank has a high internal working pressure.

MC-312/DOT 412 corrosive tank A tank that often carries aggressive (highly reactive) acids such as concentrated sulfuric and nitric acid. It is characterized by several heavy-duty reinforcing rings around the tank and holds approximately 6000 gallons of product.

MC-331 pressure cargo tank A tank that carries materials such as ammonia, propane, Freon, and butane. This type of tank is commonly constructed of steel and has rounded ends and a single open compartment inside. The liquid volume inside the tank varies, ranging from a 1000-gallon delivery truck to a full-size 11,000-gallon cargo tank.

MC-338 cryogenic tanker A low-pressure tank designed to maintain the low temperature required by the cryogens it carries. A boxlike structure containing the tank control valves is typically attached to the rear of the tanker.

Methamphetamine A psychostimulant drug manufactured illegally in illicit laboratories.

Methemoglobin Formed by a change in the iron atom in hemoglobin from the ferrous (+2) to the ferric state (+3). It's normal to have a small amount of methemoglobin in the blood, but substances such as nitrites and nitrile-containing substances convert a larger proportion of hemoglobin into methemoglobin, which does not function as an oxygen carrier.

Mitochondria Responsible for converting nutrients into energy, yielding molecules of adenosine triphosphate (ATP) to fuel the cell's activities.

Multi-gas meter A versatile detection device typically equipped with a combination of toxic gas sensors and the ability to detect flammable gases and vapors.

Muscarinic effects Effects such as runny nose, salivation, sweating, bronchoconstriction, bronchial secretions, nausea, vomiting, and diarrhea.

National Fire Protection Association (NFPA) The association that develops and maintains nationally recognized minimum consensus standards on many areas of fire safety and specific standards on hazardous materials.

National Institute for Occupational Safety and Health (NIOSH) The organization that sets the design, testing, and certification requirements for self-contained breathing apparatus in the United States.

National Response Center (NRC) An agency maintained and staffed by the U.S. Coast Guard; it should always be notified if a hazard discharges into the environment.

Nerve agent A toxic substance that attacks the central nervous system in humans.

Neutralization The method used when the corrosivity of an acid or a base needs to be minimized. This process accomplishes decontamination by way of a chemical reaction that alters the material's pH.

Neutrons Penetrating particles found in the nucleus of the atom that are removed through nuclear fusion or fission. Although neutrons are not radioactive, exposure to them can create radiation.

NFPA 704 hazard identification system A hazardous materials marking system designed for fixed-facility use. It uses a diamond-shaped symbol of any size, which is itself broken into four smaller diamonds, each representing a particular property or characteristic of the material.

Non-ionizing radiation Electromagnetic waves capable of causing a disturbance of activity at the atomic level but that do not have sufficient energy to break bonds and create ions.

Nonambulatory victim A victim who cannot walk under his or her power. These types of victims require more time and personnel when it comes to rescue.

Nonbulk storage vessel Any container other than bulk storage containers such as drums, bags, compressed gas cylinders, and cryogenic containers. Nonbulk storage vessels hold commonly used commercial and industrial chemicals such as solvents, industrial cleaners, and compounds.

Nonpressurized (general-service) rail tank car A railcar equipped with a tank that typically holds general industrial chemicals and consumer products such as corn syrup, flammable and combustible liquids, and mild corrosives.

Occupational Safety and Health Administration (OSHA) The U.S. federal agency that regulates worker safety and, in some cases, responder safety. OSHA is a part of the U.S. Department of Labor.

Operations level responders Personnel who respond to hazardous materials/WMD incidents for the purpose of implementing or supporting actions to protect nearby persons, the environment, or property from the effects of the release.

Operations section chief The general staff position responsible for managing all operations activities. It is usually assigned when complex incidents involve more than 20 single resources or when command staff cannot be involved in all details of the tactical operation.

Operations section The section within the incident command system responsible for all tactical operations at the incident. Its personnel carry out the objectives developed by the incident commander.

Oxygen (O_2) monitoring device A single-gas device intended to measure high (more than 23.5 percent) and low (less than 19.5 percent) levels of oxygen in the air.

Penetration The flow or movement of a hazardous chemical through closures such as zippers, seams, porous materials, pinholes, or other imperfections in a material. Liquids are most likely to penetrate a material, but solids (such as asbestos) can also penetrate protective clothing materials.

Permeation The process by which a hazardous chemical moves through a given material on the molecular level. Permeation differs from penetration in that permeation occurs through the material itself rather than through openings in the material.

Permissible exposure limit (PEL) The established standard limit of exposure to a hazardous material. It is based on the maximum time-weighted concentration at which 95 percent of exposed, healthy adults suffer no adverse effects over a 40-hour workweek.

Persistence The continued presence of a substance in the environment. Persistent chemicals are typically dense, oily substances with high molecular weights and low vapor pressures.

Personal dosimeter A device that measures the amount of radioactive exposure incurred by an individual.

pH An expression of the amount of dissolved hydrogen ions (H^+) in a solution.

pH paper Also called litmus paper. A type of paper used to measure the pH of corrosive liquids and gases.

Phosgene A chemical agent that causes severe pulmonary damage.

Photo-ionization detector (PID) An instrument that provides real-time measurements of airborne contaminant levels of volatile organic compounds.

Photoionization (PID) sensor A sensor that uses ultraviolet light to ionize the gases that move through the sensor; available as a stand-alone unit, or may be incorporated into a multi-gas meter.

Physical change A transformation in which a material changes its state of matter—for instance, from a liquid to a solid.

Physical evidence Items that can be observed, photographed, measured, collected, examined in a laboratory, and presented in court to prove or demonstrate a point.

Pipe bomb A device created by filling a section of pipe with an explosive material.

Pipeline A length of pipe—including pumps, valves, flanges, control devices, strainers, and/or similar equipment—for conveying fluids and gases.

Pipeline right-of-way An area, patch, or roadway that extends a certain number of feet on either side of a pipeline and that may contain warning and informational signs about hazardous materials carried in the pipeline.

Placard Signage required to be placed on all four sides of highway transport vehicles, railroad tank cars, and other forms of hazardous materials transportation; the sign identifies the hazardous contents of the vehicle, using a standardization system with 10¾-inch diamond-shaped indicators.

Plague An infectious disease caused by the bacterium *Yersinia pestis*, which is commonly found on rodents.

Planning section chief The general staff position responsible for planning functions and for tracking and logging resources. It is assigned when command staff members need assistance in managing information.

Planning section The section within the incident command system responsible for the collection, evaluation, and dissemination of tactical information related to the incident and for preparation and documentation of incident management plans.

Polycyclic aromatic hydrocarbons (PAH) A group of substances that occurs naturally in materials such as coal or crude oil. These substances also are generated during the combustion of organic materials and can be found in vehicle exhaust, tobacco smoke, and the smoke generated from structure fires, vehicle fires, wildland fires, or any other type of fire. PAHs can exist as a particle or gas.

Postal Inspection Service Established in 1737, the federal agency charged with securing and managing the U.S. mail system.

Powered air-purifying respirator (PAPR) A type of air-purifying respirator that uses a battery-powered blower to pass outside air through a filter and then to the mask via a low-pressure hose.

Pressurized rail tank car A railcar used to transport materials such as propane, ammonia, ethylene oxide, and chlorine.

Protein foam A foam that is made from organic materials. It is effective on non-alcohol-type fuels such as gasoline and diesel fuel.

Public information officer (PIO) The position within the incident command system responsible for providing information about the incident. Functions as a point of contact for the media.

Pulmonary agent Chemicals that cause severe damage to the lungs and lead to asphyxia. Chlorine and phosgene are examples. Pulmonary agents may also be referred to as choking agents.

Pulmonary edema Fluid build-up in the lungs.

Pyrolysis A chemical change brought about by the action of heat.

Radiation The combined process of emission, transmission, and absorption of energy traveling by electromagnetic wave propagation between a region of higher temperature and a region of lower temperature.

Radiation detection devices Instruments that employ a variety of technologies to detect the presence of radiation. These devices include personal dosimeters and general survey meters designed to detect alpha, beta, and/or gamma radiation.

Radiation dispersal device (RDD) Any device that causes the purposeful dissemination of radioactive material without a nuclear detonation; a dirty bomb.

Radioactive isotope A variation of an element created by an imbalance in the numbers of protons and neutrons in an atom of that element.

Radioactivity The spontaneous decay or disintegration of an unstable atomic nucleus accompanied by the emission of radiation.

Radiological agents Materials that emit radioactivity (the spontaneous decay or disintegration of an unstable atomic nucleus accompanied by the emission of radiation).

Reaction time An expression of the time from when an air sample is drawn into a detection/monitoring meter until the meter processes the sample and provides a reading. Also referred to as response time.

Recommended exposure level (REL) A value established by NIOSH that is comparable to OSHA's permissible exposure limit (PEL) and the threshold limit value/time-weighted average (TLV/TWA). The REL measures the maximum time-weighted concentration of material to which 95 percent of healthy adults can be exposed without suffering any adverse effects over a 40-hour workweek.

Recovery mode A shift in mindset and tactics whereby the incident response reflects the fact that there is no chance of rescuing live victims.

Recovery phase The stage of a hazardous materials incident after imminent danger has passed, when clean-up and the return to normalcy have begun.

Recovery time The amount of time it takes a detector/monitor to clear itself, so a new reading can be taken.

Red blood cell An oxygen-carrying cell found in mammals. Red blood cells contain hemoglobin.

Regulations Mandates issued and enforced by governmental bodies such as the U.S. Occupational Safety and Health Administration and the U.S. Environmental Protection Agency.

Relative response curve A curve that accounts for the different types of gases and vapors that might be encountered other than the one used for calibration of a flammable gas detector and/or monitor.

Relative response factor A mathematical computation that must be completed for a detector to correlate the differences between the gas that is used to calibrate the device and the gas that is being detected in the atmosphere.

Remote valve shut-off A type of valve that may be found at fixed facilities utilizing chemical processes or piped systems that carry chemicals. These remote valves provide a way to remotely shut down a system or isolate a leaking fitting or valve. Remote shut-off valves are also found on many types of cargo containers.

Rescue mode Those activities directed at locating endangered persons at an emergency incident, removing those persons from danger, treating injured victims, and providing for transport to an appropriate healthcare facility.

Retention The process of purposefully collecting hazardous materials in a defined area.

Safety data sheet (SDS) A form provided by manufacturers and compounders (blenders) of chemicals containing information about chemical composition, physical and chemical properties, health and safety hazards, emergency response, and waste disposal of a material.

Sarin A liquid nerve agent that is primarily a vapor hazard.

Secondary containment Any device or structure that prevents environmental contamination when the primary container or its appurtenances fail. Examples of secondary containment mechanisms include dikes, curbing, and double-walled tanks.

Secondary contamination The process by which a contaminant is carried out of the hot zone and contaminates people, animals, the environment, or equipment. Also referred to as cross-contamination.

Secondary device An explosive or incendiary device designed to harm emergency responders who have responded to an initial event.

Self-contained breathing apparatus (SCBA) A respirator with independent air supply used by fire fighters to enter toxic or otherwise dangerous atmospheres.

Sensitizer A chemical that causes a large percentage of people or animals to develop an allergic reaction after repeated exposure.

Sheltering-in-place A method of safeguarding people located near or in a hazardous area by keeping them in a safe atmosphere, usually inside structures.

Shipping papers A shipping order, bill of lading, manifest, or other shipping document serving a similar purpose; it usually includes the names and addresses of both the shipper and the receiver as well as a list of the shipped materials along with their quantity and weight.

Signal words Information on a pesticide label that indicates the relative toxicity of the material.

Situational awareness (SA) A term primarily used in the military, especially the Air Force, that describes the process of observing and understanding the visual cues available, orienting yourself to those inputs relative to your current situation, and making rapid decisions based on those inputs.

Size-up The rapid mental process of evaluating the critical visual indicators of the incident, processing that information based on training and experience, and arriving at a conclusion that will serve as the basis to form and implement a plan of action.

Smallpox A highly infectious disease caused by the *Variola* virus.

Smoke A collection of gaseous products of burning materials—especially organic materials—made visible by the presence of small particles of incomplete combustion (carbon).

Solidification The process of chemically treating a hazardous liquid so as to turn it into a solid material, thereby making the material easier to handle.

Solvent A substance that dissolves another substance so as to make a solution.

Soman A nerve gas that is both a contact and a vapor hazard; it has the odor of camphor.

Special-use railcar A boxcar, flat car, cryogenic tank car, or corrosive tank car.

Specialist level (OSHA/HAZWOPER only) A hazardous materials specialist who responds with, and provides support to, hazardous materials technicians. This individual's duties parallel those of the hazardous materials technician; however, the technician's duties require a more directed or specific knowledge of the various substances he or she may be called upon to contain. The hazardous materials specialist also acts as the incident-site liaison with federal, state, local, and other government authorities in regard to site activities.

Specialized detection device A unique instrument designed to specifically identify a sample substance. These devices are capable of naming the substance in question, sometimes even by its trade name.

Specific gravity The weight of a liquid as compared to water.

Standards Guidelines issued by nongovernmental entities that are generally consensus based.

State Emergency Response Commission (SERC) The liaison between local and state levels that collects and disseminates information relating to hazardous materials emergencies. SERC includes representatives from agencies such as the fire service, police services, and elected officials.

State of matter The physical state of a material—solid, liquid, or gas.

Sterilization A process using heat, chemical means, or radiation to kill microorganisms.

Sulfur mustard A clear, yellow, or amber oily liquid that may be dispersed in an aerosol form. It causes blistering of exposed skin.

Superfund Amendments and Reauthorization Act of 1986 (SARA) One of the first U.S. laws to affect how fire departments respond in a hazardous materials emergency.

Supplied-air respirator (SAR) A respirator that obtains its air through a hose from a remote source such as a compressor or storage cylinder. A hose connects the user to the air source and provides air to the face piece. SARs are useful during extended operations such as decontamination, clean-up, and remedial work. Also referred to as positive-pressure air-line respirators (with escape units).

Tabun A nerve agent that disables the chemical connections between nerves and targets organs.

Target hazards Any occupancy type or facility that presents a high potential for loss of life or serious impact to the community resulting from fire, explosion, or chemical release.

Target organ effect A situation in which specific bodily organs are typically affected by exposures to certain chemical substances.

Technical decontamination A multistep process of carefully scrubbing and washing contaminants off of a person or object, collecting runoff water, and collecting and properly handling all items. It takes place after gross decontamination.

Technical reference team A team of responders who serve as an information-gathering unit and referral point for both the incident commander and the hazardous materials safety officer (assistant safety officer).

Technician level A person who responds to hazardous materials/WMD incidents using a risk-based response process by which he or she analyzes a problem involving hazardous materials/WMD, selects applicable decontamination procedures, and controls a release using specialized protective clothing and control equipment.

Terrorism As defined by the Code of Federal Regulations, the unlawful use of force and violence against persons or property to intimidate or coerce a government, the civilian population, or any segment thereof, in furtherance of political or social objectives.

Threshold limit value (TLV) The point at which a hazardous material or weapon of mass destruction begins to affect a person.

Threshold limit value/ceiling (TLV/C) The maximum concentration of hazardous material to which a worker should not be exposed, even for an instant.

Threshold limit value/short-term exposure limit (TLV/STEL) The maximum concentration of hazardous material to which a worker can sustain a 15-minute exposure not more than four times daily without experiencing irritation or chronic or irreversible tissue damage. There should be a minimum 1-hour rest period between any exposures to this concentration of the material. The lower the TLV/STEL value, the more toxic the substance.

Threshold limit value/skin The concentration at which direct or airborne contact with a material could result in possible and significant exposure from absorption through the skin, mucous membranes, and eyes.

Threshold limit value/time-weighted average (TLV/TWA) The airborne concentration of a material to which a worker can be exposed for 8 hours a day, 40 hours a week, and not suffer any ill effects.

Toxic inhalation hazard (TIH) Any gas or volatile liquid that is extremely toxic to humans.

Toxic products of combustion Hazardous chemical compounds that are released when a material decomposes under heat.

Toxicity A measure of the degree to which something is toxic or poisonous. Toxicity can also refer to the adverse effects a substance may have on a whole organism, such as a human (or a bacterium or a plant), or to a substructure such as a cell or a specific organ.

Toxicology The study of the adverse effects of chemical or physical agents on living organisms.

Trace (transfer) evidence A minute quantity of physical evidence that is conveyed from one place to another.

TRACEMP An acronym to help remember the effects and potential exposures to a hazardous materials incident: thermal, radiation, asphyxiant, chemical, etiologic, mechanical, and psychogenic.

Train list *See* consist.

Triage The process of sorting victims based on the severity of their injuries and medical needs to establish treatment and transportation priorities.

Tube trailer A high-volume transportation device made up of several individual compressed gas cylinders banded together and affixed to a trailer. Tube trailers carry compressed gases such as hydrogen, oxygen, helium, and methane. One trailer may carry several different gases in individual tubes.

Type A packaging Packaging that is designed to protect its internal radiological contents during normal transportation and in the event of a minor accident.

Type B packaging Packaging that is far more durable than Type A packaging and is designed to prevent a release of the radiological hazard in the case of extreme accidents during transportation. Type B containers must undergo a battery of tests including those involving heavy fire, pressure from submersion, and falls onto spikes and rocky surfaces.

Type C packaging Packaging used when radioactive substances must be transported by air.

U.S. Department of Transportation (DOT) marking system A unique system of labels and placards that is used when materials are being transported from one location to another in the United States. The same marking system is used in Canada by Transport Canada.

Underground storage tank (UST) A type of tank that can hold anywhere from a few hundred gallons to several million gallons of product. USTs are usually made of aluminum, steel, or plastic.

Unified command An incident command system option that allows representatives from multiple jurisdictions and agencies to share command authority and responsibility, thereby working together as a "joint" incident command team.

Universal precautions Procedures for infection control that treat blood and certain bodily fluids as capable of transmitting bloodborne diseases.

Upper explosive limit (UEL) The maximum amount of gaseous fuel that can be present in the air if the air/fuel mixture is to be flammable or explosive.

V agent (VX) A nerve agent, principally a contact hazard; an oily liquid that can persist for several weeks.

Vacuuming The process of cleaning up dusts, particles, and some liquids using a vacuum with high efficiency particulate air (HEPA) filtration to prevent recontamination of the environment.

Vapor The gas phase of a substance, particularly of those substances that are normally liquids or solids at room temperature.

Vapor density The weight of an airborne concentration (vapor or gas) as compared to an equal volume of dry air.

Vapor dispersion The process of lowering the concentration of vapors by spreading them out, typically with a water fog from a hose line.

Vapor pressure For our purpose, the pressure associated with liquids held inside any type of closed container.

Vapor suppression The process of controlling vapors given off by hazardous materials, thereby preventing vapor ignition, by covering the product with foam or other material or by reducing the temperature of the material.

Vapor-protective clothing Fully encapsulating chemical protective clothing that offers full-body protection from highly contaminated environments and requires air-supplied respiratory protection devices such as self-contained breathing apparatus. NFPA 1991 sets the performance standards for these types of garments, which are commonly referred to as Level A ensembles.

Vent pipes Inverted *J*-shaped tubes that allow for pressure relief or natural venting of the pipeline for maintenance and repairs.

Volatile organic compound (VOC) Any organic (carbon-containing) compound that is capable of vaporizing into the atmosphere under normal environmental conditions.

VX A thick, oily liquid that is the most toxic of the weapons-grade nerve agents.

Warm zone The area located between the hot zone and the cold zone at an incident. Personal protective equipment is required for all personnel in this area. The decontamination corridor is located in the warm zone.

Washing The process of dousing contaminated victims with a simple soap-and-water solution. The victims are then rinsed using water.

Water solubility The ability of a substance to dissolve in water.

Waybill Shipping papers for railroad transport.

Weapons of mass destruction (WMD) Weapons whose use is intended to cause mass casualties, damage, and chaos. The NFPA includes the use of WMD in its definition of *hazardous materials*.

Index

© Greg Henry/ShutterStock, Inc.

Note: Page numbers followed by *f* or *t* indicate material in figures or tables respectively.

A

aboveground storage tanks (ASTs), 49, 78
absorption, 37, 39
  definition of, 255, 270
  skin, 37, 92
  technique
    for product control, 255–257
    for technical decontamination, 208, 216
acetone, 23, 23*t*
ACGIH. *See* American Conference of Governmental Industrial Hygienists
acids. *See also specific acids*
  concentrated *versus* dilute, 92
  definition of, 27, 39
  pH of, 27–28, 92, 92*f*, 329
  strong *versus* weak, 92
acute cyanide toxicity, signs and symptoms of, 164
acute exposure, 38, 39
acute health effects, 38, 39
adenosine triphosphate (ATP), 164
adsorption
  definition of, 257, 270
  technique
    for product control, 257
    for technical decontamination, 208–209, 216
aerobic metabolism, 163, 164, 173
AFFF. *See* aqueous film-forming foam
agroterrorism, 136, 137*f*, 152
AHJ. *See* authority having jurisdiction
air bill, 68, 71*f*, 78
air detection/monitoring. *See* detection/monitoring
air-purifying respirators (APRs), 105, 190, 201
  description of, 102–103, 102*f*
  disadvantages of, 103
airborne contamination, 85
alcohol-resistant concentrates, 265, 270
Alfred E. Murrah Federal Building Oklahoma City, Oklahoma, 138, 138*f*
alpha particles, 29, 39, 148, 152
ambulatory victims, 222, 231, 232
  definition of, 275, 298
  mass decontamination methods for, 225–226
  rescue of, 275
American Conference of Governmental Industrial Hygienists (ACGIH), 85
ammonia
  anhydrous, 63, 304
  exposure, health effects of, 179, 183
  outdoor release scenario, 179
ammonium nitrate fertilizer and fuel oil (ANFO), 138, 152
anaerobic metabolism, 164, 173
ANFO. *See* ammonium nitrate fertilizer and fuel oil
anhydrous ammonia lab, 304
anthrax, 36, 40, 146, 146*f*, 152
APRs. *See* air-purifying respirators
aqueous film-forming foam (AFFF), 265, 270
asphyxiants, 37, 40
assistant safety officer (ASO). *See* hazardous materials safety officer
ASTs. *See* aboveground storage tanks
atmosphere
  conditions at accident, 33
  dangerous, 86
  safe *versus* unsafe, 86
ATP. *See* adenosine triphosphate
atropine, 143
Aum Shinrikyo, 140
authority having jurisdiction (AHJ), 4, 5, 15, 111, 160, 256
autoignition temperature (ignition temperature), 23, 40
awareness level personnel
  definition of, 7, 15
  functions of, 7

B

backup team
  definition of, 116, 128, 274, 298
  responsibilities of, 116, 278–279, 279*f*
bacteria, as biologic agents, 145
bags, as storage containers, 51–52, 51*f*
bases (pH), 27, 40, 329
beta particles, 148, 152
  definition of, 29, 40
bills of lading, 66, 68*f*, 78
biological agents. *See also specific biological agents*
  anthrax, 146
  containers for, 74
  definition of, 73, 78, 145, 153
  exposure, symptoms of, 74
  indicators of, 147, 147*f*
  plague, 146–147
  protection from, 147–148
  smallpox, 147
  for terrorist incidents, 74, 145–148
biological laboratories, 306
blanket drag, 289, 290
blasting agents, 1000-pound rule, 59
BLEVE. *See* boiling liquid/expanding vapor explosion
BLEVE–safety precautions, 88*f*–89*f*
blister agents (vesicants), 144, 153. *See also specific blister agents*
  definition of, 34, 40
  exposure, health effects from, 34
blood agents, 34, 40, 145, 153
boiling liquid/expanding vapor explosion (BLEVE), 21, 40, 52
boiling point, 25, 40
bomb building instructions, on Internet, 142
bounce-off method, of foam application, 266–268
breakthrough time, 181, 182*f*
bubo, bubonic plague, 146, 146*f*
buddy system
  definition of, 116, 128
  importance of, 116
bulk storage containers, 49, 78
bump test, 317, 332
bung, 51, 78
bung wrench, 51, 51*f*

C

CAK. *See* cyanide antidote kit
Cal-OSHA, 6
calibration, 316, 317, 317*f*, 332
call centers, 91
Canadian Transport Emergency Centre (CANUTEC), 73, 78
carbon-14, 29
carbon monoxide (CO), 162–164
  detector, 315*t*, 323–324, 324*f*, 332
  exposure, health effects of, 221, 324
carboys, 52, 52*f*, 78
carcinogen, 37, 40
cardboard drums, 50
cargo tanks, 54, 78, 262
carry techniques, for victims, 284–288
CAS. *See* Chemical Abstracts Service
CFR. *See* Code of Federal Regulations
CGI. *See* combustible gas indicator
chain of custody, 242, 250
Chemical Abstracts Service (CAS), 5, 15
chemical agents, 73–74, 140–154. *See also specific chemical agents*
  blister agents, 144
  blood agents, 145
  indicators, 145
  nerve agents, 142–144
  protection from, 145
  pulmonary agents, 145
chemical and physical processes, ideal testing environment for, 24
chemical and physical properties, 20, 40
chemical, biological, radiological, and nuclear (CBRN), 100, 101, 105, 243
chemical classifier strips. *See* chemical test strips
chemical compatibility charts, 181, 181*f*
chemical containers. *See* containers
chemical degradation, 209–210, 216
chemical incidents, 37
chemical laboratories, 306

chemical-protective clothing/equipment, 96–97, 97*f*, 98*f*. *See also* personal protective equipment
  design of, 180–181
  hand signals with, 183–184, 184*f*
  ratings, 185–196
  types, 181, 183
chemical reactivity (chemical change), 21, 40
chemical references, 63–66
chemical-resistant materials, 97, 105, 181, 201
chemical test strips, 315*t*, 329, 329*f*, 332
Chemical Transportation Emergency Center (CHEMTREC), 73, 78
chlorine, 140, 153
  release, response to, 280
chloropicrin, 35
choking agents, 35, 40, 142, 153
chronic exposures, 37, 40
chronic health hazard, 37, 40
cigarette smoking, CO concentration, 324
circulatory system, 163, 163*f*
circumstantial evidence, 239, 240, 250
civilian targets, for terrorists, 136, 137*f*
clandestine drug laboratory, 303, 310. *See also* illicit laboratories
Class A foam, 264
Class B foam, 264
clean-up phase, 268
closed-circuit self-contained breathing apparatus, 102, 105
closed-head drums, 51
clothes drag, 289
clothing, 179
  chemical-protective. *See also* personal protective equipment
    design of, 180–181
    ratings for, 185–196
    types, 181, 183
clothing removal, 228
CO. *See* carbon monoxide
Code of Federal Regulations (CFR), 6, 15
cold-temperature exposures, 120
cold zone, 85, 113, 114, 128
color-coding system, 276–277
colorimetric tubes, 170, 170*f*, 173, 315*t*, 326–328, 327*f*, 332
combustible gas indicator (CGI), 314, 314*f*, 315*t*, 321–322, 321*f*, 332
combustibles, 22, 40
command staff, 121–124, 128
complete dam, 258
computer modeling, in determining size of incident, 86, 87*f*, 88, 88*f*–89*f*
concentration, of hazardous materials release, 91–92
confinement, 254, 270
consist, 68, 78
contagious, 36, 40
containers. *See also specific types of containers*
  definition of, 48, 78
  volume of, 49–50
containment, 254, 270
containment vessels, reasons for failure, 20
contaminant, physical/chemical properties of, 207
contaminated, definition of, 250
contamination, 206, 216
  definition of, 31, 40, 103, 105, 240
  of evidence, secondary, prevention of, 246–247
  secondary/cross-contamination, 32, 41, 104
contamination levels, expression of, 85
control zones, 112, 113*f*
  cold zone, 113, 114
  definition of, 104, 105, 128
  establishing, 112–114
  hot zone, 113
  warm zone, 113, 114
convulsants, 35, 40
corrosives. *See also specific corrosives*
  concentration of, 92
  skin contact, 92
  strong, 329
corrosivity, 27–28, 40
cradle-in-arms carry, 288
crank. *See* methamphetamine
crop-dusting equipment, 237*f*
  for disseminating chemical agents, 142*f*
crowd control, for mass decontamination, 231
cryogenic intermodal tanks, 50
cryogenic liquid (cryogen), 50, 52, 78
cyanide antidote kit (CAK), 165
cyanide compounds
  definition of, 34, 40, 145, 153
  exposure, health effects of, 145
cyanide exposures, 163
cyanide levels, after exposure to fire smoke, 162, 163*t*
cyanide toxicity, 164
cyberterrorism, 137–138, 153
cylinders, 52–53, 52*f*, 78
cytochrome oxidase, 164, 173

D

damming, 258–259, 270
dangerous atmosphere, 86
dangerous cargo manifest, 68, 70*f*, 78
DEA. *See* Drug Enforcement Administration
DECIDE model, 111*f*
decision-making algorithm, 111
decontamination, 206, 216
  definition of, 84, 105
  emergency. *See* emergency decontamination
  of evidence, 248–249
  of illicit laboratory, 309–310
  mass. *See* mass decontamination
  purpose of, 103, 103*f*
  standard procedures, 151
  technical. *See* technical decontamination
decontamination corridor, 103, 105, 113*f*
  stations in, 213
  in technical decontamination process, 206, 207*f*, 211*f*, 216
decontamination team, 125, 128, 207, 216
defensive actions, 94
degradation, 97, 105, 181, 201
dehydration, 196, 201
demonstrative evidence, 239, 250
denial of entry, 90, 106
Department of Transportation (DOT)
  definition of, 15
  Emergency Response Guidebook. *See Emergency Response Guidebook*
  hazardous material definition, 5
  Marking System, 58, 59*f*–60*f*
detection/monitoring
  actions for, 316
  air, 316–321, 331
  for chemical agents, 308
  concepts, 318–319
  devices, 314, 315*t*. *See also specific detection devices*
  steps in, 316
  terminology, 316–318
  types of, 321–331
  for unknown hazards, 319, 321
detonating cords, 305*f*
Dewar containers, 78
  description of, 48, 49
  low-pressure, 52, 53*f*
diesel fuel, grade #2
  flash point, 23
  ignition temperature, 23
  vapor pressure, 23
diethyl ether, vapor pressure, 24
diking, 259, 270
dilution
  definition of, 259, 270
  method
    for mass decontamination, 226, 232
    steps in, 260
    for technical decontamination, 210, 216
direct evidence, 239, 250
dirty bombs, 32, 149
disinfection, 210, 216
dissemination methods, 142
distance, in limiting exposure to radioactivity, 148
diversion
  definition of, 260, 270
  method, steps in, 260–261
divisions, 124–125, 128
documentation
  of mass decontamination, 231–232
  of personnel and scene activity, 244–246
  of sampling, 247–248
  of scene, illicit laboratory, 310
  of technical decontamination, 215
doffing, 185, 201
  level A ensemble, 187–188
  level B ensemble, 190–191
  level C ensemble, 193–194
domestic terrorism, 134
donning, 185, 201
  level A ensemble, 186–187
  level B ensemble, 189–190
  level C ensemble, 192–193
  level D ensemble, 194–195
dosimeters, personal, 331, 331*f*
DOT. *See* Department of Transportation
drags, emergency, 289–297
Drug Enforcement Administration (DEA), 238, 250
drug laboratories, clandestine, 303, 310. *See also* illicit laboratories
drums
  closed-head, 51
  definition of, 48, 50–51, 78
  open-head, 51, 51*f*
dry bulk cargo tanks, 56, 56*f*, 79
dry decontamination method, 208
"dry" foam, 265
DuoDote™, 143, 144*f*

E

ecoterrorism, 136, 153
electrochemical sensors, 168–169, 168*f*
elements, periodic table of, 29
emergency decontamination, 206, 216
  decontamination corridor, 103, 105, 113*f*
  definition of, 104, 106, 128
  performing, 114, 279, 279*f*
emergency drags, 289–297
emergency operation centers (EOCs), 150
emergency phase, 268
Emergency Planning and Community Right to Know Act (EPCRA), 12, 15

emergency responders approach, 255
Emergency Response Guidebook (*ERG*), 5
  blue section, 63
  chemical families/classes in, 63
  definition of, 4, 15, 63, 64*f*, 79
  green section, 63, 65*f*
  initial isolation and protective action distances, 86, 87*f*, 88
  orange section, 63, 64*f*
  role in mass decontamination, 228, 229*f*
  using, 63–66
  yellow section, 63
Emergency Transportation System for the Chemical Industry, Mexico (SETIQ), 73, 79
entry team
  definition of, 116, 128, 274, 298
  responsibilities of, 275–278
environmental crimes, 237
Environmental Protection Agency (EPA), 5, 15, 238, 250
EOCs. *See* emergency operation centers
EOD personnel. *See* explosive ordnance disposal personnel
EPA. *See* Environmental Protection Agency
EPCRA. *See* Emergency Planning and Community Right to Know Act
ephedrine, in methamphetamine production, 304, 304*f*
equipment
  high-temperature–protective, 96, 97*f*
  personal protective. *See* personal protective equipment
  for search, rescue and recovery efforts, 281–282, 282*f*
ERG. See Emergency Response Guidebook
etiological incidents, 32
evacuation
  definition of, 90, 106
  distances, 117
  procedures for, 116–118
evacuation chairs, 281
EVADE mnemonic, 93, 139
evaporation, 210, 216
evidence, 236, 250
  chain of custody, 242, 250
  collection, 243
    equipment for, 247
    identifying samples for, 246
    preventing secondary contamination, 246–247
    samples and preserving, 241
    standard operating procedures for, 240–241
  contaminated, 242, 250
  decontamination, 207, 248–249
  definition of, 236
  labeling, 248–249
  packaging, 248–249
  preservation, 240–243
    in mass decontamination, 232
  processing, standard operating procedures for, 240–241
  rules of, 246
  samples, identification of, 246
  sampling, 243
    and field screening protocols, 248
  types of, 238–240
evidence preservation, 240, 250
evidence sampling, 236, 243, 250
excepted packaging, 76, 79
excessive-heat disorders, 119–120
exit assists, for rescue operations, 283–285
expansion ratio, 21, 40
explosion-with-fragment hazards, military marking system for, 62, 63*f*
explosive ordnance disposal personnel (EOD personnel), 139, 153
explosives/explosive devices, 77. *See also specific types of explosive devices*
  identifying, 306
  potential, 139, 139*f*–140*f*
  potential terrorist incidents and, 73–77
  types of, 138–140
exposures. *See also specific exposures*
  concentration determination, 91–92
  definition of, 31, 40
  initial response to, 89–90
  number of, 89
  size/scope of, 91
  skin contact hazards, 92

F

FBI. *See* Federal Bureau of Investigation
federal agencies, working with, 140
Federal Bureau of Investigation (FBI), 134, 238, 250
FID. *See* flame ionization detector
field screening procedures, 246
field screening protocols, 248
finance/administration section, 127, 128
fire fighter drag, 293
Fire Fighter's Handbook of Hazardous Materials, 65
fire hazards
  mass, military marking system for, 62, 63*f*
  moderate, military marking system for, 62, 63*f*
fire point, 23
fire scene, detection and monitoring
  practices, 170–171
  technologies, 168–170
fire service, mass decontamination performed by, 222, 222*f*
fire smoke
  carbon monoxide and hydrogen cyanide, 162–164
  common fire scene detection and monitoring technologies, 168–170
  fire scene detection and monitoring practices, 170–171
  introduction to, 158, 160
  overview of, 160–162
  postfire detection and monitoring, 165–166
  smoke inhalation treatment, 165
firefighting gear, structural, protective level of, 179
FIRESCOPE, 121
fixed ammonia systems, 262
flame ionization detector (FID), 315*t*, 322, 323*f*, 332
flammable gas detectors. *See* combustible gas indicator
flammable liquids, characteristics of, 22–31
  boiling point, 25–26
  corrosivity/pH, 27–28
  flash point, 22–23
  molecular weight, 25
  specific gravity, 26–27
  toxic products of combustion, 28–29
  vapor density, 25–26
  water solubility, 27, 41
flammable range, 23, 40
flash point, 22–23, 40
fluid-chilled/water-cooled system, 199, 199*f*
fluoroprotein foam, 265, 270
foam concentrates
  application
    bounce-off method of, 266–268
    rain-down method of, 266
    roll-in method of, 266–267
  Class A, 264
  Class B, 264
  types of, 265
  vapor suppressing, 264
forced-air cooling systems, 197
forensic evidence, 236
forensic evidence team, illicit laboratory response, 309
formaldehyde, 28, 38
forward staging system, 139, 153
4H MEDIC ANNA mnemonic, 26
fourier transform infrared spectroscopy (FTIR), 330, 330*f*
Franklin, Benjamin, 171
freight bills, 66, 68*f*, 79
FTIR. *See* fourier transform infrared spectroscopy
fusible plugs, 53

G

gamma radiation, 31, 40, 148, 153
gas chromatography (GC), 315*t*, 322, 322*f*, 332
gasoline
  containers, 48
  flash point, 22–23
  ignition temperature, 23
  vapor pressure, 23
gasoline-powered positive pressure ventilation (PPV) fans, 171
Globally Harmonized System of Classification and Labelling of Chemicals (GHS), 12
gloves, removal of, 213
government agencies. *See specific government agencies*
gross decontamination, 206, 207*f*, 216

H

hand signals, 183–184, 184*f*
harm, potential, estimation of, 84–89
hazard area
  denial of entry, 90
  isolation of, 90
hazard levels, NFPA 704 hazard identification system, 61
hazardous device personnel (bomb squad), 246, 246*f*, 250, 305, 310
hazardous materials, 5–6, 6*f*, 179–192. *See also specific hazardous materials*
  chemical-protective clothing
    and equipment, 180–184
    ratings, 185–193
  containers, 48–54
  criminal use of. *See* weapons of mass destruction
  definition of, 4, 15, 47, 79
  entry routes, 35–37
  levels of training, 6–12
  locations with, 47
  marking system, of U.S. military, 62, 63*f*
  release, determining concentration of, 91–92
  respiratory protection, 184–185
  transporting, 54–57
  and WMD incident, difference between, 12–14
hazardous materials branch, 125, 125*f*, 128–129
hazardous materials group, 124, 129
hazardous materials incident, 238
  levels, 127*t*
  recovery phase of, 267
  using absorption/adsorption to manage, 257
  using dilution to manage, 260–261

hazardous materials incident (*Continued*)
using retention to manage, 262
using vapor dispersion to manage, 264
using vapor suppression to manage, 265
Hazardous Materials Information System (HMIS), 61, 62*f*, 79
hazardous materials safety officer, 118, 126, 129
hazardous materials team, response to illicit laboratories, 307
determining options, 308–310
identifying safety hazards, 308
notifying authorities, 308
hazardous materials technicians, 11, 11*f*
hazardous waste, 6, 15
HAZardous Waste OPerations and Emergency Response (HAZWOPER), 7, 15, 123
hazards
definition of, 31, 40
in illicit laboratories, 305
HAZWOPER. *See* HAZardous Waste OPerations and Emergency Response
HCN. *See* hydrogen cyanide
headspace, 49
health effects, acute and chronic, 37–39
heat exhaustion, 120, 129
heat-related illness, 196, 197
heat stroke, 120, 129
helium, cryogenic, 52
HEPA vacuum cleaner, 209
high-expansion foam, 265–266, 270
high-temperature–protective equipment, 96, 97*f*, 106, 180, 180*f*, 201
HMIS. *See* Hazardous Materials Information System
hot zone, 113, 129
hydrochloric acid, neutralization, 210
hydrofluoric acid, 256
containers, 48
hydrogen cyanide (HCN), 162–164
exposure, damage from, 183
hydrogen cyanide poisoning, 165
hydrogen sulfide (H_2S) monitor, 315*t*, 324, 324*f*, 332
hydroxocobalamin, 165
hypothermia, 120
hypoxia, 164, 174

I

IBC. *See* intermediate bulk container
IC. *See* incident commander
Ice. *See* methamphetamine
ice-cooled/gel-packed vests, 197
ICP. *See* incident command post
IDLH. *See* immediately dangerous to life and health
IEDs. *See* improvised explosive devices
ignition sources, identifying/securing, 112
ignition temperature (autoignition temperature), 23, 40
illicit laboratories, 237, 250
definition of, 302, 310
identifying, 302–305
manufacturing process, precursor chemicals, 302
for methamphetamine production, 302–305, 304*f*, 305*t*, 307
personnel/operators of, 302
potential terrorist incidents and, 76–77
responding to scene of, 308
decontamination, 309–310
responder tasks/operations, 308–310
securing/preserving, 308–309
tasks and operations, 306–310
immediately dangerous to life and health (IDLH), 85, 106, 166–167
improvised explosive devices (IEDs), 88, 88*f*–89*f*, 138, 153, 306, 309
in-suit cooling technologies, 197, 199
incendiary devices, 138–140
incident command, establishing, 150
incident command post (ICP), 123, 129
incident command system (ICS)
command, 123–124
command staff, 123–124
definition of, 121, 129
functional components, 124
general staff functions, 124–127
finance, 127
groups and divisions, 124–126
logistics, 126–127
operations, 124–126
planning, 126
staff functions of, 121*f*
incident commander (IC), 11, 15, 113, 115*f*, 119, 123, 129
incident safety officer, 118, 129
incidents, 4–5
analyzing, 237–240
classification of, 32
versus other emergencies, 12–14, 47
recognizing, 47–48
location and, 47
type of occupancy and, 47
use of senses and, 47–48
recovery phase, 267
scene of. *See* scene of incident
scope, reporting on, 91
incubation period, 146, 153
indoor safety showers, 104*f*
industrial packaging, 76, 79
infectious, 36, 40
infrastructure targets, for terrorists, 135–136, 136*f*
ingestion, 35, 37, 40
inhalation, 36–37
injection, 35, 37, 40
interagency coordination, terrorism operations, 150–152
intermediate bulk container (IBC), 53
intermodal tanks
definition of, 49–50, 79
types of, 50
International Union of Pure and Applied Chemistry (IUPAC), 24
investigative authority, 238, 251
notifying, 246
investigative jurisdictions, 238
ionization potential (IP), 321
ionizing radiation, 29, 40
irritants (riot control agents), 35, 40
isolate, 111
isolation and disposal, 211, 216
for mass decontamination, 227–228, 232
isolation, of release area, 112, 112*f*
isolation of the hazard area, 90, 106
isopropyl alcohol, vapor pressure, 24
isotopes, 29
IUPAC. *See* International Union of Pure and Applied Chemistry

J

joint agency operations, coordinating, 308
joint hazardous materials/EOD operations, 308

L

labels, 58, 79
hazardous material, 228, 229*f*
law enforcement agencies
cooperation of, 238
notification, of illicit laboratory, 308
LC. *See* lethal concentration
LD. *See* lethal dose
LEL. *See* lower explosive limit
LEPC. *See* local emergency planning committees
lethal concentration (LC), 38, 40
lethal dose (LD), 38, 41
Level A ensemble, 185, 185*f*, 201
description of, 98, 98*f*, 106
doffing, 187–188
donning, 186–187
recommendations for, 98
Level A suits, 178, 178*f*
Level B ensemble, 188–189, 188*f*, 201
description of, 98–99, 99*f*, 106
doffing, 190–191
donning, 189–190
recommendations for, 99
Level C ensemble, 190, 192*f*, 201
description of, 99, 99*f*, 106
doffing, 193–194
donning, 192–193
Level D ensemble, 193, 195*t*, 201
description of, 100, 100*f*, 106
donning, 194–195
lewisite, 34, 41
definition of, 144, 153
LFL. *See* lower flammable limit
liaison officer, 124, 129
Lilly kit, 165
linseed oil, 22
liquid splash–protective clothing, 97, 98*f*, 106, 183, 183*f*, 201
liquids
flammable. *See* flammable liquids
response to spills, 93, 93*f*
lithium, in methamphetamine production, 304
litmus paper. *See* pH paper
local emergency planning committees (LEPC), 12, 15
locations, for hazardous materials, 47
logistics section, 126–127, 129
logistics section chief, 126, 129
long backboard rescue, 295–297
low-pressure Dewar container, 52
lower explosive limit (LEL), 23, 41
definition of, 318, 332
detection/monitoring and, 319, 322
lower flammable limit (LFL)
definition of, 318
detection/monitoring and, 319, 322
Ludlum General Purpose Rate Meter, 330, 330*f*

M

magnesium, water solubility, 27
mass casualties, 151–152
mass decontamination, 206, 216, 220–222, 232
crowd control, 231
definition of, 151–153
documentation, 231–232
effectiveness, evaluation of, 224–228
evidence preservation, 232
extent of, 221
by fire service, 222, 222*f*
methods, 151–152, 222–224
for ambulatory victims, 225–226
dilution, 226–227
isolation and disposal, 227–228
for nonambulatory victims, 227
washing, 228

mass decontamination (*Continued*)
  reports, 231–232
  role of reference sources, 228–231
  success of, 220
  training, 226
mass detonation hazards, military marking system, 62, 63*f*
material safety data sheets (MSDS), 12, 15
MC-306/DOT 406 cargo tank, 54, 55*f*, 79, 263
MC-307/DOT 407 chemical hauler, 54, 55*f*, 79, 263, 269
MC-312/DOT 412 corrosive tank, 54, 55*f*, 79
MC-331 pressure cargo tank, 54, 55*f*, 79, 263
MC-338 cryogenic tank, 55, 55*f*, 79
mechanical incidents, 32
medical care
  for persons exposed to harmful substances, 84–85
  during recovery mode, 278, 278*f*
medical surveillance program, 101
methamphetamine, 237, 251, 303, 310
  production, 302
    chemicals used in, 304, 305*f*
  use, symptoms from, 302
methemoglobin, 165, 174
millimeters of mercury (mmHg), 24
mitochondria, 164, 164*f*, 174
morbidity, 163
mortality, 163
motor homes, 303
motor oil, vapor pressure, 24
multi-gas meter, 315*t*, 324–326, 324*f*, 325*f*, 332
muscarinic effects, 144, 153

N

National Fire Protection Association (NFPA)
  definition of, 5, 15
  marking system, 60–61
  versus OSHA regulations, 5
  standards of, 4–5
    NFPA 472, 6, 7
    NFPA 473, 6
National Institute for Occupational Safety and Health (NIOSH), 166, 179, 202
National Response Center (NRC), 73, 79
natural barriers, 255
natural gas, release of, 91
nerve agents, 140, 153. *See also specific nerve agents*
  definition of, 32, 41, 142
  exposure symptoms, 32, 142–144, 143*t*
  M8/M9 paper, 221, 222*f*
neutral pH, 329
neutralization, 210, 216
neutrons, 31, 41
NFPA. *See* National Fire Protection Association
NFPA 704 hazard identification system, 61, 79, 229, 229*f*
NIOSH. *See* National Institute for Occupational Safety and Health
Nithiodote, 165
Nomex jumpsuit, 96*f*, 180*f*
non-hazardous materials incidents, technical decontamination, 212
non-ionizing radiation, 31, 41
nonambulatory victims, 222, 232
  definition of, 275, 298
  mass decontamination methods for, 227
  rescue of, 275
nonbulk storage vessels, 50–54, 79
non–law enforcement personnel, 236
nonpressurized horizontal tanks, 49
nonpressurized (general-service) rail tank cars, 56, 56*f*, 79
normal temperature and pressure (NTP), 24
notes of scene, as evidence, 246
notification
  hazardous devices personnel, 246
  of investigative authority, 246
  in planned response, 111
NRC. *See* National Response Center
NTP. *See* normal temperature and pressure

O

Occupational Safety and Health Administration (OSHA), 167
  definition of, 5, 15
  HAZWOPER regulation, 7
    awareness level personnel, 8
    IC level, 11
    operations level personnel, 9–10
  versus NFPA standards regulations, 6
  non-plan states, 6
  state-plan states, 6
occupancy, type of, in recognizing type of incident, 47
offensive action, 119
Oklahoma City bombing, 212
one-person walking assist, 283, 284
one responder emergency drag from vehicle, 294
open-head drum, 46, 51, 51*f*
operations level responders, 10*f*
  definition of, 9, 15
  NFPA 472 competencies, 9–10
  role of, 112, 127
  "tasked to respond," 9
  tasks for, 9
operations section, 124–126, 129
operations section chief, 124, 129
OSHA. *See* Occupational Safety and Health Administration
OSHA HAZWOPER regulation, Level B, 189
over-the-counter drugs, in methamphetamine production, 304
overflow dam, 258
oxidizers, 1000-pound rule, 59–60
oxygen cylinders, 52
oxygen (O_2) monitoring device, 315*t*, 322–323, 323*f*, 332

P

packaging, for radiologic agents, 75–76, 76*f*
PAH. *See* polycyclic aromatic hydrocarbons
painful deposition process, 245
paper bags, 240
PAPRs. *See* powered air-purifying respirators
parts per billion (ppb), 85
parts per million (ppm), 85
parts per trillion (ppt), 85
Pasadena kit, 165
PEL. *See* permissible exposure limit
penetration, 97, 106, 181, 202
periodic table of elements, 28
permeation, 97, 106, 181, 202
permissible exposure limit (PEL), 85, 106, 167
persistence, 34, 41
personal dosimeter, 148, 153, 315*t*, 331, 331*f*, 332
personal protective equipment (PPE), 179–180, 238, 254
  chemical-protective clothing
    and equipment, 96–97, 97*f*, 98*f*, 180–184
    ratings, 185–193
  heat stress prevention and, 120
  high-temperature–protective equipment, 96, 97*f*
  illicit laboratory personnel and, 303
  levels of protection, 179, 185, 195*t*. *See also* Level A ensemble; Level B ensemble; Level C ensemble; Level D ensemble
  liquid splash-protective clothing, 97, 98*f*
  NFPA standards for, 94
  physical capability requirements, 120–121
  respiratory protection, 184–185
  for responders, at illicit laboratory scene, 309, 309*f*
  selection of, 94
  street clothing/work uniforms, 96, 96*f*
  structural firefighting gear, 96, 96*f*
  in technical decontamination process, 213
  testing of, 94
    and certification standards, 196*t*
  types of, 96–97, 96*f*–98*f*
  vapor-protective clothing, 97, 97*f*
  in warm zone, 113
personnel and scene activity, documenting, 244–246
pesticide bags, 51, 51*f*
pH, 28, 41, 329
  of acids, 27–28, 92, 92*f*, 329
  corrosivity and, 27–28, 40
  in detection/monitoring, 329
pH paper, 315*t*, 329, 329*f*, 332
phase-change cooling technology, 199, 199*f*
phenyl-2-propanone (P2P), 304
phosgene, 35, 145, 153
photoionization detector (PID), 315*t*, 321, 321*f*, 332
photographs, as evidence, 242, 246
photoionization (PID) sensor, 169–170, 174
physical capability requirements, 101
physical changes, 21, 41
physical evidence, 238, 251
physical processes, ideal testing environment for, 24
PID. *See* photoionization detector
PID sensor. *See* photoionization sensor
PIO. *See* public information officer
pipe bombs, 138–139, 139*f*, 153
pipeline right-of-way, 57, 57*f*, 79
pipelines, 57, 57*f*, 79
placards
  "Dangerous," 65*f*
  definition of, 58
  in mass decontamination process, 228
  UN number, 60
  using NFPA 704 hazard identification system, 61, 79
plague, 32, 41, 146–147, 153
planning, initial response, 92–94
planning section, 126, 129
planning section chief, 126, 129
polycyclic aromatic hydrocarbons (PAH), 161, 162,174
polyethylene drums, 50
polyurethane foam, 161
polyvinyl chloride (PVC) pipe, 258
positive-pressure self-contained breathing apparatus, 101
Postal Inspection Service, 238, 251
postfire, detection and monitoring, 165–166
powered air-purifying respirator (PAPR), 190, 202

powered air-purifying respirators (PAPRs), 103, 103*f*, 106
P2P. *See* phenyl-2-propanone
ppb. *See* parts per billion
PPE. *See* personal protective equipment
ppm. *See* parts per million
ppt. *See* parts per trillion
preplanning activities, 13–14
pressure, in cylinders, 52
pressure intermodal tanks, 50
pressurized horizontal tanks, 49
pressurized rail tank cars, 56, 57*f*, 80
primary search, 281
product control operations, 254–268
  confinement, 254
  containment, 254
  options, 254
    absorption and adsorption, 255–257
    damming, 258–259
    diking, 259
    dilution, 259–260
    diversion, 260–261
    remote valve shut-off, 262–263
    retention, 261–262
    vapor dispersion and suppression, 263–267
  recovery phase, 267–268
propane cylinders, 52
propane gas, physical change, 21
protective actions, 114–121
protein foam, 265, 270
pseudoephedrine, in methamphetamine production, 304, 304*f*
psychogenic incidents, 32
public information officer (PIO), 124, 124*f*, 129
public safety responders, 236
pulmonary agents, 142, 145, 153
pulmonary edema, 28–29, 41
pupils, responses to chemical agents, 143, 143*f*
PVC pipe. *See* polyvinyl chloride pipe
pyrolysis, 163, 174

R

radiation
  alpha particles, 29, 39, 148
  beta particles, 148
  decontamination, 148
  definition of, 29, 41
  description of, 29
  effects of, 148
  gamma, 31, 40, 148
  health hazards, 29
radiation detection devices, 315*t*, 330–331, 330*f*–331*f*, 332
radiation dispersal device (RDD), 149, 153
radioactive contamination, 29
radioactive isotope, 29, 41
radioactivity
  definition of, 29, 41
  limiting exposure to, 148
radiological agents, 148–149, 153
  definition of, 73, 80
  packaging for, 75–76, 76*f*
radiological incidents, 32
rain-down method, of foam application, 266
Raman spectroscopy, 330*f*
RDD. *See* radiation dispersal device
reaction time (response time), 317, 332
recommended exposure level (REL), 85, 106
recommended exposure limits (REL), 167–168
reconnaissance team (recon team), 150
recovery, medical care, 278, 278*f*
recovery mode, 278, 298
recovery phase, 267–268, 267*f*, 270
recovery time, 317–318, 332
red blood cells, 163, 163*f*, 174
red phosphorus ("Red P"), 304
  containers, 48
reference sources, in mass decontamination, 228–231
regulations, 5–12, 15. *See also specific regulations*
REL. *See* recommended exposure level; recommended exposure limits
relative response curve, 318, 332–333
relative response factor, 318, 333
remediation, at illicit laboratory scene, 310
remote valve shut-off, 262–263, 263*f*, 270
reporting
  of mass decontamination, 231–232
  size and scope of incident, 91
rescue mode, 118, 278, 298
rescue operations
  emergency drags, 289–297
  exit assists, 283–285
  feasibility of, 274
  tactical considerations, 274–279
rescue sleds, 281, 282*f*
residual chemicals, 74
respirators, 36, 73, 74, 102, 102*f*, 103, 106, 148
respiratory protection
  air-purifying respirators (APRs), 102–103, 102*f*
  closed-circuit self-contained breathing apparatus, 102
  levels of, 100–103
  physical capability requirements, 101
  positive-pressure self-contained breathing apparatus, 101
  powered air-purifying respirators, 103, 103*f*
  supplied-air respirators, 102, 102*f*
respiratory system
  decontamination, 213
  protection for, 184–185
responders
  awareness level, 7, 9, 112
  goal of, 13
  operations level. *See* operations level responders
  safety tips, 5, 25, 27, 31, 110
  safety, while wearing PPE, 196–197, 199
  tasks, 7
  training. *See* training
response
  to chlorine gas release, 280
  exhilaration of, 8
  initial actions, 112
  initial, planning, 92–94
  objectives, 93
  safety procedures, 112–114
response actions, implementing, 246
response time (reaction time), 317
retention
  definition of, 261, 270
  method, steps in, 262
roadway transportation, of hazardous materials, 54–56
roll-in method, of foam application, 266–267

S

SA. *See* situational awareness
safe atmosphere, 86
safety considerations
  briefings on, 118–119
  hazards. *See* hazards
  for personal protective equipment, 193–199
safety data sheets (SDS), 12, 15, 66, 104
safety officer, 123–124
safety plan, written, 119
sampling
  documentation of, 247–248
  and field screening protocols, 248
  labeling, packaging, and decontamination, 248–249
sampling teams, 243
SAR. *See* supplied-air respirators
SARA. *See* Superfund Amendments and Reauthorization Act of 1986
sarin, 32, 41, 140, 143, 143*f*, 153
SARs. *See* supplied-air respirators
SCBA. *See* self-contained breathing apparatus
scene of incident
  characterization of, 243
  scene control, 110, 112
  scene size-up, 46
  securing, characterizing, and preserving, 243–244
  viewing, from safe distance, 47
scope of incident, reporting, 91
SDS. *See* safety data sheets
search and rescue, 118
search, rescue, and recovery efforts, 281–283
  equipment for, 281–282, 282*f*
  methods, 282–283
secondary attacks and devices, 93–94
secondary containment, 49, 80
secondary contamination (cross-contamination), 32, 104, 106
  prevention, 246–247
secondary devices, 77, 80, 139, 153
secondary search, 281
self-contained breathing apparatus (SCBA), 86, 106, 158, 166, 179, 202
  closed-circuit, 102
  positive-pressure, 101, 101*f*
senses, use of, in incident recognition, 47–48
sensitizer, 38, 41
SERC. *See* State Emergency Response Commission
severity of incident, potential, estimation of, 84–89
Sevin, 274
sewer gas. *See* hydrogen sulfide monitor
sheltering-in-place, 90, 106, 118, 129, 283
shielding, in limiting exposure to radioactivity, 148
shipping papers, 66, 80
"shuns," 116
signal words, 51, 80
silent killers, 162–164
single-shell aluminum tanks, 262
situational awareness (SA), 110, 316, 333
size of incident
  determining, resources for, 86, 87*f*, 88, 88*f*–89*f*
  reporting, 91
size-up, 110, 129
skin absorption, 92
skin protection, personal protective equipment and, 94
SLUDGE mnemonic, 34, 143
smallpox, 32, 41, 147, 147*f*, 153
smoke, 158, 161, 174
  inhalation treatment, 165
  matrix, 161, 161*t*
sodium metal, in methamphetamine production, 307
sodium, metallic, water solubility, 27
soil contamination, remediation of, 210
solidification, 210, 216